36th International DAAAM Symposium on Intelligent Manufacturing and Automation 2025

Vienna, Austria
30-31 October 2025

Editor:

Branko Katalinic

ISBN: 979-8-3313-3013-2

Printed from e-media with permission by:

Curran Associates, Inc.
57 Morehouse Lane
Red Hook, NY 12571

Some format issues inherent in the e-media version may also appear in this print version.

DAAAM International Vienna
TU Wien, Getreidemarkt 9
1060 Wien
Austria, EU

Phone: +436646424398

http://www.daaam.info

Additional copies of this publication are available from:

Curran Associates, Inc.
57 Morehouse Lane
Red Hook, NY 12571 USA
Phone: 845-758-0400
Fax: 845-758-2633
Email: curran@proceedings.com
Web: www.proceedings.com

ISSN 1726-9679

DANUBE ADRIA ASSOCIATION FOR AUTOMATION & MANUFACTURING

DAAAM INTERNATIONAL VIENNA

www.daaam.info

PROCEEDINGS

OF THE 36TH INTERNATIONAL DAAAM SYMPOSIUM
"INTELLIGENT MANUFACTURING & AUTOMATION"
30-31ST OCTOBER 2025
HOSTED FROM VIENNA UNIVERSITY OF TECHNOLOGY, VIENNA, AUSTRIA, EU

ORGANIZED BY:
DAAAM INTERNATIONAL VIENNA & UNIVERSITY OF MOSTAR

IN COOPERATION WITH:
ÖIAV 1848, VIENNA UNIVERSITY OF TECHNOLOGY, INTERNATIONAL ACADEMY OF ENGINEERING AND UNIVERSITY OF APPLIED SCIENCES – TECHNIKUM WIEN

UNDER THE AUSPICES OF THE DANUBE RECTORS' CONFERENCE &
RECTORS' AND PRESIDENTS' HONOR COMMITTEE OF DAAAM INTERNATIONAL

EDITOR: B.[RANKO] KATALINIC

IMPRESSUM
2025

Proceedings of the 36th International DAAAM Symposium "Intelligent Manufacturing & Automation"
30-31st October 2025, hosted from Vienna University of Technology, Vienna, Austria, EU, Volume 36, No.1,
Abbreviated key title: ***Ann. DAAAM 2025***

ISSN 1726-9679

Ed. B. Katalinic,

Published by DAAAM International, Vienna, Austria, EU, 2025
Layout & Design: B[ranko] Katalinic
Technical Editors: I[lija] Zec / L[uka] Zec

Publisher:

DAAAM International Vienna
TU Wien, Getreidemarkt 9, 1060 Wien, Austria
E-mails: president@daaam.info / branko.katalinic@tuwien.ac.at
DAAAM homepage: http://www.daaam.info

CONTENTS

CONTENTS iii
AUTHOR INDEX vi
COMMITTEES x

Katalinic, B.: *Preface*	0001
Budic, H.; Marinac, A. & Kelic, I.: *Application of Artificial Intelligence in Tourist Destination Management*	0004
Peterka, J. & Hipp, J.: *Application of Templates in a CAM System for Geometrically Similar Components*	0013
Peterka, J. & Skursky, I.: *An Innovative Approach to Gear Manufacturing: Power Skiving on CNC Lathes*	0021
Georgieva, P.; Topalova, I.; Demirova, S. & Vasilev, I.: *Conceptual Model for Measuring and Analysing Innovation Obsolescence Through Hybrid Intelligence*	0029
Topalova, I. & Radoyska, P.: *Models for Evaluating Effective IP Traffic Management with Application of Artificial Intelligence*	0039
Zaharinov, V. & Malakov, I.: *An Approach for Automated Creation of Three-Dimensional History-Based Parametric Models*	0047
Zelenika, D.; Marusic, Z. & Markic, I.: *The Application Potential of Real-Time Human Detection Systems in Search and Rescue Operations*	0057
Ronkova, V. & Dobreva, A.: *Optimal Design Procedures in Machine Science Education*	0062
Bahrami, A. & Potekhin, V.: *Hybrid Machine Learning-Enhanced PID Controller Optimization for Industrial Process Control*	0067
Ipatov, O.; Barinova, D.; Saramakha, S.; Tsimerman, E. & Krylova, E.: *Using EduScrum Technology for Developing Academic Speaking Skills of Students Majoring in Engineering*	0074
Pandzic, A.; Hodzic, D. & Demirovic, E.: *The Influence of Annealing on the Mechanical Properties of FDM 3D Printed PET Reinforced with Carbon Fibers*	0082
Lavic, Z. & Bibic, A.: *Fuzzy AHP-Based Criteria Weight Determination for the Evaluation and Ranking of Renewable Electricity Generation Projects*	0095
Lavic, Z. & Bibic, A.: *Application of the Fuzzy TOPSIS Method for Ranking Renewable Energy Electricity Generation Projects*	0101

Galic, Z.; Trzun, Z. & Razumic, A.: *Application of Monte Carlo Simulation for Evaluating Wind Influence in External Ballistics*	0107
Al-Dailami, Z.; Elbarghthi, A. F.; Dvorak, V. & Hlava, J.: *Design-Oriented Sobol' Sensitivity Analysis of Key Parameters in Plate Heat Exchanger Optimisation*	0116
Coruzzolo, A. M.; Forgione, C.; Balugani, E.; Gamberini, R. & Roversi, A.: *Comparison of Different Motion Capture Technologies for Gait Analysis and Suggestions for Implementation in Tourism Applications*	0122
Zhao, Q.; Balugani, E.; Gamberini, R. & Lolli, F.: *A Value-Based Decision Framework for Evaluating the Integrated Efficiency of Smart Grids*	0131
Neslihan, D.: Towards Smart Factories: *The Development of Autonomous Mobile Robots for Intralogistics 4.0*	0141
Rosales Molina, E. I.: *Learning Design of Experiments through Project-Based Learning*	0147
Bisic, M.; Pandzic, A.; Jokic, N.; Bevrnja, M. & Elek, P.: *Evaluation of Firing Reliability and Penetration Depth of 3D Printed PLA Projectiles for 9 X 19 mm Ammunition*	0154
Rijkure, A.; Cerbule, K. & Megnis, J.: *Impact of Trap-Net Fishery on Marine Pollution and Solutions*	0163
Vranakova, N. & Gyurak Babelova, Z.: *Transformation of Hr Areas as a Result of Digitization and Automation: Analysis and Findings*	0169
Stastny, P.: *Impact of Remedial Interventions on Rising Damp: Insights From a Single-Case Study*	0175
Masnec, L. & Stapic, Z.: *System Dynamics and Digital Twins: A Bibliometric Insight into Real-Time Data Integration*	0180
Krause, M.; Bleicher, F. & Habersohn, C.: *Design and Evaluation of a Piezo-Driven Turning Tool for Machining Non-Circular Geometries*	0186
Fabregat-Jaén, M.; Peidro, A.; Paya, L. & Reinoso, O.: *Efficient Computation of 2D Self-Motion Manifolds: A Method Comparison in a 5-DoF Robot*	0191
Aznar Ortega, J. F.; Peidro, A.; Paya, L. & Reinoso, O.: *Design of a Hanging Parallel Robot for Plastering Brick Walls in Construction*	0201
Schneider, M.; Bobertag, M.; Glück, J.; *Hendricks, V. & Kreißl, J.: From Unbalance to Precision: Development of a Sensor-Based Balancing Concept for Grinding Technology*	0210
Dupljak, E. & Domazet, E.: *AI-Driven Approaches to Genetic Risk Stratification and Early Detection of Breast Cancer*	0217
Indra, M.; Mikulicova, M. & Kresalek, V.: *Using Luminescence Spectroscopy to Monitor the Aging of Latent Fingerprints*	0225
Nesimi, S. & Domazet, E.: *Energy-Aware Task Scheduling for Secure Autonomous Vehicles: An Analysis of Current System Designs*	0231
Milicevic, H.; Carevic, D. & Bujak, D.: *Calibration of Sediment Transport Parameters in the XBeach-G Numerical Model on Artificial Gravel Beach Ploče*	0239
Demir, B. & Domazet, E.: *Dual-View Explainable AI for 12-Lead ECG With OpenMP-Accelerated Postprocessing and External Calibration*	0244
Bendic, D.; Toric, N. & Boko, I.: *Rheological Model of Aluminium Alloys at High Temperatures*	0249
Jevtic Rundek, R. & Uros, M.: *Analysis of a Typical 1960s Large Panel Building Seismic Resistance*	0255
Straka, M.: *Machinability Comparison of Conventional Inconel and 3D Printed Inconel 718*	0260
Ivancan, D. & Gasparovic, S.: *Urban Planning Criteria for Riverfront Categorisation*	0266

Hopocky, A. & Dubek, M.: *Self-Curing Concrete: A Modern Approach to Optimising Hydration and Durability in European Conditions*	0271
Vjestica, D.; Cosic, I.; Gradojevic, N. & Dakovic, V.: *The Impact of the Ukranian Crisis on Energy Prices and Capital Market Prospects*	0276
Blabol, V. & Kutlwaser, J.: *The Impact of Changing the Sensing System on Measurement Accuracy*	0285
Sicha, M.; Stuja, K.; Aburaia, M. & Aburaia, A.: *Design and Implementation of an Additively Manufactured, Aluminium-Reinforced 4-Axis Pick-and-Place Robot*	0293
Sousedikova, L. & Adamek, M.: *Challenges of Integrating Virtual Reality into Polygraph Testing*	0298
Abdel-Wahab, Y.; Aburaia, M. & Aburaia, A.: *Evaluating Dexterity Limitations of Bionic Hands via Reinforcement Learning*	0305
Pryanichnickov, V. & Iastrebov, V.: *Technologies for Building Complex Sensor Systems for Collaborative Service Robots*	0317
Khelemendik, R. & Pryanichnickov, V.: *Cellular Environment and 4-Dimensional Cubes for Effective Representation of Robot in Labyrinth*	0321

AUTHOR INDEX

A	
Abdel-Wahab, Y.	0305
Aburaia, M.	0293, 0305
Aburaia, A.	0293, 0305
Adamek, M.	0298
Al-Dailami, Z.	0116
Aznar Ortega, J. F.	0201
B	
Bahrami, A.	0067
Balugani, E.	0122, 0131
Barinova, D.	0074
Bendic, D.	0249
Bevrnja, M.	0154
Bibic, A.	0095, 0101
Bisic, M.	0154
Blabol, V.	0285
Bleicher, F.	0186
Bobertag, M.	0210
Boko, I.	0249
Budic, H.	0004
Bujak, D.	0239
C	
Carevic, D.	0239
Cerbule, K.	0163
Coruzzolo, A. M.	0122
Cosic, I.	0276
D	
Dakovic, V.	0276
Demir, B.	0244
Demirova, S.	0029

Demirovic, E.	0082
Dobreva, A.	0062
Domazet, E.	0217, 0231, 0244
Dubek, M.	0271
Dupljak, E.	0217
Dvorak, V.	0116
E	
Elbarghthi, A. F.	0116
Elek, P.	0154
F	
Fabregat-Jaén, M.	0191
Forgione, C.	0122
G	
Galic, Z.	0107
Gamberini, R.	0122, 0131
Gasparovic, S.	0266
Georgieva, P.	0029
Glück, J.	0210
Gradojevic, N.	0276
Gyurak Babelova, Z.	0169
H	
Habersohn, C.	0186
Hendricks, V.	0210
Hipp, J.	0013
Hlava, J.	0116
Hodzic, D.	0082
Hopocky, A.	0271
I	
Iastrebov, V.	0317
Indra, M.	0225
Ipatov, O.	0074
Ivancan, D.	0266
J	

Jevtic Rundek, R.	0255
Jokic, N.	0154
K	
Katalinic, B.	0001
Kelic, I.	0004
Khelemendik, R.	0321
Krause, M.	0186
Kreißl, J.	0210
Kresalek, V.	0225
Krylova, E.	0074
Kutlwaser, J.	0285
L	
Lavic, Z.	0095, 0101
Lolli, F.	0131
M	
Malakov, I.	0047
Marinac, A.	0004
Markic, I.	0057
Marusic, Z.	0057
Masnec, L.	0180
Megnis, J.	0163
Mikulicova, M.	0225
Milicevic, H.	0239
N	
Nesimi, S.	0231
Neslihan, D.	0141
P	
Pandzic, A.	0082, 0154
Paya, L.	0191, 0201
Peidro, A.	0191, 0201
Peterka, J.	0013, 0021
Potekhin, V.	0067
Pryanichnickov, V.	0317, 0321

R	
Radoyska, P.	0039
Razumic, A.	0107
Reinoso, O.	0191, 0201
Rijkure, A.	0163
Ronkova, V.	0062
Rosales Molina, E. I.	0147
Roversi, A.	0122
S	
Saramakha, S.	0074
Schneider, M.	0210
Sicha, M.	0293
Skursky, I.	0021
Sousedikova, L.	0298
Stapic, Z.	0180
Stastny, P.	0175
Straka, M.	0260
Stuja, K.	0293
T	
Topalova, I.	0029, 0039
Toric, N.	0249
Trzun, Z.	0107
Tsimerman, E.	0074
U	
Uros, M.	0255
V	
Vasilev, I.	0029
Vjestica, D.	0276
Vranakova, N.	0169
Z	
Zaharinov, V.	0047
Zelenika, D.	0057
Zhao, Q.	0131

ORGANIZING COMMITTEE FOR 2025

KATALINIC, BRANKO – PRESIDENT OF DAAAM INTERNATIONAL
COVIC, DRAGAN – HONOR PRESIDENT OF ORGANIZING COMMITTEE
GUSEV, BORIS – PRESIDENT OF INTERNATIONAL ACADEMY OF ENGINEERING
ZEC, ILIJA - SECRETARY

ABURIA, MOHAMED
BEVANDA, IVAN
BIELIC, TONI
BOSNJAK, IGOR
BUCHMEISTER, BORUT
COLAK, IVO
COLAK, STANKO
COSIC, ILIJA
CVETKOVSKA, ROBERTA
DURAKBASA, NUMAN
FUERST, LAURENZ
GODEC, DAMIR
HOELLER, LIANE
KALENIC, MIODRAG
KNEZOVIC, NIKOLA
KOSTOLANI, DAVID
KUKUSHKIN, ILYA
MARIC, PETAR
NAUMBURGER, JOERG
PANFILOV, PETER
PARK, HONG-SEOK
PRYANICHNIKOV, VALENTIN
RAKIC, KRESIMIR
SARAVANJA, LUKA
SCHMOELLEBECK, FRITZ
SEREMET, ZELJKO
STAZHKOV, SERGEY
STAROVESKI, TOMISLAV
STOJKIC, INJA
STOJKIC, ZELJKO
STUJA, KEMAJL
TEKIC, ZELJKO
TOMIC, ZORAN
UDILJAK, TOMA
UGLESIC, ANTE
VLADOV, MAREK
ZEC, ANTE
ZEC, LUKA

RECTORS' AND PRESIDENTS' HONOR COMMITTEE

ABDRASILOV, BOLATBEK	RECTOR, KARAGANDA STATE INDUSTRIAL UNIVERSITY
ABRUDAN, IOAN VASILE	RECTOR, TRANSILVANIA UNIVERSITY OF BRASOV
ALEXANDROV, ANATOLIY	RECTOR, BAUMAN MOSCOW STATE TECHNICAL UNIVERSITY
BAAIJENS, FRANK	RECTOR, EINDHOVEN UNIVERSITY OF TECHNOLOGY
BIAŁKIEWICZ, ANDRZEJ ANTONI	RECTOR, CRACOW UNIVERSITY OF TECHNOLOGY
BONDREA, IOAN	PRESIDENT, LUCIAN BLAGA UNIVERSITY OF SIBIU
BORAS, DAMIR	RECTOR, UNIVERSITY OF ZAGREB
BOROVIKOV, YURY	RECTOR, THE YAROSLAV-THE-WISE NOVGOROD STATE UNIVERSITY
CHEON, YEON	PRESIDENT, UNIVERSITY OF ULSAN
CISNEROS PÉREZ, GUILLERMO	RECTOR, UNIVERSIDAD POLITECNICA DE MADRID
COSTOIU, MIHNEA	RECTOR, UNIVERSITY POLITEHNICA OF BUCHAREST
CZIGANY, TIBOR	RECTOR, BUDAPEST UNIV. OF TECHNOLOGY AND ECONOMICS
DAM, HENRIK	DIRECTOR, UNIVERSITY OF SOUTHERN DENMARK
DANIELS, MITCHELL	PRESIDENT, PURDUE UNIVERSITY
DI SCIASCIO, EUGENIO	RECTOR, POLYTECHNIC OF BARI
DRĂGAN, FLORIN	RECTOR, POLYTECHNIC UNIVERSITY OF TIMISOARA
G. BOUDOUVIS, ANDREAS	RECTOR, NATIONAL TECHNICAL UNIVERSITY OF ATHENS
GALIJEV, ILJHAM	PRESIDENT, OMSK STATE TRANSPORT UNIVERSITY
GARCÍA MARÍN, JUAN FRANCISCO	RECTOR, UNIVERSITY OF LEON
GARCIA, EVA FERREIRA	RECTOR, UNIVERSITY OF THE BASQUE COUNTRY
GELENCSER, ANDRAS	RECTOR, UNIVERSITY OF PANNONIA
GEYER, SYLVIA	RECTOR, UNIV. OF APPLIED SCIENCES TECHNIKUM, VIENNA
GONOKAMI, MAKOTO	PRESIDENT EMERITUS, UNIVERSITY OF TOKYO
GUBERAC, VLADO	RECTOR, STROSSMAYER UNIVERSITY OF OSIJEK

International Scientific Committee

International Program Committee

BLEY, HELMUT
BOTEZATU, CORNELIA PAULINA
BOTEZATU, MIHAI
BRAJLIH, TOMAZ
BREZAK, DANKO
BREZOCNIK, MIRAN
BUCHMEISTER, BORUT
BUCURESTEANU, ANCA
CAMPBELL, R. IAN
CEKIC, AHMET
CELAR, STIPE
CHAN, CHEE SENG
CHAUS, ALEXANDER
CHRAMCOV, BRONISLAV
COHODAR, MAIDA
COLEMAN, J. J.
COSIC, ILIJA PETAR
COSMA, SMARANDA
COTET, COSTEL EMIL
CUBONOVA, NADEZDA
CURTU, IOAN
DAMIC, VJEKOSLAV
DENKENA, BEREND
DEVAHASTIN, SAKAMON
DOLGA, LIA
DOLGA, VALER
DOLINAY, JAN
DOLINAY, VILIAM
DOSTAL, PETR
DOSTALEK, PETR
DRAGCEVIC, ZVONKO
DRAGICEVIC, MARIJA
DRSTVENSEK, IGOR
DUHOVNIK, JOZE
EJUB DZAFEROVIC,
FIRST ROGALE, SNJEZANA
FUJII, NORIO
GHEORGHIU, HORIA
GODEC, DAMIR
GOMEZ, ANTONIO
GORDIN, DOINA MARGARETA
HLAVA, JAROSLAV
HOFREITER, MILAN
HRUBINA, KAMIL
HRUSKA, FRANTISEK
HRUSKOVA, ERIKA
IBISHI, ISMET M.
JANACOVA, DAGMAR
JANSSENS, THIERRY
JARVIS, CHRISTINE W.
JAVOROVA, ANGELA
JERBIC, BOJAN
JURAGA, IVAN
JURUM-KIPKE, JASNA
KARAULOVA, TATYANA
KATALINIC, BRANKO
KAWALEC, ANDRZEJ
KHANNA, ASHUTOSH
KHOMCHENKO, VASILY
KIRK, CHRIS. S.
KISS, IMRE
KITTL, BURKHARD
KLIMENKO, S. A.
KNEGO, NIKOLA
KNEZEVIC, PETAR
KOHLERT, HELMUT
LUKIC, IVANA
LUPEA, IULIAN
LUPU, CIPRIAN
LUPULESCU, NOURAS BARBU
MACHEDON PISU, TEODOR
MACKU, LUBOMIR
MACUROVA, ANNA
MAGHETI, IOAN
MAJETIC, DUBRAVKO
MALISA, VIKTORIO
MALOPOLSKI, WALDEMAR
MAN, MARIANA
MANDIC, LIDIJA
MANDRU, DAN
MANESCU, TIBERIU STEFAN
MARASCU KLEIN, VLADIMIR
MARCOS, MARIANO
MARCU, MARIUS
MAREK, KUBALCIK
MARIAN TOLNAY
MARIN, DUMITRU
MARIN, MIHNEA
MARINESCU, NICULAE ION
MARN, FERDINAND
MARONEK, MILAN
MARSLAND, DAVID
MARTA, CONSTANTIN
MARTINEZ, SUSANA
MASTACAN, LUCIAN
MATUSOVA, MIRIAM
MATUSU, RADEK
MEHTA, NARINDER KUMAR
MEIER, HORST
MENZ, PETER
MICHAL, BLIZNAK
MICIETA, BRANISLAV
MICLOSI, VIOREL
MOHORA, CRISTINA
MOK, H.
MORARU, SORIN AUREL
MORIWAKI, T.
MORTOIU, DOINA
MOTOMANCEA, ADRIAN
MUNTEAN, MIHAELA
MUNTEAN, SEBASTIAN
MUNTEANU, ANCA
MUSCA, GAVRIL
MUSTAFIZUL, KARIM
MYSHLYAVTSEV, ALEXANDR
NAGI, MIHAI
NAGLA, K. S.
NAKAMOTO, NOBUYUKI
NEUMANN, PETR
NICA COTET, GABRIELA BEATRICE
NIKOLIC, GOJKO
NKUMBWA, RODGER
NOH, SANG DO
NOVAC, OVIDIU
NOVOSELOVIC, DANIEL
NURUL AMIN A.K.M.
OANCEA, GHEORGHE I.
OKAMOTO, J.
OLARU, ADRIAN, DUMITRU
OLARU, ONISIFOR
OLEKSIK, VALENTIN
OLTEANU, CIPRIAN
PLESA, ANGELA
PODURAEV, JURY
POKORNY, PAVEL
POMAZAN, VALENTINA
POPA, CICERONE LAURENTIU
POPPEOVA, VIERA
PRODAN, DAN
PROKOP, ROMAN
PRYANICHNIKOV, VALENTIN
PYTKO, STANISLAW
QUINN, ROGER D.
RAMAJ, VEHBI XH.
RAVAS, RUDOLF
RAY, T. S.
RIKALOVIC, ALKSANDAR
RIZESCU, SABIN VASILE
ROGALE, DUBRAVKO
ROHDE, ULRICH L.
ROMERO, FERNANDO
ROS, OLIMPIA
ROSADO, V.O.G.
RUDAS, IMRE J.
RUGESCU, DRAGOS RADU DAN
RUJA, IOAN
RUSU ANGHEL, NICOLAE
RUSU, LIGIA
RUSU-CASANDRA, AURELIA
RUTEMI, GANI BRAHIM
RUZIC, DRAGO
SAKULSKI, DUSAN
SALIHU, AVDI HAJDIN
SAMEK, DAVID
SCHREIBER, PETER
SELVAKUMAR, ARUL LAWRENCE
SEONG, POONG HYUN
SERCER, MLADEN
SHIBASAKA, TOSHIROH
SHIH, MING CHANG
SIEMIATKOWSKI, M.
SIHN, WILFRIED
SINAY, JURAJ
SISEK, BORIS
SITUM, ZELJKO
SKOCZYNSKI, WACLAW
SKOLUD, BOZENA
SLUGA, ALOJZ
SMIRNOV, ALEXANDER
STANKOVSKI, STEVAN
STAROVEŠKI, TOMISLAV
STEFANIC, NEDELJKO
STOPPER, MARKUS
STRGAR KURECIC, MAJA
SUBRAMANIAN, ARUNKUMAR
SUGAR, PETER
SYROVA, LIVIA
SYSALA, TOMAS
TADAMURA, KATSUMI
TAKALA, JOSU
TAMASILA, MATEI
TAN, KOK KIONG
THIERHEIMER, WALTER WILHELM
ULRICH, GUENTHER
VAARIO, JARI
VASEK, VLADIMIR
VELISEK, KAROL
VEZA, IVICA

REVIEWING COMMITTEE

CULJAK, EVA
CUS, FRANCI
DAR-EL, E.
DE FILIPPI, A.
DEVAHASTIN, SAKAMON
DINCMEN, M.
DINIZ, ANSELMO EDUARDO
DOHDA, KUNIAKI
DOINA MARGARETA
DONG, JIANHUA
DOVE, RICK
DURAKBASA, NUMAN
EGOROV, SERGEY
ENDTER, R.
EVERSHEIM, W.
GALANTUCCI, LUIGI MARIA
GAWLIK, JOZEF
GIACAGLIA, G. E.
GIBSON, I.
GIROT, FRANK
GOLDFARB, VENYAMIN Y.
GOMEZ, ANTONIO
GRABENWEGER, JOHANN
GUSEV, BORIS V.
HABCHI, GEORGES
HAN, QING YAO
HOBLE, DOREL ANTON
HOCK, SEO SWEE
HOFMANN, PETER
HRYU, JEE-HWAN
ILTSCHEV, VELKO
MACUTA, SILVIU-DANUTI
MARCOS, MARIANO
MARCOS, MARIANO
MARIC, PETAR
MATUSU, RADEK
MEGHDARI, ALI
MERTINS, KAI
MICHELINI, R. C.
MILOS, LAURA RAISA
MIRONEASA, SILVIA
MOMETE, DANIELA CRISTINA
MUNTEANU RADU
NASTAC, SILVIU
NEUMANN, H. J.
NICOLETTO, G.
NOVAC, CORNELIA MIHAELA
NOVAC, OVIDIU CONSTANTIN
OCZOS, KAZIMIERZ E.
OTTO, TAUNO
PALCIC, IZTOK
PANDA, ANTON
PANFILOV, PETER
PARK, HONG-SEOK
PEGDEN, DENIS
PERAKOVIC, DRAGAN
PERUTKA, KAREL
PETERKA, JOZEF
PIVONKA, PETR
PODURAEV, JURY
POP, PETRU ADRIAN
POPESCU, DANIEL
PORTER, BRIAN
TOPALOVA, IRINA
TOPLICEANU, LILIANA
TORRES-LEZA, F.
TORVINEN, S.
TSAI, B. S. Y.
UDILJAK, TOMA
UHLMANN, ECKART
UYAR, EROL
VALMOHAMMADI, CHANGIZ
VAN BRUSSEL, HENDRIK
VAN OUDHEUSDEN, DIRK L.
VARGA, GYULA
VAVRIN, PETR
VENKATESH, V. C.
VETTURI, DAVID
VITECEK, ANTONIN
VOSNIAKOS, G. CHRISTOPHER
VOUK, RUDOLF
WALDIR, P. O.
WERTHEIM, RAFI
WESSELY, EMIL
WESTKAEMPER, ENGELBERT
WIENDAHL, HANS-PETER
WILLIAMS, D. J.
WINSTON, PATRICK H.
WOZNIAK, ADAM
ZALZALA, ALI M. S.
ZEC, ILIJA
ZEC, LUKA
ZHAO, DAOZHI
ZLOTO, TADEUSZ

36TH DAAAM INTERNATIONAL SYMPOSIUM ON INTELLIGENT MANUFACTURING AND AUTOMATION

Preface

The 36th DAAAM International Symposium on Intelligent Manufacturing and Automation was organised as virtual online conference hosted by the Vienna University of Technology, Vienna, Austria, between the 30th and 31st October 2025, during the DAAAM International Week. The Symposium was organized by DAAAM International Vienna and University of Mostar in cooperation with ÖIAV 1848, Vienna University of Technology, International Academy of Engineering and University of Applied Sciences – Technikum Wien and Under the Auspices of the Danube Rectors' Conference & Rectors' and Presidents' Honor Committee of DAAAM International for 2025. The Symposium was hosted from Vienna, Austria. This year's symposium aimed at continuing the success of the previous years, focusing on the five-fold traditional objectives of the symposium: the presentation of the most recent high-quality results, support of development of young scientists and researchers, organization of international (summer) doctoral school, inauguration of new members of Central European Branch of International Academy of Engineering and the provision of the necessary setting for stimulating discussions, brainstorming and networking among European and international researchers coming both from the academia government agencies and industry

DAAAM International Symposium has been established as the leading Central European conference devoted to the cutting-edge advancements in the field of modern production and automation. Symposium provides an international forum for engineers, scientists and industrial managers to present and discuss the current status and impact of advanced manufacturing and automation. The international series of Symposium was initiated by the first meeting in Vienna (1990) and continued in Štrbske Pleso (1991), Budapest (1992), Brno (1993), Maribor (1994), Cracow (1995), Vienna (1996), Dubrovnik (1997), Cluj - Napoca (1998), Vienna (1999), Opatija (2000), Jena (2001), Vienna (2002), Sarajevo (2003), Vienna (2004), Opatija (2005), Vienna (2006), Zadar (2007), Trnava (2008), Vienna (2009), Zadar (2010), Vienna (2011), Zadar (2012), Zadar (2013), Vienna (2014), Zadar (2015), Mostar (2016), Zadar (2017), Zadar (2018), Zadar (2019), Mostar (2020), Vienna (2021), Vienna (2022), Vienna (2023), Vienna (2024), Vienna (2025).

Regular organization of the annual Symposium is one of the central activities of the DAAAM International association. The association was founded on 5. November 1990 in Vienna during a meeting organised by Professor B. Katalinic in order to celebrate the 175 Years of Vienna University of Technology. The main idea of the association was to establish a network for a new kind of international scientific and academic cooperation in the fields of intelligent manufacturing and automation. Our primary focus is on: all aspects of production of technical products made out of parts and corresponding automation and knowledge, trans-disciplinary application of methods and all kinds of education.

DAAAM International activities, during the last thirty-six years, have been supported by scientists and experts from over fifty countries all over the world. In the course of the last decade, DAAAM International has focused its activities on the regional and global long-term co-operation between (technical) universities. The result of this initiative is the founding of the University Network of DAAAM International Vienna and DAAAM International Network for Advanced Technologies which was founded on 2002-10-25 in Vienna.

The main activities of DAAAM are:

- Organisation of annual DAAAM International Symposium.
- To maintain and improve the University Network of DAAAM International Vienna as a strategic alliance and permanent international open platform for long-term co-operation and partnership in all fields of technology, research and education.
- Organisation of long-term projects, initially known as Inter-University Doctoral Studies.
- Organisation of long-term projects, initially known as the Virtual University.
- Publishing of international journals, books, and proceedings.
- Promotion of the exchange of people, scientific, and technical information.
- Promotion of international co-operation among scientists and engineers.
- Promotion of R & D, education, and industrial application.
- To co-operate with other national and international organisations in matters of common interest.
- Organisation of International Doctoral School as a platform for the support of the next generation of engineers at the highest level.

This proceeding contains 46 refereed papers from 137 authors presented at the 36th DAAAM 2025. In DAAAM proceedings, since the first symposium in 1990, we published papers from 9869 authors.

All the papers included in these proceedings are reviewed by experts in double blind peer review way, and finally revised by the authors for the improvement according to referee's reports and advices.

The papers discuss many aspects of modern manufacturing and automation such as: Algorithms, Artificial Intelligence, CAX, Computer Integration, Control, Cutting Tools, Design, FEM, Invited Lectures, Knowledge, Management, Manufacturing System, Mechatronics, Methodology, Methods, Modelling, Optimization, Robotics, Simulation, Technical Solutions, Technology and Trends.

Appreciation is forwarded to reviewers who contributed to improve the quality of papers reviewing the papers, providing constructive critical comments, improvements, and corrections to the authors. We also recognize the support of reviewers, who were in charge of acceptance/rejection process. We wish to thank the Scientific, Program and Organizing Committees, Session Chairs, student helpers and administrative assistants, for contributing to the success of this Conference. They did an excellent job.

Finally, we wish to thank all the authors who submitted papers, making this conference possible and the authors of accepted papers for making appropriate modifications and corrections with care and patience and for preparing the final manuscript. It is the quality of their presentations and their passion to communicate with other participants that really makes this conference series a success.

We would like to thank all members of the International Academy of Engineering – Central European Branch, Rector of the Vienna University of Technology Prof. Dr.-Ing. Jens Schneider, and Rector of the University of Mostar Prof. Dr. Sc. Ivo Čolak and his team, management and all those nice colleagues for their excellent support in the organization of 36th DAAAM International Symposium.

Vienna, 2025-12-15
Editor
Chair of the 36th DAAAM 2025

Branko Katalinic

Professor rtd Dr.sc. Dr.mult.h.c.Prof.h.c. Branko Katalinic
President of DAAAM International
TU Wien, IFLT-IMS, Getreidemarkt 9
1060 Wien, Austria, Europe
URL: http://www.daaam.info
E-mail: president@daaam.info
branko.katalinic@tuwien.ac.at

DOI: 10.2507/36th.daaam.proceedings.001

APPLICATION OF ARTIFICIAL INTELLIGENCE IN TOURIST DESTINATION MANAGEMENT

Hrvoje Budic, Antun Marinac & Ivan Kelic

This Publication has to be referred to as: Budic, H[rvoje]; Marinac, A[ntun] & Kelic, I[van] (2025). Application of Artificial Intelligence in Tourist Destination Management, Proceedings of the 36th DAAAM International Symposium, pp.0004-0012, B. Katalinic (Ed.), Published by DAAAM International, ISBN 978-3-902734-47-1, ISSN 1726-9679, Vienna, Austria
DOI: 10.2507/36th.daaam.proceedings.001

Abstract

This paper examines the application of Artificial Intelligence (AI) in tourist destination management, with a focus on its potential to enhance operational efficiency, decision-making, and visitor experience. The study analyses key AI tools, including recommendation systems, chatbots, virtual assistants, dynamic pricing, big data analytics, and service robots, supported by illustrative examples such as the humanoid robot Josie Pepper and the smart Hajj bracelet. Findings indicate measurable improvements in tourist satisfaction, loyalty and conversion rates, as well as in resource allocation and sustainability monitoring. AI technologies enable destinations to personalise services, predict demand and mitigate environmental impacts through real-time data processing. However, the paper also highlights potential challenges, including data privacy concerns, algorithmic bias, energy consumption and the digital divide affecting smaller destinations and enterprises. The research concludes that strategic investment in digital infrastructure, staff competencies, ethical standards and pilot implementation is essential to ensure that AI contributes to both sustainable and inclusive tourism development while maintaining the authenticity of local experiences.

Keywords: Artificial Intelligence; Tourism Management; Destination Competitiveness; Visitor Experience; Sustainable Tourism.

1. Introduction

The swift advancement in artificial intelligence (AI) is affecting many sectors, including the tourism industry, and requires transforming operations into higher technology and advanced practices. Thus, the following research question is explored: How does AI influence efficiency, decision-making, and visitor experiences in the management of tourist destinations? As tourism significantly contributes to global GDP and employment, this question is paramount for stakeholders keen to understand how this technological development is impacting the sector. The linkage between AI and tourism management has become increasingly significant. This review examines how AI technologies are transforming the management of tourist destinations by analysing how they can improve efficiency, assist in decision-making, and enhance visitor satisfaction. AI has become an integral part of tourism management innovations and developments in areas such as machine learning, natural language processing, and robotics. AI applications, such as recommendation systems and big data analytics, facilitate informed decision-making by personalising travel experiences.

The humanoid robot, Josie Pepper, at Munich airport has proven how this technology helps to improve efficiency and visitors' experiences. Also, the smart Hajj bracelet created by Saudi Arabia is another example that enhances services and visitor satisfaction. Furthermore, AI helps develop sustainability objectives, such as monitoring environmental impacts in real-time and tracking carbon emissions. Hence, the effects of AI on tourism should be thoroughly studied and considered.

The research question is addressed through a systematic review of the use of AI technologies to enhance efficiency, inform decision-making, and improve visitor experiences in the management of tourist destinations. Various recent academic works from publishers such as Wiley and Sons, Scopus and Springer that explored current trends, issues and opportunities relating to AI's impact on tourism were analysed. Moreover, the practical implications and findings were examined for various stakeholders, including tourism managers, policymakers, and tourists themselves, to gain a better understanding of how AI supports the goals and aims of the tourism sector.

The study of AI's effect on tourism management employed a systematic review and critical analysis of case studies and empirical evidence, drawing on systematic reviews, analyses of published searches and case studies in the tourism management domain. Approaches involved a comparative analysis of AI effectiveness, a critical evaluation of data sources and a thematic synthesis of findings. Furthermore, some quantitative data were integrated, such as a 30% decrease in congestion from intelligent systems, while qualitative insights, including improved satisfaction due to AI-enabled personalisation, were also reviewed.

The increasing integration of AI in tourism operations is now supported by research and development that has led to numerous applications, including AI-based recommendations, chatbots, virtual assistants, and dynamic pricing. Over the next decade, AI in the tourism market is expected to continue growing and becoming more relevant and widely utilised by the industry as innovations and technologies emerge. Challenges and limitations remain as a topic for discussion and further research. The impact of AI on tourism is reviewed in the following order:

- **Chapter 2**: AI-Technologies in Tourism Management, covering AI-based services for destinations;
- **Chapter 3**: Impact on Operational Management, analysing AI's contribution to the operations of tourism providers and its impact on destinations;
- **Chapter 4**: Impact on Visitor Experiences, evaluating how AI-based technologies will improve visitors' experiences in destinations;
- **Chapter 5**: findings, summarising findings and their implications for tourism management.

2. AI Applications in Tourism Management

The introduction of AI in tourism management has significantly enhanced customer service by offering continuous and multilingual assistance to tourists through chatbots and virtual assistants. By promptly addressing customer inquiries, fulfilling reservations, and offering personalised recommendations, the waiting time and overall convenience for customers have been significantly improved. Customer satisfaction has improved as a result, leading to enhanced customer retention (e.g., a 30% increase in customer loyalty following the implementation of a virtual assistant by WestJet). These advancements suggest that AI-led applications and resources are becoming an essential function within the organisation and replacing human roles. In contrast, while AI systems can answer basic inquiries, their understanding of culturally diverse and more complicated customer requests remains an area for development. Customer preferences can be identified by analysing customers' behaviour patterns. Thus, personalised recommendations, such as holiday destinations, accommodations, and activities, are provided to customers to increase the likelihood of making a purchase. Companies like Booking.com and Airbnb are utilising AI-generated personalisation, resulting in significant business outcomes, including a 20% rise in bookings and a 25% increase in conversion rates. [1] The collection of customer information through AI for personalisation may trigger concerns surrounding data security, privacy breaches and misuse.

AI applications also offer real-time feedback from customers to improve service quality. Quick identification of errors allows time for improvements to take place before new customers arrive. By implementing corrective measures in response to customer feedback businesses can improve their quality of service. [2] However, customer interaction cannot be left solely to artificial intelligence, as it may overlook vital aspects of human connection and feedback, aiming to achieve better service quality. Continuous and reliable customer service applications can handle a high volume of communication and services, thereby benefiting the industry by reducing costs and improving profits. [3] However, it must be noted that someone will need to control these systems and keep them updated with the changing demands of customers in different languages to maintain their use and profitability.

Data analytics are used to enhance marketing. The data and information gathered by AI-based systems include customers' purchase history, online activities, and preferences. This is used to create personalised ads, leading to higher sales and revenue returns. By tracking demand, competitor pricing policies and market trends online travel agencies can enhance their pricing and resource management strategies. For example, the Lighthouse dynamic pricing mechanism enabled automated pricing, resulting in a 20% increase in revenue generation. However, dynamic pricing may be considered exploitative for customers when prices change frequently and are determined algorithmically by the travel agencies. Machine learning (ML) is being used to analyse customer data and thus forecast emerging travel trends. [4] This allows marketing to be targeted and timely increasing the likelihood of conversion for customers and therefore remaining competitive in the market.

Nevertheless, the effectiveness of forecasting depends on accurate and representative datasets. It may thus not be appropriate in highly disruptive tourism industries due to the risk of overfitting to historical data. Predictive behaviour algorithms support the efficient distribution of marketing efforts by identifying high-value customer segments and anticipating consumer purchasing decisions. This enables more efficient and targeted campaign planning resulting in higher customer conversion rates. Another tool is sentiment analysis of customer views which involves analysing online reviews and social media posts. Tourism organisations can measure travel experiences and trends through this AI system leading to the modification of customer relationship policies. Nevertheless, the accuracy of predicting marketing trends, as well as determining travel habits, relies on the collection and processing of accurate and up-to-date data.

The use of AI and robots for service delivery introduces new approaches to operational management within the tourism industry. Hotel check-ins, room service, scheduling of housekeeping and other tourism-related tasks can be completed more efficiently and with a lower likelihood of errors. Employees, therefore, may be tasked with providing guest-centric services leading to an elevation of human connection and improved service quality.

Fig. 1. Tourism management infographic [18]

Wang and Karia estimated that within the next 10 years, up to 25% of hotel jobs will be automated. There are discussions surrounding employees losing their jobs, highlighting the need for companies to help them reskill with different skills and knowledge. AI-based systems operate continuously without requiring downtime, unlike humans. These advancements enable the service of high volumes of tourists and guests. For example, at airports or bigger hotels, there is an increase in global customers and a decrease in waiting time when AI systems are implemented for a wide variety of applications. Problem that can arise with AI-led services is that they become too impersonal and fail to emotionally or mentally connect with the consumer. Humans generally prefer human interaction, leading some consumers to dislike talking to a robot for a more personal connection; therefore, improvements are still necessary for the systems.

89% of organisations in the tourism industry that utilise digital platforms demonstrate the essentiality of their infrastructure. The digital platforms feature integrated applications for tasks such as AI-generated listings, pricing updates, and personalisation. These applications ensure efficiency, resource allocation and control within back-office operations increasing productivity. Moreover, nearly 95% of the companies agree that technology is vital to development within the tourism sector. Although digital platforms enhance efficiency and marketing, ongoing investment is necessary to improve them continually. Cybersecurity is a growing concern for all organisations, as well, with the increasing amount of customer data, systems must be kept up to date with high standards.

AI platforms connect partners within the tourism industry, often including hotels, restaurant owners, and destination management organisations (DMOs). [5] This enhances communication between stakeholders leading to increased efficiency in delivering services. Data privacy and digital connectivity can lead to increased exposure and risk including hacking and the theft of sensitive information. Cybersecurity is a concern for all organisations and companies that use digital platforms as part of their operations and thus must be emphasised during the initial stages.

The global market for AI in tourism is estimated to increase from USD 487.7 million to USD 9.8 billion from 2023-2033, growing at a compound annual growth rate (CAGR) of 35%. This statistic illustrates the increase in interest for AI in tourism, with more organizations and companies using it to create competitive advantage and thus improve operation efficiency and customer service. More companies should also strengthen their use of AI in personalisation, as most companies are currently only utilising AI on basic systems, such as personalised recommendation engines. More opportunities are found in tourism business models to be innovated, such as incorporating AI, to include real-time experiences and the ability for customers to ask an AI system to create personalised trip itineraries. [6] The ethical debates about AI, as discussed previously, need to be considered in future AI implementations for tourism. This includes data ownership and algorithmic accountability.

In summary, the use of AI in tourism management enables businesses to deliver better customer service and enhance their operational systems. It plays a significant role in customer engagement, effective marketing, and service improvements. However, it should be used and modified accordingly to achieve better benefits while ensuring that ethical concerns are addressed.

3. Impact on Tourist Destinations Operations

AI-driven systems optimise traffic and crowd management at tourist destinations by analysing real-time data from surveillance cameras, mobile applications, and GPS trackers. For instance, Hangzhou's "City Brain" system has reduced congestion rates in the West Lake region by 30%. This contributes to a higher level of visitor satisfaction and improves the overall flow of traffic within the city. The AI system accurately predicts peak hours and identifies anomalous trends, prompting timely interventions such as rerouting traffic and reallocating resources to optimise efficiency. Although AI advances efficiency, it is necessary to establish governance policies that address the ethical considerations of privacy. AI helps to coordinate tourism-related activities with transport providers, emergency services and other authorities by using algorithms to predict and act upon data. This improves safety as responses become evidence-based and quicker. [7] Nonetheless, data policies and practices that respect the privacy and consent of travellers are imperative to maintain long-term visitor assurance.

The application of AI has made tourism practices 128% more efficient than traditional methods. It enables resource management through real-time logistics and demand forecasting. Prediction models of visitor patterns allow maintenance schedules to align with needs, thereby decreasing costs. Nonetheless, the systems require continuous modification of algorithmic models and learning new information to accommodate operational shifts. Artificial intelligence has generated approximately $ 400 billion for the global tourism market. It correlates with tourism GDP contribution (correlation coefficient = 0.750, p-value = 1.55E-05) and provides economic value if included in national development strategies. However, technology-poor countries may face exclusion if international investment in AI applications is not prioritised. [8] Through real-time environmental monitoring AI can alert destination managers to contamination or the exploitation of ecological resources. This information enables the conservation of biodiversity. AI-implemented systems provide sustainability support at tourist sites, improving environmental credibility. Moreover, carbon tracking applications allow the calculation of individualised carbon emissions for visitors, providing transparency into CO_2 impact and facilitating the optimisation of travel behaviour. The usefulness of the information, however, depends on the accuracy of the environmental data and widespread behavioural changes among travel stakeholders. Artificial intelligence supports adaptive management in the natural environment. AI-implemented surveillance enables the prompt response to harmful or illegal actions in protected areas. Additionally, data-based risk mitigation measures are now used in reserve management and national park oversight. Despite the public benefits, these actions raise concerns about privacy in the context of surveillance. It is necessary to highlight the positive results of AI usage as it continues to monitor and make predictions of the state of our environment. There is a surge in consumer demand for ethical and sustainable tourism. It is therefore necessary for operators to incorporate AI-driven technologies into their efforts to meet the expectations of a market segment that prioritises environmental concerns.

Artificial intelligence can address the problems of oversaturated destinations, as seen in the Denmark pilot project of deep learning. This study utilises AI to predict where tourists are heading and then directs them to less crowded locations using digital signage, promotional platforms, and incentives. In doing so, it not only improves flow in overcrowded areas but also diversifies tourism to previously neglected areas, thereby spreading economic benefits.

Operational cost savings can be achieved with automated AI concierge services. These services can address customer requests via chat and send delivery services where required. Furthermore, AI can customise travel recommendations. These services can often deliver better service than a concierge, but depersonalization is an area that must be addressed through emotionally intelligent design. Establishing objectives of AI use and performance benchmarks is the digital foundation for effective decision-making. As part of digital transformation, tourism entities can comprehensively measure and optimise outcomes, such as visitor satisfaction. Also, AI can be applied on a pilot project basis, but iterative testing must be conducted to calibrate the optimal AI to cater to unique operational characteristics.

Despite the numerous positive aspects of AI applications within the industry, some tourism sectors are more advanced than others in terms of integration. Data from the OECD indicate that, by 2023, 11% of travel agencies and tour operators, as well as 4% of accommodation and food service activities, had adopted at least one AI technology. The digital divide presents barriers, including high initial capital costs, a lack of digital literacy, and limited infrastructure, which can hinder the growth of small to medium-sized enterprises (SMEs). To bridge this divide policymakers must focus on democratising AI to ensure its effective use by the broader sector and avoid productivity gaps.

AI adoption is accompanied by an increasingly large electricity demand with some studies predicting a doubling of data centre energy consumption by 2026. [9] The balance between improved productivity through digital applications and climate-related negative impacts is the key challenge in the ongoing digitalisation transformation. As the sector expands its efforts, AI-based solutions, energy-efficient algorithms, and data centre design must be further explored and researched for sustainable digitalisation strategies.

In a survey conducted in Egypt, the adoption of AI applications among tourists reached over 75%. When the results are applied in linear regression, the analysis highlights a significant influence of AI applications on overall service quality and assurance. [10] To maximise positive visitor assurance outcomes, tourists need to have prior experience and trust in these technologies, as well as have appropriate training on application functions.

The global market size for AI in sustainable tourism is expected to increase from USD 17.7 billion in 2022 to USD 54.5 billion by 2030. With better destination management, an integrated technology-based ecosystem, and tailored services and recommendations, destinations will be able to cater to a global market and create revenue. With data-driven analysis, destinations will be able to enhance efficiency in multiple aspects of operations while also satisfying customer preferences. This, in turn, will strengthen the relationships and bonds with the visitors, ensuring long-term loyalty and support. Proper establishment and recognition of data ownership and control are essential when introducing digital transformation. The appropriate establishment and recognition of data ownership and control are crucial when introducing digital transformation.

The 2025 Mahakumbh festival in India is expected to attract approximately 40 million visitors. It is imperative to monitor the crowds that will be moving throughout the festival areas, as density can be challenging to oversee. The application of AI surveillance, such as monitoring and mapping crowds, enables security control, monitoring and analysis during dynamic conditions. [11] Facial recognition enables faster detection and action on missing person reports, and predictive analyses using AI help facilitate real-time adjustments to facilities and operations.

4. Enhancing Visitors' Experience

The employment of AI technology enhances the visitor experience in tourism by providing highly personalised recommendations and services tailored to visitors' demographics and preferences. For example, using AI-powered recommendation tools, Booking.com and Airbnb have increased bookings by up to 20% and conversion rates by 25%. Ethical considerations must be addressed to make AI technologies more reliable, improve visitors' data privacy and prevent data bias.

The implementation of AI-powered recommendation tools has been found to influence customers' consumption by 30% significantly. [12] Marketing strategies, such as cross-selling and product bundling can be utilised in these tools to attract customers. Implementing AI technologies for product and service promotion can benefit both, destination marketers and destination operators. However, the implementation of AI in marketing must be handled with caution and transparency, as the over-personalisation of recommendations or the excessive promotion of products may make customers uncomfortable or feel pressured into purchasing unnecessary services.

The adaptability of AI-powered personalisation mechanisms which employ machine learning algorithms ensures the relevance of all recommendations based on visitor preference history, travel context, and behavioural patterns. The implementation of AI-enabled personalisation mechanisms enables catering to individual visitors' demands, thereby enhancing the appeal of destinations. Machine learning technology can be trained and adapted to travel preferences and behavioural data collected from visitors. The training dataset can consist of historical traveller information and past transaction information. However, data and machine learning models can become rigid with the accumulation of traveller information, making them inadequate for coping with emerging travel trends. Thus, there are certain shortcomings in the use of machine learning models for personalisation.

The ability of travel agencies and destination marketers to recommend individualised services attracts more customers, strengthens their brand, increases customer loyalty and leads to repeat purchases. Ultimately, travel agencies with higher satisfaction scores tend to gain a larger market share than those with lower scores. [13] AI-based personalisation mechanisms cater to customers' preferences exclusively. Thus, specific experiences that do not cater to their interests may never be recommended to them, hindering their possibilities to try something new.

Service robots and virtual assistants enhance service quality for tourists by addressing a range of visitor needs and providing them with convenience through multilingual services, real-time assistance, and prompt responses, thereby reducing service wait times and minimising language miscommunication. At Munich Airport, for example, the "Josie Pepper" robot guides passengers in various languages to the check-in gates and answers questions related to the airport. However, they cannot act out emotions, nor can they understand visitors' feelings; they are incapable of making decisions in various cases, which may result in certain limitations when it comes to coping with multiple events of the visitors.

Humanoid robots in Swiss restaurants have demonstrated their capability in operational efficiency by being used to serve food and perform repetitive tasks. They can handle complex tasks, communicate with guests, and process complex transactions quickly and accurately allowing restaurant staff to be reassigned to tasks that require creativity and innovation. Humanoid robots can also work long hours without fatigue, and they do not require additional space like other service robots. However, humanoid robots in restaurants also carry many drawbacks. They can sometimes be limited when interacting with individuals of different ages, with disabilities and with special requests. The presence of these robots may also make employees' work more competitive and stressful.

Virtual assistants are widely used for 24/7 customer support, language translation, and booking management to improve visitor convenience and ensure tourism inclusivity. [14] This has helped in increasing tourists' travel satisfaction. However, the effective use of virtual assistants in tourism greatly depends on robust natural language processing, as they need to be capable of communicating effectively in a variety of languages and must understand nuances to ensure that virtual assistants provide comprehensive and contextually accurate guidance for tourists worldwide.

AI-driven AR experiences in museums and other tourist sites significantly enhance visitor immersion, engagement, emotional arousal, knowledge transfer, and recall, ultimately impacting visitor stay length and return rates. In general, approximately 78.9% of the variance in visitor engagement scores is attributed to AI-driven AR features. Thus, the use of AI in delivering AR experiences can potentially contribute to cultural sustainability and support local economies. However, immersive technologies need to be balanced in their implementation to ensure they enhance visitors' experiences at museums and sites, rather than diverting them from the traditional and authentic culture. Generally, higher personalisation enhances visitor knowledge acquisition. Visitor engagement scores increased when the level of personalisation of the AR experience was increased. The diversity of input data must be considered to prevent alienation or marginalisation of specific groups. Additionally, AR experiences must provide value and benefits to all visitors, regardless of their preferences for immersive content.

Fig. 2. Tour planning with tickets on smartphone screen [19]

The use of AI in enhancing immersive experiences can minimise the difference between the accessible and non-accessible visits for people from varied backgrounds. AI applications, such as AR or AI-driven personalisation mechanisms significantly improve overall visitor satisfaction scores for all visitor demographics. However, despite the effectiveness and potential advantages, not all destinations or museums can afford to invest in this technology and provide it at a reasonable price. In general, AI-driven immersive experiences can be utilised to enhance visitor retention and repeat visits by encouraging greater emotional connection and knowledge transfer. The longer length of stay can, in turn, boost the local economy, thereby contributing to the community's economic development. [15] These advances have led to a new level of effectiveness in visitor retention, ultimately contributing to the increased competitiveness of destinations. However, it is also crucial to remain aware of which values should be preserved, and which elements need to be replaced with AI enhancements.

In smart hotels, AI technology is utilised to manage operational costs and environmental impacts, resulting in savings of up to 40% in labour costs and approximately 22% in energy. This aligns with sustainability goals, making smart hotels an ideal alternative for environmentally conscious and economically minded visitors, and is a significant factor in sustainable hospitality development. However, despite the promising advantages of implementing AI in hospitality, the initial investment in smart hotel infrastructure may pose substantial financial burdens, thereby limiting the feasibility for smaller hotels.

The technology of AI enables hotel operators to personalise guest rooms for individual travellers. Today, over 75% of hotels utilise online booking platforms, automated check-in systems and personalised preference mechanisms. Surveyed visitors with experience in smart hotels reported positive reviews and higher satisfaction scores compared to those without experience. [16] Thus, there are apparent advantages in employing AI applications that can be further improved through continued implementation and improvement. These smart hotels can provide much more accurate, personalised services for each customer by obtaining and memorising their preferences and can make real-time adjustments.

Increasing consumer demand for sustainability is prompting hoteliers to adopt AI-enabled systems. Thus, by implementing such technologies, smart hotels are meeting the expectations of growing traveller demand for sustainable options. With the implementation of green technologies, sustainable hospitality has undergone extensive development. Therefore, it can be inferred that consumers prefer AI technologies at hotels due to the implementation of green practices. These advancements lead not only to improved hospitality sustainability but also to increased brand loyalty. However, it is also worth noting that the implementation of certain AI practices can increase overall hotel electricity use. Moreover, evidence from the DAAAM community indicates that explainable AI (XAI) enhances transparency and user trust in AI-enabled service systems, which is highly relevant for intelligent hotel assistants and service robots. [17]

The recommendations generated by AI can be personalised to offer equal opportunities to visitors, taking into consideration factors such as accessibility and the user's language or cultural interests. Recommendations that are more accurate and helpful can be offered to a more diverse range of visitors with the implementation of newly developed visitor preference classification mechanisms. A structured framework should be incorporated into AI recommendation tools to filter out redundancies and inconsistencies, thereby increasing the accuracy of personalised recommendations and avoiding the potential exclusion of visitors from underrepresented demographic groups.

AI can also provide various real-time assistance tools that contribute to the safety, convenience and general visitor satisfaction levels. One such example of an AI-powered device for real-time visitor safety is the smart Hajj bracelet. The wearable tracking device enables the quick identification of distressed visitors based on their GPS location and medical or contact information stored within the device, thereby enhancing their safety and improving overall trip quality. The safety advantages also allow visitors to experience less stress during their trip and promote recommendations for visits from relatives and friends. However, this also raises concerns about privacy issues that need to be resolved and minimised to prevent any unintended data utilisation, as this will significantly impact the trustworthiness of tourism services to potential new customers.

The application of AI-enabled chatbots and tour guides not only enhances satisfaction for individual tourists but also helps make destinations more competitive overall, for instance, by improving the level of safety and fun tourists can expect to experience on tours and during activities. AI chatbots may be incorporated to cater to tourism services by delivering tourism information. AI tour guides and AI tour companions can increase visitor satisfaction by more than 30% compared to human tour guides. However, AI chatbots and tour guides may lack storytelling which can be crucial in allowing tourists to build a connection between the AI and the destination they are visiting.

In conclusion, AI technologies, such as those used in recommendation systems, immersive experiences and sustainable accommodations, help enhance the overall visitor experience in various ways. Addressing any ethical, operational and inclusivity challenges associated with its implementation enables the maximisation of AI technology's impact on improving tourism. These advances lead to new dimensions for global tourism management, which will be further discussed in the next chapter on concluding perspectives.

5. Conclusion

This scientific review aims to analyse the transformation of artificial intelligence in contemporary tourism destination management. The research aimed to investigate how AI has transformed operational processes and informed data-driven decisions. It also examined the ethical and practical implications for tourism stakeholders.

By reviewing a variety of scholarly articles this research attempted to illustrate the opportunities and challenges of AI in the sector. The research questions centred on how AI enhances efficiency, decision-making and visitor satisfaction. Through analysis, this research demonstrates that AI has significantly transformed the tourism industry in several key areas. Regarding operations, automation and data analysis have improved efficiency, scalability, and consistency. Customer service channels powered by AI offer continuous multilingual support to tourists. In marketing and resource management, machine learning enables dynamic pricing strategies and customised advertising leading to increased bookings and higher conversion rates. In visitor experience, AI-based recommendations improve the experience for diverse visitor demographics. Additionally, robot automation in service delivery enables staff to provide more personalised care to visitors. Finally, in destination operations, AI has enabled proactive traffic and crowd control, as well as real-time monitoring in emergencies. These elements support the position that digital transformation powered by AI is increasingly essential for competitiveness.

This scientific review suggests that the research objective has been successfully achieved. By addressing all the research questions, the synthesis indicates that AI can enhance various aspects of tourism ranging from operational management to marketing strategy. Research cited in this report demonstrates how data collection and the automation of marketing campaigns have contributed to higher revenue conversions and enhanced operational efficiency. AI-enabled improvements to operations and visitor experience are highlighted with several examples.

The synthesis conducted in this study has also helped clarify the position and relationship of this analysis within the broader scholarly and industry conversation. Several existing studies support the value of adopting AI-based technology across various sectors including tourism. They assert that these tools enhance efficiency, effectiveness and sustainability for stakeholders who embrace this shift. This research reinforces those themes while providing additional insight. This scientific review can provide a deeper understanding of the opportunities, ethical challenges and technical requirements necessary for implementing digital transformation. It can also encourage further exploration into the complex relationship between visitor behaviour and the automation of tourism practices.

With these results in mind, there are, of course, some limitations regarding the interpretation and application of the findings of this study that must be clarified. First, as a literature-based scientific review, this report relies exclusively on secondary sources to compile insights into the subject matter. As such, any conclusions or suggestions drawn by the author are based solely upon what others have studied and published. In addition, publication bias is a relevant topic to consider in this evaluation. This effect arises as publishers tend to accept only papers that display significant results. The authors may often exaggerate their findings, thereby risking overinterpretation of their data. Another limitation of this study lies in the rapid transformation and evolution of technology. The results of this review may already be obsolete within a few years. Moreover, the implementation of these technologies varies between markets. This makes it nearly impossible to apply the findings of case studies to broader trends or sector analyses.

Further research can support a greater understanding of these complex issues by undertaking primary empirical analyses. Such studies can involve the collection of long-term, cross-sectional data. This will clarify the effect of AI implementation on workforce development. This could be pursued by examining how training in these skills encourages economic opportunity or by exploring best practices for increasing the representation of women in AI-related positions. Further empirical work can focus on the social and behavioural impacts of AI in service encounters with clients, the impact of robot staff on customer experiences, and the impact of real-time visitor data feedback collection on improving tourism-based services. More research can also analyse the implications of AI technology for the sustainable development of the tourism industry.

Based on the conclusions presented in this study, we recommend that policymakers and sector leaders invest in digital infrastructure upgrades, provide training for existing employees on how to use these technologies, develop clear regulatory frameworks that allow for future flexibility, and establish incentive programs for the industry. These actions can support a more profitable, sustainable and ethically responsible tourism industry that maximises stakeholder benefit.

In conclusion, the authors reflect on the core motivation for this research study: To explore the intricate relationship between the complex service systems of tourism and the transformative power of technological innovation. To achieve this, this scientific review has incorporated multiple viewpoints, perspectives, and evidence-based insights to gain a deeper understanding of the complex phenomenon of digitalisation and its connection to the tourism sector. The findings offer direction for future sector development and support decision-makers as they navigate the ongoing transition.

6. References

[1] Zrnić, M., Stojanović, R. (2019). Application of Artificial Intelligence (AI) as a Trend in the Contemporary Tourism Market, Acad. Appl. Stud. Belgrade, Coll. Hotel Manag., Belgrade, Serbia, ISBN: 978-8682078319

[2] Erdos, F., Thinkanaran, R., Firuza, B. & Koloszár, L. (2025). The rise of AI in tourism - A systematic literature review, GeoJournal Tour. Geosites, vol. 60, no. 2spl, pp. 1254-1265, doi: 10.30892/gtg.602spl22-1498, ISSN:2065-0817, E-ISSN: 2065-1198

[3] Wang, S., Karia, N. (2024). AI Tourism: Concepts, Practices, Challenges, and Future, Glob. Bus. Manag. Res.: Int. J., vol. 16, no. 4s, pp. 475-487, ISSN: 1947-5667 (Online)

[4] Madan, V., Katicha, N. (2025). Capitalizing on the AI revolution in tourism, Strategy & PwC, Available from: https://www.strategyand.pwc.com/m1/en/strategic-foresight/sector-strategies/travel-and-tourism/ai-revolution-in-tourism/ai-in-tourism.pdf, Accessed on: 2025-01-09

[5] Kosta, A., Pazari, F., Lili, I. & Greca, S. (2025). "The role of artificial intelligence in transforming tourism services: The case of Durrës, Albania, GeoJournal Tour. Geosites, vol. 61, no. 3, pp. 1485-1494, doi: 10.30892/gtg.61308-1518, ISSN: 2065-0817, E-ISSN: 2065-1198

[6] Cristian, M. G., Tileagă, C. (2024). Challenges and Perspectives of AI in Sustainable Tourism Management. Dev., vol. 16, no. 2, pp. 1-26, doi: 10.54989/msd-2024-0012, ISSN (PRINT): 2668-2027, E-ISSN: 2668-2035

[7] Jiang, Q. (2025). Artificial Intelligence Empowers Government Tourism: Challenges, Paths, and Future Prospects, in Proc. ICEMGD Symp.: Resilient Business Strategies in Global Markets, pp. 93-99, doi: 10.54254/2754-1169/2025.CAU26260

[8] Mura, L., Stehlíková, B. (2025). Artificial Intelligence and Tourism in the EU: A Data-Driven Analysis of Adoption and Economic Contribution, Folia Geographica, vol. 67, no. 1, pp. 70-99, Available from: https://www.unipo.sk/public/media/48889/738-artificial_intelligence_and_tourism_in_the_eu.pdf, Accessed on: 2025-02-09

[9] https://www.oecd.org/content/dam/oecd/en/publications/reports/2024/12/artificial-intelligence-and tourism_41e7f157/3f9a4d8d-en.pdf, (2024). OECD Tourism Papers, no. 2024/02, pp. 1-39, Accessed on: 2025-02-09.

[10] Shalan, I. S., Gaafar, H., Al-Romedy, B. & Shawaly, A. E. (2024). The Impact of Artificial Intelligence on Improving Tourism Service Quality in the Egyptian Destination, J. Fac. Tourism Hotels-Univ. Sadat City, vol. 8, no. 2/2, pp. 1-22, ISSN: 2537-0952 E-ISSN: 2735-4202. Available from: https://mfth.journals.ekb.eg/article_385750_9eeee543d5964bbf1fe5234588ebd461.pdf?lang=en, Accessed: 2025-03-09

[11] Sharma, A., Bharti, S., Seth, S., & Guleria, S., Bhardway, V. (2025). "Role of AI and Social Media in Tourism Destination Marketing and Crisis Management," in Proc. 2nd Int. Conf. Innovation and Regenerative Trends in Tourism and Hospitality Industry, p. 343., doi: 10.2991/978-94-6463-799-1_26

[12] Wan, H., Yang, l., Yaning, W. & Chao, W. (2025). "AI-Empowered Cultural Tourism Development: Innovation Paths and Strategies for Rural Revitalisation," Front. Soc. Sci. Technol., vol. 7, no. 4, pp. 31-38, doi: 10.25236/FSST.2025.070405.

[13] Alsahafi, R., Mehmood, R. & Alqahtany, S. (2025). "A Machine Learning-Based Analysis of Tourism Recommendation Systems: Holistic Parameter Discovery and Insights," Int. J. Adv. Comput. Sci. Appl., vol. 16, no. 1, pp. 13691382, Available from: https://thesai.org/Downloads/Volume16No1/Paper_130-A_Machine_Learning_Based_Analysis_of_Tourism.pdf, Accessed: 2025-03-09

[14] Benbrahmi, F. Z. & Mehor, H. M. (2025). "The impact of artificial intelligence on enhancing the travel experience for tourists: An analysis of successful models," Bus. Excell. Manag., vol. 15, no. 5, pp. 5-13, doi: 10.24818/beman/2025. S.I.5-01

[15] Xu, J. ، & Pan, Y. (2024). "The Future Museum: Integrating Augmented Reality (AR) and Virtual-text with AI-enhanced Information Systems," J. Wireless Mobile Netw., Ubiquitous Comput., Dependable Appl. (JoWUA), vol. 15, no. 3, pp. 373-394., doi: 10.58346/JOWUA.2024.I3.025

[16] Patrichi, I. C. (2025). "Smart hospitality: the role of AI in enhancing sustainable guest experiences," J. Inf. Syst. Oper. Manag., vol. 19, no. 1, pp. 241-258, Available from: https://web.rau.ro/websites/jisom/Vol.19%20No.1%20-%202025/JISOM%2019.1_241-258.pdf, Accessed: 2025-04-09

[17] Schwaiger S., Aburaia, M., Aburaia, A. & Woeber, W. (2021). Explainable artificial intelligence for robot arm control, in Proc. 32nd DAAAM Int. Symp. on Intelligent Manufacturing and Automation, 28th-29th October 2021, Vienna, Austria, ISSN: 1726-9679, ISBN: 978-3-902734-33-4 2021, Katalinic, B. (Ed.), DAAAM International Vienna, pp. 640–647, Vienna.

[18] https://www.freepik.com/pikaso/image-editor?tool=adjust&creation=LSJL2BbswO, Accessed: 2025-07-10

[19] Tour planning with tickets on smartphone screen, Author's work

DOI: 10.2507/36th.daaam.proceedings.002

APPLICATION OF TEMPLATES IN A CAM SYSTEM FOR GEOMETRICALLY SIMILAR COMPONENTS

Jozef Peterka & Jozef Hipp

This Publication has to be referred as: Peterka, J[ozef] & Hipp, J[ozef] (2025). Application of Templates in a CAM System for Geometrically Similar Components, Proceedings of the 36th DAAAM International Symposium, pp.0013-0020, B. Katalinic (Ed.), Published by DAAAM International, ISBN 978-3-902734-47-1, ISSN 1726-9679, Vienna, Austria
DOI: 10.2507/36th.daaam.proceedings.002

Abstract

The paper addresses the application of templates in CAM systems for the development of technological operations on geometrically similar components. The introduction outlines the potential of templates as a tool for standardisation and acceleration of the programming process in environments such as Fusion 360 and SolidCAM. Subsequently, the procedure for creating templates from individual operations and their application in the machining of further similar components is described. The study demonstrates that, given sufficient similarity of parts, templates enable a significant reduction in production preparation time and the elimination of repetitive steps. The findings confirm that the principal advantages of templates lie in repeatability and efficiency, while their limitation may arise from restricted applicability to components or stock of differing geometries.

Keywords: CAM; template; geometric similarity; technological similarity

1. Introduction

Research in the field of CAM technologies has historically focused on a variety of topics, such as surface quality achieved using copy tools in CAM milling [1], [2], [3] or ground tools [4], the influence of different CAM strategies [5], wear of copy tools used in CAM [6], issues related to multi-axis milling [7], [8], and the application of additional energy sources, such as ultrasonic-assisted CAM machining [9], [10].

Contemporary CAM (Computer-Aided Manufacturing) systems no longer limit themselves to the generation of toolpaths alone but increasingly offer capabilities for process standardisation and automation. One of the tools that supports this trend is templates. In practice, templates allow proven machining settings to be saved and reused—such as tool definitions, feed rates, speeds, or entire sequences of operations [11], [12].

The advantage of this approach is that a CNC programmer does not need to repeat the same steps for every new workpiece. Instead, a prepared template is loaded, which sets the fundamental parameters and provides a consistent, verified workflow. This reduces production preparation time, minimises the risk of errors, and improves both the quality and reproducibility of results [13], [14].

The use of templates has been adopted in several professional solutions: Fusion 360 supports rapid application of CAM workflows through saved setups [13]; SolidCAM distinguishes between "operation templates" and more complex "process templates" [12]; and Siemens NX provides Manufacturing Templates to initialise the entire manufacturing process [14]. Further research indicates that systematic use of templates—such as the Automatic CAM Programming Using Machining Templates (ACPUT) methodology—can lead to significant time and cost savings in the creation of manufacturing programs [15].

An article published by Tebis presents the experience of a user implementing CAD templates with direct integration into CAM processes. The main idea is that all critical parameters—from workpiece geometry through orientation to fixturing—can be prepared directly in the CAD environment. These parameters are then automatically transferred to the CAM module, eliminating the need for repeated data entry. The author emphasises that this approach significantly accelerates programming, reduces errors, and increases consistency of manufacturing procedures, highlighting the importance of templates as a tool for automation and more efficient CAD–CAM integration [16].

An article published by Bantam Tools focuses on the use of CAM templates in Fusion 360 as a means of streamlining CNC programming. The author reports that template implementation allows standardisation of the machining operation creation process, resulting in higher productivity and a reduction in programming errors [13]. The described procedure involves multiple steps: first, a prepared CAM template is imported into Fusion 360; next, a new setup is created defining the workpiece parameters, and a specific template is selected. The template automatically predefines the technological parameters of machining, such as cut depth or feed rate. Users, however, can adjust these settings to meet specific requirements. The outcome is an NC program in the form of G-code, ready to be sent to the CNC machine [13].

From a practical perspective, the article emphasises that using CAM templates in Fusion 360 contributes to the acceleration and standardisation of the manufacturing process, while simultaneously reducing the likelihood of errors during the configuration of individual operations [13]. A contribution from the Autodesk University conference focuses on optimising repetitive workflows using templates in a CAM environment. The authors present a methodology for identifying the most frequently performed operations and subsequently creating templates to automate them. This approach reduces the time required to prepare NC programs and increases the uniformity of the resulting processes. The text highlights the benefits of templates, particularly for industrial enterprises handling large volumes of similar components and seeking to accelerate production preparation [11].

The official Siemens website introduces Solid Edge CAM Pro, which includes built-in templates designed to streamline the programming of machining operations. The software emphasises an intuitive interface and modularity, while templates simplify parameter setting and standardise repetitive procedures. The text highlights that the use of templates reduces operational complexity, speeds up NC code generation, and contributes to more accurate and stable production [17].

Based on previous research, it has been found that the process of creating operations based on similarity is very similar in both SolidCAM and Fusion 360. Initially, an operation is created for the first part and then saved as a template. This template is subsequently applied when creating operations for additional parts, significantly reducing the time required to prepare similar operations. The fundamental principle is therefore the use of templates [18]. Figure 1 shows Part 1. Machining operations for this part will be created using templates, which will then be applied to Parts 2 and 3. For clarity, surfaces/features that are similar (or different) are highlighted in red. The potential application of templates will be examined on the surfaces marked in red.

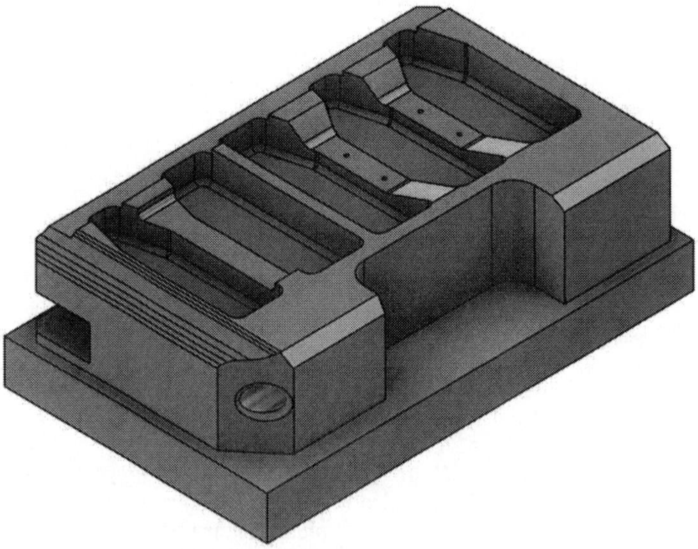

Fig. 1. Part 1

Figures 2 and 3 show similar components on which the application, advantages, and limitations of templates will be investigated across five different types of operations in the CAD/CAM software Fusion 360. In this study, it is not necessary to consider the machine, fixturing, workpiece material, cutting parameters, machining time, or other such factors, as they do not influence the aspects under investigation. Conversely, attention must be paid to the strategy employed, the tool diameter and type, and the stock material prior to milling. These parameters are critical for the creation of templates for the respective operations.

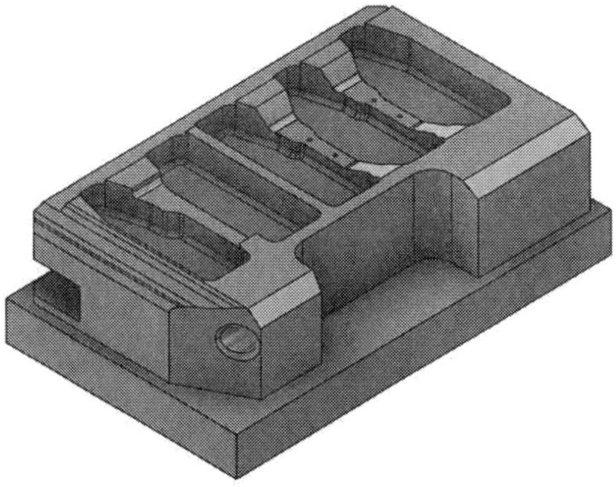

Fig. 2. Part 2

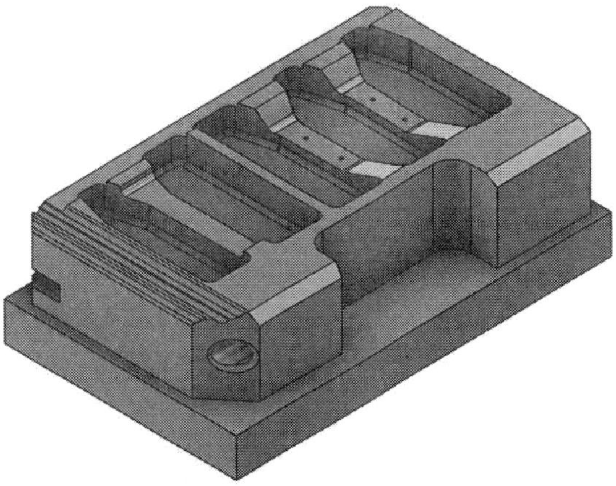

Fig. 3. Part 3

2. Creation of templates from operations

In this section, the basic operations for milling the sample component will be created. The stock for roughing the pockets from above includes a 5 mm allowance on the walls and the bottom, as illustrated in Figure 4 (the stock is shown in green). This operation accounts for these allowances, thereby eliminating air milling. The operation for roughing the pocket walls from above is performed using dynamic milling along the full length of the tool's cutting edge. A solid tool with a 16 mm diameter will be used. The machining boundaries are defined by five curves representing the pocket walls, selected directly from the model. The milling depth is set to reach the start of the corner radius at the bottom (i.e., 60 mm), and the radial step-over of the tool is set to 1.6 mm.

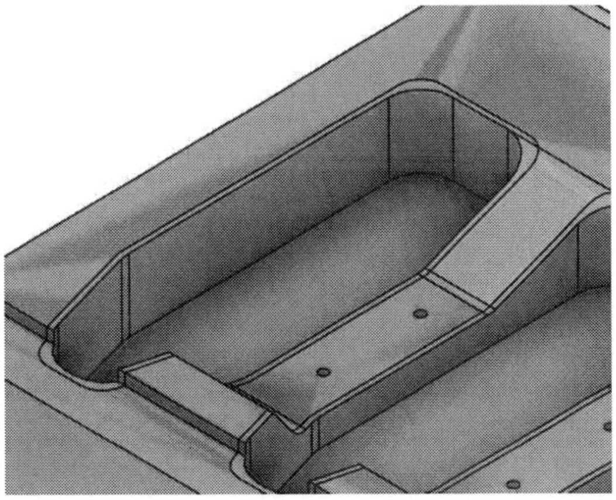

Fig. 4. Stock for the pockets from above

The operation for milling the pocket bottoms and R10 radii is performed as 3D roughing, taking into account the stock and the previous operation. A high-feed tool with a 25 mm diameter will be used, with an axial step-down of 0.6 mm and a radial step-over of 15 mm for bottom milling. The machining boundaries are identical to the previous operation, defined by the five curves representing the pocket walls. The milling depth extends from the start of the radius to the flat bottom of the pocket (i.e., 10 mm). Figure 5 shows the stock for the contoured surface. Next, the contoured surface on the model will be milled. A 1 mm allowance is applied to this surface, following its shape. The selected operation accounts for the stock and previous operations, and therefore only the contoured surface will be machined. The contoured surface will be finished using a 20 mm diameter ball-end mill, with a downward step of 0.3 mm. Toolpaths will be generated from one side to the other in both directions.

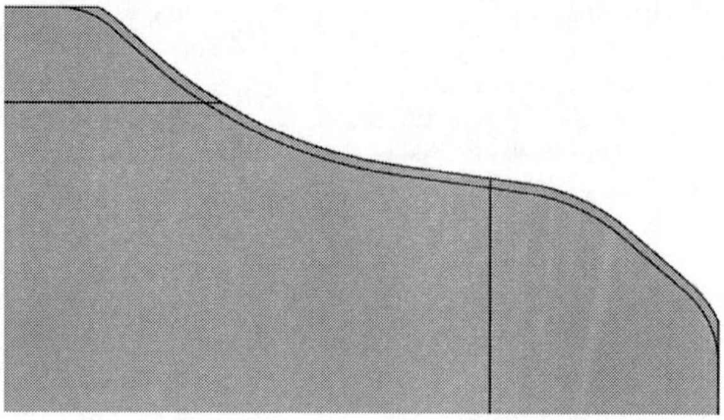

Fig. 5. Stock for the contoured surface

The stock for the fillet is defined by the previously drilled D55 hole (Figure 6). The R10 fillet on the D55 hole is finished using a 20 mm diameter ball-end mill. The downward step is set to 0.3 mm, and the toolpaths are generated in a top-down spiral.

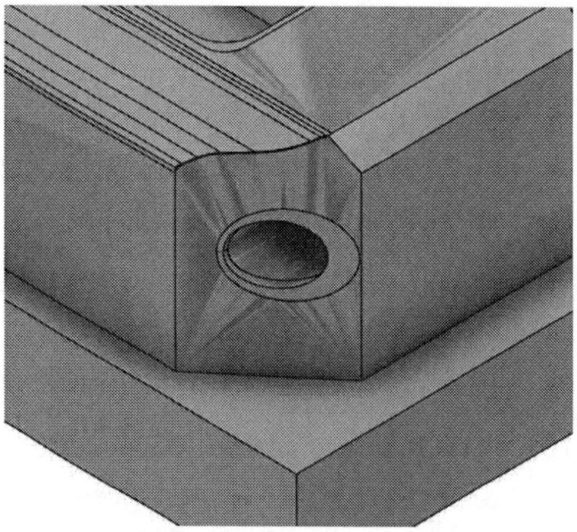

Fig. 6. Stock for the R10 fillet on the D55 hole

Figure 7 shows the stock for the slot. The operation will be created using a 2D contour. This operation does not take the stock or previous operations into account, but instead follows the defined curve. A high-feed mill with a 52 mm diameter will be used, with an axial step of 0.8 mm.

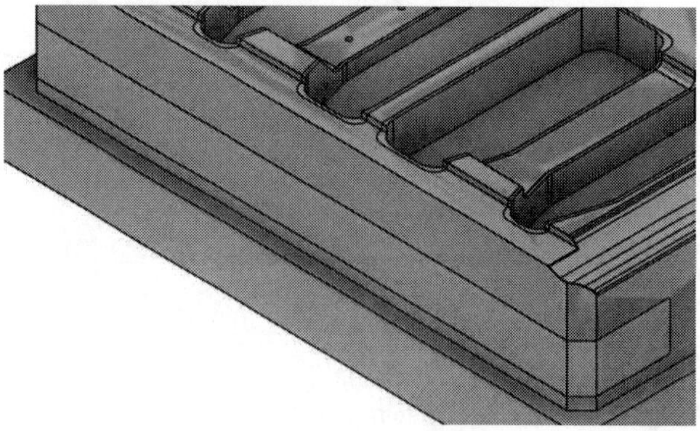

Fig. 7. Stock for the slot

After creating the program, templates will be generated from each operation. By right-clicking on a created operation, a menu will expand, providing the option "Save as Template" (Figure 8a). Upon selecting this option, a pop-up window appears, as shown in Figure 8b, where the template name is entered and the folder in which the template will be saved is specified. Clicking "Save" stores the template, and this process is repeated for all operations.

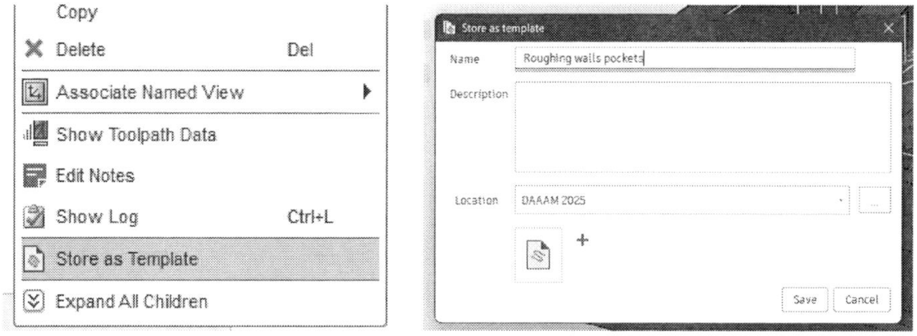

a) b)

Fig. 8. a) List of options, b) Saving the template

Next, in the template library (Figure 9), the operations are selected sequentially, and any necessary minor adjustments are made. Adding a template to a new CAM program involves selecting the desired operation and then clicking the corresponding icon (indicated by the red arrow), which adds the operation to the operation tree.

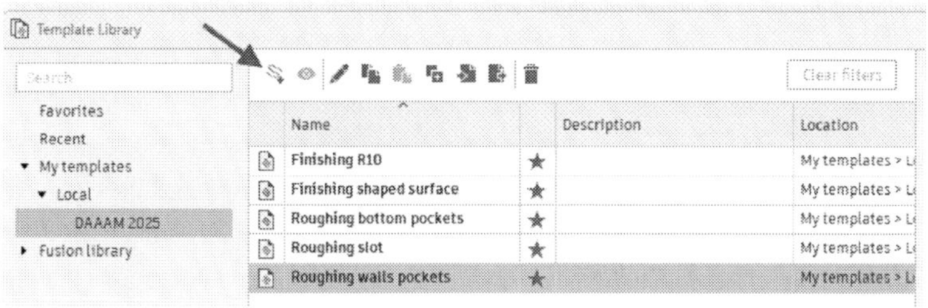

Fig. 9. Template Library

3. Creating a Program for Part 2 Using Templates

Compared to Part 1, the differences are as follows. The pockets have a different shape, while the depth and stock remain the same. The contoured surface has a different profile. The side hole has a diameter of D60, a different position, and the fillet on the wall is at a different angle. The dimensions of the front slot are unchanged. For the operations "Roughing Walls Pockets," "Roughing Bottom Pockets," "Finishing Shaped Surface," and "Roughing Slot," it is necessary to adjust and select new machining curves according to the Part 2 model. In the "Finishing R10" operation, the new fillet surface on the D60 hole must be selected. Figure 10 shows an example comparison of the machining curves for the "Roughing Walls Pockets" operation. In this way, all operations are updated.

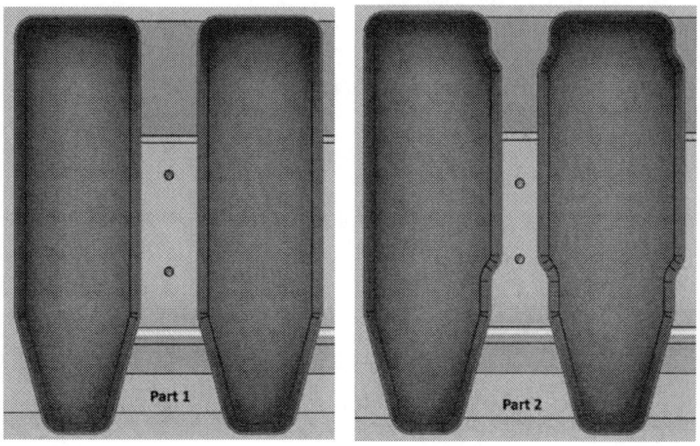

Fig. 10. Comparison of Pocket Machining Curves for Part 1 and Part 2

In each of the operations created using templates, the same tools, operation parameters (radial and axial step-over), machining planes (rapid, safe, top, and bottom), as well as entry, transition, and exit paths, are used. All these parameters are identical to those in Part 1.

For Part 2, it was possible to apply a template to each operation, and after adjusting the machining curves and surfaces, the operations were fully functional. Although the pockets have a different shape, the stock principle remains the same, and therefore the machining allowances are identical. The contoured surface has a different profile, but the machining principle remains unchanged, allowing the template to be used for this operation as well. Although there are several differences in the R10 fillet on the side hole, these do not affect the functionality or applicability of the template, so it can also be used in this case. A comparison of these two operations is shown in Figure 11. The front slot is completely identical to that in Part 1. The advantage of using templates lies in the speed of creating the operations. This approach eliminates certain repetitive tasks and saves time for operations that require more careful consideration. As seen in this case, it is necessary that the components are similar. Likewise, the operation created using the template must be correctly designed, as any errors would be replicated across subsequent components.

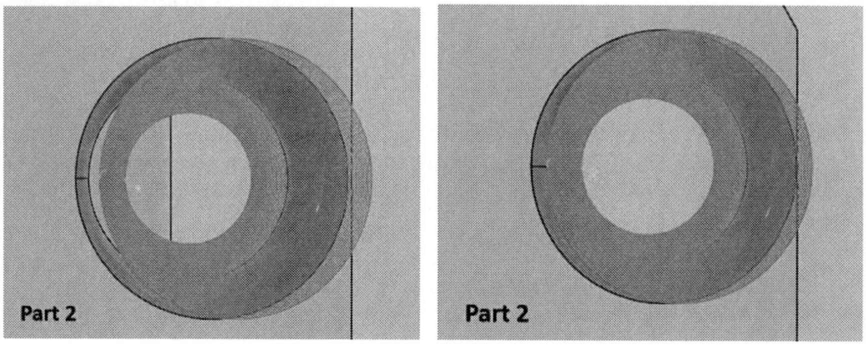

Fig. 11. Comparison of Toolpaths for the Fillet in Part 1 and Part 2

4. Creating a Program for Part 3 Using Templates

Compared to Part 1, the differences are as follows. The pockets have the same shape and depth, but the stock is different; in this case, the stock is solid. The contoured surface has a different profile. The side hole and fillet are identical, while the dimensions of the front slot differ.

In the "Roughing Walls Pockets" operation, it is possible to use a template; however, the stock from which the machining originates in Part 1 is different. In Part 1, a 5 mm allowance on the walls and bottom is considered, so only this remaining material is machined, whereas in Part 3 machining starts from solid material. A comparison of the stock for this operation is shown in Figure 12 (stock is shown in green). The template was functional when inserted into the program, but in this case, it would be advisable to consider whether a more efficient machining strategy could be employed. Therefore, it can be stated that, although the template is functional, it is not suitable from a technological process perspective, as a different stock is involved compared to Part 1.

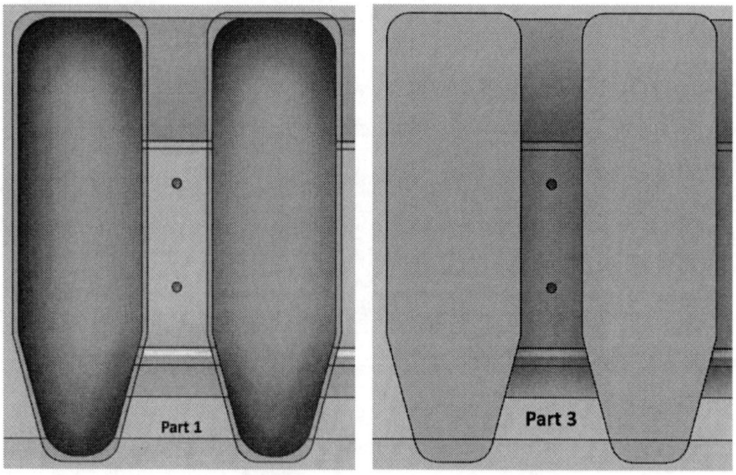

Fig. 12. Comparison of Pocket Machining Stock for Part 1 and Part 3

Figure 13 shows a comparison of the machined contoured surface for Part 1 and Part 3. A template was also used to create the "Finishing Shaped Surface" operation. In Figure 13 (left, shown in green), the entire machined contoured surface can be seen. In Figure 13 (right), certain areas of the surface (shown in white) remain unmachined.

In this case, the tool cannot reach these areas because the tool diameter is too large. Therefore, it is not advisable to use the template to create this operation, as the operation is not fully functional and does not machine the entire surface.

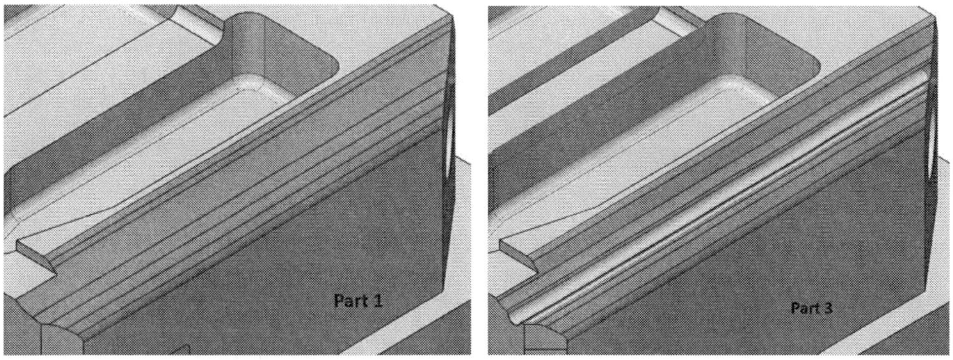

Fig. 13. Comparison of Machined Contoured Surface for Part 1 and Part 3

Figure 14 shows that the width of the front slot differs, being narrower in Part 3 compared to Part 1. It was not possible to use the template because the slot is smaller, and a 52 mm diameter tool is too large to machine the slot. This means that, in this case, the template cannot be used to create the operation.

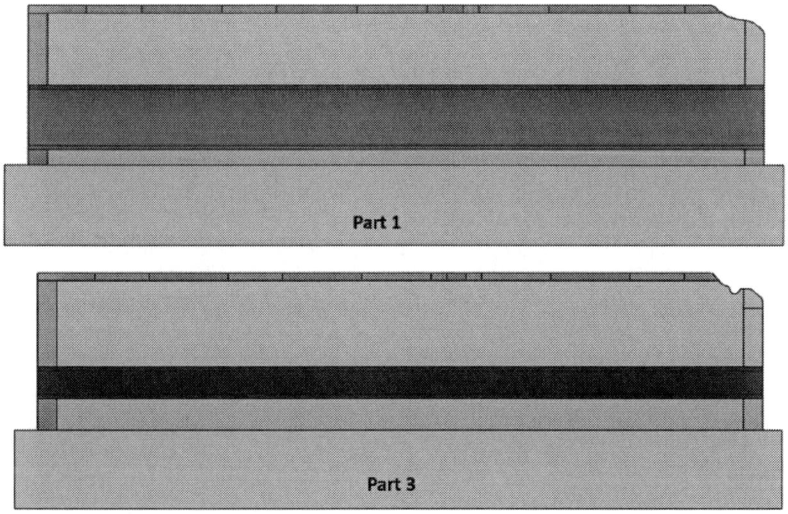

Fig. 14. Comparison of the Front Slot for Part 1 and Part 3

5. Conclusion

The conducted research confirmed that the use of templates in CAM programming enables faster and more standardised creation of operations for geometrically similar components. The results obtained in Fusion 360 demonstrated that the reuse of properly designed templates allows the CAM programmer to eliminate repetitive tasks and achieve a high level of process repeatability. Compared to manual programming, the preparation time of operations was significantly reduced. The study further showed that the functionality of templates strongly depends on the degree of geometric and technological similarity of the parts. While in the case of the second component all operations created using templates were fully functional, certain limitations were observed for the third component due to differences in the stock material. These findings are consistent with the conclusions of Kowalski et al., who also emphasised that the efficiency of automatic CAM programming based on templates rapidly decreases when the required level of similarity is not achieved. In comparison with the observations of Bantam Tools and Autodesk University, which focused primarily on increasing productivity through automation, this study extends the discussion by identifying specific situations where template-based programming no longer provides advantages. Similarly, as pointed out by Tebis and Siemens, who highlight the importance of integrating data between CAD and CAM, the results of this work underline the necessity of evaluating the technological context — particularly the stock definition, tool accessibility, and strategy compatibility — before applying templates.

The research question — to what extent can templates be effectively applied in CAM systems for geometrically similar components — has been answered by confirming that templates are highly effective when geometric similarity, machining strategy, and technological context remain consistent. Conversely, when greater deviations occur, templates lose their functionality and require manual adjustments. In conclusion, templates represent a valuable tool for improving repeatability, quality, and time efficiency in CAM programming. Their correct use, however, requires careful evaluation of geometric similarity and machining conditions. A potential next step could involve examining a more complex component in terms of template applicability, as well as analysing the technological process of such a component directly within the CAM software environment.

6. Acknowledgments

The article was supported by the VEGA 1/0391/24 and KEGA 007 STU – 4/2025 projects.

7. References

[1] Vopát, T.; Peterka, J.; Simna, V. & Kuruc, M. (2015). The Influence of Different Types of Copy Milling on the Surface Roughness and Tool Life of End Mills, Proceedings of the 25th DAAAM International Symposium on Intelligent Manufacturing and Automation, 2014, Volume100, Page 868-876, DOI10.1016/j.proeng.2015.01.443

[2] Peterka, J. (2004). A new approach to calculating the arithmetical mean deviation of a profile during copy milling, Strojniski vestnik-journal of mechanical engineering, Volume 50, Issue12, Page 594-597

[3] Polakovic, M.; Buransky, I. & Peterka, J. (2008). Simulation concept for machined surface roughness and shape deviations prediction, Annals of DAAAM for 2008 & Proceedings of the 19th International DAAAM Symposium, Page 1089-1090

[4] Kuruc, M.; Vopát, T.; Peterka, J.; Necpal, M.; Simna, V.; Milde, J. & Jurina, F. The Influence of Cutting Parameters on Plastic Deformation and Chip Compression during the Turning of C45 Medium Carbon Steel and 62SiMnCr4 Tool Steel, Materials, Volume 15, Issue 2, DOI10.3390/ma15020585

[5] Peterka, J.; Pokorny, P. & Vaclav, S. (2008). CAM strategies and surfaces accuracy, Annals of DAAAM for 2008 & Proceedings of the 19th International DAAAM Symposium, Page 1061-1062

[6] Vopát, T.; Peterka, J. & Buransky, I. (2014). The Wear Measurement Process of Ball Nose end Mill in the Copy Milling Operations, 24th DAAAM International Symposium on Intelligent Manufacturing and Automation, Volume 69, Page 1038-1047, DOI10.1016/j.proeng.2014.03.088

[7] Peterka, J.; Pokorny, P. & Vaclav, S. (2012). The task of 5-axis milling, Tehnicki Vjesnik-Technical Gazette, Volume19, Issue1, Page147-150

[8] Beno, M.; Zvoncan, M.; Kovác, M. & Peterka, J. (2013). Circular interpolation and positioning accuracy deviation measurement on five axis machine tools with different structures, Tehnicki Vjesnik-Technical Gazette, Volume 20, Issue 3, Page 479-484

[9] Kuruc, M.; Zvoncan, M. & Peterka, J. (2014). Investigation of Ultrasonic Assisted Milling of Aluminum Alloy AlMg4.5Mn, 24th DAAAM International Symposium on Intelligent Manufacturing and Automation, Volume 69, Page 1048-1053, DOI10.1016/j.proeng.2014.03.089

[10] Kuruc, M.; Vopat, T. & Peterka, J. (2015). Surface Roughness of Poly-Crystalline Cubic Boron Nitride after Rotary Ultrasonic Machining. 25th DAAAM International Symposium on Intelligent Manufacturing and Automation, 2014, Volume100, Page 877-884, DOI10.1016/j.proeng.2015.01.444

[11] https://www.autodesk.com/autodesk-university/class/Streamlining-CAM-Workflows-Templates-2019, (2019). Autodesk University, Streamlining CAM Workflows with Templates, Accessed on: 2025-09-02

[12] https://www.solidsolutions.co.uk/blog/2023/07/manufacturing-how-to-create-templates-in-solidcam, (2023). Trimech, Manufacturing: How to create templates in SoliidCAM, Accessed on: 2025-09-02

[13] https://support.bantamtools.com/hc/en-us/articles/360059977453-Fusion-360-Workflows-Basic-CAM-Templates, (2020). Bantam Tools, Fusion 360 Workflows: Basic CAM Templates, Accessed on: 2025-08-29

[14] https://community.sw.siemens.com/s/question/0D54O000061xTkjSAE/cam-templates-master-model, (2013). SIEMENS, CAM templates – Master model. Siemens Comunity, Accessed on: 2025-09-02

[15] Kowalski, M., Zawadzki, P. & Hamrol, A. (2021) Effectiveness of automatic CAM programming using machining templates for the manufacture of special production tooling. Strojniški vestnik – Journal of Mechanical Engineering, Vol. 67, No. 10, pp. 475–488., ISSN 0039-2480, DOI:10.5545/sv-jme.2021.7285

[16] https://www.tebis.com/en/blog/how-a-cad-template-truly-simplified-my-cam-programming/b2696, (2023). Tebis, How a CAD template truly simplified my CAM programming, Accessed on: 2025-08-29

[17] https://solidedge.siemens.com/cs/solutions/products/computer-aided-manufacturing-cam/solid-edge-cam/, (2025). Siemens, Solid Edge CAM Pro – NC programování, Accessed on: 2025-09-02

[18] Peterka, J. & Hipp J. (2024). Comparison of Creating Technological Operations Based on Similarity in CAM Software, Proceedings of the 35th DAAAM International Symposium, Viena University of Technology, ISSN 1726-9679, ISBN 978-3-902734-44-0, Katalinic, B. (Ed.), pp. 0026-0032, Published by DAAAM International, Viena, Austria, DOI: 10.2507/35th.daaam.proceedings.004

DOI: 10.2507/36th.daaam.proceedings.003

AN INNOVATIVE APPROACH TO GEAR MANUFACTURING: POWER SKIVING ON CNC LATHES

Jozef Peterka & Igor Skurský

This Publication has to be referred as: Peterka, J[ozef] & Skursky, I[gor] (2025). An Innovative Approach to Gear Manufacturing: Power Skiving on CNC Lathes, Proceedings of the 36th DAAAM International Symposium, pp.0021-0028, B. Katalinic (Ed.), Published by DAAAM International, ISBN 978-3-902734-47-1, ISSN 1726-9679, Vienna, Austria
DOI: 10.2507/36th.daaam.proceedings.003

Abstract

Power skiving represents a modern and highly efficient method of gear manufacturing, which in recent years has become an attractive alternative to traditional technologies such as hobbing and broaching. The principle of power skiving is based on the precise angular positioning of the tool and the simultaneous rotation of the workpiece, which makes it possible to achieve high productivity and accuracy. Its implementation on multifunctional CNC lathes enables the combination of turning and milling operations in a single setup, significantly reducing the need for subsequent technological steps in the overall production time. The article focuses on the analysis of the power skiving principle, describing its geometrical and technological aspects, the requirements for CNC lathes, and the specifics of using suitable tools.

Keywords: Power skiving; Gear manufacturing; CNC lathes; Gear hobbing; Production

1. Introduction

Wilhelm von Pittler, a German engineering pioneer and inventor, introduced in 1910 a patent that is now recognized as the basis of power skiving. This method enables the manufacturing of gears using a special gear-shaped cutting tool. The principle lies in the fact that the tool has cutting edge formed on its front face and is shaped similarly to the module of a gear. During the cutting process, the workpiece and the tool are inclined at a certain angle. This inclination causes a shearing cut to occur during their relative rotation, rather than classical hobbing. The cutting process is continuous, the chip is removed smoothly, and the method is significantly faster than gear shaping. Compared to hobbing, power skiving also allows the production of both external and internal gears [1]. Power skiving represents a continuous gear machining process in which the cutting tool engages with the workpiece in a manner like meshing of a pinion and a gear in a transmission. The principle requires extremely precise synchronization between the workpiece spindle and the tool spindle. Unlike conventional hobbing, the tool – which can be regarded as "pinion" – is simultaneously fed axially along the workpiece axis during meshing. The front face of the tool is equipped with cutting edges that remove the material during the process. The cutting speed is generated by the inclination between the axis of the gear being produced and the axis of the tool. As illustrated in the Figure 1, the tool clamped in milling unit of the multitasking CNC lathe, which is inclined relative to the internal gear; both components rotate at high speed while maintaining precise kinematic synchronization.

The cutting action takes place on the tool's face, in this case on an indexable cutting insert. The design of the fixture also incorporates openings around the spindle to ensure efficient chip evacuation [3]. The power skiving process is mainly used for production of spur gears, helical gears, internal gears, and splined shafts. It is suitable for a wide range of materials, including alloy, carbon, and steels, non-ferrous metals such as aluminium and brass, as well as certain plastics. The choice of material depends on the application requirements and the desired gear properties. This method requires specialized cutting tools capable of withstanding high cutting forces while ensuring precise machining. Such tools typically have multiple cutting edges and are commonly made from high-speed steel (HSS) or solid carbide. The selection of the appropriate tool material is influenced by the workpiece material, cutting speed, and tool life requirements [4].

Fig. 1. The working position of the milling head [2]

With power skiving technology is possible to produce whole range types of gears on the multitasking machining centres. This process is reducing machining time from several minutes, which are typical for conventional method such as Lorenz gear shaping, to only 15-20 seconds. Production with this technology is applicable for both external and internal gears with straight or helical teeth and allows machining close to shoulders (not possible with gear hobbing option). Originally, this technology was only available on dedicated gear cutting machines, which are produced by a limited number of manufacturers at high cost. The development of CNC technologies allows solving various research tasks in the theory of cutting [5], [6], [7] which could not be carried out in the past, but also to implement cutting processes that were typical for specialized machine tools. Currently, there are two options for applying "skiving" technology on CNC lathes: 1. Use special CNC hobbing lathes or 2. add "skiving" tool heads to classic CNC lathes.

The possibility to use this "power skiving" technology on CNC lathe by adding special tool holders and activate the option by purchasing it. The CNC lathes with Y-axis (needed axis) needs to be equipped with a specially designed driven toolholder which provides the necessary angular adjustment (15°- 40°) and incorporates the MTSK quick-change system for precise and rapid cutter replacement. The limit for machining on CNC lathe is the range limit of the Y-axis. The user must think about the diameter of the workpiece and of diameter of the working tool [3]. Another limit for power skiving is precise synchronization of the workpiece speed and position of the tool, which become challenging especially for large components with small gears sections. These large components require high rotational speeds (1000 – 2000 rpm) that large rotary tables or spindles often cannot achieve stably [8].

Application of production with the added toolholder to CNC lathes are achieved in terms of productivity and accuracy comparable to results obtained with dedicated gear cutting machines. Industrial applications have already been implemented in planetary gearbox manufacturing, automotive components, and packing production [9]. This study presents a practical comparison of implementing power skiving as a technology option on a multifunction CNC lathe versus conventional gear hobbing. On the lathe, the skiving cutter is clamped in a dedicated adjustable holder and the vendor-provided technology cycle automatically generates the required synchronized motions and G-code, which renders both tool setting and programming markedly simpler and more flexible. In contrast, hobbing on a CNC lathe typically requires disassembly of the live-tool holder to mount the hob arbor, precise reclamping/alignment, and predominantly manual (template-based) program creation; the machine builder merely unlocks the option that permits simultaneous rotation of spindle and the driven tool. The study documents the concrete setup steps, programming workflow, and kinematic requirements (C-axis synchronization, Y-axis constraints), and evaluates their impact on overall setup time, programming effort, and production readiness. The working hypothesis is that, for the considered gear types, power skiving on a multifunction lathe reduces setup and programming overhead relative to hobbing while maintaining the necessary synchronization accuracy for productive gear cutting.

2. Comparison of power skiving technology to gear hobbing (theory)

In practice, tool holder geometry, clamping stiffness and the selected PVD coating materially affects achievable chip load, stability and the tolerance of synchronization errors [10]. The observed balance between cycle time and synchronization accuracy is strongly influenced by tooling decisions: cutting-edge preparation and PVD coatings have measurable effects on stability, chip load and achievable output [11]. Gear hobbing process is based on the principle of generative machining, where the tooth profile is formed through the meshing engagement of the hob and the workpiece. The kinematics of hobbing correspond to the motion of a pair of gears with crossed axes. This motion consists of the rotation of the hob combined with the generative motion of the workpiece [12]. Basic schematic of the rotation of the hobbing principle is shown on the figure 2.

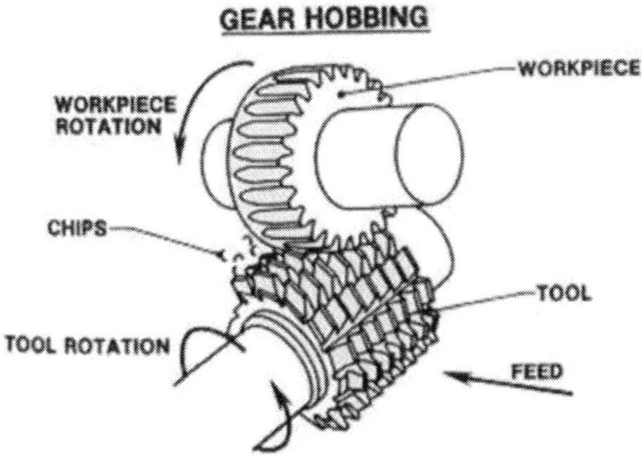

Fig. 2. Basic scheme of gear hobbing [13]

Gear hobbing is the oldest, established and most widely applied method for machining external gears. In this process, the axis of rotation of the tool and the axis of the workpiece intersect at an angle ranging from 45°- 135°, which means the axes are crossed [14]. The fundamental principle of hobbing is the rolling motion of the tool over the workpiece, where many hob teeth are engaged in cutting, resulting in high process productivity. By applying a tangential feed, new cutting edges of the tool are continuously utilized, which extends tool life and allows the production of many gears between successive resharpening operations.

In contrast to profile milling, where is a side milling cutter shaped according to the tooth gap machines each space or thread individually, hobbing enables the continuous generation of the entire gear tooth profile. Profile milling is therefore mainly applied for rough machining of large tooth gap as well as to produce worms and rotors. The process of gear hobbing can be carried out either as dry machining or with the use of cutting fluids (emulsion or oils). Modern hobbing machines are CNC controlled, equipped with a direct drive of both the table and the milling head. Their rigidity and substantial machine weight provide vibration damping, which enhances process stability and machine accuracy [15].

By using the CNC lathe machine, the tools are enabled more efficient and flexible in generating of complex tool paths to produce special gears compared to traditional dedicated gear-cutting machines. The face hobbing process is successfully applied to spiral bevel gears, but in the case of straight bevel gears, it causes profile deformation. This issue can be overcome by using six-axis CNC machines in combination with mathematical models, which allow the synchronization of tool and workpiece motions, resulting in straight tooth flanks [16]. Since five and six-axis machines are very expensive, researchers have sought technical solutions for machining bevel gears on simpler equipment. By combining three-axis CNC machines with rotary tools or rotary table, it became possible to manufacture a spiral bevel gear [17]. For performing the gear machining on a CNC lathe by hobbing, a dedicated live toolholder is required to mount the hob. Although the Y-axis is not essential, it provides the advantage of utilizing the entire tool length, as the hob can be shifted to another Y position once part of it becomes worn [18].

3. Setting comparison between gear hobbing and power skiving, and proposal of programming CNC lathe for power skiving

To use the "gear hobbing" option on a CNC lathe, the machine must be equipped with C-axis. In addition, it is necessary to purchase the rotary holder that allows the hobbing cutter to be clamped. The holder is mounted into the machine´s turret. The installation of the hobbing cutter is carried out by disassembling the shaft, onto which the required tool is mounted. Disassembling of the live tool is illustrated in figure 3. The setup for manufacturing spur gears is performed by tilting the clamped tool at angle corresponding to the tool´s module. For example, when producing a spur gear with a module of M=1.5, the tool is inclined by 1.5°.

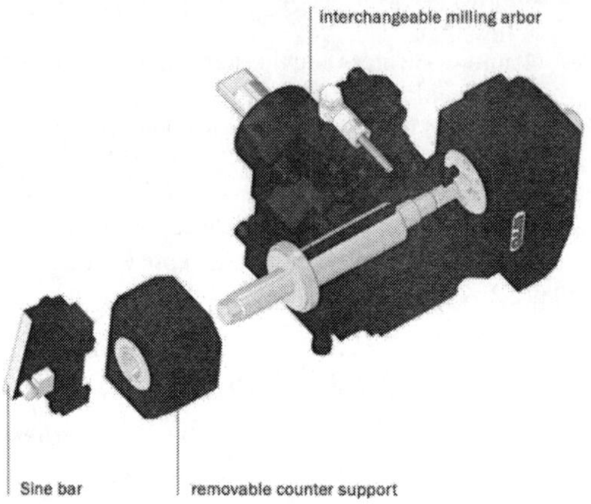

Fig. 3. Setting toolholder for hobbing [19]

The programming setup is not offered as a "cycle" but as a template. The figure shows a template with G-codes that are used in the hobbing process – see Figure 4. The marked cycles show the necessary data which must be set. The #522 code defines the tool offset along the Y-axis after the production of one workpiece. This code function is an automatic tool correction to increase tool life and ensure the tool last as long as possible. Another new code is G81.4R26.L1.Q1.5P0. the G-code starts the synchronization of the tool (hob) and the spindle (C-axis), other parameters need to be filled by the operator as diameter of the hobb and the pitch of the module. Next parameter which is marked is also filled by the operator, these parameters are the tool arrivals and departures. The #520 parameter is the parameter of Y-axis position of the tool which is raised by #522 value which is set by the machine operator. The last parameter G80.4 is the parameter which turn off the parameter of the synchronization of the tool and main spindle.

```
(***** STARTING POSITION – REFERENCE *****)
G97;
M764;
#522=0.02 (Y-AXIS MOVEMENT ON NEXT PIECE)
M8;
M69;
M5;
G330 ( V2 INTO REFERENCE)
G0G53X-110;
G0G28V0. (REFERENCE POINT Y-AXIS);
G0G53Z-350. (SAFETY POSITION IN Z AXIS);
G54G98;
G0T0505 (POSITION OF THE MOUNTED TOOL);
G0Y#520 (MOVEMENT IN Y AXIS);
G0Z-15.;
G54G98;
G81.4R26.L1.Q1.5P0. (START OF SYNCHRONIZATION OF HOBBING AND C-AXIS);
S506M13 (START OF THE RPM*VC=50/MIN);
G1X100.F1000 (THE X-AXIS APPROACH);
G1X45.F1000 (THE X-AXIS APPROACH * DIAMETER OF WORKPIECE);
G1X36.648F100. (THE HOBB APPROACH INTO CUT);
G1W-24.F25. (HOBBING REQUIRED DIAMETER);
G1U50.F1000. (TOOL EXIT FROM MATERIAL)
G0Z0;
(***** KS-OK*END OF HOBB MACHINING *****);
#520= [#520+#522](SHIFTING IN Y-AXIS);
M9;
G80.4. (ENDING THE SYNCHRONIZATION);
M5;
M46; (END OF C-AXIS);
G0G28Y0. (REFERENCE POINT OF Y-AXIS)
```

Fig. 4. Template of G-Codes for Gear hobbing

Power skiving is a technological option that can be applied on CNC lathes and is already offered as technological cycle. The operator determines the position where the G-codes generated by the cycle will be inserted, manually placing them into the program line. During tool setup, the holder is mounted into the turret of the CNC lathe, and the tool must be tilted in the holder according to the type of gear to be produced. An example of tool setup for the manufacturing of external helical gears is shown in the figure 5.

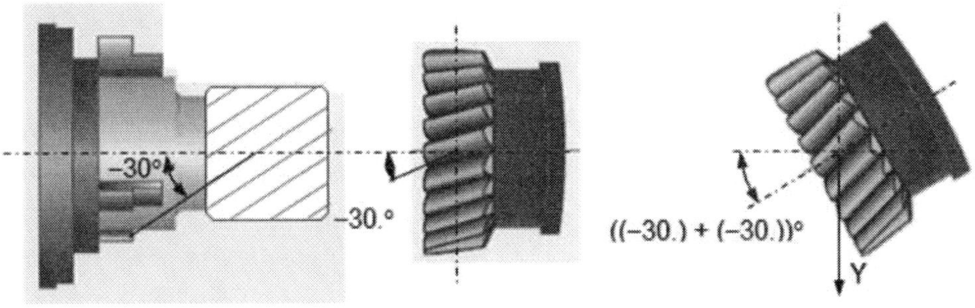

Fig. 5. Setting the angle of skiving tool [20]

The machining of external gear must be set – Tool is left-hand helix, so the helix angle βt = –30°. The workpiece is left-hand helix, so helix angle βw = –30°. Tool setting angle for helical gear machining relative to the Y-axis direction is – (βt + βw) = 60°.

The new G-codes which are generated by the operating system of the machines are:

- G900A0. – Tool definition cycle for helical gear machining
- G900A10. – Cycle for internal gear setup
- G900A11. – Cycle of machining internal gears
- G900A20. – Cycle for external gear setup
- G900A21. – Cycle of machining external gears

The Cycle of defining a tool for bevel machining of gears is shown on the Figure 6.

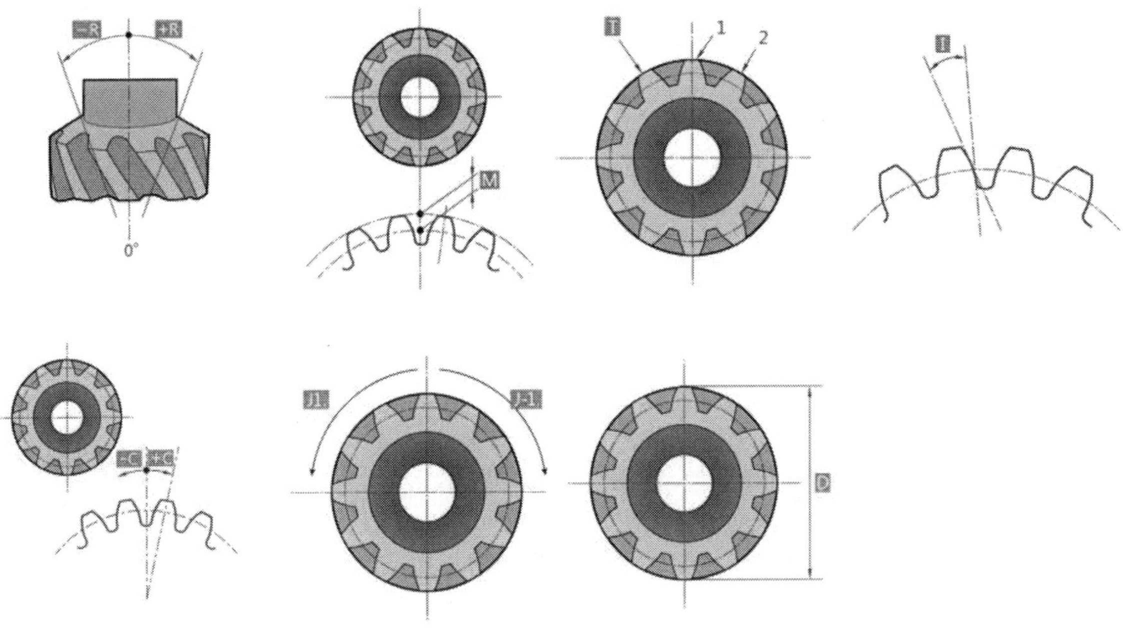

Fig. 6. Cycle definition of a tool for bevel machining gear [20]

The generated code is G900 A0 R_M_T_I_C_J_D_; where is defined:

- A0. Cycle defining a tool for bevel machining gear
- R Tool helix angle (degrees),
  where setting range is from -45. < R < 45.
  R > 0: Right-handed helix, R < 0: Left-handed helix
- M Tool module or diameter pitch
  Setting range (metric): 0,01 ≤ M ≤ 25, Setting range (inch): 0,01 ≤ M ≤ 254.
- T Number of teeth
- I Tool engagement angle, setting range: 0 < I < 90.
- C First tooth orientation angle, determines the angle for placing the first tooth on the X+ axis
  Used for phase equalization, setting range: -45 < C < 45.
- J Direction of tool rotation, J1.:M13 CW (for clockwise), J-1.: M14 CCW (for counter clockwise)
- D Diameter of the head circle of the tool

The Gear adjustment cycle is shown on Figure 7.

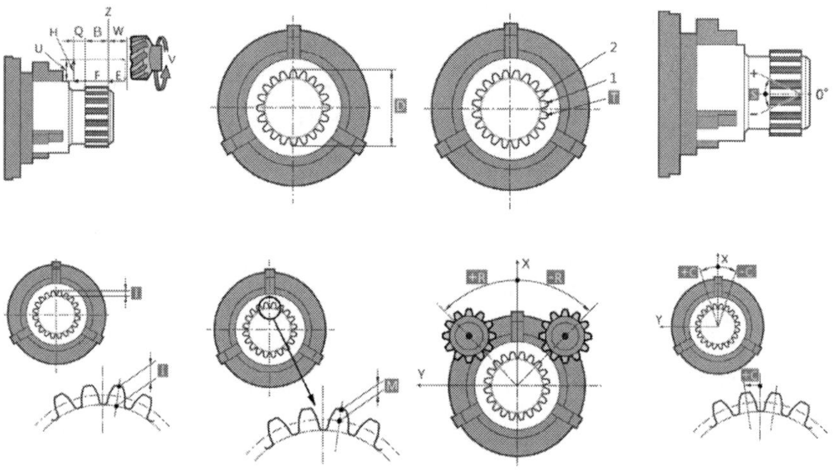

Fig. 7. Gear adjustment cycle [20]

The generated code is G900 A20. D_T_S_I_M_R_C_F_V_B_Z_Q_W_U_E_H_K1.; where is defined:

- A20…K1. Cycle for setting the machining of external gears
- D Diameter of the head circle of the workpiece (mm or inches)
- T Number of teeth
- S Helix angle (degrees),
  where setting range is from -45. ≤ S ≤ 45.
  S > 0: Right-handed helix, S < 0: Left-handed helix
- I Depth of workpiece teeth (mm or inches)
- M Tool module or diameter pitch
  Setting range (metric): 0,01 ≤ M ≤ 25, Setting range (inch): 0,01 ≤ M ≤ 254.
- R Machining position (degrees)
  Determines the angle for placing relatively to the X+ axis
  Setting range: -90 < R < 90.
- C Angle for setting the tooth position of the workpiece on the X+ axis (degrees)
  Used for phase alignment, Adjustment range: -360 < C < 360.
- F Feed rate in the Z-axis, defines the feed rate per revolution of the workpiece
- V Cutting speed (m/min or ft/min), defines the cutting speed at the tool´s pitch circle.
- B Tooth width (mm or inches)
- Z Tooth face position on the Z-axis (mm,inch)
- Q Clearance at the end of gear machining (mm or inches)
- W Clearance in the Z-axis direction at the beginning of gear machining (mm or inches)
- U Radial clearance at the beginning of gear machining (mm or inches)
- E Feed speed (mm or inches)
- H Drive feed speed (mm or inches)

4. Comparison of power skiving and gear hobbing setup and programming

In modern gear manufacturing, the choice between gear hobbing and power skiving depends not only on the final quality and geometry of the gears but also on the time and economic aspects of tool setup, tool life, and overall productivity. In the case of gear hobbing, the clamping of the hobbing cutter into the rotary holder mounted in the turret of the CNC lathe is relatively time-consuming, as it requires disassembling the shaft, mounting the cutter, and precise alignment. By contrast, power skiving tools are mounted directly into holders designed for this purpose, which makes setup significantly faster, especially when the technological cycle for skiving is integrated into CNC control, allowing automatic generation of G-codes and easier programming compared to the template-based programming used for hobbing. Table 1. presents a comparison of hobbing and skiving technologies, summarizing the differences and areas of application.

Procedure	Field of application	Particularly advantageous for	Frequency in industrial use
Hobbing	External toothing (soft)	All splines	Very high
Skiving	External gearing (hard)	Hardened tooth profiles with ISO 1328 quality 6-7	High
Power skiving	Tooth profiles (internal and external, soft)	Smaller batches, flexible production, internal gear cutting and tooth profiles “against interfering contours”	Medium

Table 1. Summary of the differences and areas of application [14]

Below I will present the advantages of both technologies in terms of: investment, cutting tool life, cycle times and input costs. From an investment point of view, the options and holders for gear hobbing are generally more affordable, often by approximately 50% compared to those required for power skiving, although the price of the hobbing cutter itself is nearly identical to that of a skiving tool. Tool life also favours power skiving, as the cutting process is smoother, chip evacuation is more efficient, and in many cases, excellent results can be achieved even without coolant, which would be impossible in hobbing due to rapid tool wear. In terms of cycle times, power skiving again demonstrates its advantages: for example, the production of a gear with a diameter of 52 mm and a module of M=1.5 takes approximately 1 minute and 45 seconds with hobbing, while the same gear can be produced in about 20 seconds using power skiving. Although hobbing offers lower entry costs, power skiving provides higher productivity, longer tool life, simpler setup, and better suitability for modern, flexible production environments.

5. Future research of power skiving, gear hobbing

In the future, it will be important for technologies that use rotating tool holders used on CNC lathes (power skiving, gear hobbing and broaching technology) to focus on:

- comparing tool life,
- assessing the quality of gears,
- energy consumption of each process,
- digitalization of the above manufacturing processes.

Comparing tool life would provide valuable data on the efficiency of each process and would allow identifying the optimal technology in terms of tool costs and production stability. Another key area of research is the evaluation of gear quality (surface integrity), where accuracy, surface roughness, microstructural changes and the presence of potential residual stresses in gears manufactured by these three technologies would be compared. The results of such research would offer a comprehensive overview of how the manufacturing technology affects the final properties of gears.

In addition to these main areas, further studies could also address other factors such as the energy consumption of each process, the feasibility of dry machining without cutting fluids, the influence of cutting parameters on process stability. Another promising topic for future research is the digitalization of manufacturing processes, where simulated results could be compared with real production data to improve the prediction of tool life and gear quality before the manufacturing process begins.

6. Acknowledgments

The article was supported by the VEGA 1/0391/24 and KEGA 007 STU – 4/2025 projects.

7. References

[1] von Pittler, W. (1910). Method for cutting gears using a gear-shaped tool with cutting edges on the tooth faces (orig. German: Verfahren zum schneiden von zahnrädern mittels eines zahnradartigen, an den stirnflächen der zähne mit schneidkanten versehenen schneidwerkzeugs). Deutsche Patentschrift Nr. 243514.

[2] https://www.mazak.com/eu-en/products/integrex-ag/, (2025). Yamazaki Mazak, Produce gears without a dedicated gear cutting machine, Accessed on: 2025-09-12.

[3] https://gearsolutions.com/Digital_Editions/2017/201704/gearsolutions-201704/AF8507FDD406DD783ED463F2C82E246E/0417-Gearsolutions.pdf, (2017). Media Solutions Inc., Understanding the basic principles of power skiving, Accessed on: 2025-09-14.

[4] https://gearsolutions.com/features/power-skiving-for-gears-advanced-solutions-and-contributions/, (2024). Briel, M.; Media Solutions Inc., Power skiving for gears: advanced solutions and contributions, Accessed on: 2025-09-09.

[5] Kuruc, M.; Vopát, T.; Peterka, J.; Necpal, M.; Simna, V.; Milde, J. & Jurina, F. (2022). The influence of cutting parameters on plastic deformation and chip compression during the turning of C45 medium carbon steel and 62SiMnCr4 tool steel. Materials, 15(2), 585, DOI: 10.3390/ma15020585.

[6] da Silva, W. T. A.; Peterka, J. & Vopát, T. (2023). Experimental research on the dynamic stability of internal turning tools for long overhangs. Journal of Manufacturing and Materials Processing, 7(2), 61, DOI: 10.3390/jmmp7020061.

[7] Sentyakov, K.; Peterka, J.; Smirnov, V.; Bozek, P. & Sviatski, V. (2020). Modeling of boring mandrel working process with vibration damper. Materials, 13(8), 1931, DOI: 10.3390/ma13081931.

[8] https://www.secotools.com/article/126125, (2024). Seco Tools, Why gear skiving can be good news for jet turbine disc manufacturers [Czech: Proč může být gear skiving dobrou zprávou pro výrobce proudových motorů], Accessed on: 2025-09-10.

[9] https://www.mtmarchetti.com/en/soluzioni/modulo-power-skiving.html, (2022). M.T. S.r.l., Power skiving module, Accessed on: 2025-09-14.

[10] Piska, M. & Holkova, K. (2023). On the effect of advanced PVD coating on performance of cutting dies in austenitic steel. Proceedings of the 34th DAAAM International Symposium, pp. 0016–0025, ISBN 978-3-902734-41-9, ISSN 1726-9679, DOI: 10.2507/34th.daaam.proceedings.003.

[11] Straka, R. & Vopát, T. (2022). Cutting edge preparation methods comparison in terms of tool life. Proceedings of the 33rd DAAAM International Symposium, pp. 0188–0192, DOI: 10.2507/33rd.daaam.proceedings.026.

[12] Dong, X.; Liao, C.; Shin, Y. C.; Zhang, H. (2016). Machinability improvement of gear hobbing via process simulation and tool wear predictions. The International Journal of Advanced Manufacturing Technology, 86, pp. 2771–2779, DOI: 10.1007/s00170-016-8400-3.

[13] Li, L.; Zhang, L.; Yu, B.; Wang, K.; Liu, F. (2015). An efficient spur gear shaping method based on homogenizing cutting area through variational circular feed rate, Proceedings of the Institution of Mechanical Engineers, Part B: Journal of Engineering Manufacture, 231, pp. 1587–1598, DOI: 10.1177/0954405415601639.

[14] https://www.emag.com/blog/en/differences-between-hobbing-skiving-and-power-skiving/?utm_source=chatgpt.com, (2025). EMAG Systems GmbH, The art of gear cutting: differences between hobbing, skiving and power skiving, Accessed on: 2025-09-13.

[15] https://www.emag.com/industries-solutions/technologies/hobbing/, (2025). EMAG Systems GmbH, Hobbing: highly productive manufacturing of gears, gear shafts, worms and rotors, Accessed on: 2025-09-13.

[16] Shih, Y.; Huang, Y.; Lee, Y.; Wu, J. (2013). Manufacture of face-hobbed straight bevel gears using a six-axis CNC bevel gear cutting machine. The International Journal of Advanced Manufacturing Technology, 68, pp. 2499–2515, DOI: 10.1007/s00170-013-4880-6.

[17] Suh, S.; Jih, W.; Hong, H. & Chung, D. (2001). Sculptured surface machining of spiral bevel gears with CNC milling. International Journal of Machine Tools and Manufacture, 41(6), pp. 833–850, DOI: 10.1016/S0890-6955(00)00104-8.

[18] https://www.cmz.com/en/applications-cnc-lathe-gear-hobbing, (2025). CMZ Machine Tool Manufacturer S.L., Gear hobbing on a lathe, Accessed on: 2025-09-09.

[19] https://www.wto-tools.com/fileadmin/user_upload/Katalog_Download/Anwendungsspezifisch/Abwaelzfraeseinheit/Gear_Hobbing_Unit_en.pdf, (2025). WTO Werkzeug-Einrichtungen GmbH, Gear hobbing unit, Accessed on: 2025-09-10.

[20] https://en.dmgmori.com/products/digitization/technology-cycles/turning-milling/machining/gearskiving, (2018). DMG MORI, gearSKIVING 2.0, Accessed on: 2025-09-11.

DOI: 10.2507/36th.daaam.proceedings.004

CONCEPTUAL MODEL FOR MEASURING AND ANALYSING INNOVATION OBSOLESCENCE THROUGH HYBRID INTELLIGENCE

Petya Georgieva, Irina Topalova*, Siyka Demirova & Iliyan Vasilev

This Publication has to be referred as: Georgieva, P[etya]; Topalova, I[rina]; Demirova, S[iyka] & Vasilev, I[liyan] (2025). Conceptual Model for Measuring and Analysing Innovation Obsolescence Through Hybrid Intelligence, Proceedings of the 36th DAAAM International Symposium, pp.0029-0038, B. Katalinic (Ed.), Published by DAAAM International, ISBN 978-3-902734-47-1, ISSN 1726-9679, Vienna, Austria
DOI: 10.2507/36th.daaam.proceedings.004

Abstract

This article presents a developed conceptual model, combining artificial intelligence and human experience to measure and analyse innovation obsolescence. By identifying key indicators and factors contributing to innovation obsolescence, a conceptual model was developed to assess the effectiveness of hybrid intelligence in detecting early signs of innovation obsolescence and supporting strategic decision-making. The hybrid model is based on machine learning with survey data focused on sales of computer devices on the Bulgarian market. In the study, the model was trained using two approaches - machine learning with a Random Forest algorithm and an MLP neural network, to compare the accuracy of the results in the test phase. The goal of the development is to provide insights and recommendations for manufacturers, retailers and policymakers on the management of innovation cycles and technological substitution in rapidly developing markets.

Keywords: Innovation Obsolescence; Hybrid Intelligence; AI-Human Collaboration; Bulgarian Market.

1. Introduction

According to an OECD (Organisation for Economic Cooperation and Development) report, systems innovation offers an integrated, multisectoral approach to addressing complex societal issues such as climate change, population aging, and resource depletion. This involves coordinated interventions across technological, social, market, and regulatory domains, promoting long-term, holistic solutions. Unlike traditional policies, systems innovation seeks to transform entire socio-technical systems-such as energy or healthcare-rather than focusing on isolated technologies or organizations [1].

Systems innovation requires collaboration among governments, businesses, civil society, and research institutions to achieve sustainable and resilient growth. It is a profound structural transformation that responds to disruptions or major breakthroughs by reshaping the components and architecture of systems. This approach transcends sectors and actors, integrating social, behavioural, and institutional change, and blending traditional innovation categories. Innovations often originate in niche technological areas and scale only when they align with broader societal needs and infrastructures. Understanding the obsolescence of systems innovation requires analysing complex transformations, enabling foresight and experimentation, and adapting strategies in times of uncertainty.

Multilevel perspective (MLP) studies [2] identify three levels:

The landscape developments layer represents macro-level trends and external pressures (e.g., climate change, societal values, and geopolitical shifts). Change here is slow and large-scale, but it can create "windows of opportunity" for innovation by putting pressure on the existing system. For example, global climate concerns are spurring the adoption of renewable energy.

The socio-technical system layer is the dominant system-existing norms, technologies, policies, infrastructure, market logics, and user practices. It is typically stable and resistant to change. Constant internal adjustments occur, but true transformation only occurs when pressure comes from above (the landscape) and below (niche innovations). Its subcomponents are markets/user preferences, science and technology, policy, and culture.

The technological niches layer represents sheltered spaces where radical innovations emerge. Innovations here are typically small-scale at first, but they are diverse, experimental, and dynamic. Through learning, communication, and development, niche innovations can grow and gain momentum. As these innovations mature and external pressures combine, they can disrupt or replace the system.

Time moves from left to right. Initially, niche innovations are isolated and experimental. As landscape pressures and system tensions increase, these niche innovations gain momentum. Eventually, a new configuration emerges, transforming the system- a shift in the socio-technical order. Once established, the new order influences future landscape developments, closing the cycle. [3]. Obsolescence management strategies are a set of proactive practices designed to mitigate risks related to innovation obsolescence throughout a product's lifecycle. These strategies incorporate innovation principles, particularly in the conceptual design phase, using model-based systems engineering to assess and respond to obsolescence risks early [4]. A modular, cross-sector framework categorizes strategies from suppliers to customers, highlighting critical obsolescence drivers and availability modelling techniques [5].

Practical approaches include establishing interfaces with suppliers and project-level processes to anticipate and address obsolescence issues [6]. Lifecycle strategies can vary from start-ups may adopt iterative obsolescence for innovation quickly, growth stages tolerate chaotic obsolescence to scale, and mature companies often adopt planned obsolescence to maintain efficiency- strategies that can be aligned using tools such as the Global Innovation Map [7]. Planned obsolescence, while sometimes profit-driven, is under increasing scrutiny under consumer protection policies, particularly as accelerating innovation cycles and global competition shorten product lifetimes [8]. Overall, effective obsolescence management balances innovation, economic value, and lifecycle considerations to ensure long-term sustainability.

2. Research Methodology and Limitations:

The study applies a multi-layered methodological framework that combines conceptual modelling, empirical field research, and machine learning techniques within a hybrid intelligence paradigm (Fig.1). Despite demonstrating high accuracy in the test phase, the model is constrained by its reliance on a small dataset from five Bulgarian companies, the short validation interval of one month, and the use of survey-based inputs that may not fully reflect market dynamics. Furthermore, the predefined strategic categories limit flexibility, while the hybrid intelligence framework requires human interpretation that introduces subjectivity. These limitations suggest that further studies with larger datasets, extended observation periods, and diversified industries are necessary to enhance the generalizability and robustness of the model.

3. State of the Art

One research direction in the literature relates to technology readiness levels to identify key components and considerations for innovation. It provides a framework for assessing technology maturity and identifying critical components related to innovation. [9]. Other research trends emphasize the need for proactive, model-based approaches to assessing and predicting system obsolescence across the lifecycle of complex systems. Salas-Cordero [4] proposes a Model-Based Systems Engineering (MBSE) approach that integrates obsolescence risk assessment early in the conceptual design phase. This not only mitigates future risks but also balances obsolescence management with innovation strategies, viewing obsolescence and innovation as opposing but interconnected phenomena. In her work, Salas-Cordero introduces two key tools that support a structured assessment: The Obsolescence Measure (HoO) – a conceptual framework for identifying obsolescence factors in relation to a system architecture; system Obsolescence Criticality Analysis (SOCA) – a method for assigning an Obsolescence Criticality Index (OCI) to risks, enabling their prioritization for mitigation during the analysis phase. Trabelsi, I. [10] develop a quantitative mathematical model to predict component obsolescence. The model considers multiple influencing factors, such as technological change, supply chain dynamics, and market trends. Trabelsi [10] also applies Monte Carlo simulation to estimate the degree of system obsolescence (SOD) by analysing the obsolescence timelines of key components over a 30-year period. This simulation approach enables probabilistic risk estimation and long-term lifecycle forecasting. Finally, obsolescence management should begin in the early stages of design and be treated as a strategic activity linked to innovation.

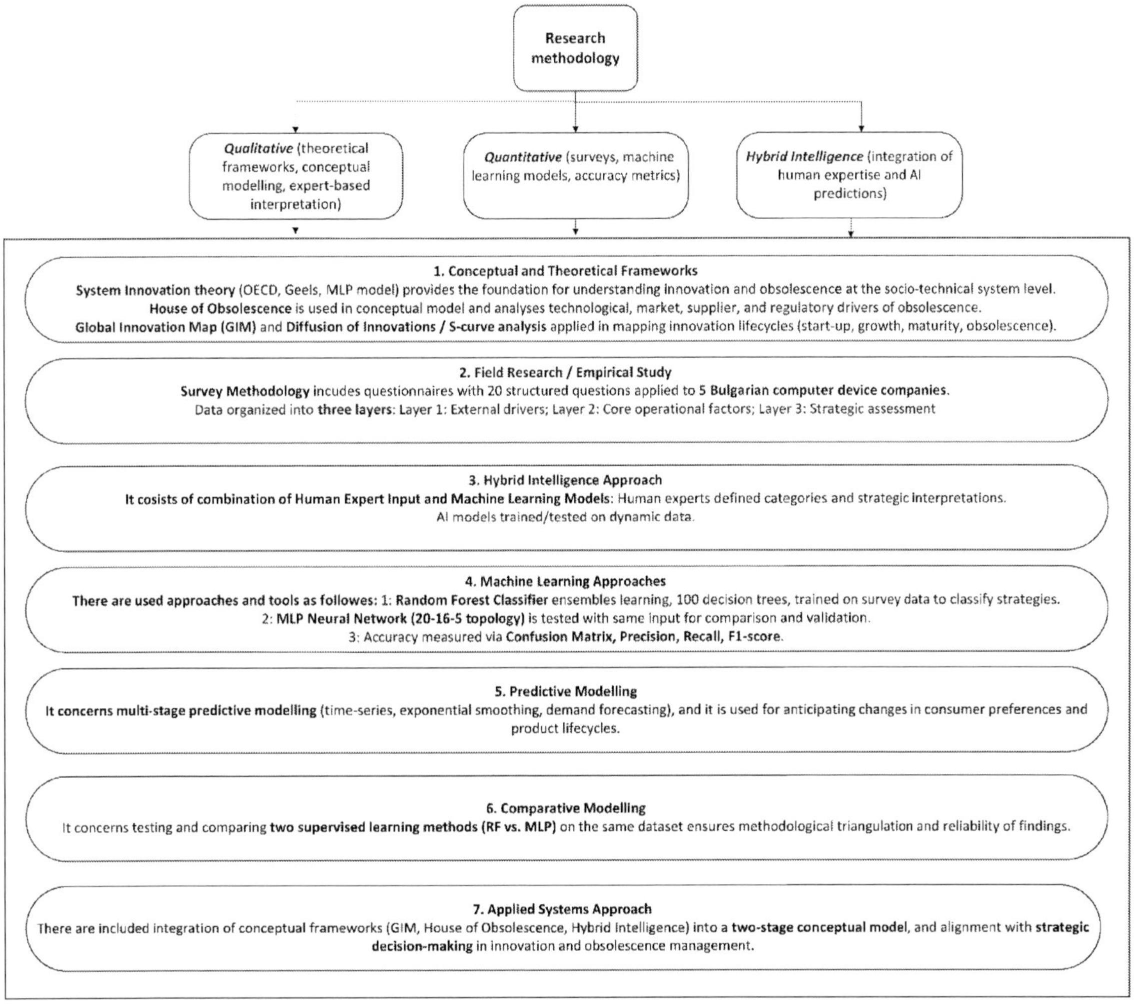

Fig. 1. Research methodology

Combining qualitative tools (such as HoO and SOCA) with quantitative models (such as Monte Carlo simulation) enhances foresight and decision-making. Integrating these models into the design and engineering processes supports more resilient and sustainable systems. The use of artificial neural networks (machine learning algorithms), applied to classify parts as active or obsolete with high accuracy, was discussed from Jennings et. al. (2016): "results have shown that machine learning algorithms (i.e., random forest, artificial neural networks, and support vector machines) can classify parts as active or obsolete with over 98% accuracy and predict obsolescence dates within a few months."[11].

The Random Forest method is a widely used machine learning algorithm that creates multiple decision trees during training and combines their results to improve prediction accuracy and robustness. It is particularly suitable for classifying components as active or obsolete and for modelling user behaviour, especially in contexts involving large amounts of data and complex, nonlinear relationships. Research shows that Random Forest, combined with other machine learning techniques such as artificial neural networks and support vector machines, can achieve classification accuracy exceeding 98% in distinguishing between active and obsolete parts. Additionally, this method can predict obsolescence timelines to within a few months of the actual date, making it a valuable tool for managing product lifecycles and planning maintenance or replacement schedules [12]. In addition to the Random Forest machine learning model, it is appropriate to compare the prediction accuracy with one of the classic neural network training models. For example, a neural network model of type Multi-Layer Perceptron (MLP) can be used to train and recognize/classify input data of the same type and format but dynamically changed over time.

The multi-stage predictive modelling approach is a methodology designed to capture the evolution of consumer preferences over time, particularly regarding product design. This approach supports design engineers in anticipating next-generation product features before they become widely adopted and potentially obsolete. This method operates in multiple stages, with each stage reflecting a different phase in the evolution of consumer preferences. A key component of this approach is the incorporation of time series analysis, specifically exponential smoothing, which is used to identify and predict emerging trends in product attributes.

The result is a dynamically evolving demand model that reflects changing consumer expectations, enabling proactive and informed product development decisions. This technique is particularly useful in rapidly changing markets, where staying ahead of consumer trends is critical to maintaining a competitive advantage and innovation relevance [13]. Hybrid intelligence is defined as the collaboration of human and machine intelligence to solve complex problems. It emphasizes conditional interdependence: tasks are divided among agents, but all contribute to a common goal. The outcome should be better than any one agent could achieve alone. Importantly, hybrid intelligence systems must continually learn and evolve through mutual interactions. These systems go beyond automation or human interaction models; they are dynamic, co-evolving systems [14].

Hybrid Intelligence (HI) represents a new and evolving paradigm, combining human cognitive abilities with AI systems to achieve superior problem-solving outcomes. Rather than seeking to replace human intelligence, HI emphasizes a collaborative partnership where each party compensates for the other's limitations. Human agents provide contextual awareness, ethical judgment, creativity, and emotional intelligence, while AI contributes speed, scalability, data processing power, and analytical accuracy [15] [16].

The literature emphasizes also the need to standardize the structure of hybrid systems and define governance frameworks that manage task allocation, data management, and incentive structures [17] [18]. The problem of the research is the high resource consumption and waste generation resulting from the production and sale of products (such as electronic devices). Measuring and analysing innovation obsolescence will support sustainable development and the application of circular economy principles. Research object is obsolescence of system innovations. The research aim is measuring and analysing system innovation obsolescence using hybrid intelligence with application in computer devices sales in Bulgaria. To address these issues, the subject of this research is to develop a machine learning-based methodology capable of forecasting obsolescence risk and product life cycle accurately while minimizing maintenance and upkeep of the forecasting system.

In this study, a hybrid supervised learning model was created to predict changes in recommended strategies based on dynamically changing input data. The hybridity of the proposed model lies in the integration between the predefined correspondence between the groups of combined input data and the desired recommended strategy. This first stage is determined by a human factor/expert. To standardize the structure of the proposed hybrid system, the second stage involves training two different well-known AI models with the prepared data. The third stage involves testing the two trained models with input data of the same type and format, but dynamically changed after a specified time, in order to correctly assign them to a specific strategy.

4. Description of the proposed conceptual system model

The diagram presents the first stage of the proposed conceptual model that integrates three key theoretical concepts: Global innovation map, the House of obsolescence, and Hybrid intelligence analysis, based on System innovation. It illustrates the interaction of these concepts in guiding strategic decision-making related to innovation obsolescence in dynamically evolving environments.

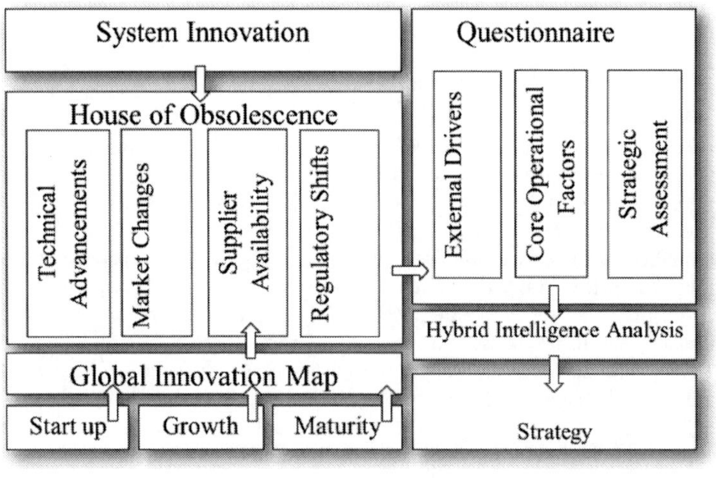

Fig. 2. First stage of the conceptual system model

4.1. System innovation block as the starting point

At the top of the diagram, System innovation captures the entire structure, focusing on a comprehensive, multi-sectoral transformation of socio-technical systems. This level establishes the premise that obsolescence and innovation should not be viewed as isolated phenomena, but rather as systemic processes embedded in evolving technological, economic, and organizational environments.

4.2. The House of obsolescence block: the multidimensional drivers behind the obsolescence of innovation

The concept of "House of obsolescence", which is based on working on the life cycle of innovation and technology turnover [19], includes four vertically aligned pillars:

- Technological developments: Refers to the pace and direction of technological progress that renders existing technologies obsolete.
- Market changes: Includes shifts in consumer demand, business models, and competitive dynamics.
- Supplier availability: Focuses on the continuity of supply of components and resources, highlighting the risks of dependency or disruption.
- Regulatory shifts: Includes legal, environmental, and political changes that enforce compliance and may lead to product or process obsolescence.

Together, these elements collectively define the obsolescence pressure affecting innovation.

4.3. Global innovation map block: mapping innovations throughout their lifecycle

The Global Innovation Map (GIM) is located below the "House of Obsolescence" and is derived from diffusion of innovation theory and S-curve analysis [20] Innovations are classified according to these stages:

- Start-up: Early-stage innovations, disruptive innovations, and niche innovations.
- Growth: Rapid expansion and market penetration.
- Maturity: Stable technologies with gradual improvement and increasing susceptibility to obsolescence.

This map provides a temporal and contextual location of innovations in the market, helping organizations visualize the vulnerabilities of each stage with respect to obsolescence.

4.4. Questionnaire block: structured inputs for Hybrid intelligence analysis block

On the right of diagram, the questionnaire section divides user or expert input into three conceptual layers:

- External factors: macro-level forces, such as politics, climate change, or global economic trends.
- Core operational factors: internal process efficiency, technological integration, and organizational capabilities.
- Strategic assessment: A high-level assessment of the alignment between innovation objectives and external/internal conditions.

This analysis aligns with systems thinking approaches to diagnosis and supports data collection for HI analysis. For our research, 5 computer sales companies in Bulgaria were surveyed, and the respondents were asked 20 questions, grouped into Layer 1: External Drivers, Layer 2: Core Operational Factors and Layer 3: Strategic Assessment. According to the number of answers, A, B C and D categories were defined in each question group. The obtained results for two of the companies are shown in Table 1. The *combinations of Layer 1 (L1), Layer 2 (L2) and Layer 3 (L3),* together with *Detailed Interpretation of results by combinations of dominant Layer, Generalized obsolescence degree*, *Detailed Strategic Actions with Examples in Sales of Electronic Devices* and *Strategic Actions Summary*, are shown in Table 2.

	Q1-Q5	**Layer 1: External Drivers**				**Q6-Q11**	**Layer 2: Core Operational Factors**				**Q12-Q20**	**Layer 3: Strategic Assessment**			
Company 1		**A**	**B**	**C**	**D**		**A**	**B**	C	**D**		**A**	B	**C**	**D**
	Q1	52	15	12	7	Q6	54	16	10	6	Q12	51	19	13	3
	Q2	54	18	13	1	Q7	52	20	11	3	Q13	55	17	9	5
	Q3	62	12	10	2	Q8	65	9	7	5	Q14	61	13	10	2
	Q4	58	10	9	9	Q9	57	11	10	8	Q15	55	20	11	0
	Q5	71	9	6	0	Q10	69	11	4	2	Q16	73	7	4	2
						Q11	55	13	12	6	Q17	53	19	13	1
											Q18	72	10	2	2
											Q19	67	11	4	4
											Q20	68	12	5	1
Copmany 2	Q1	63	18	15	9	Q6	66	20	12	7	Q12	62	23	16	4
	Q2	66	22	16	1	Q7	63	24	13	4	Q13	67	21	11	6
	Q3	76	15	12	2	Q8	79	11	9	6	Q14	74	16	12	2
	Q4	71	12	11	11	Q9	70	13	12	10	Q15	67	24	13	0
	Q5	87	11	7	0	Q10	84	13	5	2	Q16	89	9	5	2
						Q11	46	26	2	31	Q17	52	23	30	0
											Q18	57	18	24	6
											Q19	58	24	6	17
											Q20	59	22	15	9

Table 1. Questionnaire block with number of answers for two of the companies

Combinations from L1, L2, L3	Detailed Interpretation of results by combinations of dominant Layer	Generalized obscolenscence degree /GOD/	Detailed Strategic Actions with Examples in Sales of Electronic Devices	Strategic Actions Summary
AAA	Layer 1: High obsolescence risk or pressure (external/internal/strategic); Layer 2: High obsolescence risk or pressure (external/internal/strategic); Layer 3: High obsolescence risk or pressure (external/internal/strategic).	Critical obsolescence across all dimensions	Initiate an immediate redesign of outdated product lines. Allocate R&D resources toward developing next-generation PC components such as DDR5 RAM, PCIe 5.0 SSDs, and AI-accelerated GPUs. Collaborate with chipset manufacturers to ensure alignment with emerging hardware trends and address obsolescence across performance, compliance, and customer relevance.	0 Immediate product redesign and innovation investment; initiate R&D programs
BBB	Layer 1: Moderate obsolescence signals or performance issues; Layer 2: Moderate obsolescence signals or performance issues; Layer 3: Moderate obsolescence signals or performance issues.	Moderate innovation pressure; partial stability	Focus on incremental updates to existing products, e.g., bundling popular mid-tier components (e.g., RTX 4060 with NVMe drives). Refine supply chain for efficiency and consider introducing co-branded configurations with OEMs. Monitor customer usage data and feedback to identify specific needs in gaming, productivity, or creator segments.	1 Optimize operations; refresh outdated features and align product roadmap
CCC	Layer 1: Stable condition, but innovation may be lagging; Layer 2: Stable condition, but innovation may be lagging; Layer 3: Stable condition, but innovation may be lagging.	Stable but lagging in innovation	Conduct strategic foresight workshops to anticipate shifts in user preferences (e.g., demand for silent cooling systems or eco-packaging). Plan for the phased introduction of newer technologies in response to slow-moving legacy inventory. Enhance value proposition through warranty extensions, trade-in programs, or targeted marketing.	2 Consider gradual innovation upgrades; monitor customer expectations
DDD	Layer 1: Low awareness or poor alignment – high risk if ignored; Layer 2: Low awareness or poor alignment – high risk if ignored; Layer 3: Low awareness or poor alignment – high risk if ignored.	Low awareness or poor strategic fit; high risk	Reassess product-market fit and ensure offerings align with both current regulations and technological standards (e.g., energy efficiency, RoHS compliance). Diversify supplier base to reduce exposure to component shortages and integrate real-time compliance monitoring tools. Rebuild customer trust through transparency and educational campaigns on product evolution.	3 Improve strategic alignment; reassess compliance and vendor risks
ABC, ABD, ACB, ADC, BAC, BAD, BCD, BDA, CAB, CAD, DCB, DBC, DAC, CDB, DAB	Layer 1: High obsolescence risk or pressure (external/internal/strategic); Layer 2: Moderate obsolescence signals or performance issues; Layer 3: Stable condition, but innovation may be lagging.	Mixed signals; targeted improvements needed	Undertake a layered performance audit across marketing, operations, and product management functions. Deploy agile cycles for prototyping and testing updates in underperforming components. Engage B2B and enthusiast communities to crowdsource ideas and validate proposed adjustments.	4 Conduct focused diagnostics; apply agile improvement plans in weak layers

Table 2. Answer combinations and proposed strategies

4.5. Hybrid intelligence analysis block: integration of human and machine thinking

Hybrid intelligence analysis block represents a pivotal moment in the synthesis process. Hybrid intelligence refers to the combination of machine learning models and human cognition to analyse, interpret, and help to predict innovation obsolescence [21]. It combines:

- Structured insights from the survey
- Quantified pressures from the obsolescence landscape
- Lifecycle positioning within the global innovation map

The analysis enables prediction of obsolescence risks, simulation of innovation trajectories, and optimization of strategic options.

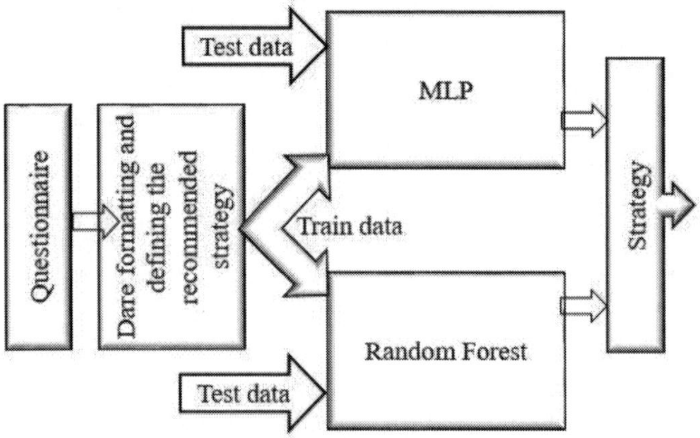

Fig. 3. Second stage of the conceptual system model - AI part of the Hybrid model

The second stage after collecting and formatting the answers to the questions, as given in Table 1, is to train the two models - Random Forest and a MLP neural network with the train data.

4.6. Strategy block: the convergent output

Essentially, the strategy emerges as a practical outcome of hybrid analysis. It may include different strategies [23, 24, 25]:

- Innovation renewal (reinvestment or re-pivot)
- Obsolescence planning
- Supplier diversification
- Regulatory compliance strategies

This last component represents the translation of complex system insights into organizational decision-making. In the test phase, the trained models are fed test data obtained after a survey with the same companies, but after one month. The system recognizes the data and classifies it into one of the five strategic categories. In this way, it responds adequately to dynamic changes in the respondents' answers.

5. Experimental results

The Random Forest model was trained with 20 samples of each one combination, defined in Table 1 (AAA, BBB, CCC, DDD, ABC, ABD, ACB, ADC, etc.). To create the model, the *Collaboratory* cloud environment was used, with access to the necessary Python libraries [22]. # RandomForestClassifier was Initialized with n_estimators=100. This parameter specifies that the Random Forest will consist of 100 decision trees, where X_train variable represents the values of input parameters Q1… to Q20. Y_train variable represents the answer category - Strategy (0,1,2,3,4 in Table 2).

X = company_data[['Q1', 'Q2', 'Q3', 'Q4', 'Q5', 'Q6', 'Q7', 'Q8', 'Q9', 'Q10', 'Q11', 'Q12', 'Q13', 'Q14', 'Q15', 'Q16', 'Q17', 'Q18', 'Q19', 'Q20']]

Y = company_data ['Answer category']

rf_classifier = RandomForestClassifier(n_estimators=100, random_state=42)

Fitting the classifier to the training data

rf_classifier.fit(X_train, Y_train)

Test data were constituted as input data of the same type and format, but dynamically changed after a month, in order to correctly classify them to a specific strategy and evaluate if the input conditions (answer of questions) represent a dynamic change. The test set consists of 10 samples of each strategy 0 to 3, and 20 samples of strategy 4, as whole 60 test samples. They are chosen randomly and fed to the model input. The obtained confusion matrix represents, that 5 samples of strategy 4 (that is 5 combinations of answers) were changed and classified/transformed as strategy 0. And 1 sample of strategy 1 is classified as strategy 4. All other instances are classified without change compared to the training phase.

To determine whether the strategy should really change according to the model's decision, we need to consider the classical estimation parameters of the model's accuracy. Namely, these are the parameters „Precision", „Recall" and „F1-score".

Real strategy categories	0 1 2 3 4
Classified in test phase 0	[[15 0 0 0 0]
1	[0 11 0 0 0]
2	[0 0 13 0 0]
3	[0 0 0 15 0]
4	[0 1 0 0 5]]

Table 3. Confusion matrix

As F1-score takes into account both precision and recall and is based on a balance of the two, it is more appropriate to calculate this parameter according to the given equation 1:

$$\mathrm{F1-score} = 2.\frac{\mathrm{preciscion.recall}}{\mathrm{preciscion + recall}} = \frac{\mathrm{TP}}{\mathrm{TP} + 1/2(\mathrm{FP} + \mathrm{FN})}, \tag{1}$$

Where:
True positives TP is an outcome where the model correctly predicts the positive strategy.
False positives FP is an outcome where the model incorrectly predicts the positive strategy.
False negatives FN is an outcome where the model incorrectly predicts the negative strategy.

Strategy	Precision	Recall	F1-score	Test samples
0	1.00	1.00	1.00	15
1	0.92	1.00	0.96	11
2	1.00	1.00	1.00	13
3	1.00	1.00	1.00	15
4	1.00	0.9	0.95	6

Table 4. Accuracy results in test phase for the 5 defined strategies

The very high value of evaluation parameter F1-score, shows that we could consider the test results as recommendation for change in strategy for these 6 different classified samples (question answers one month later). To assess the accuracy of the created machine learning model, we trained a MLP neural network model with the same set of input data, with the following topology: 20 neurons in the input layer (associated with the 20 values of the number of answers for each of the 20 questions), 16 neurons in the hidden intermediate layer and 5 neurons in the output layer (corresponding to the 5 given strategies), i.e. MLP 20-16-5. The same train and test data sets were fed to the inputs in train and test phase.

MLP Model Definition

```
model=Sequential()
model.add(Input(shape=(20,)))
model.add(Dense(16, activation='tanh')
model.add(Dense(5, activation='sigmoid'))
```

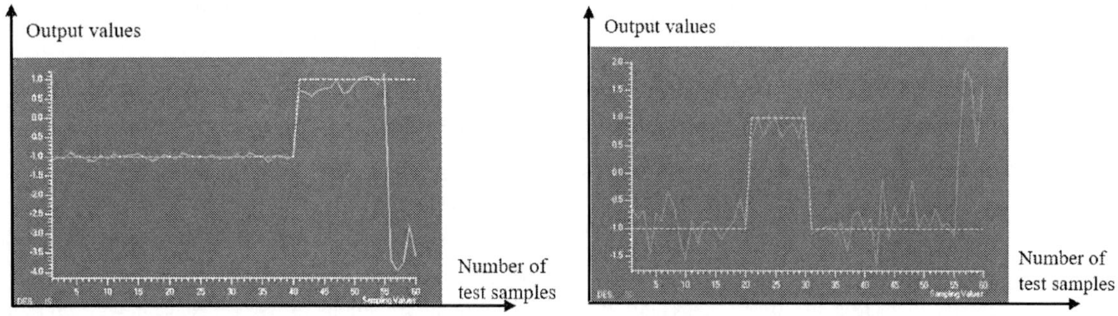

a) b)

Fig. 4. MLP output values: a/ output values of neuron 5 (strategy 4) when fed with 20 test instances of this strategy; b/ output values of neuron 2 (strategy 3) when fed with 10 test instances of this strategy

In the graphs in Figure 3, the dotted line represents the ideal values set in the training phase for the 5 neurons in the output layer, and the solid line represents the approximated values in the test phase. From the graphs in Figure 3, it can be seen that for 6 out of 20 test instances incorrect classifications are obtained - for strategy 4. That is, the number of transformed strategies for the test data obtained after one month coincides with that obtained with the Random Forest model. This is a prerequisite for considering that the methodology can claim standardization in terms of the adequacy of prescriptions for change in business strategy.

6. Conclusion and feature work

What is new in this study that distinguishes it from similar studies discussed at the beginning of the article?

- Bringing the concept of innovation to the system level in the context of innovation obsolescence.
- Studying the obsolescence of the elements of systemic innovation.
- Recommendations for strategies that are based on information about the system in which the innovation operates. In this way, the innovation can be positioned in terms of its effect on the system.
- Using Random Forest and MLP in the context of tracking innovation obsolescence.

Finally, the model represents a systems-oriented approach to managing innovation obsolescence. It presents a multidimensional, lifecycle-aware, and intelligence-enhanced model, where each theoretical component (innovation management, business architecture, and hybrid intelligence) plays a role in guiding flexible and informed innovation strategies. This integrative logic aligns well with contemporary frameworks in strategic innovation management, particularly in high-speed sectors such as electronics, software, and advanced manufacturing. Using field research, a framework was developed to measure and analyse innovation obsolescence, and a survey was developed to survey the opinions of five companies deal with Computer devices sales, from different regions in Bulgaria.

The survey was conducted over three days, and the data was used to train the AI. One month later, the survey was repeated, and the data was used to test the AI model and predict the change in strategies. In this study, a hybrid supervised learning model was created to predict changes in recommended strategies based on dynamically changing input data. The hybridity of the proposed model lies in the integration between the predefined by a human specialist correspondence between the groups of combined input data and the desired recommended strategy. Training and testing two different supervised learning models with the same input data is a prerequisite for comparing results and confirming the correctness of the choice of transformed strategy. The system recognizes the data and classifies it into one of the five strategic categories. In this way, it responds adequately to dynamic changes in the respondents' answers. An advantage of the model is the possibility of repeated re-training with different survey results, in order to build different strategies in other business sectors. Submitting data in the test phase with survey results obtained after different periods of time, leads to automated dynamic change in recommendations for changing strategies.

Research contribution concerns development of an innovative conceptual model that combines hybrid intelligence (human expertise and artificial intelligence) for innovation obsolescence analysis, applied to a real market (computer sales in Bulgaria); integration of three theoretical frameworks: Global Innovation Map, House of Obsolescence, Layered model for strategic assessment through questionnaires. The technical contributions of the research are creation and training of two machine learning models: Random Forest and MLP neural network, used to predict and classify strategies for dealing with innovation obsolescence. Use of dynamic input data collected over different time periods to validate the adaptability and accuracy of the models. Practical contribution concerns conducting a field study among 5 real companies engaged in the sale of computer devices, through 20 questions in three layers of assessment; deriving strategies that organizations can undertake: innovative renewal, planned obsolescence, supplier diversification, strategic regulatory compliance, etc. Conceptual and methodological contributions are defining hybrid intelligence as a strategic tool that combines human expert thinking with predictive analytics; forming a basis for standardization that allows the methodology to be applied across industries and adapted to the specifics of the sector.

As a future development of the method, it is planned to test it with survey data from more companies from different industries. For the analysis of predictions and recommendations for changes in strategies, the test can be further developed with data after different time periods, to establish the aging time of innovations according to the type of industry, the volume of commercial activity, etc.

7. Acknowledgments

Acknowledgements for the support of the Research Project № 3O-340/10.04.2025, of the *University of Telecommunications and Post* titled "A conceptual model for analysing innovation obsolescence through hybrid intelligence".

8. References

[1] System innovation: synthesis report, OECD 2015, https://archiwum.pte.pl/pliki/2/1/OECD%20System.pdf#page=15.12 , last access 19.07.2025

[2] Rip, Arie & Kemp, Rene. (1998). Technological Change. In: Rayner S., Malone E.L. (editors). Social Science & Medicine - SOC SCI MED. 327-399.

[3] OECD Science, Technology and Industry Scoreboard 2013, https://www.oecd.org/content/dam/oecd/en/publications/reports/2013/10/oecd-science-technology-and-industry-scoreboard-2013_g1g311d1/sti_scoreboard-2013-en.pdf, last access 19.07.2025

[4] Salas Cordero, S., Vingerhoeds, R., Baron, C., & Zolghadri, M. (2021, July). Can proactive obsolescence management and innovation go hand in hand during the conceptual design phase of complex systems? In IFIP International Conference on Product Lifecycle Management (pp. 40-54). Cham: Springer International Publishing.

[5] Cabanãs, A.A., Baguley, P., & Roy, R. (2013). Industry survey review of obsolescence management strategies in performance-based contracts.

[6] Meyer, A. (2002). Management of component obsolescence in the military electronic support environment. https://core.ac.uk/download/18221113.pdf, last access 19.07.2025

[7] Forsythe, J., & Khormaei, R. (2011). Innovation topology: The dimensions and their characteristics that drive innovation. 2011 Proceedings of PICMET '11: Technology Management in the Energy Smart World (PICMET), 1-10.

[8] Kurz, R. (2015). Quality, obsolescence and unsustainable innovation. https://hrcak.srce.hr/file/221356, last access 19.07.2025

[9] Salas Cordero, S., Vingerhoeds, R., Baron, C., & Zolghadri, M. (2021, July). Can proactive obsolescence management and innovation go hand in hand during the conceptual design phase of complex systems? In IFIP International Conference on Product Lifecycle Management (pp. 40-54). Cham: Springer International Publishing.

[10] Trabelsi, I., Barkallah, M., Zolghadri, M., Zeddini, B., & Haddar, M. (2021, March). A New Approach to Evaluate and Predict System Obsolescence: Mathematical Formulation. In International Conference on advances in Materials, Mechanics and Manufacturing (pp. 141-149). Cham: Springer International Publishing.

[11] Jennings, C., Wu, D., & Terpenny, J. (2016). Forecasting obsolescence risk and product life cycle with machine learning. IEEE Transactions on Components, Packaging and Manufacturing Technology, 6(9), 1428-1439.
[12] Jennings, C., Wu, D., & Terpenny, J.P. (2016). Forecasting Obsolescence Risk and Product Life Cycle with Machine Learning. IEEE Transactions on Components, Packaging and Manufacturing Technology, 6, 1428-1439.
[13] Tucker, C.S., & Kim, H.M. (2011). Trend Mining for Predictive Product Design. Journal of Mechanical Design, 133, 111008.
[14] Dellermann, D., Ebel, P., Söllner, M. et al. Hybrid Intelligence. Bus Inf Syst Eng 61, 637–643 (2019). https://doi.org/10.1007/s12599-019-00595-2
[15] Dellermann, Dominik and Calma, Adrian, Making AI Ready for the Wild: The Hybrid Intelligence Unicorn Hunter (March 1, 2018). Available at SSRN: https://ssrn.com/abstract=3245383 or http://dx.doi.org/10.2139/ssrn.3245383.
[16] Järvelä, S., Rummel, N., Cukurova, M., Radtke, A., Meyer, J., Yang, S., Schneider, B., Rosé, C. P., Martinez, M. O., Peppler, K., Lämsä, J., de Araujo, A., Edwards, J., Sobocinski, M., Roberts, A., Whitehead, R., & Papadopoulos, P. M. (2025). Exploring Human-Ai Interactions Across Various Learning Tasks: Diverse Perspectives on Augmenting Learner. In Oshima, J., Chen, B., Vogel, F., & Järvelä, J. (Eds.), Proceedings of the 18th International Conference on Computer-Supported Collaborative Learning - CSCL 2025 (pp. 481-489). International Society of the Learning Sciences.
[17] Tiddi, I., De Boer, V., Schlobach, S., & Meyer-Vitali, A. (2023, December). Knowledge engineering for hybrid intelligence. In Proceedings of the 12th Knowledge Capture Conference 2023 (pp. 75-82).
[18] Venigandla, K., Vemuri, N., & Vemuri, N. (2024). Hybrid intelligence systems combining human expertise and AI/RPA for complex problem solving. International Journal of Innovative Science and Research Technology, 9(30), 2066-2075.
[19] Sandborn, P. (2013). Design for obsolescence risk management. Procedia cirp, 11, 15-22.
[20] Rogers, M. (2003). A survey of economic growth. Economic Record, 79(244), 112-135. https://doi.org/10.1111/1475-4932.00082.
[21] Dellermann, D., Lipusch, N., Ebel, P. et al. Design principles for a hybrid intelligence decision support system for business model validation. Electron Markets 29, 423–441 (2019). https://doi.org/10.1007/s12525-018-0309-2.
[22] https://colab.research.google.com/notebooks/intro.ipynb
[23] Singer-Coudoux, K[atrin]; Nicolai Buxmann, H[enry] & Will, M[arkus] (2024). Strategy Development Based on Intangibles for Startups: Capacity Building in Emerging European InnovationEcosystems, Proceedings of the 35th DAAAM International Symposium, pp.0378-0386, B. Katalinic (Ed.), Published byDAAAM International, ISBN 978-3-902734-44-0, ISSN 1726-9679, Vienna, AustriaDOI: 10.2507/35th.daaam.proceedings.052
[24] Tesic, S[inisa]; Bajic, B[ojana]; Cosic, I[lija] & Rikalovic, A[leksandar] (2021). Frugal innovations in Industry 4.0: The Identification of Key Indicatiors, Proceedings of the 32nd DAAAM InternationalSymposium, pp.0262-0269, B. Katalinic (Ed.), Published by DAAAM International, ISBN 978-3-902734-33-4, ISSN1726-9679, Vienna, AustriaDOI: 10.2507/32nd.daaam.proceedings.037
[25] P. Georgieva, "Artificial Intelligence for Implementation of Business Model Innovation in an Industrial Company," 2022 International Conference on Communications, Information, Electronic and Energy Systems (CIEES), Veliko Tarnovo, Bulgaria, 2022, pp. 1-6, doi: 10.1109/CIEES55704.2022.9990677.

DOI: 10.2507/36th.daaam.proceedings.005

MODELS FOR EVALUATING EFFECTIVE IP TRAFFIC MANAGEMENT WITH APPLICATION OF ARTIFICIAL INTELLIGENCE

Irina Topalova* & Pavlinka Radoyska

This Publication has to be referred as: Topalova, I[rina] & Radoyska, P[avlinka] (2025). Models for Evaluating Effective IP Traffic Management with Application of Artificial Intelligence, Proceedings of the 36th DAAAM International Symposium, pp.0039-0046, B. Katalinic (Ed.), Published by DAAAM International, ISBN 978-3-902734-47-1, ISSN 1726-9679, Vienna, Austria
DOI: 10.2507/36th.daaam.proceedings.005

Abstract

The application of simulation models to assess the effective management of IP traffic requires the creation of appropriate network topologies, with supported functionality for building balanced and reliable traffic. The application of artificial intelligence methods for analysis and classification of statistical data determining effective management is emerging as an innovative modern approach to finding optimal solutions. But these systems do not provide a prescription for automated selection of changes in network topology/architecture to obtain better traffic efficiency. In this study, a method for automated prescription for improving the network structure is proposed, based on a database of current time parameters that determine the quality of remote connectivity. The method is developed with simulated network topologies, with the implementation of known protocols for alternative and fail-safe traffic. The measured time parameters, in different traffic regimes, serve as input training data for a proposed machine learning model. Depending on the results in the classification stage of newly received data, the model predicts the need for appropriate restructuring of the network topology and reconfiguration of the protocols. The obtained results are discussed in terms of accuracy and precision of the classification.

Keywords: IP traffic; IP network topologies; artificial Intelligence; machine learning.

1. Introduction

The most advanced network devices apply ML (Machine Learning) and AI (Artificial Intelligence) for threat detection, tracking and response, user behaviour analysis, network traffic analysis, automation in security operations and incident response. AI with the application of neural networks confidently enters the QoS mechanisms for prioritizing IP traffic to achieve better efficiency to reduce the effects of congestion [1]. In the training phase, the AI-based analytical module observes and learns the normal values of parameters. During the test phase, real-time traffic is intercepted and compared with normal behaviour. In search of plausibility in such systems, ML detects anomalies, identifies their common features and is trained on these data sets. They are analysed and classified according to the general ML model, enriching the information and creating a predictive model [2]. Various routine and repetitive tasks are intelligently designed and embedded in the modules of the software libraries.

This modern and innovative functionality does not include analysis of system behaviour in hybrid architectures, including the basic structures and protocols for balanced and fail-safe traffic, by providing alternative connections. For these reasons, it would be appropriate to implement an adaptive training system with different parameters that determine effective traffic management. This would also allow for appropriate prescriptions for reconfiguring the system according to the results of additional training of the adaptive system for different network architectures.

***State of the Art*:** Artificial intelligence (AI) methods, applied to adapt to changes in input data, dynamically updating the accumulated knowledge in order to obtain a quick and adequate response, are increasingly noticeable in many different tools in information technology. In this regard, it is reasonable, together with the various ML techniques developed in network devices, to integrate the analysis of a number of dynamically changing parameters of IP traffic such as the number of packets transmitted per second (Packets/sec), remote access time, etc. [3]. This would facilitate tracking the current state of the connection, classifying and reflecting various dynamically changing conditions. There are a number of software platforms that offer similar functionality for monitoring the quality of connections to a device on the local network or to a remote server. For example, SolarWinds Ping Sweep [4] allows you to check the connectivity to a list or range of IP addresses. This allows you to assess the performance of both DHCP and DNS servers, as well as generate warning messages about a bad or missing connection. The software offers the ability to modify ping parameters such as ICMP Timeout, ICMP packet data portion, and Packet Time-To-Lives. One of the features offered by Auvik's cloud console [5] is its Ping service, which checks for individual network endpoints from their location in the cloud infrastructure. The core network monitoring features of the Auvik system also provide continuous monitoring of SNMP-based devices. The Pingdom system [4] includes alerts that occur when ICMP response times drop. The system estimates response times for sites that are still accessible. This system also explores APIs for targeting alerts so that these notifications can be integrated into various custom applications.

An approach to proactive recommendation systems, proposed by authors in the field of road and manufacturing emergency prediction, is investigated in [6]. The information is collected from video clips, necessitating a hybrid approach that incorporates a neuro-fuzzy controller, a knowledge base, and deep learning. An algorithm for generating recommendations to prevent emergencies is proposed. This approach provides ideas for generating solutions in the design and management of communication networks based on the use of neural networks. An approach for balancing the load of parallel production lines using neural networks is discussed in. The proposed solution accounts for system variations and trains the neural network to minimize order completion time. A simulator, based on previously collected information about the production process, is used [7]. The trained neural network recommends directing the production flow to one of the production lines. The proposed idea could be adapted for predicting and reorganizing traffic in communication networks. At the same time, all these approaches do not offer automated classification of results to determine the quality of connectivity, nor do they implement options for dynamically adapting each possible classification to changes in network topology, to the assignment of different technologies and protocols, ensuring balancing and reliability of IP traffic. In this study, a method for automated prescription for improving the network structure is proposed, based on a database of current time parameters that determine the quality of remote connectivity. The method is developed with simulated network topologies, with the implementation of known protocols, such as FHRP (First Hop Redundancy Protocol) and PVST (Per VLAN Spanning-tree) for alternative and fail-safe traffic. The measured time parameters, in different traffic regimes, serve as input training data for the proposed Random Forest machine learning model. Depending on the results in the classification stage of newly received data, the model predicts the need for appropriate restructuring of the network topology and reconfiguration of the protocols. The obtained results are discussed in terms of accuracy and precision of the classification. Also feature work is proposed.

2. Description of the proposed method

The main functional model in the study is presented in Figure 1. The input training data, after appropriate formatting, is fed to the input of a Random Forest model, including a different number of training trees. The training data represent the response times of ICMP (pings) from a specific device in the network to a remote destination, at different traffic loads in the respective network. After receiving the trained data base, in the test phase, data from the response times in other situational moments are fed to it, in order to be classified by the trained Random Forest model.

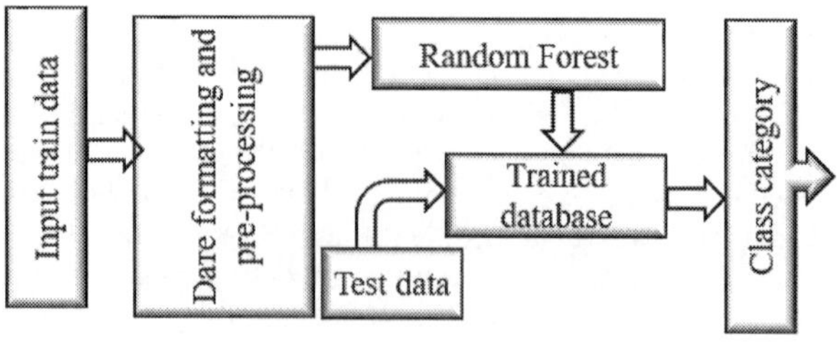

Fig. 1. Functional model flowchart

2.1. Simulation of the tested network topologies

The simulated network topologies are presented in Figure 2 as follows. In Figure 2/a - Topology1 with FHRP protocol with one group, In Figure 2/b - Topology2 with FHRP protocol with two alternative groups, In Figure 3/c - Topology3 with one VLAN and in Figure 2/e - Topology4 with two VLANs (VLAN100 & VLAN200).

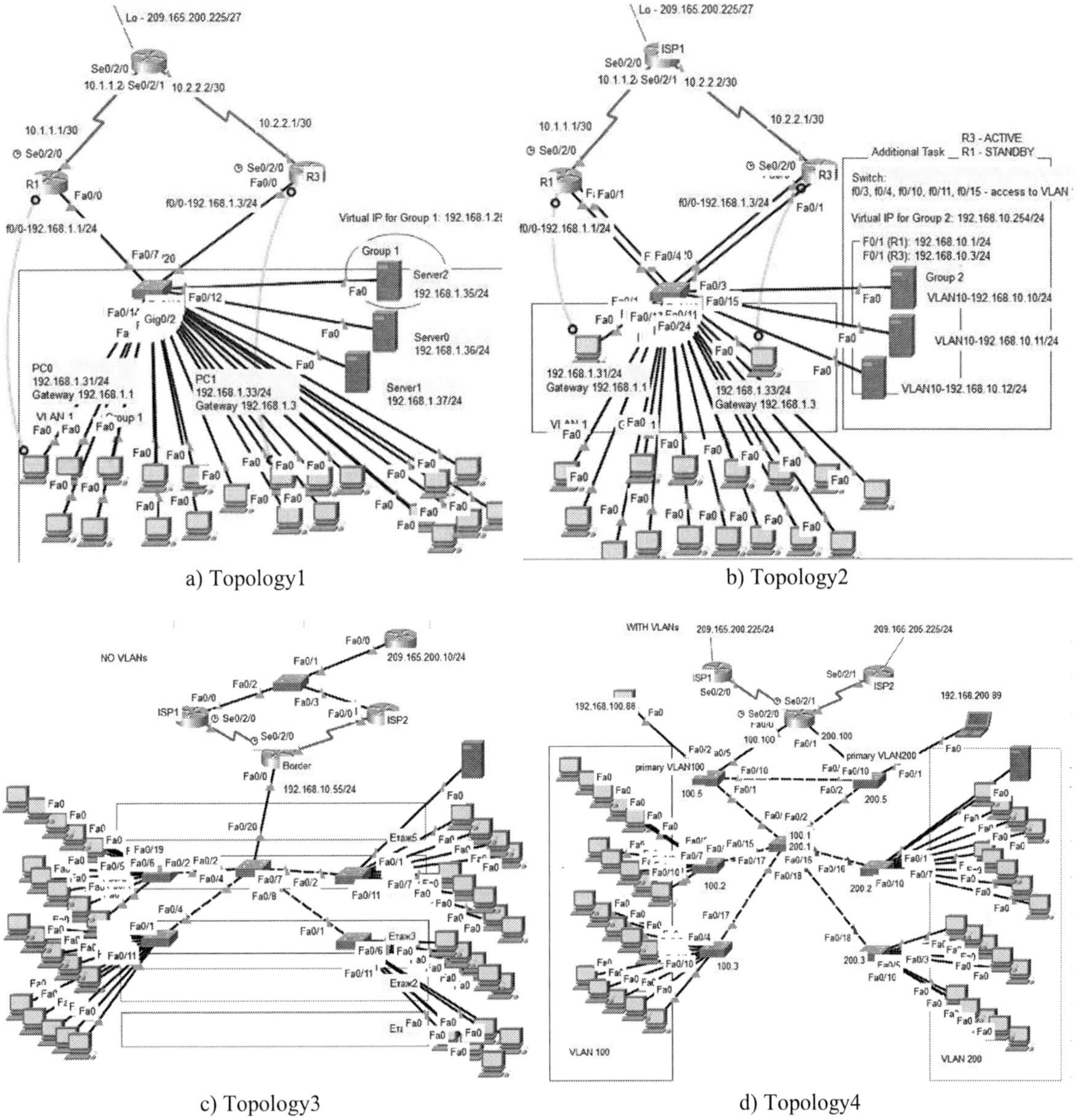

Fig. 2. Simulated network topologies: /a - Topology1 with FHRP protocol with one group, /b - Topology2 with FHRP protocol with two alternative groups, /c - Topology3 with one VLAN and /e - Topology4 with two VLANs.

The training data for the four topologies in Figure 2 is obtained and groped in *Class categories*, as follows:

- Average Ping response time for a single ping from the server in Topology 1 (Fig.2/a) to a remote address 209.165.200.225 and for a simultaneous, load-bearing, infinite broadcast ping *from 5* end devices to address 192.168.10.255 - *Class Category 1*;
- Average Ping response time for a single ping from the server in Topology 1 (Fig.2/a) to a remote address 209.165.200.225 and for a simultaneous, load-bearing, infinite broadcast ping *from 20* end devices to address 192.168.10.255 - *Class Category 2*;
- Average Ping response time for a single ping from the server in Topology 2 with two alternative FHRP groups and different roles of [(router R1 as active for Group1 - network 192.168.1.0/24 and router R3 as active for Group2-

network 192.168.10.0/24) (Fig.2/b)] to a remote address 209.165.200.225 and for a simultaneous, load-bearing, infinite broadcast ping *from 5* end devices of Group1 to address 192.168.1.255 - *Class Category 3*;

- Average Ping response time for a single ping from the server in Topology 2 with two alternative FHRP groups and different roles of [(router R1 as active for Group1 - network 192.168.1.0/24 and router R3 as active for Group2- network 192.168.10.0/24) (Fig.2/b)] to a remote address 209.165.200.225 and for a simultaneous, load-bearing, infinite broadcast ping *from 20* end devices of Group1 to address 192.168.1.255 - *Class Category 4*;
- Average Ping response time for a single ping from the server in Topology 3 (Fig.2/c) to a remote address 209.165.200.10 and for a simultaneous, load-bearing, infinite broadcast ping *from 5* end devices in the only VLAN to address 192.168.10.255 - *Class Category 1-1*;
- Average Ping response time for a single ping from the server in Topology 3 (Fig.2/c) to a remote address 209.165.200.10 and for a simultaneous, load-bearing, infinite broadcast ping *from 20* end devices in the only VLAN to address 192.168.10.255 - *Class Category 2-1*;
- Average Ping response time for a single ping from the server in Topology 4 (Fig.2/d) to a remote address 209.165.200.10 and for a simultaneous, load-bearing, infinite broadcast ping *from 3* end devices in VLAN 100 to address 192.168.100.255 and 2 end devices in VLAN 200 to address 192.168.200.255 - *Class Category 3-1*;
- Average Ping response time for a single ping from the server in Topology 4 (Fig.2/d) to a remote address 209.165.200.10 and for a simultaneous, load-bearing, infinite broadcast ping *from 10* end devices in VLAN 100 to address 192.168.100.255 and 10 end devices in VLAN 200 to address 192.168.200.255 - *Class Category 4-1*;

For each of the categories defined in this way, 10 attempts/measurements of Average response time were made. To generate a large training set, additional values were added to these values in the range +-10% relative to the average value of the 10 measurements. In this way, for each class category, 30 combinations of 10 measurements were generated. That is, for *Class categories 1,2,3 and 4*, a total of 120 combinations and for *Class categories 1-1,2-1,3-1 and 4-1*, a total of 120 combinations of measured ping responses. Table 1 represents the first rows of 10 attempts for *Class category1*.

Attempt1	Attempt2	Attempt3	Attempt4	Attempt5	Attempt6	Attempt7	Attempt8	Attempt9	Attempt10	Class category
10	11	7	8	9	11	10	8	12	9	1
10.00	14.00	4.00	11.00	12.00	10.00	10.00	8.00	14.00	11.00	1
10	12	7	5	9	9	10	8	9	6	1
8	11	10	8	11	8	9	7	13	7	1
13	13	4	10	11	8	10	6	13	11	1
13	9	5	7	10	13	10	8	15	9	1
7	12	10	11	6	11	9	6	13	9	1
11	8	5	5	9	8	12	9	11	11	1
8	11	6	8	7	10	13	9	15	6	1
12	8	6	7	12	8	7	5	9	12	1
13	11	8	5	7	10	11	9	9	9	1
11	11	7	8	7	10	12	10	14	8	1
10	11	4	9	8	11	8	5	9	6	1

Table 1. Part of training data for *Class category1*

2.2. Training the Random forest machine learning model

To create the model, the *Colaboratory* cloud environment was used, with access to the necessary Python libraries [8]. For the purposes of the experiment, two Random Forest models were trained: *Random Forest1* for class categories 1,2,3,4 and *Random Forest2* for class categories 1-1,2-1,3-1,4-1.

	Random forest 1		**Random forest 2**	
Average response time when pinging from the server to a remote point [msec]	**Topology 1**	**Topology 2**	**Topology 3**	**Topology 4**
	FHRP-One group	**FHRP-Two groups**	**One VLAN**	**Two VLANs**
Infinite broadcast ping sent from 5 end devices	9,5 msec Class category1	1,83 msec Class category3	19 msec Class category1-1	11,37 msec Class category3-1
Infinite broadcast ping sent from 20 end devices	25,8 msec Class category2	15,4 msec Class category4	45,18 msec Class category2-1	2,5 msec Class category4-1

Table 2. Average response time when pinging from server to a remote point for all Class categories

To achieve and evaluate an effective ratio between model complexity, training phase duration and achieved accuracy in the validation/testing stage, each model was trained with 1 and 10 tree structures (estimators), respectively.

```
# Features and target variable
X = data [['Attempt1', 'Attempt2', 'Attempt3', 'Attempt4', 'Attempt5', 'Attempt6', 'Attempt7', 'Attempt8', 'Attempt9', 'Attempt10']]
y = data ['Class category']
# Split data into training (55% of all 120 combinations for Random Forest1 and Random Forest2) and 45% testing sets.
X_train, X_test, y_train, y_test = train_test_split(X, y, test_size=0.45, random_state=42)
# Initialize RandomForestClassifier with n_estimators=10: This parameter specifies that the Random Forest will consist of 10 decision trees.
rf_classifier = RandomForestClassifier(n_estimators=10, random_state=42)
# Fit the classifier to the training data
rf_classifier.fit(X_train, y_train)
# Make predictions
y_pred = rf_classifier.predict(X_test)
# Calculate accuracy and classification report
accuracy = accuracy_score(y_test, y_pred)
classification_rep = classification_report(y_test, y_pred)
```

3. Experimental results

To determine whether test/validation data is effectively classified, according to the model's decision, we need to consider the classical estimation parameters of the model's accuracy. Namely, these are the parameters „Precision", „Recall" and „F1- score". As F1-score takes into account both precision and recall and is based on a balance of the two, it is more appropriate to calculate this parameter according to the given equation 1 [10]:

$$F1-\text{score} = 2.\frac{\text{preciscion.recall}}{\text{preciscion+recall}} = \frac{TP}{TP+1/2(FP+FN)}, \quad \text{where} \tag{1}$$

True positives TP is an outcome where the model *correctly* predicts the positive *Class category*;
False positives FP is an outcome where the model *incorrectly* predicts the positive *Class category*;
False negatives FN is an outcome where the model incorrectly predicts the negative *Class category*.

Random forest/ Confusion Matrix	**Class category 1,2,3,4**		**Class category 1-1;2-1;3-1;4-1**	
1 decision tree/estimator	[[13 0 0 0] [0 16 0 0] [1 0 11 0] [4 0 0 9]]	class precision recall f1-score accuracy 1 0.72 1.00 0.84 2 1.00 1.00 1.00 0.91 3 1.00 0.92 0.96 4 1.00 0.69 0.82	[[13 0 0 0] [0 16 0 0] [3 0 9 0] [4 0 0 9]]	class precision recall f1-score accuracy 1 0.65 1.00 0.79 2 1.00 1.00 1.00 0.87 3 1.00 0.75 0.86 4 1.00 0.69 0.82
10 decision trees/estimators	[[13 0 0 0] [0 16 0 0] [1 0 11 0] [3 0 0 10]]	class precision recall f1-score accuracy 1 0.76 1.00 0.87 2 1.00 1.00 1.00 0.93 3 1.00 0.92 0.96 4 1.00 0.77 0.87	[[13 0 0 0] [0 16 0 0] [1 0 11 0] [2 0 0 11]]	class precision recall f1-score accuracy 1 0.81 1.00 0.90 2 1.00 1.00 1.00 0.94 3 1.00 0.92 0.96 4 1.00 0.85 0.92

Table 3. Confusion Matrix and accuracy estimated parameters for all Class categories

The obtained high classification accuracy for all tested categories, including when the number of trees is only = 1, shows that at the lowest possible complexity of the model, an optimal ratio between the evaluation parameters for the effectiveness of the model is achieved. As 45 % of all 120 combinations are applied to test the trained model, it means that 54 combinations of Response time are randomly tested. The confusion matrix shows that only 5, 4, 7 and 3 combinations are wrong classified, respectively for Class category 1,2,3,4 (1 decision estimator), for Class category 1,2,3,4 (10 decision estimators), Class category 1-1,2-1,3-1,4-1 (1 decision estimator), Class category 1-1,2-1,3-1,4-1 (10 decision estimators). But to determine the most appropriate number of trees in the model, it is necessary to take into account the significance of each of the ten response time values/parameters in the training sample. From the graphs in Figure 3, it is clear that some parameters have almost zero significance with 1 tree in the model, while with 10 trees in the model, their significance increases. This leads to the conclusion that it is better to finally choose the model with 10 trees. All network topology simulations are held with simulator Packet Tracer [9].

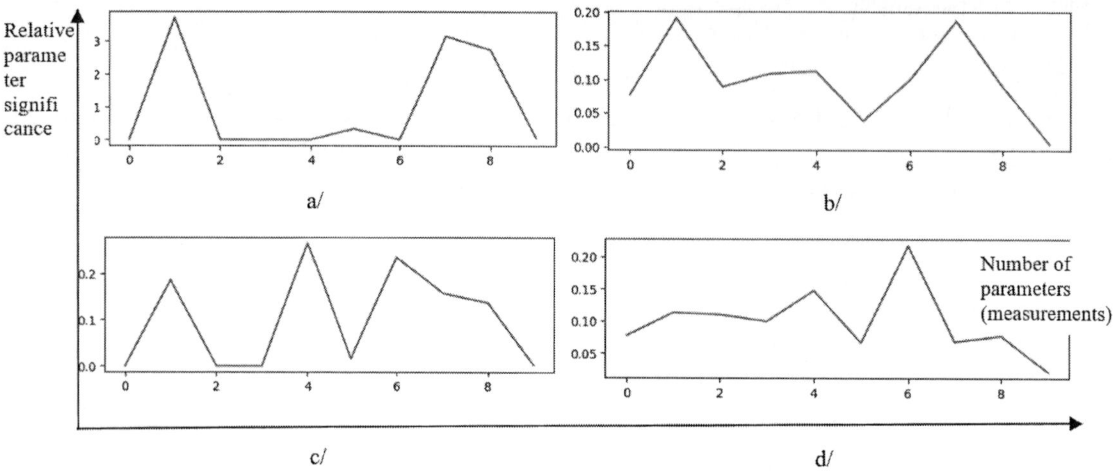

Fig. 3. Parameter significance-a/ for *Class category 1,2,3,4* with 1 tree, /b for *Class category 1,2,3,4* with 10 trees c/ for *Class category 1-1,2-1,3-1,4-1* with 1 tree, d/ for *Class category 1-1,2-1,3-1,4-1* with 10 trees

Thus, depending on the recognized *Class category*, i.e., the solution provided by the model, administrators and managers of the network infrastructure may consider it necessary to convert an existing topology from type Topology1 to type Topology 2 (with more FHRP groups), which will improve performance and lead to lower response time values, or to switch from Topology 3 to Topology 4 (with more VLANs and Per VLAN STP), which will lead to the same positive effect.

4. Comparison and interpretation of results

The two studies, the present one and the one mentioned in [3], both from the same research domain, use different machine learning models and data sets to classify ICMP connection time. The first paper, "Models for Evaluating Effective IP Traffic Management with Application of Artificial Intelligence" (Random Forest model), focuses on a network restructuring prescription based on pre-trained models., while the second, "Classification of ICMP connection time with a multi-layered neural network" (MLP model) [3], focuses on connection quality assessment and stability.

Feature	Random Forest (RF) Model	Multi-Layered Neural Network (MLP) Model [3]
Machine Learning Model	Random Forest	Multi-Layered Neural Network (MLP) / Deep Learning
Primary Research Goal	Automated prescription for improving the network structure (reconfiguring protocols and restructuring topology).	Automated classification of time variability for connection quality with focus on easy adaptation.
Input Data Features	10 measured ICMP response times (pings) from 10 attempts.	2 parameters: Local Average (LA) time and Remote Average (RA) time [ms].
Output Classes	8 Class Categories corresponding to 4 specific network topologies (FHRP, PVST) under 2 different load conditions (e.g., 5 devices vs. 20 devices).	3 Classes based on fixed connection time thresholds: Class 1 (appropriate quality), Class 0 (bad quality), and Class 2 (no connection).
Data Source	Data generated from simulated network topologies (implementing FHRP and PVST). The training set was expanded using artificial variability (±10% of the average value).	Data collected from IoT collection devices on a 2.4 GHz wireless network. Uses a larger dataset of 30,000 samples.
Classification Results	Achieved high classification accuracy even with low model complexity (1 tree). The chosen 10-tree model had a very low number of wrong classifications (e.g., 4 to 7 out of 54 tested combinations).	Achieved high training and validation accuracy (upwards of 92-93% in the best cases). The MNN structure showed stable work and absence of overfitting with a minimum number of elements.

Table 4. Comparison of results, methodology and goals of the two researches

Table 4 presents a comparison of their results, methodology, and goals. Both models achieved high accuracy, supporting the use of machine learning for ICMP classification. The MLP model was explicitly designed to achieve high accuracy with a minimum number of elements in the dense layers, which is a good prerequisite for fast performance speed when integrated into a real-time system.

The Random Forest model also demonstrated high accuracy with low complexity (1 tree) but ultimately chose a 10-tree structure for greater significance of all input features. The MLP model addresses a fundamental, diagnostic problem by classifying general connection quality based on simple LA/RA thresholds and its strength lies in its ability to show easy adaptation to different threshold requirements (using a New Data Set). While the Random Forest model in this research addresses a more complex, prescriptive problem by evaluating the ICMP response times to the physical and logical configuration of the network (topology and protocol groups). The successful classification suggests the model can serve as a decision-support tool for network restructuring.

5. Positive, negative outcomes and challenges

The core benefit is the ability to receive an automated prescription for improving the network structure. This means the network could dynamically adapt based on its performance, with the model predicting the need for appropriate restructuring of the network topology and reconfiguration of the protocols. Network managers would move away from reactive or purely manual configuration, instead relying on the high accuracy of the trained model to classify the quality of connectivity. This provides a stronger, data-backed foundation for system reconfiguration.

The model could be implemented and integrated into real-time traffic monitoring systems, allowing for a continuous cycle of data collection, classification, and prescription for newly received data. The system is inherently designed to be adaptive, allowing for additional training of the system for different network architectures and a corresponding prescription for reconfiguring the system according to the results. The implementation also presents potential challenges and negative consequences: if the simulated network does not accurately reflect the complexities, traffic patterns, and real-world latency of a live production network, the model's predictions and prescriptions for the real network may be suboptimal. Also, there will occur a risk of faulty automation.

If a new data point is misclassified (a false positive or false negative, as indicated by the wrong classifications in the confusion matrix), the system could issue a prescription for unnecessary or counterproductive changes, potentially disrupting network services or introducing new vulnerabilities. Over-reliance on artificial data generation can be pointed out as the main challenge: to generate a large training set, the we manually add values in the range of +-10% relative to the average of 10 measurements. This technique, while generating a larger set of 30 combinations per class category, introduces artificial variability rather than capturing true, organically observed network fluctuations, which could lead to a selection bias in the training.

6. Some limitations of the research and approach

The verification and testing are confined to a small set of protocols and topologies: FHRP and PVST across four specific simulated topologies. While we suggest the model is applicable to other protocols, its performance on more complex architectures, a larger number of FHRP groups, or more VLANs is proposed only as future work, not a current achievement. The input data is solely based on ICMP (ping) response times from a single device to a remote destination under two load conditions (5 and 20 end devices for Topology 1 and 2, and 5 and 20 devices split for Topology 3 and 4). This may be an insufficient representation of effective IP traffic management, which can be affected by numerous other dynamic parameters like packets/sec, jitter, packet loss, and application-layer metrics.

7. Conclusion and future work

The proposed method leverages measured time parameters (ICMP response times) from different traffic regimes as input training data for a Random Forest machine learning model. This provides a data-driven basis for management decisions rather than relying solely on manual assessment. The presented and optimized random forest model for training and classification of remote access time parameters in different network topologies shows high accuracy with +–10% variations in the input data. The model has been verified for two of the most commonly used protocols for ensuring alternative and fault-free network communication. However, it can also be applied to other network topologies, when working with other protocols. The model can be implemented and integrated into real-time traffic monitoring systems, with a dynamically created and updated database of remote access times and classification of newly received data. As a future development of the method, verification and testing of the same protocols, but with a larger number of FHRP groups, as well as with more VLANs in a more complex topology with more switches and different roles of the switches for individual VLANs, can be added. The model could be implemented and integrated into real-time traffic monitoring systems 8, allowing for a continuous cycle of data collection, classification, and prescription for newly received data. The system is inherently designed to be adaptive, allowing for additional training of the system for different network architectures and a corresponding prescription for reconfiguring the system according to the results

8. Acknowledgments

Acknowledgements for the support of the Research Project № 3O-337/10.04.2025, of the *University of Telecommunications and Post* titled "Test models for evaluating effective IP traffic management with the application of artificial intelligence".

9. References

[1] Toplalova, I. & Radoyska, P. (2018). Adjustment of the QoS Parameters on Routers with Neural Network Implementation", International Journal on Advances in Networks and Services, ISSN:1942-2644 vol. 11, no. 3 & 4, pp. 143 – 151, 2018.

[2] Hillstone Networks, (2021). AI-driven Security Solutions by Hillstone Networks", 2021, © 2021 Hillstone Networks, https://www.hillstonenet.com/.

[3] Topalova, I., (2023). Classification of ICMP connection time with a multi-layered neural network, 2022 IEEE International Black Sea Conference on Communications and Networking (BlackSeaCom) | 978-1-6654-9749 7/22/$31.00 ©2022 IEEE | DOI: 10.1109/BLACKSEACOM54372.2022.9858304.

[4] Topalova, I., Radoyska, P. & Sokolov, S. (2020). Neural Network Implementation for Detection of Denial of Service Attacks Journal of Engineering Science and Technology Reviewthis link is disabled, 2020, (Special Issue), pp. 98–102.

[5] Radoyska, P. & Atanasova, M. (2020). Free tools for testing the security of web services in the utp network, Fifth international scientific conference "Telecommunications, informatics, energy and management", October 3-4, 2020, Sofia, Bulgaria, pp.133-137

[6] Fedorov, A. & Shkodyrev, V. (2018). Recommendation System based on Neural Network for Prediction of Emergency Situations, Proceedings of the 29th DAAAM International Symposium, pp.0133-0136, B. Katalinic (Ed.), Published by DAAAM International, ISBN 978-3-902734-20-4, ISSN 1726-9679, Vienna, Austria DOI: 10.2507/29th.daaam.proceedings.018

[7] Gaku, R. (2024). Scheduling Optimization of Manufacturing Process Flows using Agent-Based Simulation Modeling and Neural Networks, Chapter 14 in DAAAM International Scientific Book 2024, pp.171-180, B. Katalinic (Ed.), Published by DAAAM International, ISBN 978-3-902734-42-6, ISSN 1726-9687, Vienna, Austria DOI: 10.2507/daaam.scibook.2024.14

[8] Radovanovic, I, (2024). Google Colab – A Step-by-step Guide

[9] Cisco.com (2025), Cisco netacad.com, Simulator Packet tracer 8.3.

[10] Capa Learning, (2025). F1 Score in Machine Learning: Formula, Precision and Recall/ Artificial Intelligence

DOI: 10.2507/36th.daaam.proceedings.006

AN APPROACH FOR AUTOMATED CREATION OF THREE-DIMENSIONAL HISTORY-BASED PARAMETRIC MODELS

Velizar Zaharinov & Ivo Malakov

This Publication has to be referred as: Zaharinov, V[elizar] & Malakov, I[vo] (2025). An Approach for Automated Creation of Three-Dimensional History-Based Parametric Models, Proceedings of the 36th DAAAM International Symposium, pp.0047-0056, B. Katalinic (Ed.), Published by DAAAM International, ISBN 978-3-902734-47-1, ISSN 1726-9679, Vienna, Austria
DOI: 10.2507/36th.daaam.proceedings.006

Abstract

The paper proposes an approach for automated creation of three-dimensional history-based parametric models from non-parametric three-dimensional models. A review and analysis of existing approaches in the literature is made. The proposed approach is presented in the form of a functional structure. For each of the ten stages constituting the approach, the problems that need to be solved, the necessary input-output information and the results that need to be achieved are formulated. An example application for creating a parametric model of a part from a non-parametric model obtained by three-dimensional scanning is given. The particularities of the practical application of the approach are indicated.

Keywords: CAD; parametric model; history-based CAD; automated model construction.

1. Introduction

The paper proposes an approach for automated creation of three-dimensional history-based parametric models from non-parametric three-dimensional models. The proposed approach is developed to be used in the context of parametric history-based CAD environments. Nowadays, these environments are extremely common as tools for creating parts and assemblies of products from mechanical engineering [1], [2], automotive engineering [3], aircraft engineering, etc. The presence of many well-known CAD environments [4], [5], [6], [7], [8], supporting this type of modeling and used in the industry by leading manufacturers such as TOYOTA, INFINITI, OMRON, Mercedes-Benz, VOLVO, MAN, General Motors, AIRBUS, FUJITSU, SIEMENS and many others, is a confirmation of the benefits of its application in the product design process. In order to take advantage of all the benefits of parametric modeling, however, it is necessary to build the model by applying the appropriate parametric modeling tools. In the design process, a number of factors require the creation of non-parametric models. Situations that necessitate such models are, for example: geometry data of an object obtained by three-dimensional scanning; transferring models between different CAD environments, which requires going through model descriptions that are external (do not fully correspond to the description used naturally by the CAD environment); the model goes through a generative design process; the model is created through direct modeling, etc. Manually creating parametric models from non-parametric models is a laborious and time-consuming process, in which geometric features can be lost if the process is not carried out methodically and accurately. Considerable flexibility and experience are required, since each model has its own geometric features, and parameterization techniques require excellent knowledge of the CAD environment used.

In addition, the work on parameterization takes up a significant amount of time for the designer, which could be used for other more creative activities related to the design and manufacturing process of the product. Taking into account the above, automating this process would lead to more accurate and faster parameterization of non-parametric models, which would relieve designers and save them time for more important and creative activities related to the added value of the product. One of the essential issues in automating the process of creating parametric models from non-parametric ones is the automated recognition of geometric features. There are a number of developments devoted to this topic, which find application not only in CAD. The first such method, described in [9], was developed for application in group technology, for coding parts, and involves the classification of the faces that make up the parts into separate groups (primary and secondary). Based on this classification, the shape of the model is analyzed and geometric features such as holes, base surfaces, channels, pockets and other connections between the shapes that make up the part are extracted. In [10] is a review of the publications made in the period from 1980 to 2016 in relation to methods for recognizing geometric features. The authors point out that the most publications, and therefore the widest scientific interest, in this period, are the methods focused on four main principles: rule/guideline-based approach, graph-based approach, volumetric decomposition approach, and artificial neural network approach. The rule-based approach [11], [12], [13] uses the idea of an expert system. This is one of the earliest researched approaches. The features are summarized in the form of templates consisting of characteristic patterns of rules. The recognition process is carried out by applying these rules in an if-then analysis. If the predefined conditions are satisfied, then the corresponding structure in the part is recognized as a geometric feature. To address some of the shortcomings of the rule-based approach, guideline-based methods have been developed. They are based on the idea that it is possible to search for an incomplete representation that can be used as a guideline for the existence of certain geometric features. The goal is to find a trace of a geometric feature that would be present in the model, even when combining several geometric features. This type of methods uses a two-step recognition procedure [10]. In the first step, guidelines are created using extraction rules based on various properties such as inferences from the geometry and topology of the model [14], [15], [16], [17], taxonomies of features [18], [19], [20], and combined probabilities for ranking potential guidelines for features [21], [22]. In the second step, guidelines are processed and can be directly compared by applying rules, with some developments also using a verification phase, after building feature volumes from guidelines and model boundary data. The graph-based approach is among the most studied methods [10] because graphs are similar to the representation of solid models in three-dimensional CAD environments. In this type of model description, nodes and arcs usually represent faces and edges, which are assigned along with some properties such as convexity and concavity of edges, face type, perpendicularity, parallelism or tangentiality of edges and faces, etc. The features are then extracted in the form of a subgraph from the full graph. In [10] 20 methods based on the graph approach are reviewed. The volume decomposition approach determines the volume of material removed from the workpiece during machining and decomposes the volume initially into intermediate volumes, after which the features are generated by combining the intermediate volumes based on certain rules. The methods implementing this approach can be broadly divided into two groups: decomposition by a described convex body and cell-based volume decomposition. The artificial neural network approach mimics human perception and the learning process in humans [23]. The main advantage is that it does not use a rigid logical procedure (algorithm) for recognizing characteristic features, but rather applies arithmetic operations to process data, training the algorithm. This avoids the need to develop a strategy (logical sequence of actions, algorithms) for each particular case and provides flexibility, allowing for exceptions or incomplete recognition patterns.

The aim of this article is to propose a solution, in the form of an approach, that would reduce the labor intensity of using modern technologies for creating CAD models and increase their usefulness by making the use and editing of the created models easier. An opportunity to achieve this goal is seen by the authors in the automated creation of parametric models from non-parametric ones.

2. Automation approach

2.1. General function and input/output information

The general function of the approach is: "Automated creation of a three-dimensional history-based parametric CAD model". The general function is the relationship between the input that the approach expects and the output from it. The input and output are informational and the main flow is information processing (Fig. 1). The approach expects two information inputs and produces one information output. One information input represents the description of the non-parametric CAD model. The second one gives the user the opportunity to adjust the process. The result of the approach is a parameterized history-based CAD model.

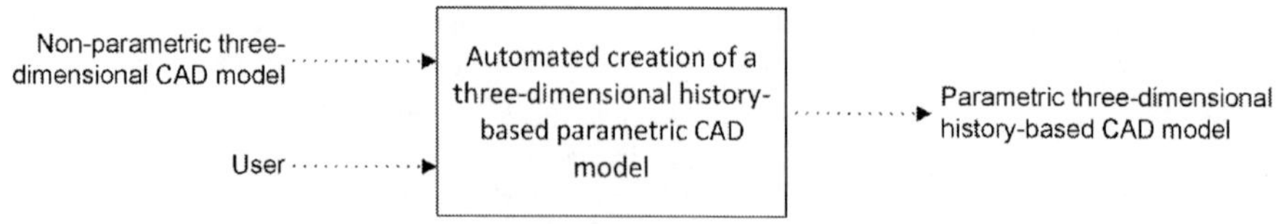

Fig. 1. General function of the approach for automated creation of three-dimensional history-based parametric models from non-parametric models

2.2. Structure of the approach

The individual stages of the proposed approach are shown in Fig. 2 in the form of a functional structure. The results obtained when performing the individual stages (functions) are also indicated. It is also shown where the inputs and outputs defined by the general function are connected (Fig. 1).

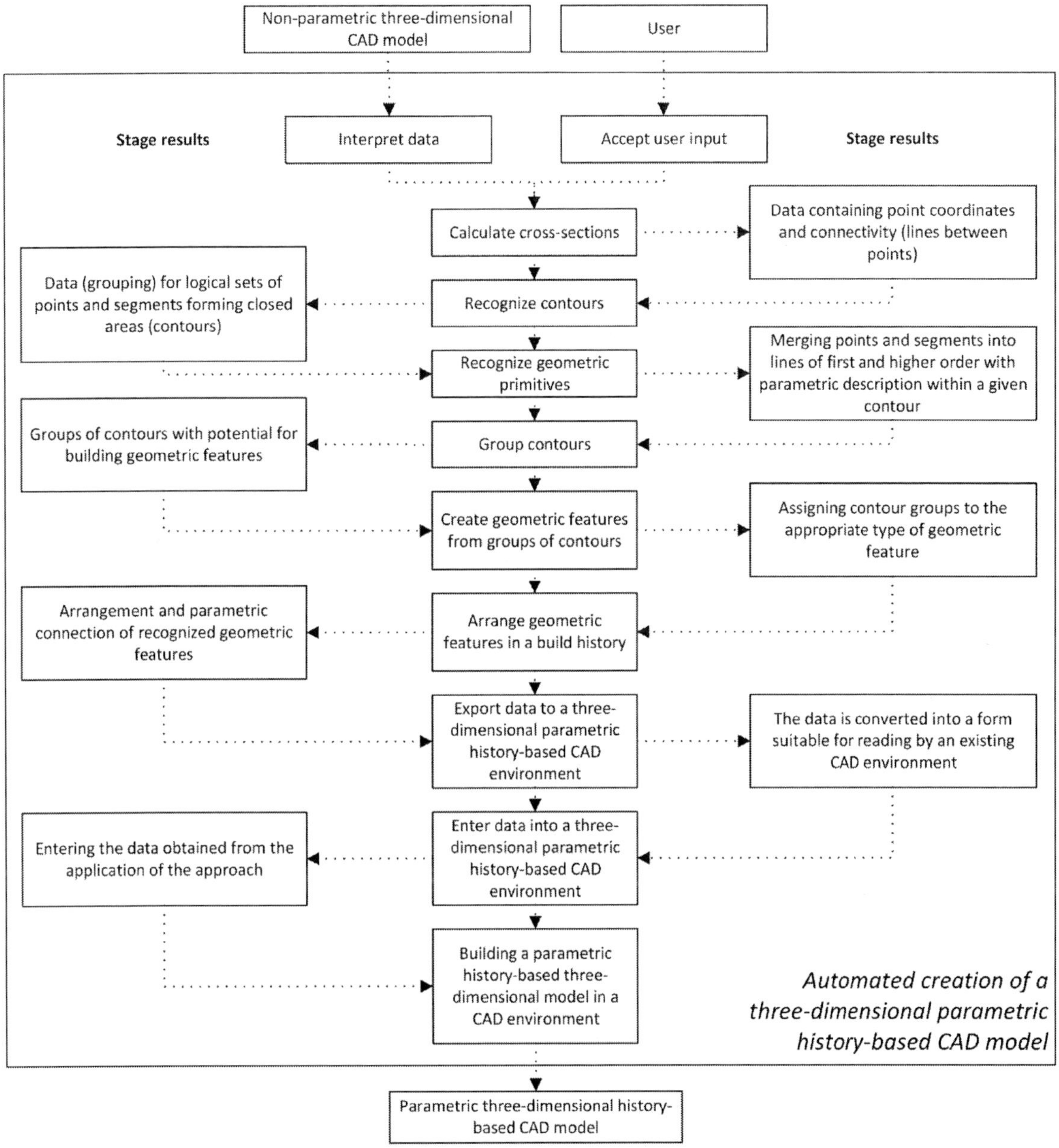

Fig. 2. Functional structure of the approach and results of the individual stages

3. Description of the approach

3.1. Stage 1. Input (includes interpreting data and accepting user input)

The transfer of data about the geometry of a three-dimensional CAD model is associated with the use of a certain format for recording the geometric information. There are a variety of CAD environments that offer a variety of ways to record data about the model. The recording is done in a file with a corresponding extension indicating the type of data recording, i.e. the file format. In general, these file formats can be classified as "native" and "neutral".

Native file formats are specific to a particular CAD environment and are usually protected by copyright. This makes it difficult to use native file formats for data transfer between competing CAD environments and requires the use of translators and validation. Another feature of native file formats is that they are not standardized, as they are developed by private companies.

An advantage of native file formats over neutral ones is that they preserve all data (including parameterization and history) that can be described by the respective CAD system and integrated into the product model. Neutral file formats offer the possibility of data transfer between different CAD environments. In addition, some of them are standardized. The disadvantage of these formats is that the parameterization and history present as information in the native format for the CAD environment are lost. However, these file formats preserve information about primitives such as cylinders, planes, points, edges, etc., related to the geometry of the model. This significantly facilitates the use of approaches and methods for parameterization, recognition of geometric features and creation of a history-based model. Table 2 shows some of the most common CAD systems for three-dimensional modeling and their native file formats (only for parametric three-dimensional models), and Table 3 shows some of the most commonly used neutral CAD file formats for data transfer for three-dimensional models.

All formats listed in Tables 1 and Table 2, except STL, include information about primitives, the so-called B-rep representation, which is standard in modern three-dimensional CAD systems. In STL files, only information about basic geometric primitives such as points and faces is stored. Due to the flexibility of the STL and OBJ file formats, they were chosen to describe the input data for the approach. This choice was made taking into account the following considerations: the ability to cover all four situations mentioned at the beginning of the introduction, in which parameterization is required; the formats are neutral and do not require the purchase of rights for use; simple data structure; the STL format is widely used and supported by a large part of modern CAD/CAM/CAE systems.

Although the OBJ format is not a traditional CAD format, it was chosen to support it in addition to the widely used STL because it allows the transfer of three-dimensional models that have been created through three-dimensional scanning, generative design methods, or direct modeling. In these situations, saving the data as an STL file is not always possible (most often due to lack of support in the model creation environment).

The main disadvantage of the chosen neutral file formats is that they offer an approximate description. This is because triangles are used to describe the geometry of the 3D model (Fig. 3), and not B-rep, as in STEP files for example. However, this disadvantage can also be seen as an advantage, since, using a single geometric primitive (triangle), the description can be generalized regardless of the type of modeling used to produce the 3D geometry (parametric, non-parametric).

The result of decoding the input data file is a set V of points v_i, each of which is defined by a triple of coordinate values $\{x_i, y_i, z_i\}$, $x_i, y_i, z_i \in \mathbb{R}$, and a set F of faces f_j, each of which is defined by the indices $\{\varphi_{j,1}, \varphi_{j,2}, \varphi_{j,3}\}$, $\varphi_{j,1}, \varphi_{j,2}, \varphi_{j,3} \in \mathbb{N}$, of three points from the set V. Further processing of the model geometry, in the course of applying the stages of the approach, will expect precisely these data as input.

CAD system	Native file format extensions
CATIA	*.CATPart; *.CATProduct
Creo	*.prt; *.asm
SolidEdge	*.par; *.asm
Inventor	*.ipt; *.iam
SolidWorks	*.sldprt; *.sldasm

Table 1. Native file formats

Name	Neutral file format extensions	Information
STEP	*.STEP; *.STP	Year of creation 1994; Standardized (ISO10303-242, 2020)
QIF	*.QIF	Year of creation 2013; Standardized (ISO23952:2020, 2020)
JT	*.JT	Year of creation 2007; Standardized (ISO14306:2017, 2017)
STL	*.STL	Year of creation 1987
IGES	*.IGS; *.IGES	Year of creation 1980; The first neutral CAD format
ACIS	*.SAT	Year of creation 1989; Created by Dassault Systemes
PARASOLID	*.X_T; *.X_B	Year of creation 1989; Created by Siemens

Table 2. Neutral file formats

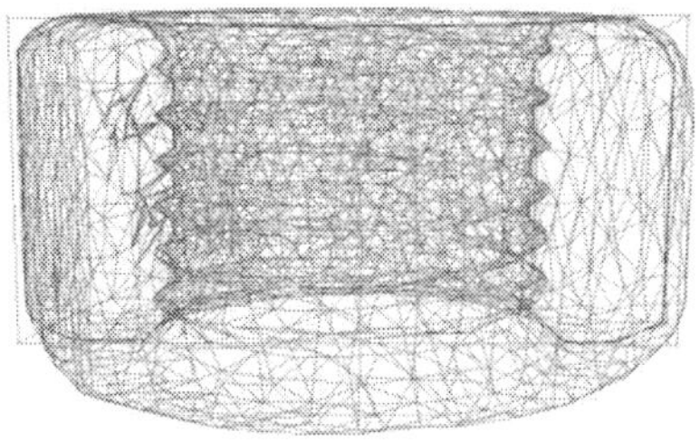

Fig. 3. A model of a part built from triangles with a depicted intersecting plane (rectangular frame) and curves obtained from the intersection of the plane with the model (depicted with overlayed red lines)

3.2. Stage 2. Calculate cross-sections

The goal of this stage is to calculate the curves obtained from the intersection of the three-dimensional model with a cutting plane. The input information required for the stage is a description of the three-dimensional model in the form of the sets V and F obtained from the decoding of an STL or OBJ file (Stage 1). The model at this stage is built from individual faces, in the form of triangles, constituting its outer surface. Each triangle consists of three points located at its vertices. They determine the intersection points of the sides of the triangles. The result of this stage is a set of points S, defined by their coordinates, a set of lines E_S, defined by pairs of indices of elements of the set S (two points each), and a set of faces (triangles) F_S, defined by triples of indices of elements of the set S (three points each). This result completely defines the resulting section. Fig. 4 shows sample results. Fig. 4a shows a section without triangles, and Fig. 4b shows a section containing triangles.

3.3. Stage 3. Recognize contours

The input for this stage is the result of the previous one, i.e. the sets S, E_S and F_S. The results of the intersection of the plane with the three-dimensional geometry of the model can be a variety of shapes – single points, points lying on a straight line, closed contours consisting of multiple points, contours containing faces (triangles), etc. In order to be able to create a parametric sketch of the section, it is necessary for these individual objects in the plane to be systematized (recognized) and separated from each other. The first step for this is the recognition of the individual contours in the set E_S, which carries the information about the connectivity of the points in the section. For the purposes of the approach, a "contour" is understood as a closed curve in the intersecting plane, composed of points, elements of S, arranged sequentially. The sequential arrangement of the points in the contour depends on the connectivity information contained in the set E_S. In practice, to create a geometric feature from a cross-section, it is necessary to produce a sketch containing only the contour, and the faces are not necessary. Moreover, they interfere with the recognition of the contour.

The obtaining (calculation) of the points from an arbitrary section does not follow a specific order. The search for the faces intersecting with the plane from the three-dimensional model is performed according to their order recorded in the STL or OBJ file. This order does not necessarily follow a specific regularity (if such a thing is assumed, it falls into a special case). Therefore, the points from the section are obtained in a random order and are arranged in the same way in the set S (in the order of their calculation). The set E_S, containing the lines from the section, and the set F_S, containing the faces, are also not arranged in any regular way. On the other hand, in order for the recognition of the contour to be possible, the points from it must be arranged one after the other (sequentially). It does not matter which is the starting point, or in which direction the contour will be traversed when recording the points, but the sequential arrangement is a condition for recognizing the contour as such and distinguishing it from the other objects in the intersecting plane.

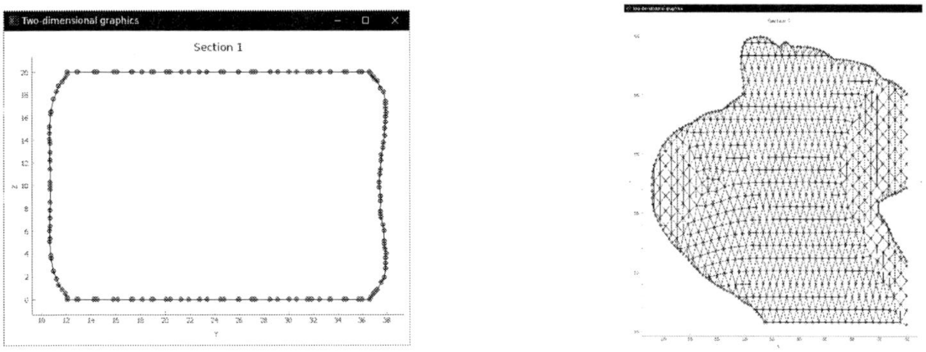

(a) (b)

Fig. 4. Two types of common sections: (a) containing contour(s) filled with faces (triangles) and (b) empty contour(s)

The result of this step is a set C containing all the contours of the section S, both those that contain faces in their interior and those without faces. If the section contains contours filled with faces (Fig. 4b) these must be removed. In addition, the set C also contains objects that are not contours - lines and single points, if such exist in the section. These objects are easily distinguished from contours, since the last element of the set does not point to the first (contours are closed, but lines and points are not).

3.4. Stage 4. Recognize geometric primitives

The goal of this stage is to parameterize the recognized contours. The idea is for the parameterized contours to be the basis for parametric sketches, from which geometric features can be created. The parameterization of three types of primitives is proposed: segments, arcs, and circles. The reason for choosing these three primitives is their widespread use in creating parametric sketches. The goal is also to minimize the number of parametric primitives used for simplicity and efficiency. Of course, based on the recognized contours, it is possible to parameterize other types of lines in the plane, which is only a matter of developing additional recognition algorithms. To implement the recognition, it is proposed to perform "segmentation" of the contours in the section. By "contour segmentation" is meant the division of the contour into separate sections, which are subsets of the contour, i.e. sequences of points belonging to the contour. All points of the segment are sequentially connected, i.e. the first point is connected to the second, the second to the third, etc. (Fig. 5).

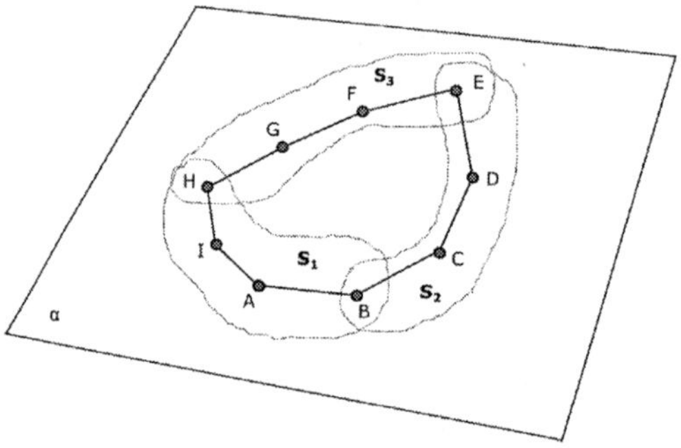

Fig. 5. Example of segmenting a contour. The segments are labelled S_1, S_2, and S_3 and are enclosed by dashed lines. The segmentation points are B, E, and H

It is possible to use multiple segmentation criteria, for example, an interval of permissible values for the angle between two adjacent contour segments. Multi-criteria decision-making for segmentation is also possible. The result of this stage is three sets for each recognized contour in the cross-section, which contain the recognized lines, arcs, and circles for the specific contour.

3.5. Step 5. Group contours

The grouping of contours is a stage that is important for the correct determination of the types of geometric features that can be constructed from the obtained cross-sections. To solve the problems at this stage, the following requirements can be formulated: it is necessary to obtain a sufficient number (sufficient "density") of cross-sections; it is necessary to formulate criteria and/or constraints for grouping; it is necessary to formulate a decision criterion. The problem of the sufficiency or density of the sections produced in stage 2 of the approach is significantly related to the scale of the model and the accuracy with which it is desirable to parameterize it. A larger number of sections increases, in addition to the accuracy of parameterization, the computational time. Intuitively, it can be assumed that the optimization of the computational time is largely related to adaptive decision-making about the number of sections in certain regions of the three-dimensional model - in some parts, which need to be parameterized more precisely, the density of sections is high, and in others - low. This conclusion can also be reached after visual analysis by a human. The full automation, i.e., without human guidance, of the adaptive determination of the number of sections, has to be approached iteratively. It is necessary to initially calculate a small number of sections in order to be able to analyze the existing geometry of the model. The number can be determined from the overall dimensions of the three-dimensional model in the direction of the intersecting planes (normal to them). Based on the change in geometry between two adjacent sections, a decision can be made whether more or fewer sections are needed between them. The result of this stage is groups of contours, grouped according to their similarity and/or suitability for use as feature sketches.

3.6. Stage 6. Create geometric features from groups of contours

Different parametric history-based CAD environments offer the creation of different geometric features. Depending on the target CAD system, there may be some differences in the types of geometric features and their parameterization.

However, there are some common geometric features, such as those created by extruding a sketch and by rotating a sketch about an axis of symmetry. The mentioned common ways of creating geometric features also come in two varieties, those that add material to the model and those that remove material from it (producing cavities or holes). Even with just these four parametric geometry creation tools, it is possible to build complex 3D models. Therefore, assigning geometric features to groups of contours is a complex task. In many cases, combining several geometric features produces geometry that is difficult to determine its elementary components. The problem of classifying groups of contours can be approached in the following way: selection of a set of characteristic geometric features to which the groups of contours will be classified; analysis of the minimum number of parameters necessary for parameterization of the selected characteristic geometric features; formulation of classification criteria for each characteristic geometric feature; development of tools to automate the process. Many common geometric features require a two-dimensional drawing – a sketch – on which they are built. The sketch can be created from the contours in the group, with one of them being chosen as representative of the entire group. It is also possible to use averaging.

3.7. Stage 7. Arrange characteristic geometric features in a build history

The stage of arranging the recognized characteristic geometric features in a build history can be considered as a combinatorial problem, in which one must find such a combination of characteristic geometric features that it is possible to parameterize the overall model and that it has a certain meaning. A certain combination of the recognized characteristic geometric features represents a possible construction sequence. Some of the possible combinations will lead to invalid geometry. Depending on the number of recognized geometric features, for some models it will be possible to analyze all combinations, while for others the analysis time will be too long (combinatorial explosion). In these cases, a directed search for a solution will be necessary, as a possible approach is the use of characteristic ways of building models used in practice: for example, a certain direction of construction - from one initial plane to another final one, establishing a basic geometric feature, symmetrical construction, etc.

3.8. Stage 8. Export data to a three-dimensional parametric history-based CAD environment

At this stage, the sketches and geometric features are saved in a file format that allows the transfer of the data obtained from the analysis. The goals of this format are: to be open, so that it can be used by CAD environment manufacturers; to transfer the information in a compact and lossless way.

3.9. Stage 9. Enter data into a three-dimensional parametric history-based CAD environment

This is the stage in which an existing CAD environment reads and interprets the data from the file format for transferring the analysis made according to the proposed approach. The necessary preliminary preparation is carried out before executing the automated construction process. It is necessary to determine a sequence for interpreting the transferred data.

3.10. Stage 10. Building a parametric three-dimensional history-based model in a CAD environment

At this stage, using the capabilities of the CAD environment, a sequence of commands is executed to create a parametric model with build history, following the analysis results obtained within the approach. It is necessary to perform interpretation of the results of the approach in an existing CAD environment and automated execution of a sequence of commands in an existing CAD environment.

4. Application example

The input data (Stage 1) is a model obtained by 3D scanning (Artec 3D LTD, 2024) (Fig. 6a). This model has a relatively simple geometry, but due to its obtaining by 3D scanning, even the relatively simple geometry is significantly "weighted" due to the fact that an extremely large amount of geometric information is generated during scanning. In addition, modern 3D scanners achieve resolutions of 0.01 mm, which means that the model may also contain geometry that is not relevant to the technical documentation of the object (for example, traces of wear and imperfections, as can be seen in Fig. 6a). Usually, this type of geometry is not desirable to be restored during parameterization.

The model consists of 446 350 points defining 1 339 450 segments that make up 892 947 triangles. Sections (Stage 2) are calculated in the horizontal (Fig. 6b) and vertical (Fig. 9) axes of the model. The points that define the contours of the section in the horizontal axis (Fig. 6b) are very densely located, at small distances from each other. The geometry is not ideal, i.e. as in an engineering drawing, but reflects the real object with its imperfections. The chamfers on the edges of the model are not straight lines, not even parts of arcs, but combinations of straight sections and curves. These anomalies may be due to wear and/or the post-processing that is necessary after the three-dimensional scan to remove noise and unwanted geometry. In Stage 3, the contours in the sections are recognized – in this case, there are two in the horizontal and vertical axes (Fig. 7b and Fig. 9b). Then, segmentation and recognition of primitives (Stage 4) is applied – segments and arcs.

Fig. 7b shows the left contour from Fig. 6b and the parameterized result after segmentation (Fig. 7c). Fig. 7c shows that the parameterized contour is far from following the shape of the original contour well. To improve the obtained results, a reduction of the triangles making up the three-dimensional model is applied. The reduction is performed in the non-parametric modeling application Blender 3.5 (Blender Foundation, 2024). The tool implemented in the cited application aims for minimal changes to the shape when performing the reduction. Fig. 8 shows the result of the same procedure for obtaining the parameterized contour from Fig. 7, but applied to the reduced model. From Fig. 8 it can be seen that the similarity between the parametric contour and the non-parametric one is significantly improved compared to Fig. 7. Stage 5 is performed after visual analysis of the obtained cross-sections in both directions. Three groups of contours are identified that are suitable for parametric construction of the part. This is the profile shown in Fig. 8c and the circle and pentagon from the cross-section shown in Fig. 9a (the parametrized pentagon is shown in Fig. 9c). Three characteristic features can be built from these three groups of contours (Stage 6). These are: extrusion of the pentagon (feature A), removal of material by extruding it in the shape of the circle from Fig. 9a (feature B) and removal of material by rotation around the central axis of the contour from Fig. 8c (feature C). The logical sequence (Stage 7) for building the model is to create first feature A, then B and finally C (Fig. 10a and b). Data output to a CAD environment (Stage 8) is performed via a DXF file. The DXF file contains the parametric curves - segments, arcs and circles, segmented and recorded with their basic parameters. The DXF file is introduced into the CAD environment (Stage 9) as separate sketches, the basis for building features. The parametric model is built using the tools of the CAD environment (Fig. 10c).

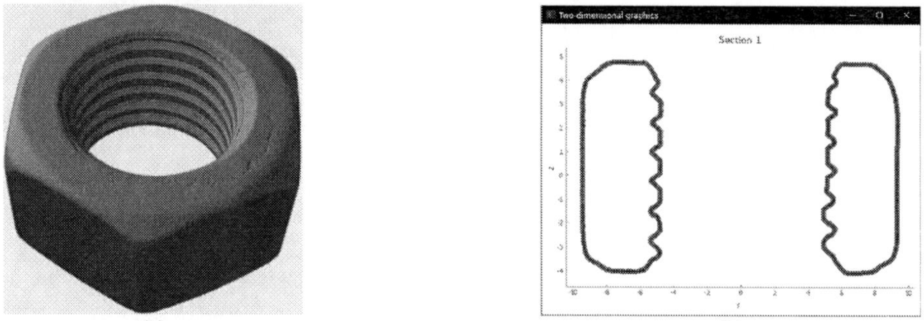

(a) (b)

Fig. 6. Model of a three-dimensional scanned part

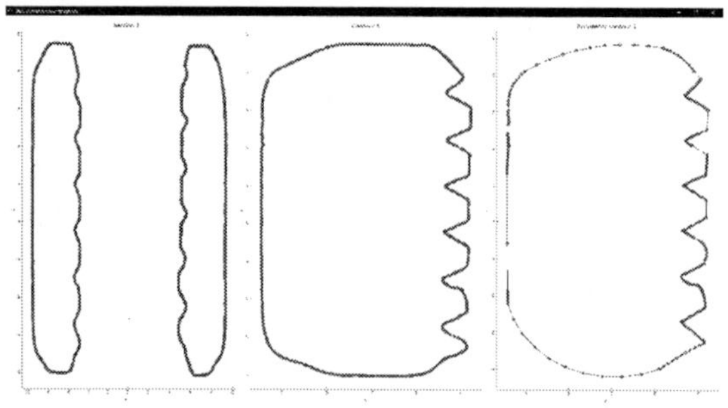

(a) (b) (c)

Fig. 7. Parameterization of a contour. (a) section along the horizontal axis, (b) contour; (c) parameterized contour

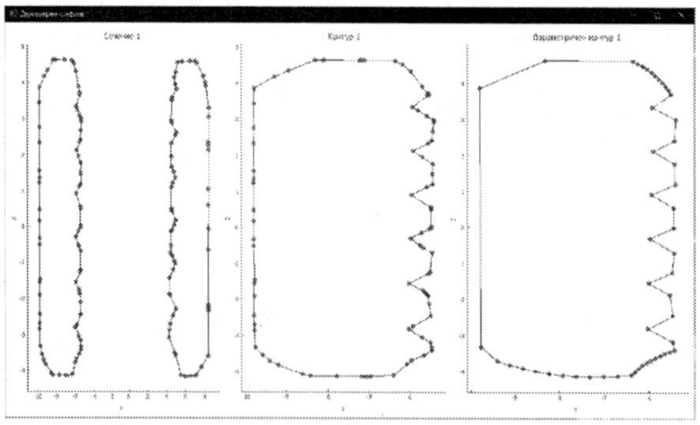

(a) (b) (c)

Fig. 8. Parameterization of a contour after reduction

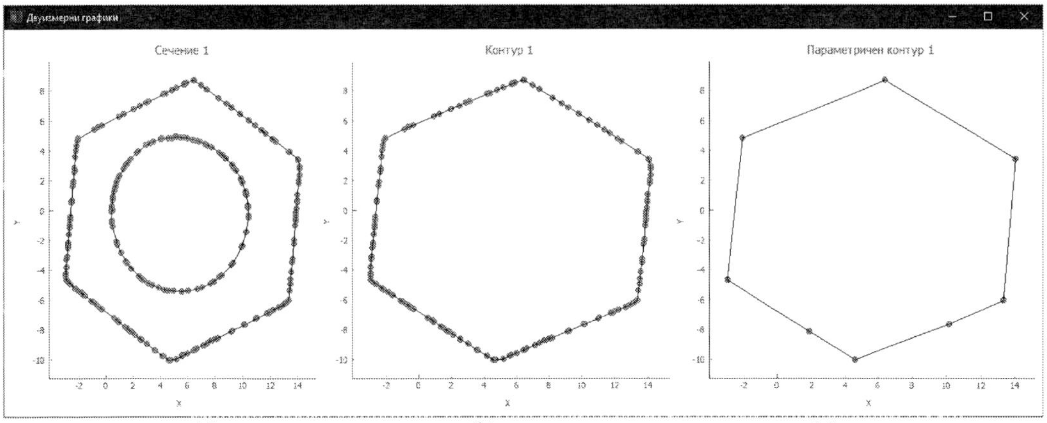

(a) (b) (c)

Fig. 9. Section and contour along the vertical axis of the model

(a) (b) (c)

Fig. 10. Building the parametric model: (a) characteristics A and B, (c) characteristic B, (d) resulting model

5. Conclusion

This article proposes an approach to automate the process of obtaining a parametric history-based model with a from a non-parametric one. The focus of the approach is on three-dimensional models of technical products, which are used in a CAD environment to document the design and production process of a given product. The approach is described in the form of a functional structure, with each function being a separate stage. The expected results of each stage are indicated, as well as the problems that each stage must solve. An example application of the approach is shown, indicating the problems when working with real models. The report describes the principles of the approach implementation, but does not go into depth of the mathematical and algorithmic foundations for implementing the proposed ideas, since each stage of the approach includes multiple mathematical models and algorithmic support, and the volume is limited. The authors believe that it is important to first present the overall approach as a working framework, and then to go into depth (the principle from general to particular). A large part of the problems identified in the first four stages have been solved by the authors' developments, and another part is yet to be solved (from the fifth stage to the last stage of the approach). Work is underway to publish the algorithmic and software support for the first four stages and develop mathematical models, methods and algorithms for solving the formulated problems in the remaining stages. In addition to this work, testing the developed approach and tools with a more diverse set of models and ways to obtain them is envisaged as a future development.

6. Acknowledgments

Both authors are affiliated to Technical University Sofia and National Centre of Excellence „Mechatronics and Clean Technologies", Sofia, Bulgaria. This work was supported by European Regional Development Fund under "Research Innovation and Digitization for Smart Transformation" program 2021-2027 under the Project BG16RFPR002-1.014-0006 "National Centre of Excellence Mechatronics and Clean Technologies". This work was accomplished by the Center of Competence for Mechatronics and Clean Technologies "Mechatronics, Innovation, Robotics, Automation and Clean Technologies" – MIRACle, with the financial support of contract No. BG16RFPR002-1.014-0019-C01, funded by the European Regional Development Fund (ERDF) through the Programme "Research, Innovation and Digitalisation for Smart Transformation" (PRIDST) 2021–2027.

7. References

[1] Dichev, D., Zhelezarov, I., Georgiev, B., Karadzhov, T., Dicheva, R., & Hasanov, H. (2025). A Method for Measuring Angular Orientation with Adaptive Compensation of Dynamic Errors. Sensors, Vol. 25, No. 16, 4922, DOI: https://doi.org/10.3390/s25164922

[2] Todorov, T.T.; Todorov, G.; Sofronov, Y. (2025) Process Parameter Optimization through Virtual Simulations for Enhanced Durability of Artificial Ski Slope Tiles, DOI: https://doi.org/10.1088/1742-6596/3061/1/012008.

[3] Kamberov K. H., Todorov G. D., Ivanov Ts. T. (2024) Automotive Product Validation through Virtual Prototyping at Early Design Stage, AIP Conference Proceedings, Vol. 3064, No. 1, art. no. 030007, DOI: https://doi.org/10.1063/5.0199194.

[4] https://www.ptc.com/en/products/creo, (2025). PTC Inc. Creo. Accessed on: 2025-09-23.

[5] https://www.3ds.com/products/catia (2025) Dassault Systemes CATIA. Accessed on: 2025-09-23.

[6] https://www.solidworks.com/ (2025) Dassault Systemes SOLIDWORKS. Accessed on: 2025-09-23.

[7] https://www.autodesk.com/products/inventor/overview (2025) Autodesk Inventor. Accessed on: 2025-09-23.

[8] https://solidedge.siemens.com/en/ (2025) SIEMENS Solid Edge. Accessed on: 2025-09-23.

[9] Kyprianou, L. (1980). Shape classification in Computer-Aided Design, Ph.D. Dissertation, Cambridge University.

[10] Shi, Y. Z. (2020). A critical review of feature recognition techniques, Computer-Aided Design and Applications, Vol. 17, No. 5, pp. 861-899, DOI: https://doi.org/10.14733/cadaps.2020.861-899.

[11] Donaldson, I. C. (1993). Rule-based feature recognition for 2.5D machined components, International Journal of Computer Integrated Manufacturing, Vol. 6, No. 1-2, pp. 51-64, DOI:https://doi.org/10.1080/09511929308944555.

[12] Vosniakos, G. D. (1993). A shape feature recognition framework and its application to holes in prismatic parts. The International Journal of Advanced Manufacturing Technology, Vol. 8, No. 6, pp. 345-351, DOI:https://doi.org/10.1007/bf01751095.

[13] Chan, A. C. (1994). Process planning by recognizing and learning machining features. International Journal of Computer Integrated Manufacturing, Vol. 7, No. 2, pp. 77-99, DOI:https://doi.org/10.1080/09511929408944597

[14] Bhandarkar, M. N. (2000). STEP-based feature extraction from STEP geometry for agile manufacturing. Computers in Industry, Vol. 41, No. 1, pp. 3-24.

[15] Gao, J. Z. (2004). Extraction of machining features for CAD/CAM integration. The International Journal of Advanced Manufacturing Technology, Vol. 24, No. 7-8, pp. 573-581, DOI:https://doi.org/10.1007/s00170-003-1882-9.

[16] Dimov, S. B. (2007). A hybrid method for feature recognition in computer aided design models. Journal of Engineering Manufacture, Vol. 221, No. 1, pp. 79-96, DOI:https://doi.org/10.1243/09544054JEM437.

[17] Verma, A. R. (2008). A hint-based machining feature recognition system for 2.5 D parts. International Journal of Production Research, Vol. 46, No. 6, pp. 1515-1537, DOI:https://doi.org/10.1080/00207540600919373.

[18] Xu, X. H. (1998). Recognition of rough machining features in 212D components. Computer-Aided Design, Vol. 30, No. 7, pp. 503-516, DOI:https://doi.org/10.1016/S0010-4485(97)00090-0.

[19] Fu, M. O. (2003). An approach to identify design and manufacturing features from a data exchanged part model. Computer-Aided Design, Vol. 35, No. 11, pp. 979-993, DOI:https://doi.org/10.1016/s0010-4485(02)00160-4.

[20] Nasr, E. K. (2006). A new methodology for extracting manufacturing features from CAD system. Computers & Industrial Engineering, Vol. 51, No. 3, pp. 389-415, DOI:https://doi.org/10.1016/j.cie.2006.08.004.

[21] Vandenbrande, J. R. (1993). Spatial reasoning for the automatic recognition of machinable features in solid models. IEEE Transactions on Pattern Analysis and Machine Intelligence, Vol. 15, No. 12, pp. 1269-1285, DOI:https://doi.org/10.1109/34.250845.

[22] Han, J. R. (1997). Integration of feature based design and feature recognition. Computer-Aided Design, Vol. 29, No. 5, pp. 393-403, DOI:https://doi.org/10.1016/S0010-4485(96)00079-6.

[23] Topalova, I. & Staeva, Y. (2022). Classification of Two-dimensional Mechanical Parts Using a Convolutional Neural Network, Proceedings of the 33rd DAAAM International Symposium, pp.0042-0049, B. Katalinic (Ed.), Published by DAAAM International, ISBN 978-3-902734-36-5, ISSN 1726-9679, Vienna, Austria DOI: 10.2507/33rd.daaam.proceedings.007

DOI: 10.2507/36th.daaam.proceedings.007

APPLICATION POTENTIAL OF YOLO-BASED REAL-TIME HUMAN DETECTION SYSTEMS IN SEARCH AND RESCUE OPERATIONS

Danijel Zelenika, Željko Marušić & Markić Ivan

This Publication has to be referred as: Zelenika, D[anijel]; Marusic, Z[eljko] & Markic, I[van] (2025). The Application Potential of Real-Time Human Detection Systems in Search and Rescue Operations, Proceedings of the 36th DAAAM International Symposium, pp.0057-0061, B. Katalinic (Ed.), Published by DAAAM International, ISBN 978-3-902734-47-1, ISSN 1726-9679, Vienna, Austria
DOI: 10.2507/36th.daaam.proceedings.007

Abstract

In Search and Rescue (SAR) operations, rapid and accurate human detection in remote or hazardous areas is critical for saving lives. This study explores the application of modern computer vision algorithms that enable automated, real-time human detection. The focus is placed on deep learning methods, particularly convolutional neural networks (CNNs), with an emphasis on models such as YOLO (You Only Look Once). Although other models may offer higher accuracy, their computational demands pose challenges for real-time processing. The proposed approach was tested on aerial imagery from well-known HERIDAL dataset. The HERIDAL dataset primarily contains simulated rescue scenarios in mountainous and rural areas. The results demonstrate that the model achieves a recall of 85.6% at an input image resolution of 1024px, despite the input image resolution being reduced by nearly four times. Furthermore, the real-time detection can be achieved without significant loss of accuracy, even under low-visibility conditions and complex terrain. Overall, the findings confirm the potential of such systems for practical use in real-world SAR missions.

Keywords: computer vision; human detection; real-time processing; search and rescue missions; convolutional neural networks.

1. Introduction

In emergency situations such as natural disasters, accidents, or missing person incidents, the ability to rapidly and accurately locate individuals can be the determining factor between life and death. Traditional search methods rely on the deployment of large numbers of rescuers, search dogs, and various ground and aerial vehicles, making such operations logistically demanding, costly, and time-consuming. In response to these challenges, unmanned aerial vehicles (UAVs) have become increasingly utilized in SAR missions due to their capability to swiftly survey extensive areas and capture aerial imagery that supports the localization of victims. However, a significant challenge remains, as the analysis of footage captured by UAVs still largely depends on human interpretation, which makes the process slow and prone to errors [1]. This creates an opportunity for the integration of artificial intelligence, particularly deep learning techniques, which enable the automation of human detection in complex environments. Notably, models from the YOLO family, known for their high-speed and high-accuracy object detection capabilities [2], have demonstrated substantial potential for application in SAR operations.

Fig. 1. Examples of human detection on the HERIDAL dataset using the YOLOv11 model.

The objective of this paper is to present the concept of real-time automatic human detection, examine existing methodologies, and discuss the advantages and limitations of such systems in practical SAR scenarios. This paper is structured as follows. Chapter 2 provides an overview of previous research in the field of human detection for SAR operations, presenting related works that address this challenge, the methodologies applied, and the results achieved by various authors. Chapter 3 introduces the YOLOv11 model architecture and describes the datasets used in this study, which serve as the foundation for model training and evaluation. Chapter 4 presents and analyzes the experimental results, focusing on detection accuracy, model performance, and limitations observed during testing. Finally, Chapter 5 summarizes the key findings and provides conclusions and recommendations for future research in the domain of UAV-based human detection systems for SAR applications.

2. Related work

Human detection from aerial imagery represents a highly complex problem in the field of computer vision, primarily due to the significant variability of objects caused by different poses in which a missing or injured person may be found, variations in clothing color, lighting conditions, and background complexity. Traditional object detection approaches have generally been designed for relatively large objects with clearly defined edges and contours. However, in images captured by UAVs, the objects of interest are typically very small, and image instability caused by drone movement further complicates the task.

Conventional object detection methods mainly rely on object size, texture, and shape, whereas modern approaches often employ bimodal systems that combine thermal and optical imagery to improve detection accuracy. Examples include the works of Gaszczak and Breckon [3] and Rudol and Doherty [4], where thermal images are used to identify regions of elevated temperature corresponding to human silhouettes. At the same time, the optical spectrum is analyzed using cascades of boosted classifiers with Haar-like features. Similar approaches have also been applied to the detection of swimmers in open water environments using unsupervised learning techniques [5], whereas studies [6], [7] have proposed models that utilize pyramidal feature extraction in SSD architectures or a combination of color and depth information for human detection.

The use of transfer learning has proven to be an extremely effective strategy in situations where the number of labeled images is limited, which is a common challenge in the analysis of satellite and aerial imagery. Since collecting and manually annotating such data is both time-consuming and resource-intensive, transfer learning enables the reuse of existing pre-trained neural networks as a foundation for further training on domain-specific datasets. This approach significantly improves model accuracy and generalization capabilities without the need to generate large quantities of new annotations or to train models from scratch [8]. For instance, in [9], a model for human detection on aerial images was first trained on a large, general-purpose dataset encompassing a wide range of visual characteristics, and then fine-tuned on a smaller, domain-specific dataset, achieving high performance and better adaptation to the target environment.

Similarly, the study in [10] demonstrates that models utilizing transfer learning on larger and more generic datasets can outperform those trained solely on smaller, specialized datasets. These findings confirm that transferring knowledge from large-scale models enables more efficient learning in resource-constrained scenarios, establishing transfer learning as a key technique in the field of aerial and satellite image analysis.

In Mediterranean regions, the application of thermal cameras during summer months is often constrained by high ambient temperatures, which significantly reduce the temperature contrast between the human body and the surrounding environment. As a result, within the IPSAR project, optical cameras were employed as the primary sensing modality for image acquisition. The image processing workflow incorporated a two-stage mean-shift segmentation algorithm optimized for small image segments, followed by a heuristic analysis stage designed to enhance object distinction and reduce noise. This approach effectively minimizes computational complexity while maintaining a high level of detection accuracy.

Subsequent research efforts expanded upon this foundation by comparing system performance on compressively reconstructed versus original images[11], as well as evaluating the impact of various salient detection algorithms [12] on detection reliability. Additionally, hybrid approaches combining salient detection techniques with convolutional neural network (CNN)-based models [13] have demonstrated promising potential for improving detection robustness and accuracy. However, these methods still face challenges related to false alarm rates and sensitivity under variable environmental conditions, indicating the need for further optimization and adaptive model design in real-world SAR scenarios. Studies [14] present comparative analyses of the Score Map to ROI algorithm based on deep neural network performance and human expert accuracy across various types of SAR missions, where the Score Map to ROI algorithm have achieved substantially higher accuracy than human evaluators. In the same study, the best results across the four evaluated missions were achieved in experiments where human experts performed detection with the assistance of the Score Map to ROI algorithm. This hybrid approach demonstrated that combining human intuition with algorithmic guidance can significantly enhance detection accuracy and reduce false positives in complex SAR scenarios.

3. Model architecture and Datasets

This chapter focuses on the fundamental components that underpin this research's experimental framework: the YOLOv11 object detection architecture and the aerial HERIDAL image datasets used for model training and evaluation. Collectively, these elements form the technical foundation for developing and accessing automated human detection systems in SAR scenarios.

The YOLOv11 model represents the latest generation of the YOLO family of real-time object detection algorithms. Building on the architecture of YOLOv8, this version integrates several structural and performance improvements aimed at enhancing detection accuracy, computational efficiency, and adaptability across diverse application domains. The model is available in several size variants allowing researchers to balance accuracy and efficiency based on the available computational resources. In this study, the YOLOv11n (nano) model was employed. The model processes an input image of 640×640 pixels in only 8.64 milliseconds on the NVIDIA Jetson Orin NX (16 GB), allowing the system to achieve approximately 116 frames per second (FPS). This exceptional performance underscores the model's suitability for deployment in time-critical applications, such as Search and Rescue (SAR) missions, where rapid detection and low latency are essential for real-time decision-making.

As the foundation for model training and testing, the Heridal [13] dataset was employed. This database consists of a collection of aerial photographs captured by drones, depicting both rural and urban locations across Bosnia and Herzegovina. For the purposes of this study, a subset of images was selected from 11 specific locations: Blidinje, Čapljina, Goranci, Kupres, Ljubuški, Međugorje, Posušje, Rakitno, Široki Brijeg, Stolac, and Velež. The dataset's diversity, including both rural and urban environments, and its high-resolution imagery, make it a valuable resource for developing effective and reliable detection systems for real-world SAR scenarios.

The Heridal dataset contains approximately 1,700 high-resolution images (4000 × 3000 pixels), which allows for detailed analysis of complex scenes. The complete dataset was divided into training and validation sets. To ensure reliability and objective evaluation, the data were randomly split in an 80:20 ratio. This approach ensures that most of the data is used for model learning, while maintaining a representative and unbiased subset for assessing model performance on unseen samples. This dataset has been widely used in research focused on human detection in aerial imagery, particularly in the context of SAR operations. It enables the development and evaluation of various deep learning models, including YOLO-based models for real-time detection, ensemble approaches to improve detection robustness, and transfer learning techniques that leverage pre-trained models adapted to the specific characteristics of aerial images.

4. Results

The proposed model was trained on images with resolutions of 640px and 1024px. The experimental results show that, in both experiments, a stable and prosperous learning process was achieved over 50 epochs, as evidenced by the continuous reduction in the loss function and a significant improvement in performance metrics, both in precision and recall, as summarized in Table 1. The key metric, mAP@0.5, reached a value of 66.11% at the end of training on 640px images, and 78.33% on 1024px images, indicating a solid foundational ability of the model to identify objects and place bounding boxes with acceptable accuracy.

Despite this success, a more detailed analysis reveals notable weaknesses. The model's performance drops sharply at stricter IoU thresholds, as reflected by the low mAP@0.5-0.95 values of only 26.60% on 640px images and 40.66% on 1024px images. This indicates a significant issue with object localization precision, i.e., placing highly accurate bounding boxes. However, for application in SAR missions, precise bounding box placement is not critical.

Epoch	Precision		Recall		mAP50		mAP50-95	
	640px	1024px	640px	1024px	640px	1024px	640px	1024px
10	51.20	52.65	44.69	54.07	41.64	47.12	13.37	19.91
20	56.01	56.88	51.85	56.60	47.92	62.60	16.16	27.70
30	54.92	53.03	55.18	55.93	49.38	65.57	17.13	20.94
40	57.08	72.69	61.11	70.74	58.46	71.65	19.97	30.41
50	64.53	74.85	68.05	78.89	66.12	78.33	26.60	40.66

Table 1. Performance metrics during training on images with resolutions of 640 and 1024px per epoch.

For model validation, a set of 147 images from the HERIDAL database was selected, containing 270 instances of the "Human" class. Fig. 2 shows an example of successful human detections in a very complex environment, demonstrating the model's ability to accurately identify targets despite challenging background conditions. Two experiments were conducted. In the first experiment, the input image to the model was 640 px, and in the second, 1024 px, given that the original images from the HERIDAL database have dimensions of 4000×3000 pixels. The results of these experiments are presented in Table 2.

Precision		Recall		F1	
640px	1024px	640px	1024px	640px	1024px
71.9	68.8	63.3	85.6	67.3	76.2

Table 2. Comparison of performance metrics for image resolutions of 640 and 1024px.

The analysis of the results on validation set indicates a notably high proportion of false negative detections. Specifically, 99 out of 270 actual instances of the "Human" class were missed in the 640px images. This elevated miss rate, which directly corresponds to a recall of 63.3%, represents the primary limitation of the model. In contrast, the recall on the 1024px images reaches 85.6%, marking a substantial improvement of 22.3%. This significant increase demonstrates that the model benefits considerably from higher-resolution input, leading to more accurate detection of actual objects.

Fig. 2. An example of successful detections in a very complex environment

While the transition to higher-resolution images improved recall, it also increased false positive detections, rising from 67 to 105. However, the overall detection performance, as measured by the F1 score, and the considerable improvement in recall outweighs this drawback. The tendency toward false positives suggests that the model adopts a more "aggressive" detection strategy, aiming to identify every potential object, a behavior commonly observed in models optimized for high recall. The disparity in false negatives, 39 for 1024px images versus 99 for 640px images, highlights one of the model's key challenges. A detailed examination reveals that the root cause of these limitations lies in the dataset's inherent characteristics. Analysis of the bounding box size distribution confirms that the model is trained predominantly on extremely small objects. Detecting and localizing such small objects is an inherently difficult task, which partially explains the reduced recall at lower resolutions and the increased number of false negatives. These observations underscore the importance of image resolution in object detection tasks and suggest that higher-resolution inputs enable the model to capture fine-grained features necessary for accurate localization and classification better.

5. Conclusion

This study highlights the critical role of input image resolution in achieving accurate human detection for Search and Rescue (SAR) operations. The results demonstrate that the model's recall improves markedly from 63.3% at 640 px to 85.6% at 1024 px, underscoring the importance of high-resolution inputs for reliable detection. While higher resolutions led to a slight increase in false positive detections, the overall F1 score of 85.6% and substantial gain in recall of 22.3% validate the effectiveness of this approach. Error analysis reveals that the primary challenge lies in detecting very small objects, as represented in the HERIDAL dataset. These limitations are particularly evident at lower resolutions, where a higher incidence of false negatives reduces overall performance.

Despite the increase in false positives, the model adopts a more "aggressive" detection strategy aimed at identifying all relevant objects—a strategy particularly beneficial in SAR applications, where maximizing the detection of actual targets outweighs the cost of occasional false alarms. In summary, these findings confirm that optimizing input resolution and carefully balancing recall and precision can significantly enhance model performance, supporting the practical deployment of human detection systems in complex and hazardous environments.

One of the future goals of this research is to create a more robust image dataset that includes diverse SAR scenarios, such as poor daytime light conditions or snow and other challenging environments, which are currently underrepresented in the official HERIDAL dataset. Another goal is to integrate multiple sensors to improve detection in difficult conditions, such as low-light or poor daytime illumination.

6. References

[1] A. A. Bany Abdelnabi and G. Rabadi, "Human Detection From Unmanned Aerial Vehicles' Images for Search and Rescue Missions: A State-of-the-Art Review," IEEE Access, vol. 12, pp. 152009–152035, 2024, doi: 10.1109/ACCESS.2024.3479988.

[2] M. Abdank, M. Aburaia, and W. Woeber, "USING COLOUR-BASED OBJECT DETECTION FOR PICK AND PLACE APPLICATIONS," in Annals of DAAAM and Proceedings of the International DAAAM Symposium, DAAAM International Vienna, 2021, pp. 536–541. doi: 10.2507/32nd.daaam.proceedings.077.

[3] A. Gaszczak, T. P. Breckon, and J. Han, "Real-time people and vehicle detection from UAV imagery," in Intelligent Robots and Computer Vision XXVIII: Algorithms and Techniques, 2011, p. 78780B. doi: 10.1117/12.876663.

[4] P. Rudol and P. Doherty, "Human body detection and geolocalization for UAV search and rescue missions using color and thermal imagery," in IEEE Aerospace Conference Proceedings, 2008. doi: 10.1109/AERO.2008.4526559.

[5] E. Lygouras, N. Santavas, A. Taitzoglou, K. Tarchanidis, A. Mitropoulos, and A. Gasteratos, "Unsupervised Human Detection with an Embedded Vision System on a Fully Autonomous UAV for Search and Rescue Operations," Sensors, vol. 19, no. 16, 2019, doi: 10.3390/s19163542.

[6] A. Al-Kaff, M. J. Gómez-Silva, F. M. Moreno, A. de la Escalera, and J. M. Armingol, "An Appearance-Based Tracking Algorithm for Aerial Search and Rescue Purposes," Sensors, vol. 19, no. 3, 2019, doi: 10.3390/s19030652.

[7] B. Mishra, D. Garg, P. Narang, and V. Mishra, "Drone-surveillance for search and rescue in natural disaster," Comput Commun, vol. 156, pp. 1–10, Apr. 2020, doi: 10.1016/J.COMCOM.2020.03.012.

[8] A. Bochkovskiy, C.-Y. Wang, and H.-Y. M. Liao, "YOLOv4: Optimal Speed and Accuracy of Object Detection," Apr. 2020, [Online]. Available: http://arxiv.org/abs/2004.10934

[9] A. J. Mantau, I. W. Widayat, J.-S. Leu, and M. Köppen, "A Human-Detection Method Based on YOLOv5 and Transfer Learning Using Thermal Image Data from UAV Perspective for Surveillance System," Drones, vol. 6, no. 10, 2022, doi: 10.3390/drones6100290.

[10] A. Majidizadeh, H. Hasani, and M. Jafari, "Transfer Learning Framework for Semantic Segmentation of High-Resolution UAV-based images in Urban Area," Journal of Geospatial Information Technology, vol. 10, no. 4, 2023, doi: 10.61186/jgit.10.4.87.

[11] J. Musić, I. Orović, T. Marasović, V. Papić, and S. Stanković, "Gradient Compressive Sensing for Image Data Reduction in UAV Based Search and Rescue in the Wild," Math Probl Eng, vol. 2016, p. 6827414, 2016, doi: 10.1155/2016/6827414.

[12] S. Gotovac, V. Papić, and Ž. Marušic, "Analysis of saliency object detection algorithms for search and rescue operations," in 2016 24th International Conference on Software, Telecommunications and Computer Networks, SoftCOM 2016, 2016. doi: 10.1109/SOFTCOM.2016.7772118.

[13] D. Božić-Štulić, Ž. Marušić, and S. Gotovac, "Deep Learning Approach in Aerial Imagery for Supporting Land Search and Rescue Missions," Int J Comput Vis, 2019, doi: 10.1007/s11263-019-01177-1.

[14] S. Gotovac, D. Zelenika, Ž. Marušić, and D. Božić-Štulić, "Visual-Based Person Detection for Search-and-Rescue with UAS: Humans vs. Machine Learning Algorithm," Remote Sens (Basel), vol. 12, no. 20, 2020, doi: 10.3390/rs12203295.

DOI: 10.2507/36th.daaam.proceedings.008

OPTIMAL DESIGN PROCEDURES IN MACHINE SCIENCE EDUCATION

Vyarka Ronkova & Antoaneta Dobreva

This Publication has to be referred as: Ronkova, V[yarka] & Dobreva, A[ntoaneta] (2025). Optimal Design Procedures in Machine Science Education, Proceedings of the 36th DAAAM International Symposium, pp.0062-0066, B. Katalinic (Ed.), Published by DAAAM International, ISBN 978-3-902734-47-1, ISSN 1726-9679, Vienna, Austria
DOI: 10.2507/36th.daaam.proceedings.008

Abstract

A theoretical survey in the area of application of optimal educational technologies and design in Machine Science Education has been implemented. Based upon this analysis, the main objectives of the study have been derived: to define updated strategies for developing and improving collaborative teaching practices, to create an appropriate model that meets educational objectives and combines relevant co-teaching approaches; to indicate opportunities for improving the acquisition of specific competencies of engineering students corresponding to the challenges of Industry 4.0. A modified collaborative teaching model has been created. The research has been carried out based upon a methodology containing six stages. Practical applications of the model created have been described. Based on the results of the study, a system of additional activities is proposed to in order to improve the quality of training in the higher education system. Conclusions and recommendations are made, and the authors' vision for future work in this area is presented.

Keywords: Optimal Design procedures; Machine Science Education; Modified co-teaching model; Industry 4.0.

1. Introduction

The application of optimal educational technologies and training design in the area of Machine Science depends on access to modern study materials and on the opportunities for students to communicate with well-known professionals in a certain scientific field, [1]. A significant number of studies have investigated the impact of study process management based upon outcomes and innovation in engineering higher education [2], [3]. The authors of [4] emphasize the extremely important role of creating a sustainable functional and theoretical approach concerning teaching and learning by applying a fundamental framework related to knowledge. According to the authors of [5], the appropriate decision-making in managing the flow and volume of knowledge in fundamental engineering subjects involves critical evaluation of learning outcomes, which requires time and additional effort. In [6], the importance of knowledge management, providing a link between universities and engineering practice, has been emphasized. In addition, methods for managing time for independent study of students have been discussed. The most important results of this study are relevant to the possibilities for applying optimization procedures in the field of Machine Science and Engineering Design Training. Besides, interactive work with 3D models is important for clarifying spatial representations and relationships, especially if sufficient effort is made to combine graphic drawings, models and textual descriptions according to [7].

Models of collaborative teaching have been discussed in various literature sources, but the degree of responsibility of the academic staff in a co-teaching environment and the appropriate way to define effective collaborative training require further study according to the authors of [8]. The investigation, described in [9], defines co-teaching models that include: teamwork in a large group, simultaneous lectures, presentation of different perspectives and different methods for solving problems. The selection of different approaches and responsibilities for each lecturer depends on the aims and objectives of the study contents to be fulfilled, the individual needs of the students, the expertise of the academic staff and the results which are to be obtained. It is particularly important to emphasize that no single model of co-teaching should be used quite often. It is recommended to develop different modifications or combinations of co-teaching approaches for certain periods of time [10], [11].

Based upon a theoretical analysis of European university practices, the following specific indicators can be summarized: important prerequisites for implementing optimization procedures in the education technology in Machine Science is the availability of curricula in engineering Bachelor degree courses synchronized with the requirements of the Bologna Process. A particularly significant condition for the successful creation of an appropriate modified model of co-teaching is the perfect and transparent approach for introducing international dimensions into the university experience of the academic staff, applied by the Vice-Rectors for International Cooperation at the University of Ruse over the past 12 years: [12], [13] and [14].

Due to the beneficial academic environment created by the Rectors' Management teams of the University of Ruse, almost all members of the academic staff show awareness of the need to introduce new, innovative models of education. The most important reasons for facing this challenge are: students' chances of finding a job in the internationally oriented labour market are increasing, [15]; students need important competencies to work successfully in a multicultural society, [16]; global challenges in the labour market require a multidisciplinary and international approach, [17], [18]. It becomes clear that at the University of Ruse, Bulgaria there is a sustainable trend towards creating beneficial conditions for the implementing new and modified lecturing and training approaches.

The main objective of the study can be summarized in the following way: to define updated strategies for developing and improving collaborative teaching practices, to create an appropriate model for different purposes that meet educational objectives and to combine relevant co-teaching approaches. Therefore, the second objective of the study presented is to indicate opportunities for improving the acquisition of specific competencies of engineering students corresponding to the challenges of Industry 4.0.

2. Creating a modified collaborative teaching model

In order to achieve the research objectives, a modified model of co-teaching has been created based upon the professional experience of the authors' team [19], [20] and [21]. The model is characterized by the following features:

A system of additional educational activities has been created, including the integration of specially developed training for co-teaching in the field of Machine Science. The authors' team integrated newly developed training strategies for co-teaching, providing individual additional training programs. The following components of the modified co-teaching method have been provided: control of the analytical part of the main components of the product developed; ensuring co-teaching combined with consultations regarding the elaboration of the representation from the assembly drawing of the product, containing as much information as possible; organizing additional co-teaching aiming to explain and correct errors, to indicate omissions; continuing co-teaching and consulting activities for completing the project task.

This approach achieves the objectives of the individual educational programs, considering the prior level of preparation of each student in the target group. Therefore, the significant spectrum of academic and behavioural diversity in a student group is taken into account.

The authors' team strives to ensure the credibility of the research by using methods for data collection and analysis, described in details in [22] and [23]. The transferability of the research results can be established by identifying the academic study environment and describing the activities carried out by the participants in the research process. Within the framework of the investigation preparatory work, information on the equipment facilities is updated, the types of training modules are planned in appropriate time intervals. Administrative support is provided if necessary. Audit options are described with the aim of: guaranteed verification of the stages used and the decisions made within these research activities.

3. Practical application of the developed model in a new BIP

The authors' team developed a new project for the Blendid Intensive Program (BIP): "Smart Design Case Studies". The work on the project will be carried out within the framework of a joint international group of several European universities, aiming to foster a deep understanding of creative projects in the fields of Machine Science, Design and Digital Products, [24], [25]. These initiatives are closely related to the requirements and needs of local markets, which guarantees their relevance. The project envisages a preliminary two-month virtual training, followed by a one-week face-to-face work. Students participating in the intensive program will have the opportunity to: build a network of contacts internationally; learn and work in a team, participating in interesting challenges supervised by leading scientists. The created modified model of co-teaching will be implemented within the framework of this BIP.

4. Research Methodology

Aiming to achieve the objectives of the study and based upon the professional experience of the authors' team [26], an advanced methodology has been developed, which includes the following stages:

1. Determining recommendations for a system of teaching methods, training and additional activities aimed at improving the quality of education in the higher education system based on the results of the study and the practical application of the study in international educational and scientific activities.
2. Determination of an academic staff team engaged in the implementation of the modified collaborative teaching model. The definition of this target group of simultaneously teaching lecturers depends on the well-known characteristics of effectiveness in carrying out the relevant training: they should share their vision and commitment to co-teaching; participate in planning the study process on an equality of opportunities; responsibilities are to be defined in advance clearly and precisely, [27];
3. Selection of the research target group in order to identify the challenges in the training of students from different Bachelor degree courses;
4. Implementation in stages of collaborative teaching with the target group, including the described components of the modified co-teaching model. The integration and application of the described collaborative teaching strategies aim to adapt of the education system to the generation of students with a digital orientation in the field of engineering design;
5. Conducting a pilot study, discussing the advantages and features of a modified model for co-teaching with students from the target group;
6. Analysis and discussion of the research results obtained;
7. Determining recommendations for a system of teaching methods, training and additional activities aimed at improving the quality of education in the higher education system based on the research results and the practical application of the study in international educational and scientific activities.
8. Determining recommendations for a system of teaching methods, training and additional activities aimed at improving the quality of education in the higher education system based on the results of the study and the practical application of the study in international educational and scientific activities.

5. Analysis and discussion of results

The academic staff team, engaged in the creation and implementation of the modified model of collaborative teaching in the field of Machine science, consists of three habilitated lecturers in this scientific field. The target group has been selected among students from different Bachelor degree courses in three engineering faculties of the University of Ruse. The prerequisite for participation in this group, set by the academic staff team, was compliance with the principle of voluntary participation in a "project".

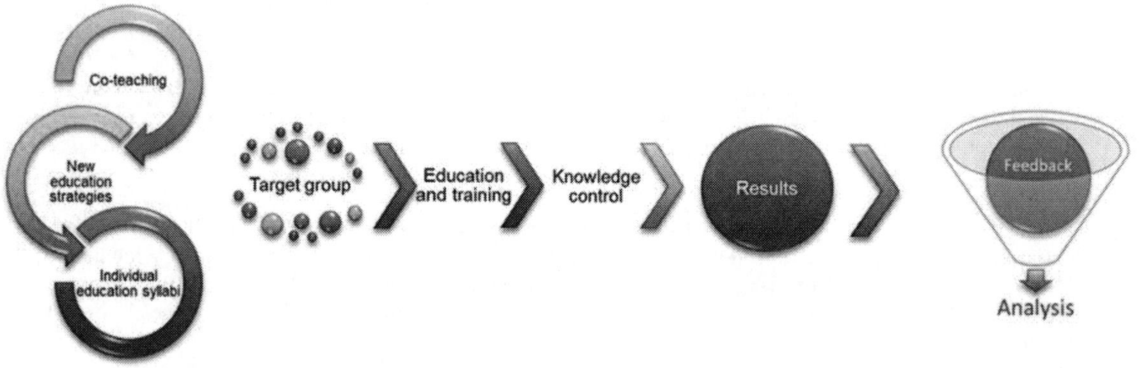

Fig. 1. Research implementation stages

Students have been informed about the expected benefits: teamwork, development of creative thinking, upgrading of spatial vision skills and improvement of design abilities. After successful completion of the project, students receive a certificate for a completed elective course with the corresponding ECTS. This achievement can be included in their Bachelor's Degree diploma and professional CV. A phased implementation of collaborative teaching with the target group within the framework of the research work has been carried out, Fig. 1. After the implementation of the described model of collaborative teaching, the students' motivation for independent learning improved. An additional survey has been conducted, discussing the advantages and features of the modified co-teaching model with students from the target group. Particularly good feedback was received when discussing the individual supervising support provided to almost all students from the target group. Additional positive results have been obtained in improving communication and professional development of the academic staff members who participated in the study.

The following challenges had to be faced: organizing preliminary meetings to ensure good planning; managing time for carrying out the relevant activities for teachers and students. It became clear that the differences in the teaching styles of each teacher should be taken into account and achieving their unification to a certain extent is quite necessary. The positive consequences of the research can be summarized in the following way: improving the quality of education, increasing the motivation of students to do independent work. The expected negative consequences could be: not all lecturers are to be equally loaded; creative contradictions may arise regarding education technology and teaching methods; the organization of the learning process between lecturers is complex, which would require an additional investment of time.

6. Conclusions and future work

The research of the scientific problem has been imposed by the need to optimize and modernize educational practices in the area of Machine sciences. Special attention has been dedicated to the development of collaborative teaching models that would meet the requirements of the European educational systems. According to the authors of the paper the solution to this challenge is the elaboration and implementation of a modified co-teaching models based upon modern pedagogical approaches and practical experience of the authors' team. Based upon the research implemented, the following conclusions can be formulated:

The main objectives of the study have been achieved through the implementation of a methodology containing six stages. Updated strategies for developing and improving co-teaching practices have been defined. The solutions included the creation of a system of additional educational activities, the integration of collaborative learning strategies and adaptation to the students' digital environment. The practical application of the model within a new Blended Intensive Program (BIP) demonstrates its effectiveness and the possibilities for improving students' skills and competencies corresponding to the challenges of modern industry. An appropriate modified collaborative teaching model has been created which corresponds to the educational objectives and combines the relevant co-teaching approaches. The practical applications of this model have been described.

The results of the study showed that the implementation of the modified model improved motivation, communication and independent learning among students. In addition, the professional experience of the lecturers has been significantly enriched. The results confirmed the prospects for further improvement of co-teaching practices and integration of international dimensions. The possibilities for improving the acquisition of specific competencies of engineering students corresponding to the challenges of Industry 4.0 are indicated.

Future scientific work in this area include expanding the application of the developed model to other academic subjects and to other Bachelor and Master degree courses, improving learning management strategies, and continuing international cooperation. The authors' team intent to systematize and popularize good practices, ensuring their sustainability and adaptability to the dynamically changing requirements of industry and education.

7. Acknowledgments

The authors express gratitude and acknowledgements for the support of the scientific fund at the University of Ruse, Bulgaria.

8. References

[1] Mogos, R., Bodea, C., Dascalu, M., Safonkina, O., Lazarou, E., Trifan, E. et al. (2018). Technology enhanced learning for Industry 4.0 engineering education. Rev. Roum. Des. Sci. Tech. Ser. Electrotech. Energy 63, pp. 429–435.

[2] Abubakar, A., Elrehail, H., Alatailat, M., & Elci, A. (2019). Knowledge management, decision-making style and organizational performance. Journal of Innovation & Knowledge, 4(2), pp.104-114.

[3] Brix, J. (2017). Exploring knowledge creation processes as a source of organizational learning: A longitudinal case study of a public innovation project. Scandinavian Journal of Management, 33(2), pp. 113–127.

[4] Stoyanov S., Dobreva. A. (2021). Systems Analysis and Design of Gear Drives through Innovative Software Approach IEEE Xplore: 2021 5th International Symposium on Multidisciplinary Studies and Innovative Technologies (ISMSIT), pp 366-371, doi: 10.1109/ISMSIT52890.2021.9604575

[5] Fitzgerald, D. R., Mohammed, S., & Kremer, G. O. (2017). Differences in the way we decide: The effect of decision style diversity on process conflict in design teams. Personality and Individual Differences, 104, pp. 339–344.

[6] Katalinic, B. (2010). Engineers for Knowledge Based Society, Proceedings of the 21st DAAAM International Symposium, B. Katalinic (Ed.), Published by DAAAM International, ISBN 978-3-901509-73-5, ISSN 1726-9679, Vienna, Austria

[7] Simion, I., Dobre, D., Pascu, N., Adir, V. & Adir, G. (2009). Learning System for Engineering Graphics and Design. Annals of DAAAM for 2009 & Proceedings of the 20th International DAAAM Symposium, Vol 20 (1), ISSN 1726-9679, ISBN 978-3-901509-70-4, Editor B. Katalinic, Published by DAAAM International, Vienna, Austria.

[8] Mastropieri, M. A., Scruggs, T. E., Graetz, J., Norland, J., Gardizi, W., & McDuffie, K. (2005). Case Studies in Co-Teaching in the Content Areas Successes, Failures, and Challenges. Intervention in School and Clinic, 40 (5), 260-270.

[9] Friend, M. & Bursuck, W. (2009). Including Students with Special Needs: A Practical Guide for Classroom Teachers (5th Ed.). Columbus, OH: Merrill.

[10] Ronkova, V. & Dobreva, A. (2024). Improving Design Parameters of Driving Systems, Proceedings of the 35th DAAAM International Symposium, pp.0091-0095, B. Katalinic (Ed.), Published by DAAAM International, ISBN 978-3-902734-44-0, ISSN 1726-9679, Vienna, Austria, DOI: 10.2507/35th.daaam.proceedings.012

[11] Tomsu, M. (2024). Information Overload as a Problem for IT Specialist, Proceedings of the 35th DAAAM International Symposium, pp.0177-0182, B. Katalinic (Ed.), Published by DAAAM International, ISBN 978-3-902734-44-0, ISSN 1726-9679, Vienna, Austria, DOI: 10.2507/35th.daaam.proceedings.023

[12] Temole F., D. Atanasova (2025). The Transformative Impact of Container Technology in Public Sector Organizations Science, Technology & Public Policy, 9 (1), 27-35, doi: 10.11648/j.stpp.20250901.13

[13] Ivanova, G., Velikova, M. (2024). Application and assessment of digital resources in the education of future pedagogues, Strategies for Policy in Science & Education, 5s, 197 – 206, doi: 10.53656/str2024-5s-20-app

[14] Popova, J., Koev, K., Popova, A. (2020). The Istanbul Convention and its Political, Social and Cultural Reflections in the Danube Space: The Case with Bulgaria, Journal of Danubian Studies and Research, 1, Vol. 10, 528-540

[15] Beloev H., V. Pencheva, J. Popova (2018). The academic values in the changing world (the University of ruse, Bulgaria as a signatory of MAGNA CHARTA UNIVERSITATUM), Proceedings of UoR, Vol. 57 (9).

[16] Koev, K., A. Popova (2024). The Risks for the Global Security in the Context of the Global Risks Report, Proceedings of UoR, Vol. 63 (5.4), pp 42-45

[17] Beloev, H., A. Smrikarov, V. Voinohovska, G. Ivanova. (2023). Determining the Degree of Digitalization of a Higher education institution Strategies for Policy in Science and Education. Education in the Information Society., Vol. 31, N.4s, pp 9-21, doi: 10.53656/str2023-4s-1-det

[18] Dobrev, V., Dobreva, A. (2023). Approaches to training and support for doctoral students, Proceedings of the University of Ruse, Vol. 62, (4.1.), pp 30-33

[19] Kamenov, K., A. Dobreva, Ronkova, V. (2017). Advanced Engineering Methods in Design and Education, Material Science and Engineeering, IOP Publishing, 252, 012033 - 37, doi: 10.1088/1757-899X/252/1/012033

[20] Popova, J., A. Dobreva, A. Ahmed (2014). Cooperation with Industry and Work Placements at the University of Ruse, 4TH VALENCIA GLOBAL 2014, pp 296-299

[21] Ronkova, V., Dobreva, A. (2022). Personal effectiveness and scientific productivity of doctoral students, Proceedings of the University of Ruse, Vol 61 (4.1), pp 12-17

[22] Berg, B. (2001). Qualitative research methods for the social sciences. Needham Heights, MA:Allyn & Bacon.

[23] Shenton, A. (2004). Strategies for ensuring trustworthiness in qualitative research projects. Education for Information, 22, pp. 63-75.

[24] Ivanova B., K. Shoilekova, D. Atanasova, R. Rusev (2024). Application of Zspace Technology in the Disciplines of the STEM cycle, Strategies for Policy in Science and Education, Vol.32, N 5s, 161-171, doi: 10.53656/str2024-5s-16-app

[25] Beloev, H.; Smrikarov, A.; Ivanova, A.; VassileV, C.; Georgiev, C.; Smrikarova, S.; Ivanova, G.; Stoykova, V.; Ibryamova, E.; Aliev, Y. & ZlataroV, P. (2020). A Vision of the University of the Future. Proceedings of the 21st International Conference on Computer Systems and Technologies (CompSysTech '20). Association for Computing Machinery, pp 307–312, New York, NY, USA.

[26] Dobreva, A., Dobrev, V. (2022). Inclusive Design in Engineering education, 33rd International DAAAM Symposium on Intelligent Manufacturing and Automation, pp 193-197, doi: 10.2507/33rd.daaam.proceedings.027

[27] Dobreva, A., Dobrev, V. (2020). Digital Transformation Dynamics in Higher Education PROCEEDINGS OF UNIVERSITY OF RUSE - 2020, volume 59 (4.1), pp. 12-16

DOI: 10.2507/36th.daaam.proceedings.009

HYBRID MACHINE LEARNING-ENHANCED PID CONTROLLER OPTIMIZATION FOR INDUSTRIAL PROCESS CONTROL

Amirhossein Bahrami & Vyacheslav Potekhin
Peter the Great St. Petersburg Polytechnic University

This Publication has to be referred as: Bahrami, A[mirhossein] & Potekhin, V[yacheslav] (2025). Hybrid Machine Learning-Enhanced PID Controller Optimization for Industrial Process Control, Proceedings of the 36th DAAAM International Symposium, pp.0067-0073, B. Katalinic (Ed.), Published by DAAAM International, ISBN 978-3-902734-47-1, ISSN 1726-9679, Vienna, Austria
DOI: 10.2507/36th.daaam.proceedings.009

Abstract

This article proposes a new hybrid method to optimize Proportional-Integral-Derivative (PID) controllers in industrial processes using the combination of machine learning techniques and classical control theory. This method couples reinforcement learning algorithms with metaheuristic optimization for achieving adaptive PID parameter adjustment in real-time industrial processes. The hybrid architecture is composed of three modules that interact with each other: a Reinforcement Learning Module using Q-learning, a Metaheuristic Optimization Module using hybrid particle swarm optimization and differential evolution, and an Adaptive Parameter Selection Module using fuzzy logic inference. The architecture illustrates better performance than standard tuning techniques with up to 49% enhancement in settling time and much lower control signal variation over several industrial process conditions. The work is a contribution to the development of intelligent control systems by marrying classical control theory and contemporary artificial intelligence strategies, providing pragmatic solutions for Industry 4.0 applications without compromising on computational efficiency to enable real-time implementation.

Keywords: PID control; machine learning; reinforcement learning; industrial automation; artificial intelligence

1. Introduction

Industrial automation has altered dramatically in recent years due to the growing integration of artificial intelligence (AI) into conventional control frameworks. Despite constant advancements, proportional-integral-derivative (PID) controllers—which comprise about 95% of all control loops in production systems—remain the industry standard [1]. Nonetheless, as industrial processes become more complex and diverse, there is an increasing need for more advanced and flexible control systems. Historically, PID tuning has relied on methods such as Ziegler-Nichols, Cohen-Coon, and pole placement. While these techniques have proven effective in stable environments, they usually perform poorly in dynamic settings where operating conditions are constantly changing. Specifically, manual tuning takes a lot of time and heavily depends on domain-specific expertise. Moreover, fixed controller parameters cannot cope with the unpredictability of modern industrial operations [2]. Due to these challenges, the focus has shifted to intelligent tuning methods based on machine learning.

As AI and control engineering continue to converge, opportunities have emerged for creating controllers that can tune and adjust themselves in real time. Machine learning approaches, especially reinforcement learning and evolutionary algorithms, have shown great promise for optimizing PID parameters in industrial settings that demand precision and adaptability [3]. In this work, we present a comprehensive framework that integrates the basic concepts of classical control theory with the flexible capabilities of modern machine learning to meet the evolving demands of process automation.

2. Related Work

Recent developments in machine learning-enhanced control systems have the potential to greatly improve the performance of industrial operations. Patel et al. [4] introduced a reinforcement learning-based approach for process control and optimization by extending the concept of automatically changing SISO PID controllers to MIMO MPC techniques. Their research demonstrated the effectiveness of RL algorithms in dealing with scenarios involving multiple control variables. Metaheuristic methods for PID optimization have recently become popular. Differential evolution, particle swarm optimization, and genetic algorithms have all been used successfully to solve PID tuning problems [5]. However, these methods may not be suitable for real-time applications and frequently necessitate a large amount of processing power. Evolutionary programming methods for PID parameter optimization have shown promise in a variety of settings [6]. These algorithms can avoid local optima while exploring large parameter spaces, making them ideal for tackling complex optimization problems. However, incorporating online learning features remains a challenge.

Neural networks have been investigated recently for adaptive PID control [7], [8], [9]. Complex nonlinear correlations between process variables and ideal controller settings can be learned using deep learning techniques. Neural networks' interpretability and usefulness in safety-critical industrial applications are, however, limited by their black-box nature. Adaptive control systems have been extensively studied for responsive manufacturing environments, with emphasis on the design of controllers which are able to respond quickly to disturbances and production process changes [10].

3. Methodology

3.1. Hybrid Architecture Design

The proposed hybrid machine learning-enhanced PID optimization framework is made up of three interconnected modules: the Adaptive Parameter Selection Module (APSM), the Reinforcement Learning Module (RLM), and the Metaheuristic Optimization Module (MOM). This architecture maintains the clarity and reliability of conventional PID controllers while allowing for continuous learning and adaptation. The RLM employs a Q-learning algorithm with function approximation to learn optimal control policies based on system state observations and performance feedback. The state space is defined as:

$$S = \left\{e(t), \frac{de(t)}{dt}, \int_0^t e(\tau)d\tau, u(t), y(t)\right\} \tag{1}$$

where e(t) represents the error signal, de(t)/dt is the error derivative, ∫e(τ)dτ is the error integral, u(t) is the control signal, and y(t) is the process output.

The action space consists of incremental adjustments to PID parameters:

$$A = \{\Delta K_p, \Delta K_i, \Delta K_d\} \tag{2}$$

The reward function is designed to minimize the integral of time-weighted absolute error (ITAE) while penalizing excessive control effort:

$$R(t) = -\alpha \cdot ITAE(t) - \beta \cdot |u(t)| - \gamma \cdot |\Delta u(t)| \tag{3}$$

where α, β, and γ are weighting factors.

3.2. Metaheuristic Optimization Integration

The MOM implements a hybrid particle swarm optimization (PSO) and differential evolution (DE) algorithm to explore the PID parameter space efficiently. The optimization objective function combines multiple performance criteria:

$$J = w_1 \cdot ISE + w_2 \cdot IAE + w_3 \cdot ITAE + w_4 \cdot TV \tag{4}$$

where:

- ISE = $\int_0^\infty [e(t)]^2 dt$ (integral of squared error)
- IAE = $\int_0^\infty |e(t)| dt$ (integral of absolute error)
- ITAE = $\int_0^\infty t|e(t)| dt$ (integral of time-weighted absolute error)
- TV represents the total variation of the control signal

The hybrid PSO-DE algorithm updates particle positions using:

$$\overrightarrow{v_i^{(t+1)}} = w \cdot \vec{v}i^{(t)} + c_1 \cdot r_1 \cdot (\vec{p}best, i - \vec{x}i^{(t)}) + c_2 \cdot r_2 \cdot \left(\vec{g}best - \overrightarrow{x_i^{(t)}}\right) \quad (5)$$

$$\overrightarrow{x_i^{(t+1)}} = \vec{x}i^{(t)} + \vec{v}i^{(t+1)} + F \cdot (\vec{x}r1^{(t)} - \vec{x}r2^{(t)}) \quad (6)$$

where the differential evolution component introduces additional exploration capability through the mutation factor F and randomly selected particles $\vec{x}_{r1}$ and $\vec{x}_{r2}$.

3.3. Adaptive Parameter Selection

To achieve the best balance between RL-based adjustments and metaheuristic optimization results, the APSM employs a fuzzy logic inference system. [11] utilized fuzzy logic methods in automatic control systems, illustrating the applicability of fuzzy inference systems to parameter selection in industry. The fuzzy system selects appropriate parameters based on performance requirements, disturbance levels, and process characteristics. The fuzzy rule base includes linguistic variables for:

- Process dynamics (slow, medium, fast)
- Disturbance magnitude (low, medium, high)
- Performance priority (stability, speed, accuracy)

A typical fuzzy rule example: "IF process dynamics is fast AND disturbance magnitude is high THEN use RL-based parameters with high confidence".

4. Experimental Setup and Implementation

4.1. Simulation Environment

The proposed methodology was evaluated using a comprehensive simulation environment that models various industrial processes including:

- First-order plus dead time (FOPDT) systems
- Second-order systems with varying damping ratios

The simulation platform implements realistic noise models and disturbance patterns commonly found in industrial settings. Process models are described using the following transfer functions:

$$G(s) = \frac{K \cdot e^{-\tau s}}{Ts+1} \quad (7)$$

for first-order plus dead time (FOPDT) systems, where:

- K is the process gain
- τ is the dead time
- T is the time constant

And for second-order systems:

$$G(s) = \frac{K \cdot \omega_n^2}{s^2+2\zeta\omega_n s+\omega_n^2} \quad (8)$$

where:

- ω_n is the natural frequency
- ζ is the damping ratio
- K is the steady-state gain

4.2. Performance Metrics

The evaluation framework employs multiple performance indices to assess controller effectiveness:

- Settling Time (T_s): Time required for the response to reach and stay within 2% of the final value
- Overshoot (M_p): Maximum percentage overshoot in the step response
- Steady-State Error (e_{ss}): Permanent error between desired and actual output
- Integral of Absolute Error (IAE): $IAE = \int_0^{\infty}|e(t)|\,,dt$
- Total Variation (TV): Measure of control signal smoothness defined as: $TV = \sum_{k=1}^{N-1}|u(k+1) - u(k)|$

4.3. Comparative Analysis

The proposed hybrid approach was compared against several benchmark methods:

- Traditional Ziegler-Nichols tuning
- Particle Swarm Optimization (PSO) alone
- Manual expert tuning

5. Results

Tables 1,2 present comparative results across different process types and operating conditions. The hybrid ML-enhanced approach demonstrates consistent improvements in key performance metrics.

Method	**Settling Time (s)**	**Overshoot (%)**	**IAE**	**Steady-State Error**	**Total Variation**
Ziegler-Nichols	2.800	0.00	19.972	0.9968	1680.976
PSO Only	19.990	0.00	19.988	1.0001	120.231
Manual expert tuning	12.940	0.00	19.990	1.0010	741.847
Hybrid ML-Enhanced	2.330	0.00	19.992	0.9983	90.169

Table 1. FOPDT process

Method	**Settling Time (s)**	**Overshoot (%)**	**IAE**	**Steady-State Error**	**Total Variation**
Ziegler-Nichols	2.020	0.00	19.952	0.9984	1776.554
PSO Only	19.990	0.00	19.996	0.9998	387.230
Manual expert tuning	19.990	0.00	19.966	0.9997	781.345
Hybrid ML-Enhanced	1.020	0.00	19.964	0.9983	390.039

Table 2. Second order process

Based on the results presented in the tables 1,2, the ranking of the methods based on weighted composite score is as follows (lower is better):

1. Hybrid ML-Enhanced: 28.7081
2. PSO Only: 35.5685
3. Manual Expert: 85.2947
4. Ziegler-Nichols: 177.7915

The comparison of control signals for the FOPDT process in figure 1 indicates a very important benefit of the Hybrid ML-Enhanced technique. Although the Ziegler-Nichols and Manual Expert methods have very oscillatory and erratic control signals with high variations, the Hybrid ML-Enhanced technique results in an extremely smooth and stable control signal. Such smooth control action is highly desirable in industrial practice since it lowers actuator wear and tear, reduces energy usage, and avoids mechanical stress on control hardware. The PSO Only technique demonstrates moderate control activity but still has higher variation compared to the suggested hybrid technique.

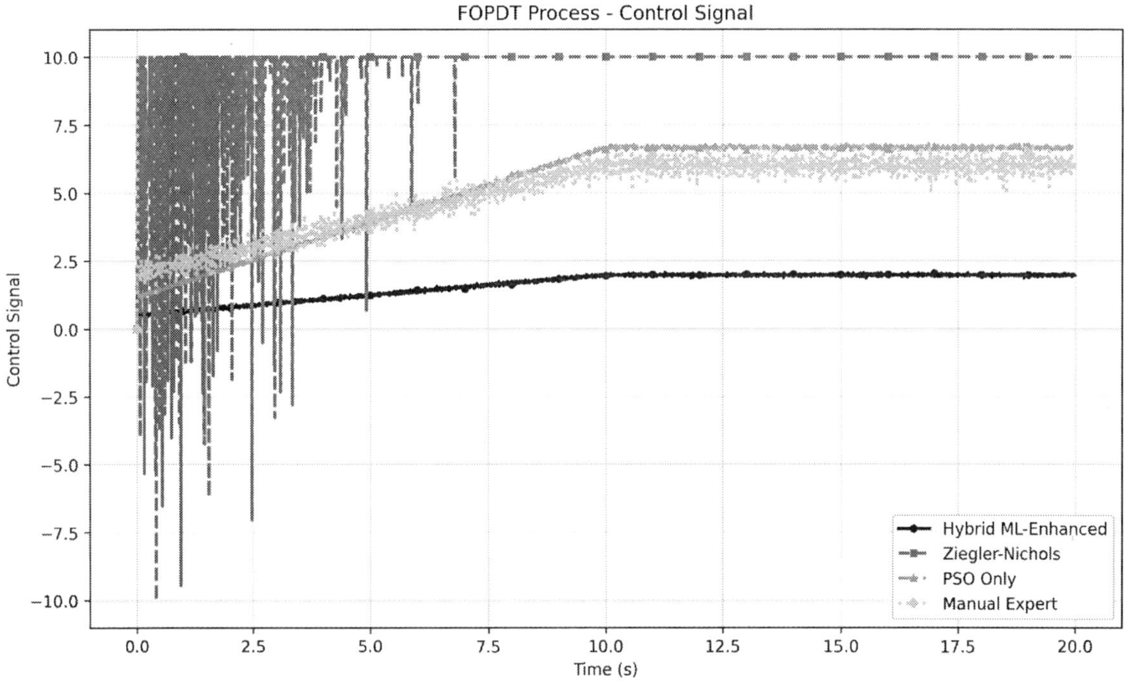

Fig. 1. Control Signal Analysis (FOPDT)

The comparison of settling time in figure 2 illustrates the remarkable speed of the Hybrid ML-Enhanced approach. For the FOPDT process, it reaches a settling time of merely 2.33 seconds, considerably outperforming all alternatives. The Ziegler-Nichols, PSO Only, and Manual Expert approaches all exhibit settling times at or close to the maximum simulation time (19.99 seconds), reflecting poor transient response. For second-order systems, the Hybrid ML-Enhanced approach reaches 1.02 seconds, once more considerably ahead of rivals. Such a fast-settling feature is vital in industrial processes where speedy response to setpoint change has a direct effect on productivity and product quality.

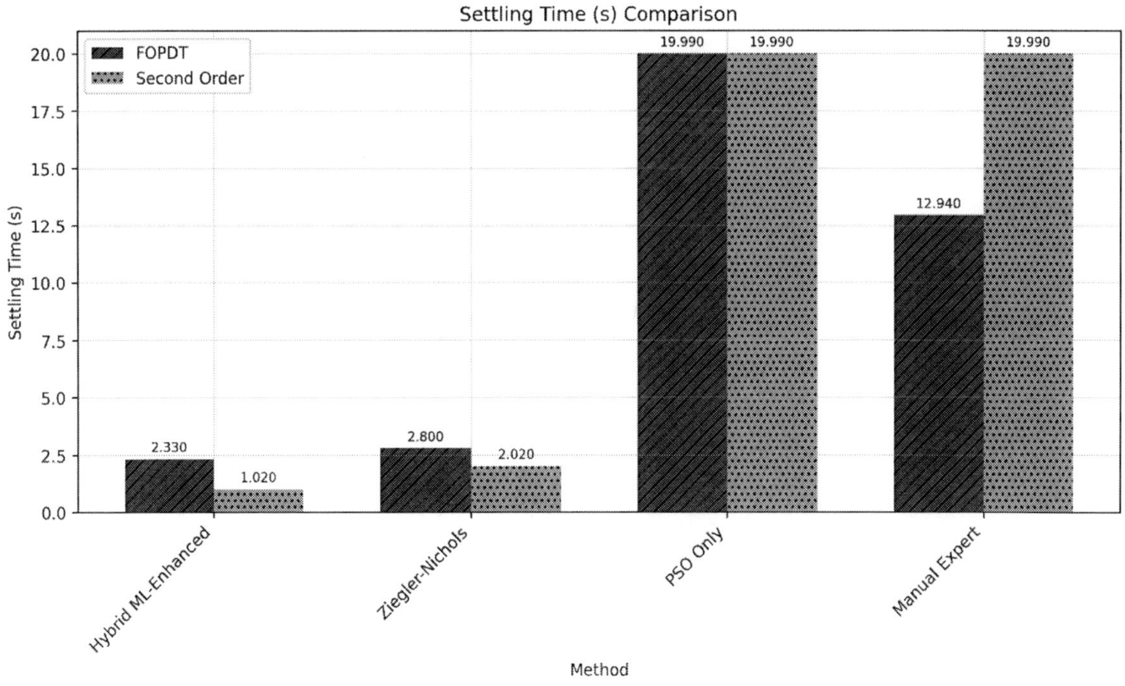

Fig. 2. Settling Time Performance

The comparison of total variation in figure 3, emphasizes the arguably most important benefit of the Hybrid ML-Enhanced method. With total variations of just 90.17 for FOPDT and 390.04 for second-order systems, it far outperforms the other methods. The Ziegler-Nichols method is seen to have too much control activity (1680.98 and 1776.55), while the Manual Expert method also has high variation (741.85 and 781.35). This minimal control variation shows that the ML-enhanced method obtains its performance without taking overly aggressive control actions that would destabilize the system or lead to premature equipment failure.

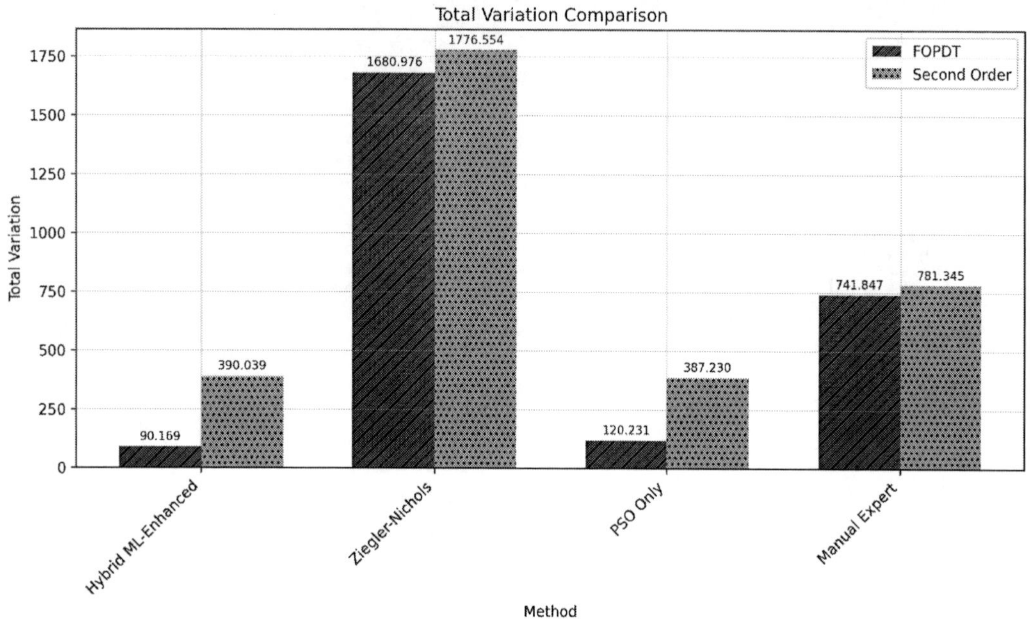

Fig. 3. Control Effort and Smoothness

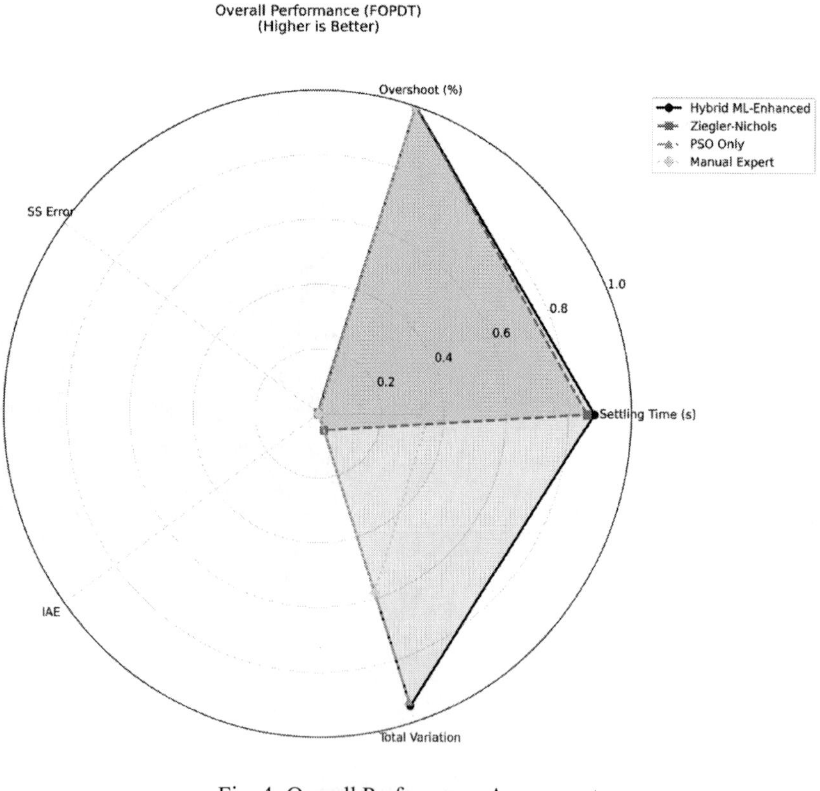

Fig. 4. Overall Performance Assessment

The FOPDT performance radar chart, in figure 4, reveals the Hybrid ML-Enhanced method delivering near-optimal performance in all the metrics. It performs exceptionally well in settling time and total variation while having very good steady-state accuracy and low overshoot. The well-balanced performance profile illustrates the method's capability in optimizing several competing objectives simultaneously, a characteristic of high-end optimization methods.

6. Future Directions and Limitations

Future research should focus on:

- Development of theoretical stability guarantees for RL-based control systems
- Integration with digital twin technologies for enhanced learning
- Application to distributed control systems and multi-agent scenarios
- Extension to model predictive control frameworks

7. Conclusion

This study describes a novel hybrid machine learning-enhanced method for optimizing PID controllers in industrial processes. Adaptive parameter selection, metaheuristic optimization, and reinforcement learning outperform traditional tuning techniques in several ways. The proposed method addresses major issues in industrial control systems by incorporating self-tuning, adaptive features capable of responding to changing process conditions. The hybrid design minimizes the drawbacks of each optimization approach while boosting their advantages.

The work enhances intelligent control systems and offers a strong basis for using AI-enhanced controllers in Industry 4.0 settings. The modular architecture allows for integration with existing industrial infrastructure and supports standard protocols. Future studies will focus on expanding the framework to more complex control scenarios and fortifying the theoretical underpinnings of stability guarantees in adaptive control systems.

8. References

[1] Zhao, D., Gao, C., Li, J., Fu, H., & Ding, D. (2025). PID control and PI state estimation for complex networked systems: A survey. International Journal of Systems Science, 1–16. DOI: 10.1080/00207721.2025.2456001.

[2] Lawrence, N. P., Damarla, S. K., Kim, J. W., Tulsyan, A., Amjad, F., Wang, K., Chachuat, B., Lee, J. M., Huang, B., & Gopaluni, R. B. (2024). Machine learning for industrial sensing and control: A survey and practical perspective. Control Engineering Practice, 145, 105841. DOI: 10.1016/j.conengprac.2024.105841.

[3] Bujgoi, G., & Sendrescu, D. (2025). Tuning of PID controllers using reinforcement learning for nonlinear system control. Processes, 13(3), 735. DOI: 10.3390/pr13030735.

[4] Raven, D. B., Chikkula, Y., Patel, K. M., Al Ghazal, A. H., Salloum, H. S., Bakhurji, A. S., & Patwardhan, R. S. (2024). Machine learning & conventional approaches to process control & optimization: Industrial applications & perspectives. Computers & Chemical Engineering, 189, 108789. DOI: 10.1016/j.compchemeng.2024.108789.

[5] Borase, R. P., Maghade, D. K., Sondkar, S. Y., & Pawar, S. N. (2021). A review of PID control, tuning methods and applications. International Journal of Dynamics and Control, 9(5), 818–827. DOI: 10.1007/s40435-020-00665-4.

[6] Ahmed, G., Eltayeb, A., Alyazidi, N. M., Imran, I. H., Sheltami, T., & El-Ferik, S. (2024). Improved particle swarm optimization for fractional order PID control design in robotic manipulator system: A performance analysis. Results in Engineering, 18, 103089. DOI: 10.1016/j.rineng.2024.103089.

[7] Bloor, M., Ahmed, A., Kotecha, N., Mercangöz, M., Tsay, C., & Del Rio Chanona, E. A. (2024). Control-informed reinforcement learning for chemical processes. Control Engineering Practice (in press). DOI: 10.48550/arXiv.2408.13566.

[8] Chen, L., Yin, Q., Zhao, J., Xia, H., & Gu, J. (2024). Accelerated optimization in deep learning with a proportional-integral-derivative controller. Nature Communications, 15, 10543. DOI: 10.1038/s41467-024-54451-3.

[9] Dogru, O., Velswamy, K., Ibrahim, F., Wu, Y., & et al. (2022). Reinforcement learning approach to autonomous PID tuning. Computers & Chemical Engineering, 161, 107760. DOI: 10.1016/j.compchemeng.2022.107760.

[10] Buchmeister, B., Polajnar, A., Palcic, I., Brezocnik, M., & Acko, B. (2010). Adaptive Control Systems for Responsive Factories. Proceedings of the 21st DAAAM International Symposium, B. Katalinic (Ed.), Published by DAAAM International, Vienna, Austria. Available: https://www.researchgate.net/publication/290075186_Adaptive_control_systems_for_responsive_factories

[11] Veselov, O., Veselov, A., & Svetushenko, S. (2019). The Use of Fuzzy Logic in Automatic Control Systems. Proceedings of the 30th DAAAM International Symposium, pp. 0897-0903, B. Katalinic (Ed.), Published by DAAAM International, Vienna, Austria. DOI: 10.2507/30th.daaam.proceedings.125

DOI: 10.2507/36th.daaam.proceedings.010

USING EDUSCRUM TECHNOLOGY FOR DEVELOPING ACADEMIC SPEAKING SKILLS OF STUDENTS MAJORING IN ENGINEERING

Oleg Ipatov, Darina Barinova, Sofia Saramakha,
Evgenia Tsimerman & Elena Krylova

Peter the Great St.Petersburg Polytechnic University

This Publication has to be referred as: "patov, O[leg]; Barinova, D[arina]; Saramakha, S[ofia]; Tsimerman, E[vgenia] & Krylova, E[lena] (2025). Using EduScrum Technology for Developing Academic Speaking Skills of Students Majoring in Engineering, Proceedings of the 36th DAAAM International Symposium, pp.0074-0081, B. Katalinic (Ed.), Published by DAAAM International, ISBN 978-3-902734-47-1, ISSN 1726-9679, Vienna, Austria
DOI: 10.2507/36th.daaam.proceedings.010

Abstract

The paper presents the didactic possibilities of EduScrum technology application in teaching as an effective tool for increasing students' motivation, improving the quality of learning, and integrating an interdisciplinary approach into learning. The purpose of the research is to create a theoretical and methodological basis for the formation of academic communication competencies among students in engineering fields of study, achieved by introducing the EduScrum methodology into the academic disciplines. A methodical analysis of the application of EduScrum technology tools was carried out and the tool that is the most effective for the development of communicative skills was singled out. Algorithms for using of the introduced innovative EduScrum technology for integral indicators of oral academic speech are given in the paper, the results of the experimental research have been analysed and conclusions have been drawn. EduScrum has a high degree of applicability for the development of meta-subject and professional competencies necessary for communication in the field of engineering technology. The article will serve the interest for those who research new trends in education as well as those who are interested in introducing integration of academic speaking skills development with the acquisition of professional terminology, including Agile and EduScrum concepts.

Keywords: EduScrum; engineers; higher education; agile methodology; project-based learning.

1. Introduction

Modern technologies are developing rapidly, and knowledge is rapidly becoming obsolete, especially in such dynamic areas where constant updating of skills is required at the junction of the humanities and digital sciences. In these conditions, a specialist who is able to quickly adapt to changes, master new tools and work in conditions of uncertainty is especially valuable – that is, he has flexible skills [1]. This work is an analysis of the objectives and principles of the Agile methodology (and specifically EduScrum), widely used in the field of IT and digital humanities for project management and teamwork. Studying these approaches within academic courses will contribute not only to enhancing communicative competencies but also to forming ideas about organizing effective project activities for future engineers.

Project-based learning and the method of projects are not fundamentally new in global educational practice. This teaching method was developed in the early 20th century in the USA by American philosopher and educator John Dewey [2], along with his disciple William H. Kilpatrick, and it was closely associated with the ideas of the humanistic movement in philosophy and education. According to J. Dewey's opinion, the learning process should be based on active participation of learners, where their personal interest in the subject matter plays a key role. Knowledge acquires value when it can be applied to solve real-world problems relevant to the student.

Students, in addition to developing technical and professional skills, acquire abilities necessary for working on projects—the so-called "21st-century skills." Lampert describes project-based learning as an approach grounded in the assumption that students are capable of utilizing a wide range of resources to create new meanings, ideas, and products. The Project-Based Learning (PBL) methodology is widely employed in many universities, and its effectiveness has already been studied. Zhang and Ma conducted research on the impact of PBL and concluded that its effectiveness depends on various factors such as region, discipline, course type, academic period, group size, and duration of the experiment [3]. They found that project-based learning particularly engages engineering and technological students, especially during laboratory sessions, in small groups of 4–5 people, and over periods lasting between 9 and 18 weeks. In our view, this is equally relevant for the specialization "computational linguistics," which lies at the intersection of linguistics and computer science. Students in this field work with real data, develop natural language processing algorithms, build machine learning models, requiring both technical skills and teamwork capabilities, the ability to analyze tasks from different perspectives, and adaptability to rapidly changing industry demands.

Traditional project-based learning typically follows a linear model, where students first immerse themselves in theoretical material before proceeding to planning, development, implementation, testing, and presenting the final result. Despite its logical structure, this approach has several drawbacks. Firstly, it often separates learning from practical context, reducing student motivation. Secondly, the rigid sequence of steps limits flexibility—errors made early may go unnoticed until too late, and adaptation to changed conditions or ideas becomes difficult. Additionally, students might struggle with engagement because feedback appears only near the end of the project.

To address these issues, new teaching methods have emerged, some inspired by predictive ("waterfall") or adaptive project management methodologies. In 2001, a group of 17 software developers published the "Manifesto for Agile Software Development," which outlined core values and principles of the Agile approach [4]. According to this manifesto, the main idea is to adopt iterative rather than extensive upfront planning in software projects. The term "Agile methodology" originates from the world of IT technologies. Translated from English, "agile" means "nimble, quick," but in Russian IT terminology, it has come to mean "flexible."

Agile methodology is a modern and flexible approach to organizing project activities, which is becoming increasingly popular in the educational environment, especially in areas such as engineering. In the context of rapid technological changes and constant updating of knowledge, Agile offers tools for step-by-step and adaptive learning, focusing on the value of the result, teamwork, feedback and self-organization. In addition, in the process of project activities, students learn future professional vocabulary. Agile gave rise to several tools (frameworks). A framework is a set of rules that helps organize the work of a project team according to the principles and values of Agile in order to create a specific product. For the field of education, the adaptation of Scrum technology became preferable.

The founders of Scrum are K. Schwaber and J. Sutherland, who argue that any organization across various fields can apply Scrum. K. Schwaber and J. Sutherland define Scrum as one of the tools of Agile methodology, where products are developed iteratively through a series of fixed-length iterations called "sprints" [5]. At the end of each sprint, progress becomes tangible, and using short sprints enhances the importance of good estimation and fast feedback. This technology helps understand what you want to achieve, how to achieve it, and how to track progress toward completing the project.

EduScrum (education + scrum) became the educational version of Scrum. The idea of creating an educational version of Scrum originated in 2011 thanks to W. Wijnands — a Dutch teacher who was inspired by a story about J. Sutherland's training session and the concept of applying Scrum in company operations. W. Wijnands wanted to change traditional learning systems and create an environment where learners would act as active participants rather than passive listeners and content consumers. He believed that focus should shift away from rigid plans towards flexibility, taking into account student feedback and individual characteristics (interests, difficulties, experiences) while aiming to reveal their strengths [6]. Emphasis is placed on maximizing value both in terms of specific subject outcomes and broader skills like organization, planning, collaboration, and teamwork.

It is important to note that Scrum does not replace project-based learning (Project-Based Learning); instead, they function as frameworks (platforms) for its organization. Scrum is used in conjunction with project-based learning (PBL), where the primary goal is to produce a final product (the result of the project method) by simulating teamwork in small groups following a flexible approach. Initially, Scrum was adopted in education within the field of software engineering. Later, it expanded to other STEM (Science, Technology, Engineering, Mathematics) disciplines, such as mathematics, and currently spans interdisciplinary areas. Since Scrum is implemented in diverse forms of education, its adoption is accompanied by modifications tailored to the unique features of the educational process in particular institutions or levels of education. It should be noted that the development of eduScrum technology in Russia began in 2019 under the supervision of official representatives—certified trainers. According to the official website eduScrum.ru, there are three internship platforms: schools in Saratov, Nizhny Tagil, and a gymnasium in Elektrostal. As reported by Skillbox Media journal in 2019, approximately 850–900 individuals attended Russian eduScrum trainings, with around 60% being school teachers, 30% university lecturers, and 10% corporate trainers [7].

Muller–Amthor et al. provided examples of technology adaptation for higher education [8]:

- Learning objectives are formulated as user stories included in the Product Backlog;
- Sprints can last a week or more, depending on the frequency of classes;
- Daily SCRUM sessions can take place without the physical presence of participants.

The main principles of the eduScrum educational framework are:

- Process transparency: significant aspects of the process should be agreed upon by all participants in the process and posted in a common access, which will allow all participants to have a common understanding of them. For example: terminology related to the process should be understandable to all participants; understanding of the project readiness criteria for the start should be the same for those who perform the work and those who accept it.
- Inspection (reviews): all participants in the eduScrum project should periodically independently inspect (inspect) the eduScrum elements and the progress towards achieving the educational goals to detect unwanted deviations.
- Adaptation: if the teacher or students determine that one or more aspects of the process are not feasible and/or the result will be unacceptable, it is necessary to partially adjust the plan or tasks in the work. Changes should be made as quickly as possible in order to limit further deviations.

The eduScrum trainer has a mixed role of Product Owner and eduScrum Master (with the goal of gradually handing it over to the learner team captain). As Product Owner and eduScrum Master, the trainer is responsible for defining the learning objectives to be achieved and assessing them; monitoring the eduScrum process; and facilitating the learning of individual team members; for example, by addressing subject matter, answering questions, and providing examples. During the learning process, no global goals are set, such as "learning a foreign language." Instead, the entire process is divided into sprints with deadlines established depending on the curriculum. Applying eduScrum in foreign language classes offers the following advantages:

- Developing independence in students: self-education, self-reflection, and self-control.
- Enhancing students' "4C" skills: critical thinking, creativity, team communication, and coordination.
- Boosting student motivation by enabling experiments rather than strict adherence to lesson plans.
- Providing the opportunity to use the foreign language directly in communication with team members.
- Organizational methods for this technology are implemented step-by-step, allowing complex tasks to be broken down into smaller sprints. By dividing large goals into short sprints, even in a multifaceted process like learning a foreign language, progress can be seen clearly.
- Enabling the integration of cross-disciplinary knowledge, which is particularly relevant for teaching foreign language for special purposes.

Engineering students will need to present their developments, participate in international conferences, defend startup projects, and conduct technical demonstrations in the future. To successfully interact in these formats, they must not only master academic vocabulary but also navigate modern IT-project design terminology, including concepts related to Agile and its Scrum framework. Understanding this terminology should become part of the specialist's active academic lexicon. The authors of this study conducted a pedagogical experiment to test the hypothesis: the process of developing academic spoken foreign language proficiency among students majoring in computational engineering will be more effective when using the EduScrum technology, since it promotes active involvement of learners in communication-oriented project activities involving regular use of current professional terminology and situations close to future professional contexts.

The authors examined the didactic potential of implementing the eduScrum technology in foreign language instruction for professional purposes as an approach based on Agile principles adapted to the educational environment. The main components of the framework, its structure, implementation stages, participant roles, and fundamental values underlying the methodology were analyzed, as well as opportunities for applying the technology in teaching English for professional purposes, particularly for preparing engineering students for communication in a professional setting. It was concluded that the use of eduScrum contributes to the formation of metacognitive and professional skills among learners, such as self-regulation, teamwork, critical thinking, and adaptability. Furthermore, eduScrum represents an effective tool for increasing student motivation, improving the quality of material assimilation, and integrating an interdisciplinary approach into education. In the context of modern language education, the main focus is the integration of innovative educational technologies. EduScrum, based on the principles of the flexible Agile methodology, can be effectively used in teaching a foreign language for special purposes to future engineers. Analysis of the framework structure, its stages and the roles of the participants allows us to conclude that it is highly applicable for developing students' meta-subject and professional competencies necessary for communication in the field of information technology. The EduScrum methodology promotes the formation of self-regulation skills, teamwork and critical thinking and creates conditions for the implementation of interdisciplinary interaction.

2. Methods

The data for the research were collected through the implementation of an experimental teaching process, where the effectiveness of the EduScrum technology was assessed. To address the set objectives the following research methods were chosen:

1. Theoretical Methods were employed during the study and analysis of psychological, pedagogical, methodological, and didactic literature, both domestic and foreign, as well as the experience of leading pedagogical researchers. Additionally, this stage involved familiarizing with Federal State Educational Standards (FSES) and other relevant documentation.
2. Experimental Methods included a pedagogical experiment, surveys.
3. Observational Methods involved observing the process and outcomes of students' learning of the course content.
4. Statistical Methods: Both quantitative and qualitative analyses of the collected data were carried out.

3. Experiment

The experiment was conducted at Peter the Great St. Petersburg Polytechnic University. The experiment involved second-year undergraduate students (24 people) in the field of study (specialty) 15.05.01 Design of Technological Machines and Complexes, studying within the framework of the discipline "Foreign Language: Professionally Oriented Course". The students were divided into an experimental (12 people) and a control (12 people) group. The level of foreign language proficiency of the students in the experimental and control groups was known to us in advance; this level is B2–C1, Upper–Intermediate – Advanced according to the CEFR. To visualize the results of EduScrum's work in the educational process, the Russian online board Holst was chosen, which provides a convenient and functional interface for teamwork. Holst allows you to create interactive boards on which students can visualize the stages of completing tasks, distribute roles in the team, record key ideas and adjust the work plan in real time. The experimental training took place in 2025 for four weeks. The number of classroom hours allocated to studying the section on academic oral speech was the same for both groups: four academic hours devoted to teaching oral academic speaking skills using EduScrum technology. In addition, the students were given two academic hours to prepare a presentation before the control stage of the experiment during independent work. Students in the experimental group studied using EduScrum technology.

To implement the experiment, a set of exercises was prepared (using EduScrum technology) aimed at mastering speaking skills in an academic environment, with seven assessment criteria:

- Criterion 1. Level of speech fluency;
- Criterion 2. Level of consistency and coherence of speech;
- Criterion 3. Level of proficiency in the lexical aspect of the language;
- Criterion 4. Level of proficiency in the grammatical aspect of the language;
- Criterion 5. Adequate development of the presentation topic;
- Criterion 6. Use of vocal characteristics;
- Criterion 7. Use of visual accompaniment;
- Criterion 8. Use of non-verbal behavior (gestures, facial expressions, posture, etc.);
- Criterion 9. Implementation of communicative intent.

The following table outlines the stages of work using the EduScrum methodology, their alignment with selected evaluation criteria, as well as the instructional materials and exercises utilized in the course.

Skills	Evaluation Criteria	Exercises	Stage
Consistently and logically convey information	Level of coherence and logic in speech	• Listening comprehension with content analysis • Focused listening on structure and content of presentations; • Analysis of text coherence and consistency for speeches; • Sentence rearrangement to improve logical flow; • Rewriting paragraphs to enhance coherence and logic	Sprint 1 Listening
Use varied and accurate lexical resource	Mastery of lexical aspect of language	• Text analysis demonstrating frequently used linguistic units in public speaking; • Reading comprehension focusing on hypotheses; • Summarization of key points from texts	Sprint 2 Reading
Speak fluently and spontaneously with clear structure without excessive repetitions or corrections	Fluency level	Practice spontaneous oral speech using structural elements	Sprint 3 Presentation Aspects
Apply grammatical constructions accurately	Command of grammar aspects of language	• Analysis of grammatical structures suitable for oral presentations.	Sprint 3 Presentation Aspects

		• Matching exercise linking grammatical unit, example, purpose in text	
Logically and thoroughly expand upon topic	Adequate development of presentation theme	• Draft writing (Draft Writing) • Editing and refining (Editing and Refining)	Sprint 3 Presentation Aspects
Utilize pace, intonation, volume effectively	Use of vocal characteristics	Analyzing presentations to identify appropriate nonverbal strategies for public speaking	Sprint 3 Presentation Aspects
Effectively utilize visual aids	Effective use of visual support	Materials providing examples of correct slide layouts for presentations	Sprint 3 Presentation Aspects
Employ nonverbal behavior appropriately	Physical behavior usage	Authentic video analysis to determine appropriateness of voice intonations, gestures, facial expressions	Sprint 3 Presentation Aspects
Answer questions, engage in discussion	Realization of communicative intent	• Role-play: Preparing questions and response templates. • Small presentation preparation by students on selected topics	Sprint 4 Presentation

Table 1. Stages of work using EduScrum technology aimed at teaching oral academic speech

During the scientific experiment, students work in teams to prepare and deliver a presentation on a given topic. The process is divided into sprints, each of which is aimed at developing specific language and communication skills. Each sprint is accompanied by EduScrum elements: planning, completing tasks, discussing progress, demonstrating results, and reflecting.

The experiment was conducted in five stages:

1. preparatory
2. preliminary;
3. formative;
4. intermediate;
5. control.

During the preparatory stage, the experimental and control groups were determined, a set of exercises aimed at developing academic speaking skills was developed, and tasks were prepared on the Holst online digital resource "eduScrum–board" that the team needed to complete to achieve the educational goal. At the preliminary stage, students' presentations were heard on topics included in the list of practical classes conducted within the framework of the discipline "Foreign Language: Professionally Oriented Course". The input results were recorded. Then the students were divided into teams for project activities, and roles were assigned within the teams. At the formative stage, 4 classes were held, each lasting one academic hour, with each of the small groups of students included in the experimental group within the framework of our study. During the classes, a set of exercises developed by the authors using the EduScrum technology was used and an attempt was made to develop the skills of giving an oral presentation on scientific topics in English. With the students of the control group, a set of exercises related to scientific presentations was used, using the existing textbook on the discipline program without the use of the EduScrum technology. During the intermediate stage, students made presentations in their groups in accordance with Sprint 4. Students answer questions for group reflection (teamwork) and for personal reflection, using a specially developed questionnaire according to the EduScrum technology. During the control stage, we listened to students' oral presentations on one of the proposed topics. Then the level of formation of criteria related to academic speaking skills was assessed. At the end of the experiment, the obtained data were statistically processed and analyzed. The assessment was carried out on a three-point scale. The gradation is shown in Table 2.

Criteria	**Satisfactory**	**Good**	**Excellent**
Speech coherence and sequence	Speech is often incoherent, making it difficult to follow the logic with unclear transitions.	Speech is coherent but occasionally lacks smoothness or clarity in transitions between ideas.	Speech is clear, logical, and smoothly transitions from one idea to another without significant disruptions.
Fluency level	Slow speech with long pauses and frequent difficulties.	Speech has pauses and repetitions but remains generally comprehensible.	Fluent and confident speech almost devoid of pauses, easily understood.
Lexical aspect proficiency	Limited vocabulary using simple and repetitive words.	Vocabulary expands but still limited in complex contexts, frequently uses basic words.	Broad lexical repertoire with correct and varied use of words and expressions, minimal errors.

Grammatical aspect proficiency	Frequent grammatical errors and limitation in constructions usage.	Few grammatical errors that do not significantly affect understanding; variety of grammatical structures used.	Almost error-free grammar with diverse grammatical constructs used with high precision.
Adequate development of presentation topic	Topic either not covered or very superficially developed, weak idea development.	Topic adequately addressed but material not fully connected.	Well-developed theme with precise and deep analysis, all aspects thoroughly explored.
Use of vocal characteristics (tempo, tone, volume)	Exceptional use of tempo, intonation, and volume to maintain interest; natural and lively speech.	Acceptable voice usage; minor weaknesses that don't greatly impact perception.	Frequent mistakes in tempo, volume, and tone reducing speech quality.
Use of visual aids	High-quality and diverse supporting materials enhancing credibility and clarity of the subject.	Supporting materials logically linked to thesis make speech more engaging.	Poorly made materials weakly related to thesis, negatively affecting speech perception.
Nonverbal behavior (gestures, facial expression, posture)	Outstanding posture, gestures, facial expressions, eye contact, and appearance complementing verbal message.	Acceptable use of gestures, facial expressions, posture; slight mismatches don't hinder perception.	Bad posture, distracting gestures and facial expressions reduce trustworthiness and impair perception.
Achievement of communicative intent	Difficulty answering questions, incomplete or off-topic responses.	Understandable answers though may be incomplete or somewhat inaccurate.	Precise, complete, detailed answers free of errors covering all aspects clearly.

Table 2. Criteria for assessing academic speaking skills

The maximum points that a student could score was 45. In total, it was proposed to assess 9 components of competence in the field of public academic speaking. Figure 1 presents comparative results of the assessment of academic oral speech of students from two groups according to nine criteria during preliminary assessment.

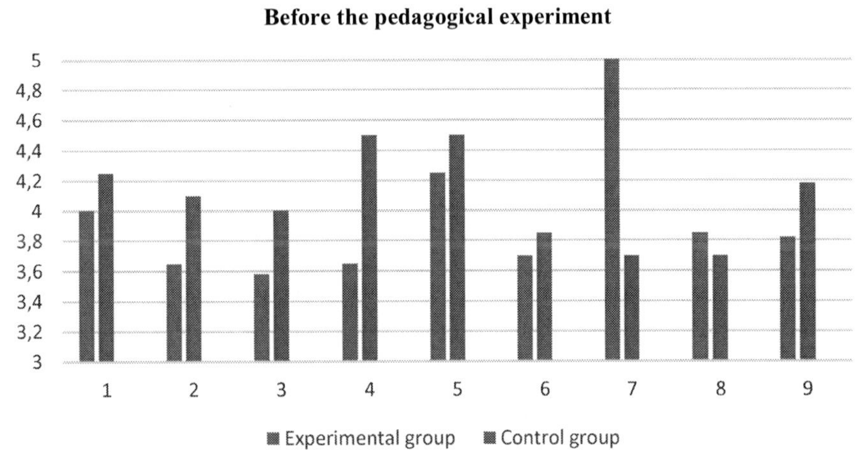

Fig. 1. Comparative results of the assessment of academic oral speech during preliminary assessment

From the analysis of the input survey, it was found that the students in the control group demonstrated higher results in almost all the criteria of academic oral speech. This is especially noticeable in such parameters as the level of consistency and coherence of speech, proficiency in the lexical and grammatical aspects of the language, as well as the development of the presentation topic. In addition, the control group showed better results in the use of voice characteristics and the implementation of communicative intention. The experimental group, on the contrary, demonstrated less pronounced indicators at the initial stage of the experiment, which may indicate a lower starting level of training. However, for some criteria - for example, the use of physical behavior and visual accompaniment - the difference between the groups was insignificant. We concluded that at the input stage the control group had a higher level of proficiency in academic oral speech, which creates favorable conditions for further monitoring the dynamics of the development of these skills as a result of the use of EduScrum technology in the experimental group.
Figer 2presents the students' results for nine criteria after the pedagogical experiment.

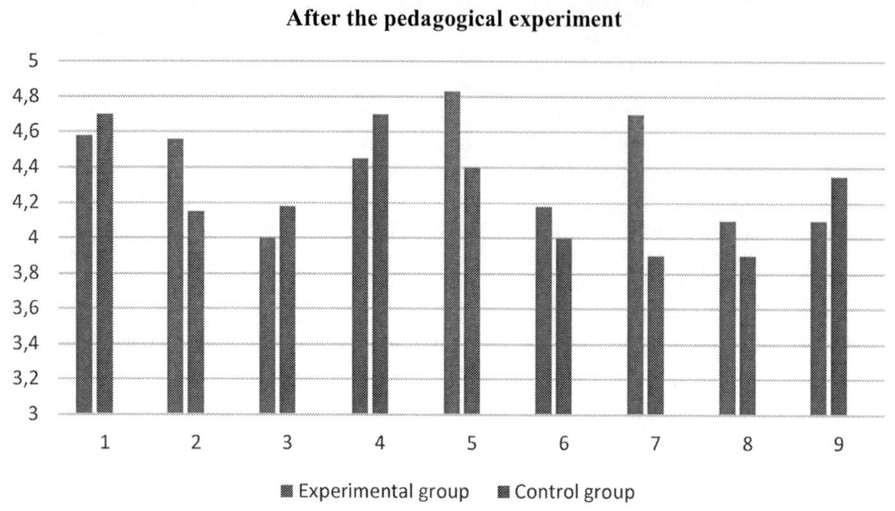

Fig. 2. Comparative results of the assessment of academic oral speech after pedagogical intervention

From the above data, reflecting the level of academic oral speech of students after the pedagogical experiment, it is evident that the indicators of students in the experimental group for most criteria exceed the results of the control group. Therefore, it indicates the effectiveness of the application of EduScrum technology. The next task involves determining whether differences have emerged between initially matched experimental and control groups after conducting sessions utilizing the EduScrum technology within the experimental group. To achieve this goal, the chi-square test (a statistical criterion facilitating conclusions regarding how random coincidences and discrepancies are in the data) was used. In this research, an ordinal scale with three gradation levels was applied for assessing the degree of formation of criteria: a satisfactory result, a good result, and an excellent result. A significance level of $\alpha = 0.05$ as the critical value was chosen, meaning that the acceptable probability of error does not exceed 5%. The critical value of the criterion for the existing number of gradations and given significance level equals 5.99. Figure 3 presents the formula for calculating the empirical value for the selected chi-square criterion.

$$\chi^2_{emp} = N \cdot M \cdot \sum_{i=1}^{L} \frac{\left(\frac{n_i}{N} - \frac{m_i}{M}\right)^2}{n_i + m_i}$$

Fig. 2. Comparative results of the assessment of academic oral speech after pedagogical intervention

The following notations are used in this formula: N – number of members of the experimental group; M – number of members of the control group; L – number of gradation levels; n_i – number of members of the experimental group who received the $_i$ score; m_i – number of members of the control group who received the $_i$ score. The obtained data are presented in Table 3.

Criterion	**Empirical value of the χ^2 criterion when comparing experimental and control groups**	
	χ^2 before the experiment	**χ^2 after the experiment**
Speech fluency	6.316	2.182
Coherence and sequencing	3.429	0.178
Lexical aspect	5.474	4.444
Grammatical aspect	9.600	1.510
Adequate development of presentation topic	1.600	12.000
Vocal characteristics	0.889	2.182
Use of visual aids	0.000	12.267
Use of physical behavior	0.889	2.000
Realization of communicative intention	4.000	0.202

Table 3. Calculation of the χ^2 criterion value

Having analyzed the obtained results, it can be noted that before the experiment, the values of the χ^2 criterion for all diagnostic parameters, with the exception of the grammatical aspect ($\chi^2 = 9.600$) and speech fluency ($\chi^2 = 6.316$), were below the critical value of 5.99 at a significance level of $\alpha = 0.05$. This indicates the absence of statistically significant differences between the experimental and control groups, which confirms their comparability and the correctness of the conditions of the pedagogical experiment. After the experiment, significant differences between the groups were recorded for the following criteria: "Adequate development of the presentation topic" ($\chi^2 = 12.000$) and "Use of visual accompaniment" ($\chi^2 = 12.267$). These values significantly exceed the critical threshold and indicate a reliable improvement in the students of the experimental group. Positive changes are observed for the remaining criteria, but their differences from the results of the control group did not reach the level of statistical significance.

Thus, it can be concluded that the students included in the experimental group demonstrated higher results in a number of key parameters characterizing the level of development of academic oral speech skills. The developed set of exercises has proven its effectiveness and can be recommended for use in the educational process. All participants in the experiment demonstrated, to one degree or another, the development of skills necessary for a successful presentation with a scientific presentation. Consequently, the developed set of exercises using the EduScrum technology can be successfully used in the learning process.

4. Conclusion

The integration of EduScrum into the educational process contributed to active student engagement in project activities and fostered key skills such as teamwork, independent planning, analysis, and reflection. Working in sprints allowed for logical structuring of learning, sequentially developing academic oral communication skills. Elements of EduScrum like planning, visualization of goals, regular feedback, and team evaluation ensured a high level of involvement. The conducted pedagogical experiment demonstrated statistically significant improvement in the level of students' academic oral skills in the experimental group taught through EduScrum compared to the control group. This confirms the hypothesis that using EduScrum enhances effective development of academic foreign language speaking abilities. The obtained results indicate the feasibility of implementing this technology in university teaching practice, especially for engineering students. The paper is useful to researchers dealing with pedagogics, teaching methodology and linguistics. The theoretical significance of the study is due to the possibility of further use of the obtained results as recommendations to teachers to use the EduScrum technology in the educational process.

5. References

[1] Ipatov, O.; Barinova, D.; Ni, O.; Saramaha, S. & Odinokaya, M. (2022). New Teaching Approaches in Conditions of Globalisation in Digitized Higher Engineering Education, Proceedings of the 33rd DAAAM International Symposium, pp.0215-0220, B. Katalinic (Ed.), Published by DAAAM International, ISBN 978-3-902734-36-5, ISSN 1726-9679, Vienna, Austria DOI: 10.2507/33rd.daaam.proceedings.030

[2] Lampert, M. (2010). Learning teaching in, from, and for practice: What do we mean? Journal of Teacher Education, 61(1–2). p.21–34

[3] Zhang, L., & Ma, Y. (2023). A study of the impact of project–based learning on student learning effects: A metaanalysis study. Frontiers in Psychology, 14, 1202728. https://doi.org/10.3389/fpsyg.2023.1202728.

[4] Beck, K. (2001). The Agile Manifesto. Agile Alliance. http://agilemanifesto.org/

[5] Sutherland, J. (2014). Scrum: the art of doing twice the work in half the time. New York, NY: Crown Publishing Group.

[6] Wijnands, W., & Stolze, A. (2019). Transforming education with eduScrum. InD. Parsons, K. MacCallum (Eds.), Agile and Lean Concepts for Teaching and Learning 100 P. Singapore: Springer. doi:10.1007/978–981–13–2751–3_5. P.100

[7] https://eduscrum.ru (2025). EduScrum Education. Accessed on: 2025-08-30

[8] Müller–Amthor, M., Hagel, G., Gensheimer M. & Huber, F. (2020). Scrum Higher Education – The Scrum Master Supports as Solution–focused Coach. IEEE Global Engineering Education Conference (EDUCON), Porto, Portugal, 2020. pp. 948–952, https://doi.org/10.1109/EDUCON45650.2020.9125304

DOI: 10.2507/36th.daaam.proceedings.011

THE INFLUENCE OF ANNEALING ON THE MECHANICAL PROPERTIES OF FDM 3D PRINTED PET REINFORCED WITH CARBON FIBERS

Adi Pandzic, Damir Hodzic & Ermin Demirovic

This Publication has to be referred as: Pandzic, A[di]; Hodzic, D[amir] & Demirovic, E[rmin] (2025). The Influence of Annealing on the Mechanical Properties of FDM 3D Printed PET Reinforced with Carbon Fibers, Proceedings of the 36th DAAAM International Symposium, pp.0082-0094, B. Katalinic (Ed.), Published by DAAAM International, ISBN 978-3-902734-47-1, ISSN 1726-9679, Vienna, Austria
DOI: 10.2507/36th.daaam.proceedings.011

Abstract

This study investigates the influence of the annealing process on the mechanical properties of FDM 3D-printed PET reinforced with carbon fibers (PET-CF), with a particular focus on tensile strength, flexural strength, and the corresponding elastic moduli. Specimens were thermally treated at three different temperatures (80 °C, 130 °C, and 180 °C) and for three different durations (30 min, 120 min, and 210 min). The obtained results were compared with the material specifications provided by the manufacturer to determine the conditions under which the mechanical properties are maximized. To assess the significance of temperature and annealing time on the material behavior, an analysis of variance (ANOVA) was conducted. This statistical method enabled the identification of key factors and their interactions affecting the tensile and flexural mechanical properties of the PET-CF material. Based on the experimental data, predictive mathematical models were developed to describe the material's behavior as a function of annealing temperature, time, and their interaction.

Keywords: FDM; PET-CF; 3D printing; annealing; ANOVA.

1. Introduction

In modern manufacturing environments, a wide range of technologies are used to produce functional products. Traditional approaches such as subtractive manufacturing, which removes material from a solid block, and formative manufacturing, which reshapes materials using force or heat without adding or removing material, have been foundational for decades. However, additive manufacturing (AM) has recently emerged as a transformative technology based on the principle of adding material layer by layer [1], [2]. Additive Manufacturing, commonly referred to as 3D printing, encompasses a set of technologies that build physical, three-dimensional objects from digital CAD (Computer-Aided Design) models. These models are translated into layers, which are deposited sequentially, offering unique advantages such as design freedom, material efficiency, and fast prototyping. AM supports the use of advanced materials, enables the production of complex geometries, and significantly reduces lead time and material waste. Consequently, AM has found wide application in aerospace, medical, automotive, architectural, and consumer product industries [3], [4], [5].

Among AM technologies, Fused Deposition Modeling (FDM), also known as Fused Filament Fabrication (FFF), is the most widely used. First patented in 1989 by Stratasys, FDM belongs to the material extrusion category of AM technologies. In this process, thermoplastic filament is heated in a print head and extruded through a nozzle in a semi-molten state to form layers along a predefined path generated by slicing software (e.g., Ultimaker Cura) [6]. The workflow typically starts with the creation of a 3D CAD model using software like SolidWorks, CATIA, or Autodesk Fusion 360, or through 3D scanning of a physical object (Fig. 1). The model is converted into the STL format and imported into slicer software, which divides it into layers and generates G-code instructions that control the printer's movements [7], [8]. Once the G-code is transferred to the printer, and both the nozzle and build platform reach the desired temperatures, the material is extruded layer by layer until the part is complete. FDM printers are user-friendly and capable of printing multiple parts with varying infill patterns and geometries in a single operation. However, challenges include internal porosity, surface roughness, interlayer adhesion issues, and anisotropy, especially in the Z-direction, which can lead to delamination [9].

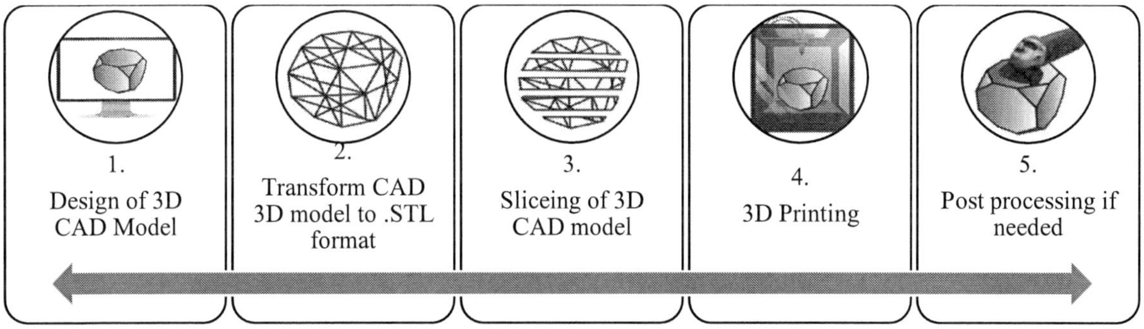

Fig. 1. The principle of product manufacturing using additive technologies [5]

These limitations can be mitigated by optimizing print parameters, but further improvements can be achieved using fiber-reinforced composite filaments. One of the most promising materials is polyethylene terephthalate reinforced with carbon fibers (PET-CF). Carbon fibers significantly enhance the mechanical properties of thermoplastics while maintaining low density. PET-CF filaments are produced by combining PET as the matrix and chopped carbon fibers as reinforcement, through an extrusion process that ensures homogeneous distribution of fibers within the polymer matrix [10]. The resulting composite exhibits high tensile strength, stiffness, and thermal resistance, capable of withstanding temperatures up to 76 °C. These properties make PET-CF suitable for functional prototyping and end-use applications in the automotive, aerospace, and tooling industries. Due to its high wear resistance and rigidity, PET-CF is also used to fabricate replacement parts on demand, reducing downtime in industrial operations. Ultimaker, a leading 3D printer manufacturer, states that PET-CF properties can be further enhanced by thermal post-processing such as annealing [11].

Annealing is a thermal treatment process traditionally used in metallurgy to reduce internal stresses, improve ductility, and increase structural uniformity. When applied to polymer materials, the principle remains similar. In injection molding, for example, controlled cooling improves part strength and thermal resistance. The same technique can be applied to FDM-printed components, which are prone to internal stress and poor interlayer bonding due to rapid cooling between deposited layers [12], [13]. In FDM, layers solidify immediately upon deposition, often resulting in stress accumulation, poor layer adhesion, microcracks, and dimensional distortion. Smaller features cool faster than larger ones, leading to thermal gradients and localized shrinkage. Moreover, the layered structure inherently contains microvoids and imperfections that reduce load-bearing capacity compared to injection-molded counterparts [13], [14]. Annealing improves mechanical and thermal properties by relieving internal stresses, enhancing interlayer diffusion, and promoting the formation of stable crystalline regions. The key parameters in the annealing process are temperature, exposure time, and the cooling method. The annealing temperature must lie between the material's glass transition temperature (Tg) and melting temperature (Tm) to enable molecular mobility without melting the polymer. Prolonged exposure at optimal temperature facilitates molecular reorganization and interlayer fusion [13].

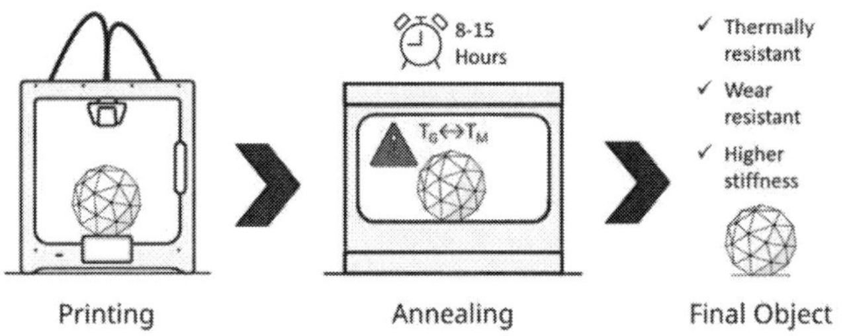

Fig. 2. Schematic representation of the annealing process applied to FDM 3D printed PET-CF material [14]

Controlled cooling, whether inside the oven or in ambient air, is essential for minimizing warping and dimensional inaccuracies. Gradual temperature decline promotes molecular alignment into energetically favorable crystalline patterns, enhancing density and structural integrity. However, excessive crystallization may reduce material toughness, and uneven heating can lead to internal stress buildup, deformation, or degradation of the polymer chains [13], [15]. While annealing offers notable advantages, such as increased stiffness, thermal resistance, and dimensional stability, it must be carefully controlled to avoid brittleness and geometric distortion. In this study, a detailed experimental investigation was carried out to evaluate the influence of annealing temperature and time on the mechanical properties of FDM-printed PET-CF composite material. Samples were thermally treated at three different temperatures (80 °C, 130 °C, and 180 °C) and for three different durations (30, 120, and 210 minutes). The focus was placed on analyzing the resulting changes in tensile and flexural strength, as well as in the corresponding modulus of elasticity. Based on the experimental data, predictive mathematical models were developed to describe the relationship between annealing conditions and mechanical properties, contributing to optimized post-processing strategies for FDM 3D printed PET-CF materials.

2. Literature review

Several studies have investigated how annealing affects the mechanical behavior of FDM-printed materials. Annealing has been shown to improve tensile and flexural strength, hardness, and dimensional stability, although outcomes depend on the material type, annealing temperature, and exposure time. Rengiseti et al. [13] studied the mechanical behavior of PLA, ABS, PETG, and their carbon-fiber-reinforced variants (PLA-CF, CF-ABS, and PETG-CF). All samples were annealed at a temperature 5 °C above their glass transition temperature (Tg) for 60 minutes. The samples were heated in a cold oven to the target temperature, then allowed to cool inside the oven after heating. PETG and PETG-CF showed significant improvements in tensile strength and Young's modulus, while flexural strength slightly decreased. This suggests a complex balance between improved stiffness and reduced flexibility due to heat treatment [16].

Bart DE Wijk [12] evaluated the effects of annealing on PLA and PET-CF materials printed under controlled conditions. PLA samples were annealed at 80 °C and 110 °C, and PET-CF samples at 120 °C, 170 °C, and 220 °C, each for a duration of 2 hours. PET-CF specimens annealed at lower temperatures showed up to an 18.5% increase in flexural strength, while higher temperatures led to increased stiffness and thermal resistance. However, interlayer adhesion decreased at higher temperatures, indicating a potential trade-off between thermal performance and structural integrity [13]. Valvez et al. [14] examined PETG and its reinforced versions (PETG-CF and PETG-KF), using a B2X300 printer and a 0.6 mm nozzle. Samples were annealed at 90 °C, 110 °C, and 130 °C for 30, 240, and 480 minutes. Increased temperature and duration generally enhanced flexural strength and elastic modulus, but dimensional deformation was also observed. For parts where dimensional precision is critical, careful selection of annealing parameters is necessary [17].

Stojković et al. [15] investigated the effects of annealing temperature, exposure time, and print layer height (0.1 mm, 0.2 mm, 0.3 mm) on PLA, PETG, and PETG-CF. Samples were annealed at temperatures ranging from 60 °C to 100 °C for 30, 60, and 90 minutes. The results showed that increasing layer height reduced tensile strength, while longer annealing times improved mechanical properties, especially for PETG-CF. This material also showed the smallest dimensional change and highest modulus of elasticity, indicating its robustness under thermal treatment [18].

Sathish Kumar et al. [16] studied the influence of infill density and annealing on PETG and PETG-CF samples. Infill densities of 25%, 50%, 75%, and 100% were analyzed. Samples were annealed at temperatures 5 °C above their Tg for 60 minutes. Increasing the infill density resulted in higher hardness, tensile and flexural strength, and toughness. For PETG-CF with 100% infill, flexural strength increased by 18% and tensile strength by 25%, compared to annealed PETG, confirming the benefit of carbon fiber reinforcement and thermal treatment [19].

Across these studies, researchers have explored how annealing improves mechanical properties in various thermoplastic composites, including PLA, PETG, ABS, and their carbon and Kevlar-reinforced variants. A common approach involves combining annealing with optimization of print parameters such as infill density and layer height. While improvements in tensile and flexural properties are well documented, outcomes can vary significantly depending on the specific material formulation and annealing protocol. Building upon this background, the current study focuses on Ultimaker's PET-CF composite filament and aims to determine the optimal annealing parameters—temperature and time—for improving tensile and flexural mechanical properties. Experimental tests were conducted using nine different annealing conditions to evaluate mechanical behavior and develop mathematical models that correlate treatment conditions with performance outcomes. This approach enables a more systematic understanding of post-processing effects on high-performance 3D printed composites.

3. Methodology

The experimental part of this research was conducted to investigate the influence of annealing temperature and time on the mechanical properties of FDM 3D printed PET-CF material. The experiments were fully carried out in the Laboratory for Testing of Polymeric Materials of the Kingdom of Norway (IPMlab), located at the Faculty of Mechanical Engineering, University of Sarajevo. Samples were subjected to thermal treatment at three different temperatures and three different time intervals, followed by tensile and flexural mechanical testing. The complete procedure is illustrated in Figure 3.

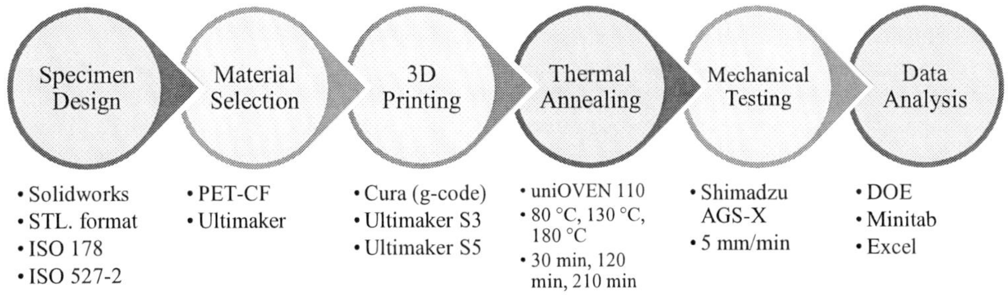

Fig. 3. Overview of the experimental procedure

Test samples for evaluating tensile and flexural mechanical properties were designed using SolidWorks 2025 software. The geometry for tensile testing followed the ISO 527-2 standard [20], while the geometry for flexural testing was based on ISO 178 [21]. The dimensional characteristics of both types of specimens are shown in Figure 4.

Fig. 4. Geometric characteristics of tensile and flexural test specimens

The CAD models were exported to STL format for slicing and G-code generation. Model preparation for 3D printing was performed using UltiMaker Cura 5.6.0. For the Ultimaker PET-CF (black) filament, predefined parameters within Cura (Normal Profile 0.15 mm, Shell Thickness 1.8 mm, Infill Density 100%) were used to ensure minimal influence of slicing settings on the final mechanical results. The printing parameters used are listed in Table 1.

3D Printing Parameters	Value
Nozzle diameter, [mm]	0,6
Layer height, [mm]	0,15
Infill density, [%]	100%
Printing temperature, [°C]	270
Build plate temperature, [°C]	80
Printing speed, [mm/s]	25
Fan speedora, [%]	10

Table 1. 3D printing parameters for PET-CF specimens

After slicing, all samples were printed in a flat orientation on desktop FDM 3D printers: Ultimaker S3 and Ultimaker S5. A total of 100 samples were printed—50 for flexural testing and 50 for tensile testing. After 3D printing, the specimens were annealed using a uniOVEN 110 laboratory oven, shown in Figure 5.

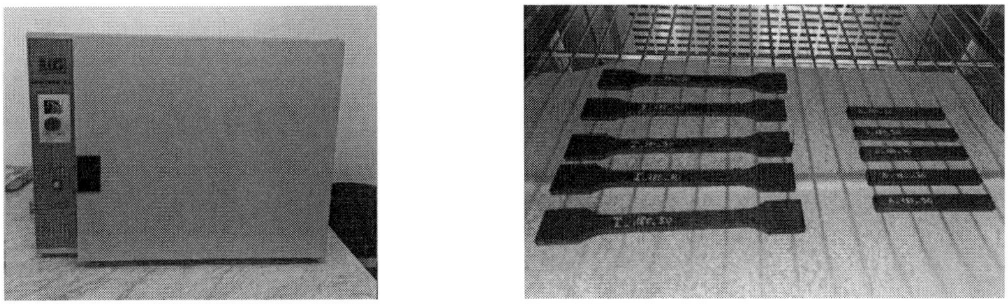

Fig. 5. Laboratory oven uniOVEN 110 used for thermal annealing and test specimens inside

For both tensile and flexural tests, 45 samples were thermally treated under nine different combinations of temperature (80 °C, 130 °C, and 180 °C) and time (30 min, 120 min, and 210 min). For each combination, five specimens were used. Additionally, five untreated samples for each test type were preserved as a control group to evaluate the influence of annealing by comparison. After the annealing, the samples were removed from the oven and left to cool at room temperature. Mechanical testing of the annealed and non-annealed samples was performed using a Shimadzu AGS-X universal testing machine, with a maximum force capacity of 10 kN. Tensile testing was conducted in accordance with ISO 527-2, while flexural testing followed ISO 178. The crosshead speed for both types of testing was set to 5 mm/min, as shown in Figure 6.

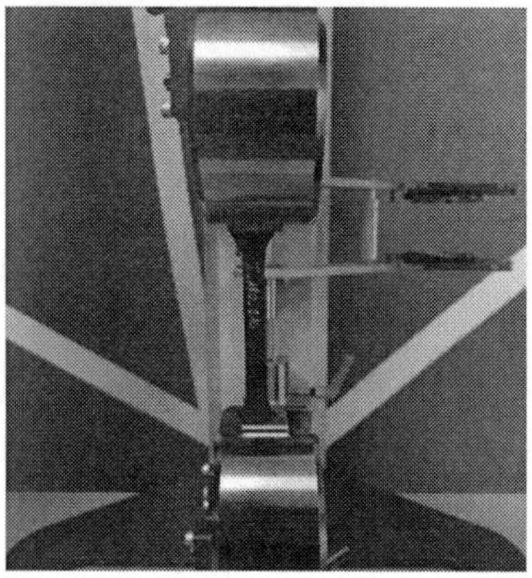
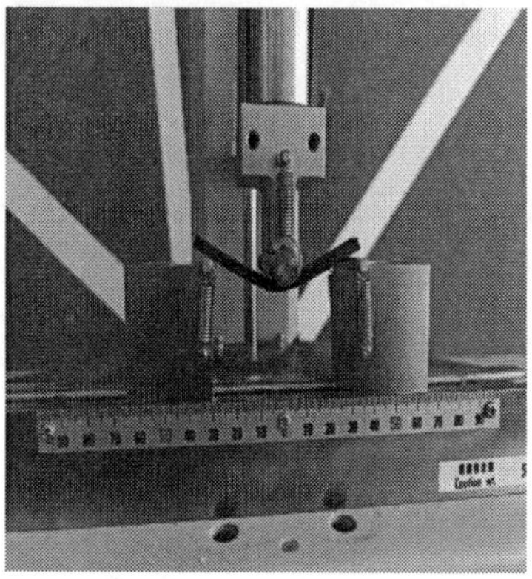

Fig. 6. Tensile (left) and flexural (right) testing setup

Test data acquisition and logging were performed using the TrapeziumX software, developed by Shimadzu. Initially, the mechanical results of the untreated specimens were compared to the values provided in the material's Technical Data Sheet (TDS) to verify the consistency of the printed specimens with the manufacturer's specifications. Subsequently, the results of the annealed samples were compared to the untreated control group to evaluate the effect of annealing on the mechanical properties. Collected data were analyzed statistically using Python programming language along with Microsoft Excel and Minitab software. The results and discussion derived from this analysis are presented in the following chapters.

3.1. Material

The material used in this study is UltiMaker PET CF — a carbon fiber-reinforced version of polyethylene terephthalate (PET), designed for high-performance FDM 3D printing. It is supplied as 2.85 mm filament spools (750 g) and is optimized for use with UltiMaker S series printers. UltiMaker PET CF combines the easy printability of PETG with significantly improved stiffness, dimensional stability, and chemical and thermal resistance. After annealing, it can withstand temperatures up to 180 °C, making it suitable for functional prototypes, tools, replacement parts, and manufacturing aids. Compared to other composites like PA CF, PET CF offers better ease of use, lower warpage, excellent surface finish, and reliable print quality. In this work, all specimens were printed using predefined settings in UltiMaker Cura 5.3 (layer height 0.15 mm, infill 100%, nozzle 0.6 mm) and conditioned at room temperature for 24 hours before testing. All samples were printed individually using new filament spools on UltiMaker S7 with the UltiMaker Material Station to ensure dry storage and optimal results. Table below present the mechanical properties of UltiMaker PET CF before and after annealing, as specified by the manufacturer.

Mechanical properties	**Method**	**No annealing**	**After annealing**
Tensile (Young's) modulus	ASTM D3039 (1mm / min)	4342 ± 89 MPa	5530 ± 124
Tensile stress at yield	ASTM D3039 (5 mm / min)	50.6 ± 0.6 MPa	-
Tensile stress at break	ASTM D3039 (5 mm / min)	75.2 ± 0.8 MPa	72,8 ± 2,0
Elongation at yield	ASTM D3039 (5 mm / min)	3.9 ± 0.1%	-
Elongation at break	ASTM D3039 (5 mm / min)	5.5 ± 0.6%	3,4 ± 0,2
Flexural modulus	ISO 178 (1 mm / min)	5743 ± 150 MPa	6280 ± 114
Flexural strength	ISO 178 (5 mm / min)	102.8 ± 2.6 MPa at 4.4%	136,6 ± 2,8
Flexural strain at break	ISO 178 (5 mm / min)	No break (>10%)	3,0 ± 0,1

Table 2. Mechanical properties of UltiMaker PET CF before and after annealing, as specified by the manufacturer [14]

According to UltiMaker's technical documentation, annealing at 120 °C for 2 hours can improve strength by up to 30%, stiffness by 10%, and increase heat resistance from 76 °C to 181 °C [14].

4. Results and discussion

Following the mechanical testing of FDM 3D printed PET-CF material, the obtained results were analyzed and are presented in this chapter. A total of 100 specimens were tested, consisting of 50 samples for tensile testing and 50 samples for flexural testing. The average values for all tested groups are summarized in tabular format. Table 3. summarizes the results from the tensile and flexural tests, showing average values of tensile strength (Rm, MPa) and Young's modulus (E, GPa) for each condition.

Temperature (°C)	Time (min)	Tensile properties		Flexural properties	
		Rm [MPa]	E [GPa]	Rm [MPa]	E [GPa]
Nontreated (non-annealed)		47.5	4.0	84.6	4.8
80	30	46.8	4.0	85.0	4.9
80	120	45.6	4.0	84.3	4.8
80	210	45.4	3.9	83.2	4.7
130	30	51.8	4.5	109.8	5.5
130	120	50.0	4.5	108.1	5.5
130	210	56.3	4.6	109.5	5.5
180	30	43.8	4.6	110.7	5.5
180	120	47.1	4.6	105.9	5.5
180	210	47.8	4.7	100.6	5.5

Table 3. Average tensile and flexural properties for PET-CF samples under different annealing conditions

The variation in tensile strength of PET-CF samples after annealing is presented in Figure 7, where all annealing conditions are shown in a single comparative chart. The results indicate that annealing at 80 °C leads to a decreasing trend in tensile strength. Longer exposure times result in more pronounced degradation, with the most significant reduction observed after 210 minutes—showing a 4.42% decrease compared to the untreated reference. In contrast, samples annealed at 130 °C exhibited an overall improvement in tensile strength at all time intervals. The most notable increase was recorded after 210 minutes, with an improvement of 18.34% over untreated specimens. This represents the highest tensile strength achieved in the entire experiment and highlights the 130 °C / 210 min combination as the most favorable condition for enhancing tensile performance. Annealing at 180 °C showed a variable influence. At 30 minutes, tensile strength dropped significantly by 7.95%. However, with increased exposure time, the strength gradually recovered. Still, no substantial improvement was recorded compared to untreated samples, indicating that excessive temperature may initially weaken the material before structural stabilization occurs.

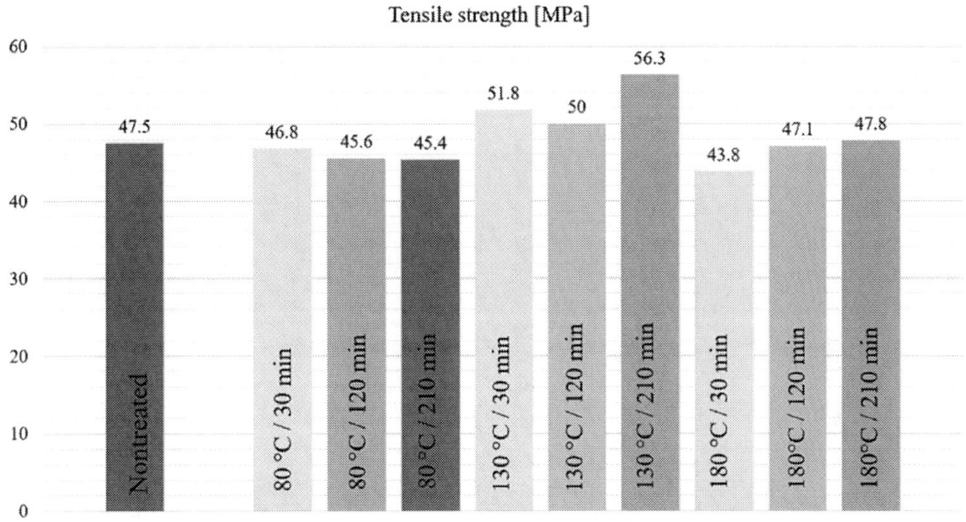

Fig. 7. Influence of annealing in different conditions on tensile strength of PET-CF samples

The results for tensile modulus are presented in Figure 8, which combines all data points across the three annealing temperatures into a single chart. At 80 °C, a slight initial improvement of 1.51% was observed after 30 minutes of treatment. However, prolonged annealing times (120 min and 210 min) led to a decrease in tensile modulus, with a 1.5% drop recorded after 210 minutes.

At 130 °C, a continuous increase in tensile modulus was observed with longer annealing times. The improvement was most evident at 210 minutes, suggesting that the exposure duration positively correlates with material stiffness at this temperature. Annealing at 180 °C produced the most significant increase in tensile modulus. Even the shortest exposure time (30 min) resulted in a 15.67% enhancement. The trend continued with increasing time, reaching a maximum improvement of 18.7% after 210 minutes. This clearly indicates that higher annealing temperatures, when combined with longer durations, can considerably enhance the stiffness of PET-CF under tensile loading.

3.3

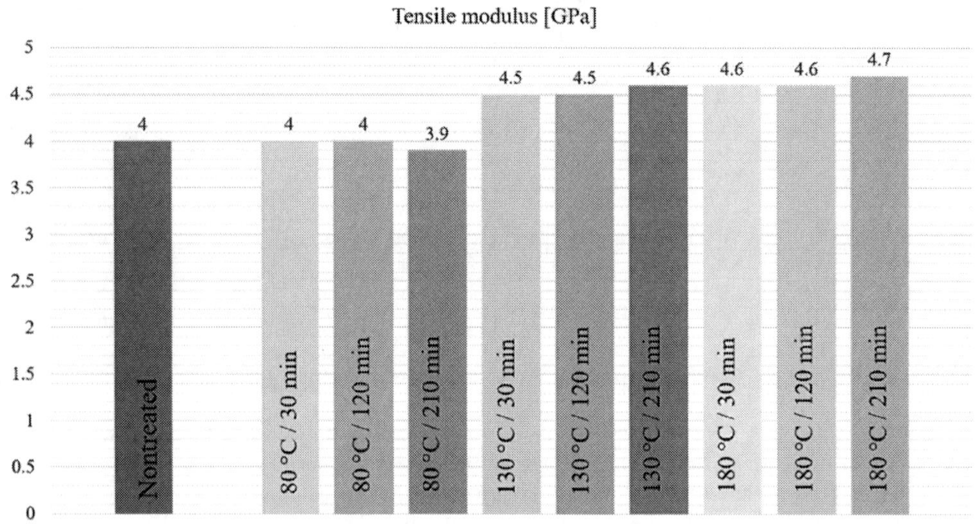

Fig. 8. Influence of annealing in different conditions on tensile moduluss of PET-CF samples

The analysis of flexural strength results indicates that annealing has an even more pronounced impact on the flexural mechanical properties of PET-CF compared to tensile properties. All variations across different temperatures and durations are consolidated in Figure 9.

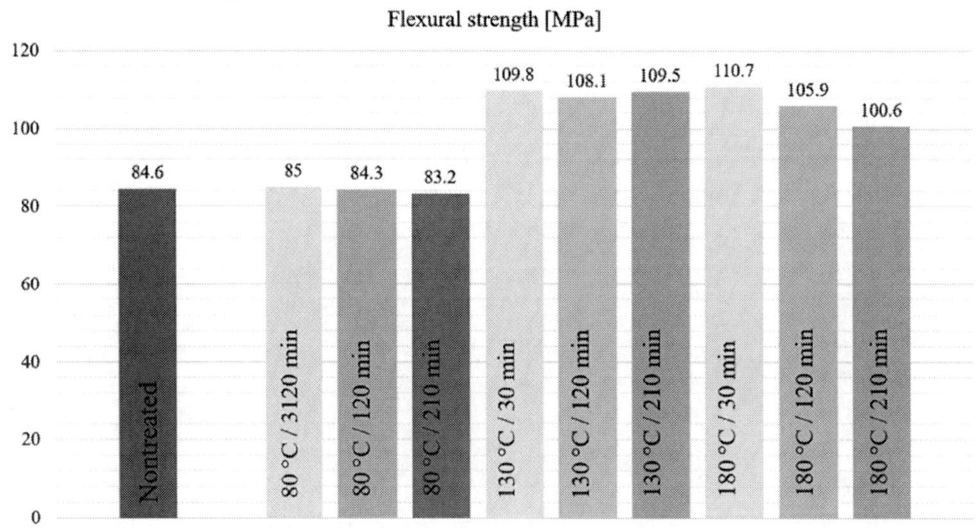

Fig. 9. Influence of annealing in different conditions on Flexural strength of PET-CF samples

At 80 °C, a slight improvement of 0.4% was observed after 30 minutes of treatment. However, with longer exposure times, the flexural strength began to decrease. A reduction of 0.35% was recorded after 120 minutes, and a more noticeable decrease of 1.7% occurred after 210 minutes, compared to the untreated reference samples. Annealing at 130 °C produced significant improvements in flexural strength. After just 30 minutes, strength increased by 29.8%. Although a slight decline to 27.7% was noted at 120 minutes, the strength rose again to 29.4% after 210 minutes. These results suggest that annealing at 130 °C consistently enhances flexural performance, though the effect is somewhat influenced by treatment duration.

At 180 °C, the initial 30-minute treatment resulted in the highest increase in flexural strength recorded in this study—30.85%. With longer exposure times, the improvement gradually declined: 25.12% after 120 minutes and 18.22% after 210 minutes. This trend indicates that while high temperatures promote strong initial gains, prolonged treatment may lead to diminishing returns. The evolution of flexural modulus under different annealing parameters is shown in Figure 10. At 80 °C, flexural modulus increased by 2.4% after 30 minutes, but longer treatments reduced this benefit. After 210 minutes, the modulus dropped by 1.05% compared to the non-annealed samples, suggesting that prolonged low-temperature annealing may not be beneficial for flexural stiffness. At 130 °C, a consistent improvement was observed with increasing treatment time. The flexural modulus rose by 13.13% after 30 minutes, reached 13.64% after 120 minutes, and peaked at 15.2% after 210 minutes, indicating strong positive effects from moderate-temperature annealing. At 180 °C, flexural modulus exhibited the highest recorded increases. The value improved by 15.14% after 30 minutes, peaked at 16.47% after 120 minutes, and slightly decreased to 15.87% after 210 minutes. Despite the small drop at longer durations, the results clearly demonstrate the beneficial effect of high-temperature annealing on flexural stiffness.

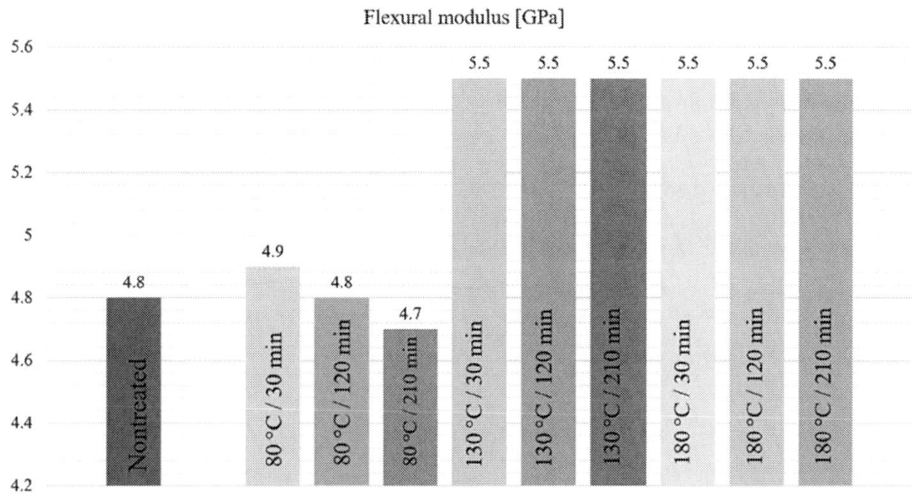

Fig. 10. Influence of annealing in different conditions on Flexural modulus of PET-CF samples

To better understand how annealing parameters influence the mechanical behavior of PET-CF samples, a statistical analysis was performed using Minitab software. The Design of Experiments (DOE) methodology was applied, specifically a full factorial design with two input factors: annealing temperature (80 °C, 130 °C, 180 °C) and annealing time (30, 120, 210 minutes). Table 4. summarizes the factors and their corresponding levels used in the experimental design. For each combination of input parameters, five test specimens were produced and measured. To minimize the impact of outliers, the maximum and minimum values were excluded, and the ANOVA analysis was conducted using the remaining three data points.

Factor	**Levels**	**Values**
Temperature (°C)	3	80, 130, 180
Time (min)	3	30, 120, 210

Table 4. Factors and levels used in the experimental design

Analysis of variance (ANOVA) was used to determine the statistical significance of each factor and their interactions on the mechanical properties. A significance level of 95% was used, and results with a p-value less than 0.05 were considered statistically significant. In addition, regression models were developed to predict mechanical responses based on input parameters. The model accuracy and reliability were assessed using R^2 (R-sq, coefficient of determination), adjusted R^2, and predicted R^2. A model was considered acceptable if the adjusted R^2 exceeded 75% and the difference between adjusted and predicted R^2 was less than 20%. The final mathematical model, developed to predict the tensile strength of the FDM-printed PET-CF composite material, is provided in Table 5. This model enables the estimation of tensile performance based on specific annealing parameters.

R_m (MPa)	=	47.019 - 1.10 T_80 + 4.11 T_130 - 3.02 T_180 - 0.24 t_30 - 1.62 t_120 + 1.86 t_210 + 1.35 T_80 * t_30 + 1.23 T_80 * t_120 - 2.58 T_80 * t_210 - 1.73 T_130 * t_30 - 1.68 T_130 * t_120 + 3.41 T_130 * t_210 + 0.37 T_180 * t_30 + 0.45 T_180 * t_120 - 0.83 T_180 * t_210

Table 5. Equation for predicting tensile strength

The analysis of variance (ANOVA) and corresponding results for tensile strength are presented in Table 6.3. Based on the data, the contribution of each individual input factor, as well as their interactions, to the variation in tensile strength within the developed model can be clearly observed. This analysis provides valuable insight into the most influential parameters in the annealing process and confirms the statistical validity and reliability of the tensile strength prediction model.

Source	DF	Seq SS	Contribution	Adj SS	Adj MS	F	P
Model	8	385.75	56.73%	385.75	48.22	2.95	0.027
Temperature (°C)	2	245.21	36.06%	245.21	122.6	7.5	0.004
Time (min)	2	55.21	8.12%	55.21	27.6	1.69	0.213
Temperature * Time	4	85.34	12.55%	85.34	21.33	1.31	0.306
Error	18	294.21	43.27%	294.21	16.34		
Total	26	679.96	100.00%				

Table 6. ANOVA results for tensile strength

The ANOVA results presented in Table 6. indicate that only the temperature is a statistically significant factor ($p = 0.004$), contributing 36.06% to the model. The other factors—time and the temperature–time interaction—are not statistically significant (p-values greater than 0.05). Time contributes 8.12%, while the two-way interaction between temperature and time contributes 12.55% to the model. These results show that the model explains only 56.73% of the total variation (R-sq), leading to the conclusion that the model is not satisfactory. Additionally, the results shown in Table 7. indicate that the majority of the variation in tensile strength (Rm) cannot be reliably predicted using the developed mathematical model.

S	R-sq	R-sq (adj)	R-sq (pred)
4.04287	56.73%	37.50%	2.65%

Table 7. Coefficient of determination (R-sq) for the model – Tensile strength

The final mathematical model developed to predict the tensile modulus of FDM-printed PET-CF material is presented in Table 8.

E (GPa)	=	4.36667 - 0.3889 T_80 + 0.1333 T_130 + 0.2556 T_180 + 0.0111 t_30 - 0.0111 t_ 120 - 0.0000 t_210 + 0.0444 T_80 * t_30 + 0.0333 T_80 * t_120 - 0.0778 T_130 * t_210 - 0.0111 T_130 * t_30 - 0.0222 T_130 * t_120 + 0.0333 T_130 * t_210 - 0.0333 T_180 * t_30 - 0.0111 T_180 * t_120 + 0.0444 T_180 * t_210

Table 8. Equation for predicting tensile modulus

The ANOVA results for the tensile modulus model, shown in Table 9., indicate a strong model fit with 97.88% of the total variation explained. Temperature is the most influential factor, with a significant p-value (0.000) and a contribution of 95.86%. The interaction between temperature and time is also statistically significant ($p = 0.016$), contributing 1.92%, while time alone is not significant ($p = 0.658$) and has a minimal effect. With only 2.12% of unexplained variation, the model is considered highly accurate, emphasizing the importance of temperature and its interaction with time.

Source	DF	Seq SS	Contribution	Adj SS	Adj MS	F	P
Model	8	2.15333	97.88%	2.15333	0.26917	103.82	0
Temperature (°C)	2	2.10889	95.86%	2.10889	1.05444	406.71	0
Time (min)	2	0.00222	0.10%	0.00222	0.00111	0.43	0.658
Temperature * Time	4	0,04222	1.92%	0.04222	0.01056	4.07	0.016
Error	18	0.04667	2.12%	0.04667	0.00259		
Total	26	2.2	100.00%				

Table 9. ANOVA results for tensile modulus

The ANOVA results presented in Table 10. indicate that the majority of variation in the tensile modulus (E) can be accurately predicted by the developed mathematical model. The coefficient of determination (R-sq) shows that the model explains 97.88% of the total variation, while the adjusted R-sq value of 96.94% confirms the model's robustness without overfitting. Additionally, the predicted R-sq of 95.23% demonstrates the model's strong ability to forecast outcomes reliably. Together, these indicators suggest that the model is highly reliable, with minimal error and excellent predictive performance.

S	R-sq	R-sq (adj)	R-sq (pred)
0.0509175	97.88%	96.94%	95.23%

Table 10. Coefficient of determination (R-sq) for the model – Tensile modulus

The significance of the input factors is also illustrated in the Pareto chart shown in Figure 11., clearly identifying which factors are statistically significant. Temperature stands out as the dominant factor with the greatest influence, while time has no significant effect. The interaction between temperature and time shows statistical significance but with a smaller impact compared to temperature. This trend is further confirmed in the 3D surface plot in Figure 11., which demonstrates how the tensile modulus (E) is affected by annealing temperature and time. The diagram shows that increasing the temperature (up to 180 °C) significantly raises the tensile modulus, while longer annealing durations (210 minutes) lead to only minor improvements. The highest values are observed at the combination of high temperature and extended time, confirming that temperature is the key factor in maximizing tensile modulus.

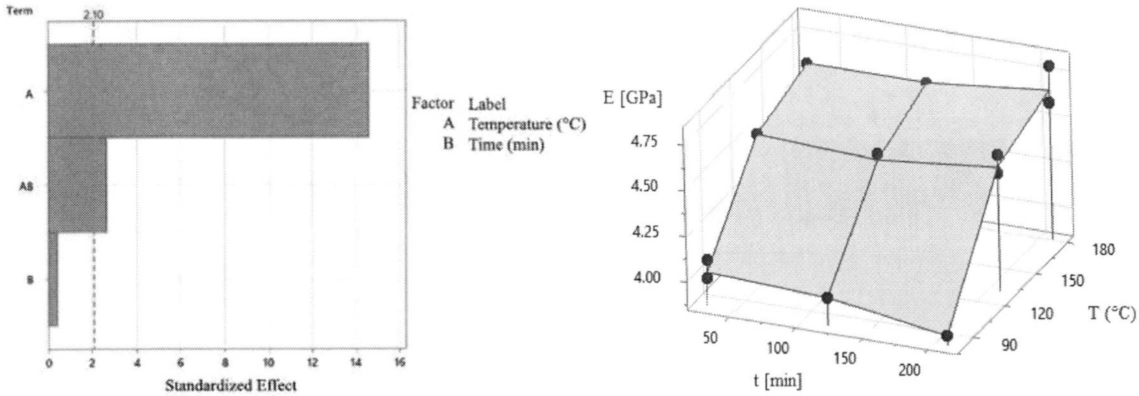

Fig. 11. Significance analysis of factors for tensile modulus – Pareto chart (left) and 3D plot showing the influence of temperature and annealing time on tensile modulus (right)

Based on the experimentally obtained values of flexural strength and flexural modulus for FDM-printed PET-CF material subjected to various annealing temperatures and exposure times, a statistical evaluation was performed to identify the key influencing factors. The final mathematical model developed to predict the flexural strength of the material is presented in Table 11.

R (MPa)	=	$99.567 - 15.484\,T_{80} + 9.158\,T_{130} + 6.326\,T_{210} + 1.689\,t_{30} - 0.156\,t_{120} - 1.533\,t_{210} - 1.013\,T_{30} * t_{30} + 0.307\,T_{80} * t_{120} + 0.706\,T_{130} * t_{210} - 1.966\,T_{130} * t_{30} - 0.526\,T_{130} * t_{120} + 2.492\,T_{130} * t_{210} + 2.979\,T_{180} * t_{30} + 0.220\,T_{180} * t_{120} - 3.198\,T_{180} * t_{210}$

Table 11. Equation for predicting flexural strength

ANOVA results presented in Table 12. show that the model explains 96.68% of the total variation, indicating a high level of fit. Temperature is the dominant factor, with a significant p-value ($p = 0.000$) and a contribution of 92.71%. Time is also statistically significant ($p = 0.048$), but with a low contribution of 1.33%, making it less influential. The interaction between temperature and time is significant ($p = 0.025$) with a contribution of 2.64%. The remaining unexplained variation is only 3.32%, confirming the model's reliability. Optimization should focus primarily on temperature, while time and interaction effects are relevant but less impactful.

Source	DF	Seq SS	Contribution	Adj SS	Adj MS	F	P
Model	8	3413.31	96.68%	3413.31	426.66	65.59	0,000
Temperature (°C)	2	3272.88	92.71%	3272.88	1636.44	251.58	0,000
Time (min)	2	47.05	1.33%	47.05	23.53	3.62	0.048
Temperature * Time	4	93.37	2.64%	93.37	23.34	3.59	0.025
Error	18	117.09	3.32%	117.09	6.5		
Total	26	3530.4	100.00%				

Table 12. ANOVA results for flexural strength

The ANOVA results presented in Table 13 indicate that the majority of variation in flexural strength can be predicted using the developed mathematical models. The coefficient of determination (R-sq) shows that the model explains 96.68% of the total variation. The adjusted R-sq is 95.21%, confirming a good model fit without overfitting, while the predicted R-sq of 92.54% demonstrates strong predictive capability. Together, these values suggest that the model is reliable, accurately represents the data relationships, and maintains minimal error with high predictive performance.

S	R-sq	R-sq (adj)	R-sq (pred)
2.55044	96.68%	95.21%	92.54%

Table 13. Coefficient of determination (R-sq) for the model – Flexural strength

The significance of individual factors is also illustrated in the Pareto chart (Figure 12.), where temperature clearly emerges as the dominant factor with the highest statistical influence. While time and the interaction between temperature and time also exhibit statistical significance, their effects are notably less pronounced compared to temperature. The 3D plot shown in Figure 12. further illustrates how flexural strength (Rm) is influenced by both annealing temperature and time. Based on the diagram, the highest flexural strength is achieved at an annealing temperature of 130 °C and a duration of 120 minutes.

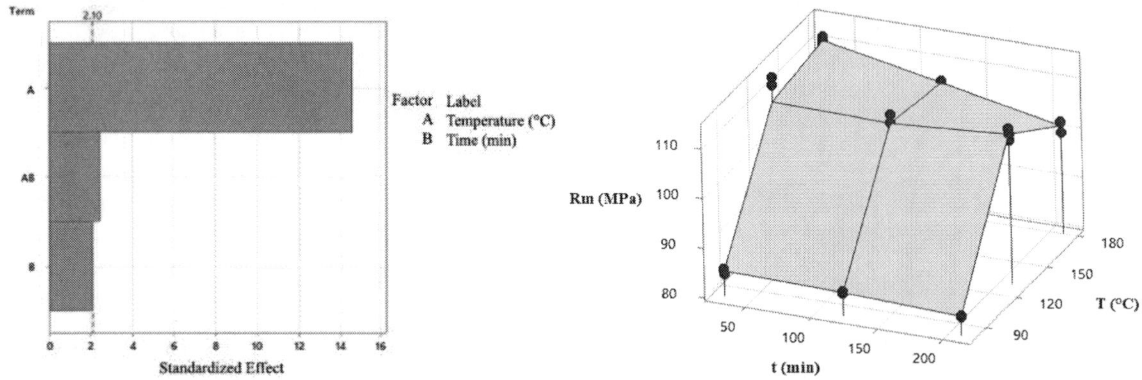

Fig. 12. Significance analysis of factors for flexural strength – Pareto chart (left) and 3D plot showing the influence of temperature and annealing time on flexural strength (right)

The mathematical model developed to predict the flexural modulus of FDM-printed PET-CF material, based on experimental results obtained under different annealing temperatures and exposure times, is presented in Table 14.

E (GPa)	=	5.26327 - 0.4826 T_80 + 0.2688 T_130 + 0.2139 T_180 - 0.0015 t_30 + 0.0083 t_120 - 0.0068 t_210 + 0.0546 T_80 * t_80 30 + 0.0238 T_80 * t_120 - 0.0784 T_80 * t_210 - 0.0332 T_130 * t_30 - 0.0091 T_130 * t_120 + 0.0423 T_130 * t_210 - 0.0213 T_180 * t_30 - 0.0147 T_180 * t_120 + 0.0360 T_180 * t_210

Table 14. Equation for predicting flexural modulus

The ANOVA analysis and results presented in Table 15. show the contribution of input factors to the model. The analysis indicates that the model explains 98.27% of the total variation, demonstrating a high level of fit to the data. Temperature stands out as the primary influencing factor (p = 0.000), contributing 97.34% to the output (flexural modulus). Time is not statistically significant (p = 0.798) and contributes only 0.03%. In contrast, the interaction between temperature and time is statistically significant (p = 0.008), contributing 1.35%, suggesting a relevant but much smaller effect compared to temperature alone. The model's unexplained variation is 1.28%, further confirming its adequacy.

Source	DF	Seq SS	Contribution	Adj SS	Adj MS	F	P
Model	8	3.20322	98.72%	3.20322	0.4004	174.06	0,000
Temperature (°C)	2	3.15827	97.34%	3.15827	1.57914	686.49	0,000
Time (min)	2	0.00105	0.03%	0.00105	0.00052	0.23	0.798
Temperature * Time	4	0.0439	1.35%	0.0439	0.01097	4.77	0.008
Error	18	0.04141	1.28%	0.04141	0.0023		
Total	26	3.24462	100.00%				

Table 15. ANOVA results for flexural modulus

The ANOVA results presented in Table 16. indicate that the model explains 98.27% of the total variation. The adjusted coefficient of determination (R-sq adj) is 98.16%, while the predicted coefficient of determination (R-sq pred) is 97.13%. Together, these values suggest that the model is reliable and accurately describes the relationships within the data, with minimal error and strong predictive capability.

S	R-sq	R-sq (adj)	R-sq (pred)
0.0479616	98.72%	98.16%	97.13%

Table 16. Coefficient of determination (R-sq) for the model – Flexural modulus

The significance of the input factors is also illustrated in the Pareto chart (Figure 13), clearly showing which variables have a statistically significant influence. Temperature demonstrates a dominant effect, while the interaction between temperature and time is also statistically significant, though with a much smaller contribution. Time alone does not show statistical significance and has no meaningful impact. The 3D surface plot in Figure 13 reveals that temperature is the primary factor influencing the increase in flexural modulus. At elevated temperatures (130 °C and 180 °C), the modulus significantly improves, while variations in annealing time do not notably affect the outcome. The highest recorded flexural modulus was 5.6 GPa, achieved at 130 °C after 120 minutes of annealing.

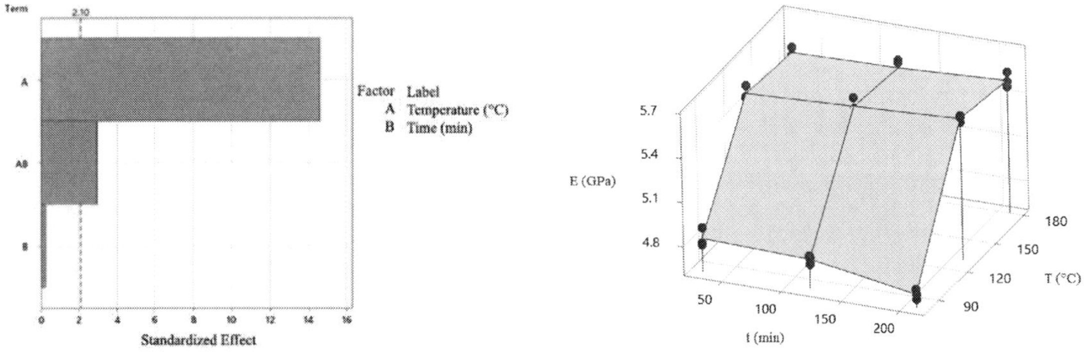

Fig. 13. Significance analysis of factors for flexural modulus – Pareto chart (left) and 3D plot showing the influence of temperature and annealing time on flexural modulus (right)

5. Conclusion

This study investigated the effect of annealing on the mechanical properties of FDM-printed PET-CF material, with a focus on tensile and flexural strength, as well as tensile and flexural modulus. Specimens were annealed at three temperatures (80 °C, 130 °C, and 180 °C) and three durations (30, 120, and 210 minutes), and the results were analyzed using ANOVA and linear regression models. The findings confirmed that annealing can significantly improve the mechanical performance of PET-CF. The most notable improvements were achieved at higher temperatures and longer durations. The maximum observed increases were 30.85% in flexural strength, 18.34% in tensile strength, 18.7% in tensile modulus, and 15.87% in flexural modulus. These results support the claim that thermal post-processing enhances mechanical performance but also highlight differences compared to manufacturer specifications—primarily due to differences in printer models, filament storage conditions, and moisture sensitivity of the PET-CF material. ANOVA results showed that temperature is the dominant factor affecting all mechanical properties, while time and the interaction between temperature and time had limited or no statistical significance. The model for predicting tensile strength was found to be inadequate, likely due to outliers and inconsistent cooling conditions. Improving the model's accuracy could be achieved by increasing the number of replicates and introducing controlled post-annealing cooling. Future work should focus on studying the influence of moisture, the impact of controlled cooling on mechanical properties, exploring annealing effects on different FDM filaments, and developing software tools for optimizing annealing conditions.

6. Acknowledgments

The authors would like to express their sincere gratitude to Ultimaker for their valuable support and contribution to this research. Special thanks also go to the Faculty of Mechanical Engineering, University of Sarajevo, and the Laboratory for Testing of Polymeric Materials (IPM Lab) for their continued support, resources, and collaboration throughout the experimental work.

7. References

[1] Abdulhameed, O.; Al-Ahmari, A.; Ameen, W. & Mian, S. H. (2019). Additive Manufacturing: Challenges, Trends, and Applications, Advances in Mechanical Engineering, Vol. 11, No. 2, pp. 1–27, ISSN 1687-8140, https://doi.org/10.1177/1687814018822880

[2] Al-Tamimi, A. A.; Pandžić, A. & Kadrić, E. (2023). Investigation and Prediction of Tensile, Flexural, and Compressive Properties of Tough PLA Material Using Definitive Screening Design, Polymers, Vol. 15, No. 20, Article 4169, ISSN 2073-4360, https://doi.org/10.3390/polym15204169

[3] Singh, S.; Prakash, C. & Ramakrishna, S. (2020). Additive Manufacturing: Innovations, Advances, and Applications, World Scientific Publishing, ISBN 978-9811218527, Singapore, https://doi.org/10.1142/11953

[4] Zhou, L.; Miller, J.; Vezza, J.; Mayster, M.; Raffay, M. et al. (2024). Additive Manufacturing: A Comprehensive Review, Sensors, Vol. 24, No. 9, Article 2668, ISSN 1424-8220, https://doi.org/10.3390/s24092668

[5] Pandžić, A. (2022). Doktorska disertacija: Prilog procjeni mehaničkih osobina FDM printanih polimernih materijala sa različitim dizajnom ispune, Mašinski fakultet Sarajevo, Sarajevo, Oktobar 2022, https://doi.org/10.13140/RG.2.2.25939.86563

[6] Godec, D.; Gonzalez-Gutierrez, J.; Nordin, A.; Pei, E. & Erena Alcaraz, J. (2022). A Guide to Additive Manufacturing, Springer Nature Switzerland AG, ISBN 978-3-031-05863-9, https://doi.org/10.1007/978-3-031-05863-9

[7] Hodžić, D.; Pandžić, A.; Hajro, I. & Tasić, P. (2020). Strain Rate Influence on Mechanical Characteristics of FDM 3D Printed Materials, Proceedings of the 31st DAAAM International Symposium on Intelligent Manufacturing and Automation 2020, Vienna, Austria, ISSN 1726-9679, ISBN 978-3-902734-29-1, Katalinić, B. (Ed.), pp. 022–027, DAAAM International Vienna, https://doi.org/10.2507/31st.daaam.proceedings.022

[8] Hodžić, D. & Pandžić, A. (2021). Influence of Infill Design on Compressive and Flexural Mechanical Properties of FDM Printed PLA Material, Proceedings of the 32nd DAAAM International Symposium on Intelligent Manufacturing and Automation 2021, Vienna, Austria, ISSN 1726-9679, ISBN 978-3-902734-32-1, Katalinić, B. (Ed.), pp. 029–034, DAAAM International Vienna, https://doi.org/10.2507/32nd.daaam.proceedings.029

[9] Krisiawan, R. B.; Imaduddin, F.; Ariawan, D.; Ubaidillah & Arifin, Z. (2021). A Review on the Fused Deposition Modelling 3D Printing: Filament Processing, Materials and Printing Parameters, Open Engineering, Vol. 11, No. 1, pp. 522–536, ISSN 2391-5439, https://doi.org/10.1515/eng-2021-0063

[10] Alarifi, I. M. (2023). PETG/Carbon Fiber Composites with Different Structures Produced by 3D Printing, Polymer Testing, Vol. 124, Article 107949, ISSN 0142-9418, https://doi.org/10.1016/j.polymertesting.2023.107949

[11] Pandžić, A.; Hodžić, D.; Tasić, P. & Hajro, I. (2024). Influence of Nozzle Diameter on Tensile and Flexural Mechanical Properties of FDM 3D Printed PET-CF Material, Proceedings of the 35th DAAAM International Symposium on Intelligent Manufacturing and Automation 2024, Vienna, Austria, ISSN 1726-9679, Katalinić, B. (Ed.), DAAAM International Vienna, https://doi.org/10.2507/35th.daaam.proceedings.017

[12] Zhodi, N. & Yang, R. (2021). Material Anisotropy in Additively Manufactured Polymer Composites, Polymers, Vol. 13, No. 19, Article 3368, ISSN 2073-4360, https://doi.org/10.3390/polym13193368

[13] De Wijk, B. (2022). Annealing of 3D Printed Parts, Internship Report, University of Groningen, Faculty of Science and Engineering, Netherlands, February 2022, Available from: https://fse.studenttheses.ub.rug.nl/26639/1/B.de.Wijk_Internship.pdf

[14] Chapman, A. (2023). How to Anneal Your PET CF Parts for Better Performance, Ultimaker – Learn, September 12, 2023, Available from: https://ultimaker.com/learn/how-to-anneal-pet-cf-for-better-performance/, Accessed: 2025-07-03

[15] Guo, Q. (2016). Polymer Morphology, Canada, Available from: http://kinampark.com/PL/files/Books/Guo%202016%2C%20Polymer%20Morphology.pdf, Accessed: 2025-07-03

[16] Rangisetty, S. & Peel, L. D. (2017). The Effect of Infill Patterns and Annealing on Mechanical Properties of Additively Manufactured Thermoplastic Composites, Proceedings of the ASME 2017 Conference on Smart Materials, Adaptive Structures and Intelligent Systems (SMASIS 2017), Snowbird, Utah, USA, September 18–20, 2017, Paper No. SMASIS2017-4011, https://doi.org/10.1115/SMASIS2017-4011

[17] Valvez, S.; Silva, A. P.; Reis, P. N. B. & Berto, F. (2022). Annealing Effect on Mechanical Properties of 3D Printed Composites, Proceedings of ICSI 2021 – The 4th International Conference on Structural Integrity, Procedia Structural Integrity, Vol. 37, pp. 1307–1312, Elsevier B.V., https://doi.org/10.1016/j.prostr.2022.02.004

[18] Stojković, J. R.; Turudija, R.; Vitković, N.; Gorski, F.; Pacurar, A.; Plesa, A.; Ianosi-Andreeva-Dimitrova, A. & Pacurar, R. (2023). An Experimental Study on the Impact of Layer Height and Annealing Parameters on the Tensile Strength and Dimensional Accuracy of FDM 3D Printed Parts, Materials, Vol. 16, No. 13, Article 4574, MDPI, ISSN 1996-1944, https://doi.org/10.3390/ma16134574

[19] Kumar, K. S.; Soundararajan, R.; Shanthosh, G.; Saravanakumar, P. & Ratteesh, M. (2020). Augmenting Effect of Infill Density and Annealing on Mechanical Properties of PETG and CFPETG Composites Fabricated by FDM, Sri Krishna College of Engineering and Technology & National Institute of Technology, India, October 2020

[20] International Organization for Standardization (2012). ISO 527-2:2012 – Plastics – Determination of tensile properties – Part 2: Test conditions for moulding and extrusion plastics, Available from: https://www.iso.org/standard/56046.html, Accessed: 2025-07-03

[21] International Organization for Standardization (2019). ISO 178:2019 – Plastics – Determination of flexural properties, Available from: https://www.iso.org/standard/70513.html, Accessed: 2025-07-03

DOI: 10.2507/36th.daaam.proceedings.012

FUZZY AHP-BASED CRITERIA WEIGHT DETERMINATION FOR THE EVALUATION AND RANKING OF RENEWABLE ELECTRICITY GENERATION PROJECTS

Zedina Lavic & Amina Bibic

This Publication has to be referred as: Lavic, Z[edina] & Bibic, A[mina] (2025). Fuzzy AHP-Based Criteria Weight Determination for the Evaluation and Ranking of Renewable Electricity Generation Projects, Proceedings of the 36th DAAAM International Symposium, pp.0095-0100, B. Katalinic (Ed.), Published by DAAAM International, ISBN 978-3-902734-47-1, ISSN 1726-9679, Vienna, Austria
DOI: 10.2507/36th.daaam.proceedings.012

Abstract

Due to persistent challenges in the implementation of planned investments for the development and construction of renewable electricity generation facilities, particularly hydro, wind, and solar power plants, public utility companies are steering to project acquisitions from privately owned companies through public tenders. This paper offers a set of evaluation and ranking criteria for projects submitted through public tenders. The criteria are founded on both qualitative and quantitative aspects and can be divided into three major categories: technical, economic, and environmental. Their importance was determined by using the Fuzzy Analytic Hierarchy Process (FAHP). Overall, this paper provides a critical reflection on the results as well as suggestions for untapped research opportunities.

Keywords: Renewable Electricity Generation Projects; Evaluation and Ranking Criteria; Fuzzy Analytic Hierarchy Process.

1. Introduction

The electricity sector in Bosnia and Herzegovina, as well as in Croatia, has adopted the view that environmental pollution is an urgent issue, not only in these countries but also the region. Public utility companies in the electricity sector aim to address this challenge by focusing on increasing the production of electricity from renewable energy sources and reducing pollutant emissions. With this aim, the largest public utility company in Bosnia and Herzegovina purchases renewable electricity generation projects through open tenders, where independent producers can propose their offers. The significant number of submitted projects reflects an increased interest and active participation of local companies in promoting the energy transition. These projects include hydroelectric, wind, and photovoltaic power plants, spanning a diverse range of locations, providing tremendous options for production portfolio optimization and enhancing the resource base. In general, before issuing a public tender call, companies need to establish the criteria for evaluation and assign weights to each criterion. Such criteria should include technical, economic and environmental aspects. Although it is possible to measure some of these criteria, others may be either difficult to quantify or lack sufficient data. Moreover, in the case of analysing these less measurable criteria, companies often depend on expert judgement.

Each of the criteria should be assigned a weight based on the specific objectives it aims to achieve in the project acquisition, which may include increasing capacity, improving production, optimizing investment efficiency, or selecting the best geographical locations for power plants. When using expert judgement, assigning these weights can be challenging because experts are subjective and imprecise, and the various factors (quantitative and qualitative) involved in the decision making are difficult to balance and measure accurately. This presents a multicriteria decision-making issue in a "fuzzy" environment. Based on Fuzzy set theory [1], fuzzy logic is an appropriate approach to handle a vast array of problems that involve a form of uncertainty, or imprecision or incomplete information.

In the paper [2] a security system application for outdoor cabinets is designed using fuzzy logic. Methodology for selecting a source of electrical energy suitable for the site, based on fuzzy logic is presented in [3]. Also, among the numerous examples in the literature, fuzzy logic has been used to solve problems in the dairy industry [4], wind farm site selection [5], and scheduling problems in flexible robotic assembly cells [6]. Fuzzy Analytic Hierarchy Process (FAHP) which is based on the extent analysis technique proposed by Chang [7] has been extensively used in many studies to deal successfully with problems of complex multicriteria decision making. Although the method of extent analysis has proven to have several strengths, it is not devoid of weakness such as zero-wights [8]. This paper, therefore, takes the FAHP method given by Buckley [9] which is a more theoretically strong method of dealing with the fuzzy pairwise comparisons, and obtaining the priority vectors.

2. Criteria for evaluation and ranking of projects

Various criteria may be applied to rank projects for electricity generation from renewable energy sources submitted through the acquisition process via a public tender. The criteria can be grouped according to their nature, and in this research, the criteria are divided into three different groups:

- group of technical criteria,
- group of economic criteria,
- group of environmental criteria.

Each group of criteria consists of three selected criteria, which may be quantitative or qualitative, as shown in Table 1.

No.	Group of criteria	Criterion	Quant./Qual.	Unit of measure	Max/ Min
1.	Technical group	Project implementation status	Qualitative	-	Max
2.		Installed capacity	Quantitative	MW	Max
3.		Projected annual production	Quantitative	GWh/ year	Max
4.	Economic group	CAPEX	Quantitative	Million EUR	Min
5.		Payback period	Quantitative	years	Min
6.		Efficiency	Qualitative	-	Max
7.	Environmental group	Impact on ecosystem	Qualitative	-	Min
8.		Location	Qualitative	-	Max
9.		Spatial requirements	Qualitative	-	Min

Table 1. Criteria groups and criteria

The technical criteria group has three criteria including the status of project implementation, the installed capacity and the projected annual energy production. Project implementation status is an indicator that provides information about the progress and current state of the project in relation to defined goals, deadlines, budget, and resources. It shows whether the project is still in the planning phase, in the process of obtaining permits, under construction, or already in operation. Installed capacity is the capacity maximum of an individual power plant to produce electricity at a certain time unit, and it is expressed in MW or GW. Projected annual energy production is an estimation of the level of energy production of the power plant over a period of one year. It is among the most significant indices in assessing potential and performance of energy projects.

The economic criteria group is composed of three evaluation and ranking criteria: capital expenditure (CAPEX), investment payback period, and project efficiency. CAPEX reflects the total capital expenditure necessary to construct, maintain, and possibly upgrade the generation facility of power. The cost is dependent on various construction factors such as the scope of the project, the type of technology, geographical area and compliance standards. The investment payback period represents the period needed to cover the initial capital investments via net cash inflows or operational cost reductions. It is considered as one of the most important parameters to determine the financial viability and the payback potential of a project. Efficiency can be described as the extent to which the available resources are systematically and optimally employed in ensuring that predetermined objectives on a given project are met. Efficiency will be measured within this framework based on a qualitative evaluation and quantitative measures (percentages).

The environmental criteria group has three indicators: impact on ecosystem, project location, and spatial requirements. The main environmental effects of photovoltaic power plants are the change in land use pattern, as well as lost habitats, water pollution, during manufacturing, and disposable processes of solar panels, and water-use patterns. The potential effects of wind power projects on ecosystems may include land-use change, alteration of habitats, and bird and bat collisions with turbines, and noise and turbulence concerns [10]. Large-scale hydropower has major environmental effects, such as alteration of habitats, change in river regimes, sedimentation patterns and disturbance to fish populations and migration channels. Moreover, dam constructions may cause changes in land-use and the relocation of the people living locally [11]. Project locations are actual geographical locations of the planned energy plants. The appropriateness of site is evaluated regarding natural conditions, vicinity to the distribution and/or transmission grid, presence of transformer substations, presence of road infrastructure, and so forth. Spatial needs refer to measuring of land landscape to cover the needs of implementation and operational running of specific project.

3. FAHP – Buckley's Geometric Mean Method

The classical Analytic Hierarchy Process (AHP) is not able to handle the ambiguity and imprecision of human judgment. To overcome this weakness, FAHP methodology is used for solving of many decision-making problems in such environment. The FAHP allows employing the idea of fuzzy logic and represents the preferences of decision-makers more qualitative and closer to reality. In this frame, the criteria are also compared pairwise and linguistic variable values, considered as triangular fuzzy numbers, are used in the comparison [12]. Table 2[13] provides example of these variable values (from Saaty's scale) and their related fuzzy values. Among other contributions, Buckley significantly improved the FAHP method: he suggested a way to aggregate local weights from fuzzy comparison matrices. Applying the Buckley FAHP method [13] removes the „zero-weight "weakness, which may, in certain cases, occur in applying the FAHP method based on Chang's fuzzy extent analysis.

Linguistic variable values	**Triangular fuzzy scale**	**Triangular fuzzy reciprocal scale**
Equally important	(1,1,1)	(1,1,1)
Weakly important	(2,3,4)	(1/4,1/3,1/2)
Fairly important	(4,5,6)	(1/6,1/5,1/4)
Strongly important	(6,7,8)	(1/8,1/7,1/6)
Absolutely important	(9,9,9)	(1/9,1/9,1/9)
The intermittent values between two adjacent scales	(1,2,3)	(1/3,1/2,1)
	(3,4,5)	(1/5,1/4,1/3)
	(5,6,7)	(1/7,1/6,1/5)
	(7,8,9)	(1/9,1/8,1/7)

Table 2. Linguistic variable values and corresponding triangular fuzzy numbers

The following text provides a description of Buckley's method. Before applying the steps of Buckley's method, it is necessary to construct the reciprocal fuzzy matrix $\tilde{A}$ (1) of size $n\ x\ n$, which represents the comparison between elements i and j.

$$\tilde{A}=\begin{bmatrix}(1,1,1) & \tilde{a}_{12} & \cdots & \tilde{a}_{1n}\\ \tilde{a}_{21} & (1,1,1) & & \tilde{a}_{2n}\\ \vdots & \vdots & \cdots & \vdots\\ \tilde{a}_{n2} & \tilde{a}_{n2} & \cdots & (1,1,1)\end{bmatrix} \tag{1}$$

Triangular fuzzy number $\tilde{a}=(l,m,u)$ where is $l \leq m \leq u$ is defined on $\Re$ using the membership function $\mu_{\tilde{a}}(x): R \rightarrow [0,1]$:

$$\mu_{\tilde{a}}(x)=\begin{cases}\frac{x}{m-l}-\frac{l}{m-l}, & x\in[l,m]\\ \frac{x}{m-u}-\frac{u}{m-u}, & x\in[m,u]\\ 0, & otherwise\end{cases} \tag{2}$$

Buckley's method is carried out through the following steps:

1. calculation of the fuzzy value of the geometric mean $\tilde{r}_i$ (3),
2. calculation of the fuzzy weight $\tilde{w}_i$ (4),
3. defuzzification (centre of area method) and normalization (by dividing each value by the summation of all values [13].

$$\tilde{r}_i=[\tilde{a}_{i1}\otimes\tilde{a}_{i2}\otimes...\otimes\tilde{a}_{in}]^{\frac{1}{n}}\ \forall\ i=1,2,\dots,n \tag{3}$$

$$\tilde{w}_i=\frac{\tilde{r}_i}{(\tilde{r}_1\oplus\tilde{r}_2\oplus...\oplus\tilde{r}_n)} \tag{4}$$

4. Data and results

Comparison matrices of the criterion groups and the individual criteria are formed using expert judgments about the relative significance of the elements compared. After the consistency is confirmed, the Fuzzy Analytic Hierarchy Process (FAHP) of Buckley is implemented. The criteria groups comparison matrix, fuzzy geometric means as well as fuzzy weights are presented in Table 3. In Table 4 calculated weights and normalized weight values are given.

	Technical group	Economic group	Environmental group	Fuzzy geometric mean	Fuzzy weight
Technical group	(1,1,1)	(4,5,6)	(2,3,4)	(2.00,2.47,2.88)	(0.44,0.66,0.96)
Economic group	(1/6,1/5,1/4)	(1,1,1)	(1,2,3)	(0.55,0.74,0.91)	(0.12,0.20,0.30)
Environmental group	(1/4,1/3,1/2)	(1/3,1/2,1)	(1,1,1)	(0.44,0.55,0.79)	(0.10,0.15,0.27)

Table 3. Comparison matrix of the criteria groups, fuzzy geometric means and fuzzy weights

	Weight	Normalized weight
Technical group	0.69	0.65
Economic group	0.21	0.19
Environmental group	0.17	0.16
	Sum	1

Table 4. Criteria group weights and normalized weights values

In Table 5 the comparison matrix of the technical criteria, as well as the fuzzy geometric mean and fuzzy weights are given. The Table 6 shows corresponding calculated weights and normalized weights.

	Project implementation status	Installed capacity	Projected annual production	Fuzzy geometric mean	Fuzzy weight
Project implementation status	(1,1,1)	(2,3,4)	(1,2,3)	(1.26,1.82,2.29)	(0.28,0.51,0.83)
Installed capacity	(1/4,1/3,1/2)	(1,1,1)	(6,7,8)	(1.14,1.33,1.59)	(0.26,0.37,0.58)
Projected annual production	(1/3,1/2,1)	(1/8,1/7,1/6)	(1,1,1)	(0.35,0.41,0.55)	(0.08,0.12,0.20)

Table 5. Comparison matrix of the technical criteria, fuzzy geometric means and fuzzy weights

	Weight	Normalized weight
Technical group	0.54	0.50
Economic group	0.40	0.37
Environmental group	0.13	0.13
	Sum	1

Table 6. Calculated weights and normalized weights of the technical criteria

In Table 7 the comparison matrix of the economic criteria are given, as well as the fuzzy geometric mean and fuzzy weights. Table 8 shows calculated weight and normalized weight of each economic criterion.

	CAPEX	Payback period	Efficiency	Fuzzy geometric mean	Fuzzy weights
CAPEX	(1,1,1)	(5,6,7)	(3,4,5)	(2.47,2.88,3.27)	(0.51,0.68,0.88)
Payback period	(1/7,1/6,1/5)	(1,1,1)	(6,7,8)	(0.95,1.05,1.17)	(0.20,0.25,0.32)
Efficiency	(1/5,1/4,1/3)	(1/8,1/7,1/6)	(1,1,1)	(0.29,0.33,0.38)	(0.06,0.08,0.10)

Table 7. Comparison matrix of the economic criteria, fuzzy geometric means and fuzzy weights

	Weight	Normalized weight
CAPEX	0.69	0.67
Payback period	0.25	0.25
Efficiency	0.08	0.08
	Sum	1

Table 8. Calculated weights and normalized weights of the economic criteria

In Table 9 the comparison matrix of the environmental criteria is given, as well as the fuzzy geometric mean and fuzzy weights. Table 10 shows calculated weight and normalized weight of each that criterion.

	Impact on ecosystem	Location	Spatial requirements	Fuzzy geometric mean	Fuzzy weights
Impact on ecosystem	(1,1,1)	(5,6,7)	(3,4,5)	(2.47,2.88,3.27)	(0.51,0.67,0.87)
Location	(1/7,1/6,1/5)	(1,1,1)	(1/9,1/8,1/7)	(1,1.10,1.22)	(0.51,0.67,0.87)
Spatial requirements	(1/5,1/4,1/3)	(7,8,9,)	(1,1,1)	(0.28,0.31,0.36)	(0.06,0.07,0.10)

Table 9. Comparison matrix of the environmental criteria, fuzzy geometric means and fuzzy weights

	Weight	Normalized weight
CAPEX	0.68	0.67
Payback period	0.26	0.26
Efficiency	0.08	0.07
	Sum	1

Table 10. Calculated weights and normalized weights of the environmental criteria

In the Table 11, the weighting structure is displayed: weights of criteria groups as well as weight (local and global) of every individual criterion.

No.	**Group of criteria**	**Weight of criteria group**	**Criterion**	**Local weight of criterion**	**Global weight of criterion***
1.	Technical group	0.65	Project implementation status	0.50	0.33
2.			Installed capacity	0.37	0.24
3.			Projected annual production	0.13	0.08
4.	Economic group	0.19	CAPEX	0.67	0.13
5.			Payback period	0.25	0.05
6.			Efficiency	0.08	0.02
7.	Environmental group	0.16	Impact on ecosystem	0.67	0.11
8.			Location	0.26	0.04
9.			Spatial requirements	0.07	0.01

Table 11. Global weights of criteria (*global weights of criteria are rounded on two decimal places)

The technical criteria group in this analysis has the highest weight (0.65) which implies that it has predominant importance among the other criteria groups. The environmental criteria group comes next with its weight value 0.19, followed by the economic group whose relative importance is the lowest (0.16). Out of all individual criteria, the implementation status of a project has the greatest value of global weight (0.33), meaning that there may be a high preference of projects that are in implementation stages or in operation as opposed to those not so close to implementation.

(operation). These are followed by installed capacity (0.24), CAPEX (0.13) and impact on the ecosystem (0.11). The other five criteria together constitute global weight of 0.20. These results indicate an obvious strategic focus on ramping up renewable energy projects, particularly the projects with the highest technical maturity and capacity as well as trying to keep capital outlays and the negative impact on the environment to a minimum.

5. Conclusion

The purpose of this paper is to improve the acquisition process of renewable electricity generation projects through public procurement. Nine evaluation criteria were determined indicating a careful consideration of technical, economic, and environmental aspects of renewable energy projects. This set of criteria was systematically arranged into three broad categories that relate to each of the three aspects. The technical criteria group has three criteria including the status of project implementation, the installed capacity and the projected annual energy production. The economic criteria group is composed of three criteria including the capital expenditure (CAPEX), the investment payback period, and the project efficiency, while the environmental criteria group consists of the impact on ecosystem, the project location, and the spatial requirements. The evaluation framework encompasses quantitative and qualitative criteria thus making it possible to include expert judgment where the data is imprecise, not readily available or may be hard to find. To calculate relative importance (weights) of the criteria from the consistent fuzzy pairwise comparison matrices, formed by expert, the study uses the FAHP suggested by Buckley, with the local weights calculated using the geometric mean. The results (global weight of each criterion) are aggregated according to the principles of the classical AHP method given by Saaty. The obtained weight values of criteria reflect the strategic orientation of Bosnia and Herzegovina in the direction of the faster realization of decarbonization goals in the electricity generation sector, trough economically feasible and environmentally sustainable means. The proposed criteria and the weighting method could be successfully applied in the context of other countries aiming to accelerate the process of power sector decarbonisation by leveraging project acquisition mechanisms. The results obtained in this study can, in the context of future research, be compared with the results that would be obtained by applying another multicriteria decision-making method in a fuzzy environment.

6. References

[1] Zadeh, L. A. (1965). Fuzzy Sets. Information and Control, 8(3), 338–353.

[2] Aydin, Selahattin Mert; Keklik, Emre; Akdogan, Anil & Vanila, Ali Serdar (2024). Fuzzy Logic-Based Security System for Remote Outdoor Cabinets, Proceedings of the 35th DAAAM International Symposium, pp.0204-0209, B. Katalinic (Ed.), Published by DAAAM International, ISBN 978-3-902734- 44-0, ISSN 1726-9679, Vienna, Austria

[3] Prochazka, Jan; Prochazkova, Dana; Rusko, Miroslav; Rockel, Georg & Ilko, Jan (2023). Methodology for Selecting a Source of Electrical Energy Suitable for the Site, Proceedings of the 34th DAAAM International Symposium, pp.0172-0182, B. Katalinic (Ed.), Published by DAAAM International, ISBN 978- 3-902734-41-9, ISSN 1726-9679, Vienna, Austria

[4] Jamegh, Riyadh; Kassam, Alla Eldin & Sabih, Sawsan (2018). Dynamic Fuzzy Logic to Identify Safety Stock in Dairy Industries / Empirical Case Study, Proceedings of the 29th DAAAM International Symposium, pp.0983-0992, B. Katalinic (Ed.), Published by DAAAM International, ISBN 978-3-902734- 20-4, ISSN 1726-9679, Vienna, Austria

[5] Aljicevic, Zajim; Kostic, Aleksandra; Dautbasic, Nedis & Karli, Gunay (2016). Model of Fuzzy Logic for Selection Infrastructural Investment Project of Wind Farm Locations, Proceedings of the 27th DAAAM International Symposium, pp.0743-0748, B. Katalinic (Ed.), Published by DAAAM International, ISBN 978-3-902734-08-2, ISSN 1726-9679, Vienna, Austria

[6] Abd, Khalid; Abhary, Kazem & Marian, Romeo (2017). Application of Fuzzy AHP and TOPSIS Approach for Multi- objective Scheduling Problems in Robotic Flexible Assembly Cells, Chapter 21 in DAAAM International Scientific Book 2017, pp.279-290, B. Katalinic (Ed.), Published by DAAAM International, ISBN 978-3-902734-12-9, ISSN 1726-9687, Vienna, Austria

[7] Chang, D.-Y. (1996). Applications of the extent analysis method on fuzzy AHP. European Journal of Operational Research, 95(3), 649–655

[8] Wang, Y.-M., Luo, Y., Hua, Z., (2008). On the extent analysis method for fuzzy AHP and its applications. European Journal of Operational Research, 186(2), pp.735-747

[9] Buckley, J. J. (1985). Fuzzy Hierarchical Analysis, Fuzzy Sets and Systems, 17(3), 233–247.

[10] NREL. (2021). Environmental Impacts of Wind Energy Projects. National Renewable Energy Laboratory. Available from: www.nrel.gov , Accessed: 17. January 2024

[11] Iha. (2021). Hydropower Status Report. International hydropower association. Available from: www.hydropower.org , Accessed: 07. February 2024

[12] Emrouznejad, A. and Ho, W. (2017). Fuzzy Analytic Hierarchy Process. CRC Press.

[13] Ayhan, M. B. (2013). A fuzzy AHP approach for supplier selection problem: a case study in a gearmotor company. International Journal of Managing Value and Supply Chains (IJMVSC) Vol.4, No.3. Available from: https://www.airccse.org/journal/mvsc/papers/4313ijmvsc02.pdf, Accessed: 07. February 2024.

DOI: 10.2507/36th.daaam.proceedings.013

APPLICATION OF THE FUZZY TOPSIS METHOD FOR RANKING RENEWABLE ENERGY ELECTRICITY GENERATION PROJECTS

Zedina Lavic & Amina Bibic

This Publication has to be referred as: Lavic, Z[edina] & Bibic, A[mina] (2025). Application of the Fuzzy TOPSIS Method for Ranking Renewable Energy Electricity Generation Projects, Proceedings of the 36th DAAAM International Symposium, pp.0101-0106, B. Katalinic (Ed.), Published by DAAAM International, ISBN 978-3-902734-47-1, ISSN 1726-9679, Vienna, Austria
DOI: 10.2507/36th.daaam.proceedings.013

Abstract

An imperative placed on companies in the electricity power sector is to invest in new capacities for electricity generation from renewable energy sources, with the aim of increasing energy portfolio flexibility, improving energy efficiency, and reducing emission of pollutants and greenhouse gases. These objectives are also pursued through the acquisition of projects via public tender, based on predefined evaluation criteria. Due to budget constraints for this purpose, it is necessary to rank the offered renewable energy electricity generation projects (hydropower, wind, and photovoltaic). The paper proposes an effective solution to project ranking problem based on the use of the Fuzzy Technique for Order Preference by Similarity to Ideal Solution (FTOPSIS) method with an illustrative case study. The findings are presented, and additional research implications outlined.

Keywords: Renewable energy projects; Ranking problem; Fuzzy Technique for Order Preference by Similarity to Ideal Solution.

1. Introduction

The Energy Community Treaty signed in 2006, between the European Union (EU) and the countries of Southeast Europe, aims to generate an integrated energy market and assist in the integration of these nations into the energy structure of the EU. Another major goal of the Treaty is to improve the environmental situation in terms of energy sources provision and promote renewable energy sources (RES) and energy efficiency. The EU has announced its intentions of becoming climate-neutral by 2050, as part of the European Green Deal. The way to accomplish this objective is by reducing dependency on fossil fuel, and by using more clean technologies, as well as by making energy more efficient. Potential candidate and candidate countries to EU membership like Bosnia and Herzegovina, are expected to embrace and follow EU policies on climate and energy. Slow progress in this process may hurt their EU membership endeavours. The Indicative Energy Strategy of Bosnia and Herzegovina up to 2035 envisages priorities in the development of the energy sector, grouped around three pillars - supply security, competitive pricing and decarbonization [1]. The business strategy of the leading electric utility in the country Elektroprivreda Bosne i Hercegovine (EPBiH) uses these principles as well.

However, EPBiH works in a difficult political and legal environment which often creates delays in the planning and implementation of renewable energy projects. Such an environment slows the rate of energy transition. To increase renewable energy holdings, EPBiH was inspired by the example of Hrvatska elektroprivreda (HEP) that has attempted to develop superior renewable energy sources via acquisitions. The level of use of RES, especially solar energy, is still not at a satisfactory level in Bosnia and Herzegovina [2]. In 2020, EPBiH announced its first public tender to buy RES projects to generate electricity from independent producers. Since funding on such investments is limited, the evaluation criteria should be defined to maintain transparency in the selection process, supported by a suitable ranking method. EPBIH ranked the projects by deploying a linear combination of predefined, fixed weight criteria, considering levelized cost of electricity, installed capacity, and the project implementation stage [3].

The paper [4] illustrates application of multi-criteria decision analysis, which is a combination of Analytic Hierarchy Process (AHP) and Preference Ranking Organization Method for Enrichment Evaluation (PROMETHEE). AHP is used as a tool to generate relative importance of the criteria and PROMETHEE is applied to evaluate and rank alternatives (small hydropower projects). In the paper [5], the criteria for RES electricity generation projects ranking are comprehensively defined, and their relative weights were computed using the Fuzzy Analytic Hierarchy Process (FAHP), with local weights derived through Buckley's geometric mean method.

This paper addresses the problem of ranking renewable energy projects for electricity generation in the acquisition process through a public tender. Project ranking is based on nine criteria categorized into three groups: technical, economic, and environmental criteria [5]. Due the nature of the criteria and the incompleteness of available information, a fuzzy approach is applied because it can handle uncertain and imprecise data [6]. The Fuzzy Technique for Order Preference by Similarity to Ideal Solution (FTOPSIS) method is used in this paper for ranking problem solving in fuzzy environment.

2. FTOPSIS method

Fuzzy TOPSIS is a multi-criteria method that integrates fuzzy approach and TOPSIS method to be more effective with imprecise data. This method allows ranking alternatives when the data is either unclear or not precise enough. In contrast to the traditional TOPSIS method which operates only using the correct and precise information (classical numbers), fuzzy TOPSIS considers uncertainty and imprecision of the information based on the concepts of fuzzy logic [7]. Chen extended the TOPSIS method with triangular fuzzy numbers which in general can be from various fuzzy scales [8]. The fuzzy scale (linguistic variable values and corresponding fuzzy numbers) used by Chen are shown in Table 1. The application of Chen's fuzzy TOPSIS method consists of six steps [9].

Linguistic variable values	Fuzzy number
Very low importance	(0,0,0.1)
Low importance	(0,0.1,0.3)
Medium-low importance	(0.1,0.3,0.5)
Medium importance	(0.3,0.5,0.7)
Medium-high importance	(0.5,0.7,0.9)
High importance	(0.7,0.9,1)
Very high importance	(0.9,1,1)

Table 1. Fuzzy linguistic variable values

Step 1. Construction of the fuzzy decision matrix

Considering m alternatives, n criteria, and k decision-makers, a multi-criteria group fuzzy decision-making problem can be formally expressed in matrix form (1):

$$\tilde{D}=\begin{matrix} \\ A_1 \\ A_2 \\ \vdots \\ A_m \end{matrix}\begin{matrix} C_1 & C_2 & & C_n \\ \left[\begin{matrix} \tilde{x}_{11} \\ \tilde{x}_{21} \\ \vdots \\ \tilde{x}_{m1} \end{matrix}\right. & \begin{matrix} \tilde{x}_{12} \\ \tilde{x}_{22} \\ \vdots \\ \tilde{x}_{m2} \end{matrix} & \begin{matrix} \dots \\ \\ \dots \\ \dots \end{matrix} & \left.\begin{matrix} \tilde{x}_{1n} \\ \tilde{x}_{2n} \\ \vdots \\ \tilde{x}_{mn} \end{matrix}\right] \end{matrix}, i = 1,2,\dots,m;\ j = 1,2,\dots,n \tag{1}$$

where $A_1,A_2,...A_m$ represent set of the alternatives, $C_1,C_2,...,C_m$ are criteria, and $\tilde{x}_{ij}$ corresponds to the evaluation of alternative A_i with respect to criterion C_j as assessed by the decision-makers.

The mean value method for aggregating the fuzzy performance evaluations $\tilde{x}_{ij}$ of decision-maker k with respect to the same criterion is calculated according (2):

$$\tilde{x}_{ij}=\frac{1}{k}\left(\tilde{x}_{ij}^{1}+\tilde{x}_{ij}^{2}+\dots\tilde{x}_{ij}^{k}\right) \tag{2}$$

where $\tilde{x}_{ij}$ is the evaluation of alternative A_i with respect to criterion C_j given by the k-th decision-maker, and is expressed as a triangular fuzzy number $\tilde{x}_{ij}^k=\left(\tilde{a}_{ij}^k, \tilde{b}_{ij}^k, \tilde{c}_{ij}^k\right)$.

Step 2. Normalization of the fuzzy decision matrix given by (1)
Normalization of the fuzzy decision matrix is given by (3) [10].

$$\tilde{R}=[r_{ij}]_{mxn}, i = 1,2, \dots, m; j = 1,2, \dots, n \tag{3}$$

The normalization of the criteria to be maximized is given by (4):

$$r_{ij}=\left(\frac{x_{1ij}}{x_{3j}^*}, \frac{x_{2ij}}{x_{3j}^*}, \frac{x_{3ij}}{x_{3j}^*}\right), x_{3j}^*=max x_{3ij}. \tag{4}$$

The equation (5) is used for the criteria to be minimized:

$$r_{ij}=\left(\frac{x_{1j}^-}{x_{3ij}}, \frac{x_{1j}^-}{x_{2ij}}, \frac{x_{1j}^-}{x_{1ij}}\right), x_{1j}^-=min x_{1ij}. \tag{5}$$

Step 3. Forming of the weighted normalized fuzzy decision matrix
The values of the elements in the weighted fuzzy decision matrix $\tilde{V}$ (6) are calculated according to (7):

$$\tilde{V}=\left[\tilde{v}_{ij}\right]_{mxn}, \quad i = 1,2, \dots, m; j = 1,2, \dots, n \tag{6}$$

$$\tilde{v}_{ij}=\tilde{r}_{ij} \otimes \tilde{w}_j. \tag{7}$$

Values $\tilde{w}_j$ represent the weighting factors.

Step 4. Determination of the positive and negative ideal solution
The positive ideal solution A^* is given by (8), and the negative ideal solution A^- is given by (9):

$$A^*=(v_1^*, v_2^*, \dots, v_n^*) \tag{8}$$

$$A^-=(v_1^-, v_2^-, \dots, v_n^-) \tag{9}$$

where are $v_j^*=(1,1,1)$, $v_j^-=(0,0,0)$, and $j=1,2, \dots, n$.

Step 5. Determination of the distance of alternatives from the A^* and A^-
The distance of alternatives from the A^* and A^- are given by (10) and (11) respectively:

$$d_i^*=\sum_{j=1}^{n} d(\tilde{v}_{ij}, v_j^*), i = 1,2,3, \dots, m, j = 1,2,3, \dots, n \tag{10}$$

$$d_i^-=\sum_{j=1}^{n} d(\tilde{v}_{ij}, v_j^-), i = 1,2,3, \dots, m, j = 1,2,3, \dots, n \tag{11}$$

Step 6. Determination of the relative closeness of the alternatives to the ideal solution
The relative distance of each alternative is calculated according to (12):

$$CC_i=\frac{d_i^-}{d_i^*+d_i^-}, i = 1,2,3, \dots, m \tag{12}$$

The alternative with the index CC_i approaching the value 1 indicates that this alternative is close to the fuzzy positive ideal solution and far from the fuzzy negative ideal solution. In other words, a high value of the index CC_i indicates good performance of the alternative A_i.

3. Data for practical example and results

The FAHP method was used on a case study to five energy projects (two hydroelectric power plants (HPPs), two wind farms (WFs), and one photovoltaic power plant (PVPP)) in Bosnia and Herzegovina. To respect confidentiality, the raw data was a little changed yet in a manner that maintained their overall integrity. Table 2, Table 3, and Table 4 contain the data on these projects.

Alternative	Project	Project implementation status	Projected annual production (GWh/year)	Installed capacity (MW)
1	HPP1	Construction	95	20
2	HPP2	Permitting process	45	10
3	WF1	Construction	165.17	50.6
4	WF2	Planning stage	145	54
5	PVPP1	Planned	17.2	13.54

Table 2. Project implementation status, projected annual production and installed capacity of the projects

Alternative	Project	CAPEX (mill. EUR)	Payback period (years)	Efficiency
1	HPP1	70	7	80%
2	HPP2	40	6	70%
3	WF1	81	6	55%
4	WF2	80	5	53%
5	PVPP1	11.1	8	20%

Table 3. CAPEX, payback period and efficiency of the projects

Alternative	Project	Impact on ecosystem	Location	Spatial requirements (ha/MW)
1	HPP1	Large impact	L1	160
2	HPP2	Very large impact	L2	80
3	WF1	Large impact	L3	66
4	WF2	Moderate impact	L4	78
5	PVPP1	Moderate impact	L5	25

Table 4. Impact on ecosystem, location and spatial requirements of the projects

One decision-maker participated in the evaluation, and the fuzzy ratings of the alternatives according to individual criteria from the technical, economic, and environmental groups are given in Table 5, Table 6, and Table 7, respectively.

Alternative	Project	Project implementation status			Projected annual production			Installed capacity		
1	HPP1	0.90	1.00	1.00	0.50	0.70	0.90	0.70	0.90	1.00
2	HPP2	0.10	0.30	0.50	0.00	0.10	0.30	0.10	0.30	0.50
3	WF1	0.70	0.90	1.00	0.50	0.70	0.90	0.50	0.70	0.90
4	WF2	0.30	0.50	0.70	0.50	0.70	0.90	0.50	0.70	0.90
5	PVPP1	0.10	0.30	0.50	0.10	0.30	0.50	0.10	0.30	0.50

Table 5. The fuzzy ratings of the alternatives according to individual criteria from the technical group

Alternative	Project	CAPEX			Payback period			Efficiency		
1	HPP1	0.30	0.50	0.70	0.50	0.70	0.90	0.90	1.00	1.00
2	HPP2	0.30	0.50	0.70	0.10	0.30	0.50	0.70	0.90	1.00
3	WF1	0.50	0.70	0.90	0.30	0.50	0.70	0.50	0.70	0.90
4	WF2	0.50	0.70	0.90	0.70	0.90	1.00	0.50	0.70	0.90
5	PVPP1	0.70	0.90	1.00	0.50	0.70	0.90	0.50	0.70	0.90

Table 6. The fuzzy ratings of the alternatives according to individual criteria from the economic group

Alternative	Project	Impact on ecosystem			Location			Spatial requirements		
1	HPP1	0.50	0.70	0.90	0.70	0.90	1.00	0.90	1.00	1.00
2	HPP2	0.70	0.90	1.00	0.30	0.50	0.70	0.70	0.90	1.00
3	WF1	0.50	0.70	0.90	0.50	0.70	0.90	0.50	0.70	0.90
4	WF2	0.30	0.50	0.70	0.70	0.90	0.10	0.70	0.90	1.00
5	PVPP1	0.30	0.50	0.70	0.50	0.70	0.90	0.30	0.50	0.70

Table 7. The fuzzy ratings of the alternatives according to individual criteria from the environmental group

Table 8 contains relative proximities of each alternative to the ideal solution (CC_i values), and their normalized values.

Alternative	Project	CC_i	Normalized value
1	HPP1	0.0924	0.2130
2	HPP2	0.0865	0.1993
3	WF1	0.0857	0.1977
4	WF2	0.0876	0.2021
5	PVPP1	0.0815	0.1879

Table 8. Relative closeness of each alternative to the ideal solution

In Table 9, the alternatives are presented, ranked from the best to the worst.

Rank	Alternative	Project	CC_i	Normalized value
1	1	HPP1	0.092	0.2130
2	4	WF2	0.087	0.2021
3	2	HPP2	0.086	0.1993
4	3	WF1	0.085	0.1977
5	5	PVPP1	0.081	0.1879

Table 9. Alternatives ranked from the best to the worst

The findings of the implemented ranking process showed that the score of Project 1 (HPP1) is the highest in the way of relative proximity to the ideal solution, and, therefore, the most desirable among the given projects. The following three ranked projects, Projects 4 (WF2), 2 (HPP2), and 3 (WF1), indicate a high index of alignment to the criteria as well but not as strongly as the project with the highest rank. Project 5 (PVPP1) has the lowest value of relative closeness, making it the least favourable option in this evaluation process. The obtained results were influenced by the established criteria and their assigned values. Looking at the technical group of criteria, which carries the greatest weight, a noticeable difference among the alternatives is evident. HPP1 and WF2 are in construction phase, HPP2 is in the permitting phase, WF1 is in the planning stage, while PVPP1 is still only planned for construction. Additionally, WF1 and WF2 have higher projected annual output than both HPP1 and HPP2, while the projected annual output off HPP1 is higher than that of HPP2. In addition to the technical criteria group, an important criterion is CAPEX, where HPP1 has a higher CAPEX than HPP2, but significantly lower than WF1 and WF2.

4. Conclusion

The problem of ranking electricity generation projects acquired through a public tender is considered in this paper as a fuzzy multicriteria decision-making problem. The criteria (both quantitative and qualitative) and their corresponding weights were predetermined, and the fuzzy TOPSIS method is selected for ranking the projects. The method is implemented on case study of five projects in Bosnia and Herzegovina (two hydropower plants, two wind farms and one photovoltaic power plant) at different levels of development. The findings clearly show the differences in project performance as per the stated criteria with some important information in influencing decisions in the renewable energy project acquisitions process. Project evaluations were provided by a single decision-maker, so future research could involve experts for each specific aspect: technical, economic, and environmental.

5. References

[1] Okvirna energetska strategija Bosne i Hercegovine do 2035. godine. [Framework Energy Strategy of Bosnia and Herzegovina until 2035], Council of Ministers of B&H, 2018.

[2] Lulic, Haris; Gadara, Merima; Metovic, Sadjit & Softic, Almira (2024). Optimum Tilt Angle of Flat Plate Solar Collectors for Different Locations in Bosnia and Herzegovina, Proceedings of the 35th DAAAM International Symposium, pp.0238-0244, B. Katalinic (Ed.), Published by DAAAM International, ISBN 978-3-902734-44-0, ISSN 1726-9679, Vienna, Austria

[3] Lavic, Z.; Dacic-Lepara, S. & Salkic, S. (2021). Ranking method in the project acquisition process, 15. Conference of BH K/O CIGRE Neum, 17-20. october 2021., Neum, Bosna and Herrcegovina

[4] Urosevic Gvozdenac, B & Marinovic, B. (2021). Ranking construction of small hydro power plants using multi-criteria decision analysis. Renewable Energy, Elsevier, vol. 172(C), 1174-1183

[5] Lavic, Zedina & Bibic, Amina (2025). Criteria for the Evaluation and Ranking of Renewable Electricity Generation Projects, unpublished

[6] Aydin, Selahattin Mert; Keklik, Emre; Akdogan, Anil & Vanli, Ali Serdar (2024). Fuzzy Logic-Based Security System for Remote Outdoor Cabinets, Proceedings of the 35th DAAAM International Symposium, pp.0204-0209,

B. Katalinic (Ed.), Published by DAAAM International, ISBN 978-3-902734- 44-0, ISSN 1726-9679, Vienna, Austria

[7] Petrovic, B. (2015). Primena fuzzy TOPSIS metodologije u višekriterijumskoj analizi. [Application of the fuzzy TOPSIS methodology in multi-criteria analysis]. Engineering Management 1(1), 42-56

[8] Sorin, N. (2016). Fuzzy TOPSIS: A General View. Procedia Computer Science, 91, 823 – 831.

[9] Wang, T.C, Chang, T.-H. (2007). "Application of TOPSIS in evaluating initial training aircraft under a fuzzy environment", Expert Systems with Applications, 33, 870-880

[10] Solangi, Y. A., Tan, Q., Mirjat, N. H. and Ali, S. (2019). Evaluating the strategies for sustainable energy planning in Pakistan: An integrated SWOT-AHP and Fuzzy-TOPSIS approach. Journal of Cleaner Production 236. 117655. ISSN 0959-6526.

DOI: 10.2507/36th.daaam.proceedings.014

APPLICATION OF MONTE CARLO SIMULATION FOR EVALUATING WIND INFLUENCE IN EXTERNAL BALLISTICS

Antonia Galić, Zvonko Trzun & Andrej Razumić

This Publication has to be referred as: Galic, Z[vonko]; Trzun, Z[vonko] & Razumic, A[ndrej] (2025). Application of Monte Carlo Simulation for Evaluating Wind Influence in External Ballistics, Proceedings of the 36th DAAAM International Symposium, pp.0107-0115, B. Katalinic (Ed.), Published by DAAAM International, ISBN 978-3-902734-47-1, ISSN 1726-9679, Vienna, Austria
DOI: 10.2507/36th.daaam.proceedings.014

Abstract

This study investigates the application of Monte Carlo simulation to evaluate the impact of wind on the accuracy and precision of unguided rockets. The influence of constant stochastic wind is modelled with wind direction and intensity as variable parameters. Given that wind has the most significant effect on projectile trajectory deviations and impact point dispersion among atmospheric conditions, accurate modelling and parameter selection are critical for reliable simulation outcomes. The study further explores the effect of different probability distributions on the dispersion of rocket impact points. To this end, two commonly used distributions in external ballistics—the Gaussian (normal) and Rayleigh distributions—are employed to generate random wind direction and intensity. Following Monte Carlo simulations, the resulting data are analysed and compared both analytically and graphically. Results show that the Gaussian distribution yields a more concentrated grouping of impact points around the target compared to the Rayleigh distribution, and that wind indeed has a significant influence on the dispersion of unguided projectile impact points.

Keywords: Monte Carlo simulation; unguided projectile precision; stochastic wind modelling; Gaussian vs. Rayleigh probability distribution; rocket precision.

1. Introduction

In modern military technology and aviation systems, understanding and predicting the influence of weather conditions on rocket performance is of utmost importance. One of the key factors that can significantly affect the trajectory and precision of unguided rockets is wind. Wind is a complex variable that can vary in intensity and direction during the rocket's flight [1]. Therefore, it is interesting to compare simulations of the impact of constant stochastic wind on unguided rockets using different probability distributions. This study addresses two research questions. First, do different probability distributions (used to generate random wind) introduce significant differences in the estimated dispersion of impact points for unguided projectiles [2], [3]. The second research question is whether Monte Carlo simulation can be successfully applied to this specific problem (evaluating the dispersion of trajectories and impact points of unguided projectiles due to wind influence) and whether Monte Carlo provides additional insights compared to a deterministically formulated problem.

Monte Carlo simulation is a widely used method for analysing the effects of multiple disturbances and resulting changes in the observed variable [4], [5]. It is employed when it is impractical to establish a clear functional relationship between multiple input parameters and outcomes that depend on their variations. The method relies on repeated nondeterministic sampling to obtain numerical results, leveraging the stochastic nature of the sample to draw conclusions about processes that may be deterministic in nature. This method has already been used in several studies examining the influence of disturbances on rocket flight [6], [7], but these works did not focus on the impact of wind. Of particular interest are studies [8] and [9], which link wind effects to rocket trajectories, and our paper aims to expand on those findings.

2. Methodology

To address the research questions posed in the Introduction, a Monte Carlo simulation of rocket firings will be conducted. The rocket's characteristics will remain constant, while only the wind speed and direction will vary. The Monte Carlo method is especially suitable when the nature and probability distribution of the variables affecting the outcome (in this case, projectile flight) are unknown [10]. It is also useful for cases where simplifying assumptions (e.g., linearization or normality of the sample) cannot be made in advance, or for analysing samples that exhibit asymmetry, skewness, or kurtosis in their frequency distributions [11].

In this study, the Monte Carlo method is used to generate a large number of random wind samples under different probability distributions. Each simulation produces data on the rocket's trajectory and impact point, including deviation from the target, trajectory elements, and flight time. The results are compared with previously verified data from controlled test firings published in the Firing Table for the specified rocket, serving as a control for the simulation results.

Wind force is scaled using the Beaufort Wind Scale (Table 1).

Force	Description	Wind Speed	Main Effects on Land
0	Calm	< 1 km/h (< 0.3 m/s)	Smoke rises vertically
1	Light Air	1–5 km/h (0.3–1.5 m/s)	Wind direction seen from smoke
2	Light Breeze	6–11 km/h (1.6–3.3 m/s)	Wind felt on face, weather vanes move
3	Gentle Breeze	12–19 km/h (3.4–5.5 m/s)	Leaves and twigs in constant motion
4	Moderate Breeze	20–28 km/h (5.5–7.9 m/s)	Dust raised, small branches move
5	Fresh Breeze	29–38 km/h (8–10.7 m/s)	Small leafy trees begin to sway
6	Strong Breeze	39–49 km/h (10.8–13.8 m/s)	Large branches move, whistling in wires
7	Near Gale	50–61 km/h (13.9–17.1 m/s)	Whole trees move, walking is difficult
8	Gale	62–74 km/h (17.2–20.7 m/s)	Twigs break off trees
9	Strong Gale	75–88 km/h (20.8–24.4 m/s)	Light structural damage
10	Storm	89–102 km/h (24.5–28.4 m/s)	Trees uprooted, significant building damage
11	Violent Storm	103–117 km/h (28.5–32.6 m/s)	Widespread destruction
12	Hurricane	≥ 118 km/h (≥ 32.7 m/s)	Catastrophic destruction

Table 1. Beaufort Wind Scale

For generating random samples, the wind speed is constrained to $|W| < 9.8\ \mathrm{m/s}$, which approximately corresponds to the upper limit of Force 5 on the Beaufort scale. Beyond this threshold, simulation results may become unrealistic, especially if strong crosswinds act on the rocket's trajectory, causing significant deviations from the target. Depending on the type of artillery projectile, wind speed, and the angle at which it acts, the wind's influence can vary. In ballistics, wind is typically decomposed into longitudinal (head/tail) and lateral (left/right) components. Longitudinal wind follows intuitive logic: tailwinds increase range, while headwinds decrease it.

Lateral wind is more complex, affecting the firing direction depending on the projectile type and specific conditions:

- Classical projectiles (and rockets in passive flight) drift downwind (e.g., leftward wind pushes them right).
- Rockets in active flight (with thrust) turn into the wind (e.g., leftward wind causes a leftward turn). Once the motor stops, they behave like classical projectiles. This is due to the centre of pressure being behind the centre of mass, creating a moment that turns the rocket into the wind.

Two series of rocket firings are simulated using Monte Carlo. Each series consists of 1,000 nondeterministically generated initial conditions (ballistic winds). In the first series, wind speed follows a normal distribution, while in the second, it follows a Rayleigh distribution. The Rayleigh distribution is a special case of the Weibull distribution (shape parameter $k=2$) and is commonly used in external ballistics to model wind dispersion.

2.1. Wind Speed Dispersed According to the Normal Distribution

For normal distribution the probability density is calculated as:

$$f(W,\sigma) = \frac{1}{\sigma\sqrt{2\pi}} e^{\frac{(W-\mu)^2}{2\sigma^2}} \tag{1}$$

where, in our problem, the independent variable is the wind speed *W*. This implies that the probability that the wind speed magnitude |*W*| is less than or equal to a certain threshold W_1 is given by:

$$P(|W| \le W_1) = \frac{1}{\sigma\sqrt{2\pi}} \int_{-W_1}^{W_1} e^{\frac{W^2}{2\sigma^2}} dW \tag{2}$$

It is evident that the probability in Equation (2) is directly determined by the selected standard deviation σ. Based on the Beaufort scale (Table 1) and empirical data, a value of $\sigma = 5\ \mathrm{m/s}$ is selected for simulation purposes, assuming a normal distribution. By choosing this parameter, it is ensured that, with 95% confidence, the wind speed remains within the range of interest $|W| \le 9.8\ \mathrm{m/s}$. This also maintains compatibility with results derived from firing tables, which are typically constructed based on test firings conducted under low-wind conditions.

The generated input dataset consisted of 1,000 wind speed values, and the dispersion parameters indicate that the sample was appropriately selected and satisfies the conditions of normality: the mean is -0.11, standard deviation 5.04, kurtosis 0.028, and skewness -0.009. These statistics, together with the histogram confirming visual similarity to the expected frequency distribution, verify that the sample is normally distributed. Figure 1 presents the histogram of total wind speed.

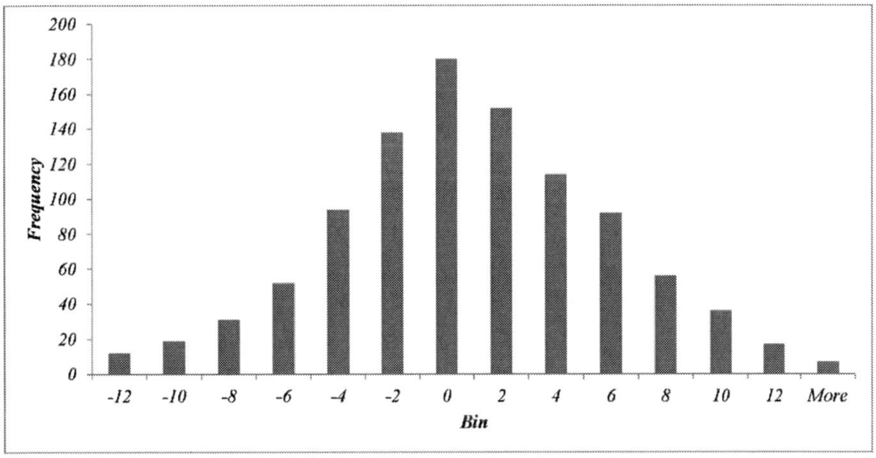

Fig. 1. Histogram of a wind speed sample, distributed according to the normal distribution

In the generated sample, some values significantly exceed the predefined threshold of 9.8 m/s. Such outliers are removed from the input dataset and are not included in the final simulation. A total of 62 such outliers were identified, which slightly exceeds the remaining 5% probability. After outlier removal, the input dataset consists of 938 wind speed values, which are then uniformly decomposed into longitudinal and lateral wind components (as will be explained in Section 2.3).

2.2. Wind Speed Dispersed According to the Rayleigh Distribution

The Rayleigh distribution is a continuous, non-negative probability distribution (it is actually a special case of the Weibull distribution, with shape parameter *k = 2*). It arises in various applications, particularly in the context of signal processing, communications, and physics, where it describes the magnitude of a vector whose components are independent and identically distributed Gaussian random variables. The Rayleigh probability density function is given as:

$$f(W,\sigma) = \frac{W}{\sigma^2} e^{-\frac{W^2}{2\sigma^2}} \tag{3}$$

where the probability that the wind speed *W* will be less than a certain threshold value W_1 is equal to:

$$P(W \leq W_1) = \int_0^{W_1} \frac{W}{\sigma^2} e^{-\frac{W^2}{2\sigma^2}} dW = 1 - e^{-\frac{W_1^2}{2\sigma^2}} \tag{4}$$

It can be seen that, as with the normal (Gaussian) distribution, the shape of the probability density function and cumulative distribution function directly depends on the choice of the parameter σ. It should be emphasized that in the Rayleigh distribution, $\sigma$is a scale parameter and not the standard deviation of the function. The maximum probability density of the above function occurs at the value W_m which is obtained by differentiating the probability density function:

$$f'(W_m) = -\frac{(W_m^2-\sigma^2)^2}{\sigma^2} e^{-\frac{W_m^2}{2\sigma^2}} = 0 \quad \rightarrow \quad W_m = \sigma \tag{5}$$

while the expected value of this distribution is defined by the following integral:

$$\mu = \int_0^{\infty} W \cdot f(W) dW = \int_0^{\infty} \frac{W^2}{W_m^2} e^{-\frac{W_m^2}{2\sigma^2}} dW = W_m \sqrt{\frac{\pi}{2}} \tag{6}$$

from which it follows that the expected value of the Rayleigh distribution can also be expressed in terms of W_m or equivalently σ as the scale parameter ($\mu \approx 1.253\, W_m$).

The Rayleigh distribution is frequently used in wind energy and meteorological studies to model wind speed, especially in regions where wind direction is uniformly distributed (and this is one of assumptions for our analysis) and speed follows a positively skewed pattern. It is particularly suited for modelling horizontal wind speed when the wind has random direction and its two orthogonal components are normally distributed with zero mean and equal variance σ.

For easier comparison of the effects of two selected probability density function, the scale parameter was set to $\sigma =$ 4 m/s. This choice ensures an approximately equal 95% confidence that the wind speed remains within the range of interest $W \leq 9.8$ m/s. The generated dataset, consisting of 1000 values, was evaluated through descriptive statistics to assess its alignment with the theoretical Rayleigh distribution with a chosen scale parameter. Key statistical indicators were calculated, and the mean (4.96) and standard deviation (2.59) were consistent with the expected theoretical values:

$$\mu[X] = \sigma\sqrt{\frac{\pi}{2}} \approx 5.013, \quad STD[X] = \sigma\sqrt{\frac{(4-\pi)}{2}} \approx 2.538 \tag{7}$$

The sample also exhibited a skewness of 0.537 which closely align with the theoretical values for a Rayleigh distribution, namely 0.631. These results suggest a distribution that is positively skewed and moderately peaked, in line with the Rayleigh profile. The histogram analysis further confirmed the visual similarity between the empirical and theoretical probability density functions (Figure 2).

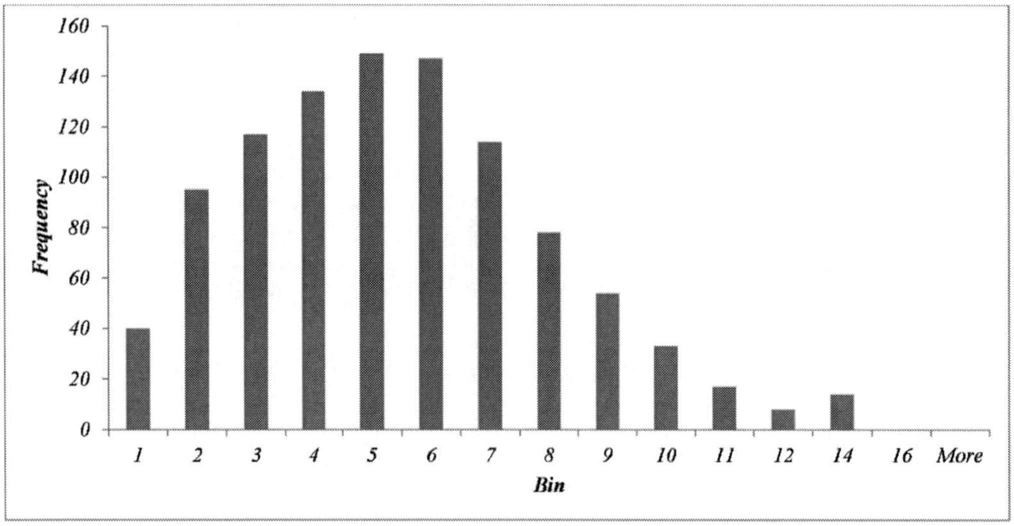

Fig. 2. Histogram of a wind speed sample, distributed according to the Rayleigh distribution

As for outliers (e.g. values where $W > 9.8$ m/s) in the generated wind sample 49 cases have been identified (which corresponds very well with the remaining probability of 5%); these winds have been removed from the input wind dataset, leaving the final input set for 951 simulated firing of rockets.

2.3. Wind direction and other simulation parameters

The wind direction in both series is dispersed according to a uniform distribution over the full circle $\phi_W = 0 - 360°$ Since the normal distribution includes negative values, it can be also used with $\phi_W = 0 - 180°$. An alternative approach is to use a half-normal distribution, which contains only non-negative values, while the wind direction is dispersed over the full circle. These two methods ultimately produce very similar or identical results, especially with sufficiently large sample sizes (empirically: $n > 300$). As previously mentioned, lateral wind is particularly hazardous for rocket accuracy, although strong tailwinds or headwinds can also significantly affect the magnitude of the miss distance.

Overall, considering both wind speed and direction, the wind is modelled as a constant stochastic wind. The term "stochastic" is used here due to the nondeterministic nature of the two parameters defining the wind—both variables change from one simulated rocket to another. The term "constant" refers to the fact that once defined, the wind maintains the same direction and magnitude throughout the flight of a single rocket; no changes are introduced with atmospheric layers or as the rocket approaches the target during flight. In external ballistics, this type of wind is also referred to as ballistic wind and is calculated to produce the same effect on the projectile's trajectory as actual wind varying with altitude and distance to the target would. Regarding other atmospheric parameters, both simulations use the standard ICAO atmosphere (T = 15 °C; p = 1013 hPa; relative humidity Φ = 0%). The only deviation from the standard atmosphere is the presence of wind.

A mathematical 6DOF (six degrees of freedom) rocket flight model was used, which accounts for all influential parameters such as the rocket's inertial properties, aerodynamic coefficients, initial angles, and others [12].

$$
\begin{aligned}
&\begin{bmatrix} \dot{x} \\ \dot{z} \\ \dot{z} \end{bmatrix} = \boldsymbol{L}_{LP} \begin{bmatrix} u_k \\ \tilde{v}_k \\ \widetilde{w}_k \end{bmatrix} \\
&\dot{u}_k = -\tilde{q}\widetilde{w}_k + \tilde{r}\tilde{v}_k + \frac{X + F_x}{m} - g \sin\vartheta - a_{Kx} \\
&\dot{\tilde{v}}_k = p_P\widetilde{w}_k - \tilde{r}u_k + \frac{\tilde{Y} + F_{\tilde{y}}}{m} - a_{K\tilde{z}} \\
&\dot{\widetilde{w}}_k = -p_P\tilde{v}_k + \tilde{q}u_k + \frac{\tilde{Z} + F_{\tilde{z}}}{m} + g\cos\vartheta - a_{K\tilde{z}} \\
&\dot{p} = \frac{(L + L^F)}{I_x} \\
&\dot{\tilde{q}} = p_P\tilde{r} + \frac{-I_x p\tilde{r} + \widetilde{M} + \widetilde{M}^F}{I_y} \\
&\dot{\tilde{r}} = -p_{PA}\tilde{q} + \frac{I_x p\tilde{q} + \widetilde{N} + \widetilde{N}^F}{I_y} \\
&\dot{\varphi} = p + \tilde{r}tg\vartheta \\
&\dot{\vartheta} = \tilde{q} \\
&\dot{\psi} = \frac{\tilde{r}}{\cos\vartheta}
\end{aligned}
\tag{8}
$$

This mathematical model also allows the conversion of constant (ballistic) wind into real wind conditions using meteorological bulletins (e.g., Standard Computer Met Message STANAG 4082). However, ballistic wind was intentionally used here to simplify the correlation between a given wind condition and accuracy relative to the target. The simulation modelled the firing of an unguided rocket with a diameter of 122 mm, a length of 2.7 m, and a maximum range of 20,500 m. The target was set at 13,700 m, representing two-thirds of the maximum range, which is a common practice for verifying the correctness of firing tables and mathematical models.

3. Results and Discussion

A sample generated according to the normal distribution is expected to typically show greater clustering around the ideal (mean) value compared to a sample from the Rayleigh distribution. This is because the normal distribution is symmetric around the mean μ, which also corresponds to the highest probability density. Therefore, most of the sample is expected to be tightly clustered around the target location (e.g., approximately 68% of hits are expected, given a sufficiently large sample, within the confidence interval $\mu \pm \sigma$). The dispersion plot of impact points indeed shows significant clustering around the target (Figure 3):

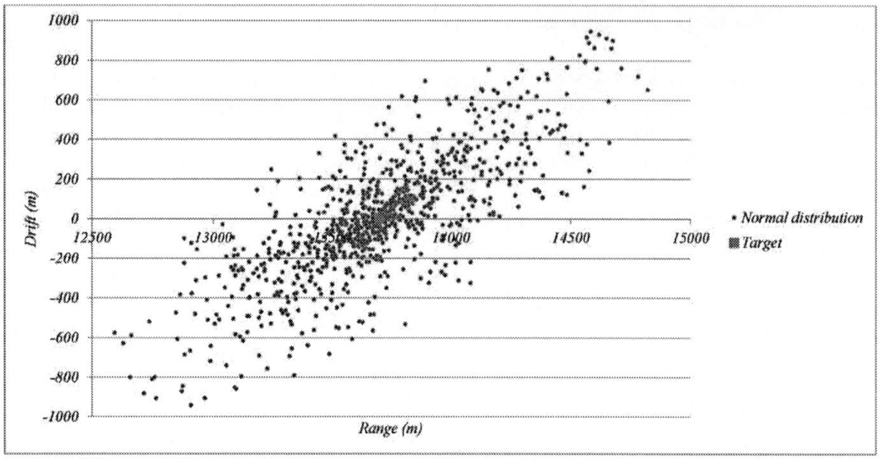

Fig. 3. Effect of normally distributed wind on the dispersion of projectile impact points

The dispersion pattern of impact points appears rotated relative to the firing line (which corresponds to the x-axis on the graph). This is a result of the asymmetry of forces and moments acting on the rocket. The fins at the rear of the rocket are rotated relative to the projectile's axis of rotation for design reasons, which introduces an additional component of Magnus force causing an induced moment and rotation of the dispersion pattern, as shown in Figure 3. The average range is 13,707 m, the minimum range is 12,594 m, and the maximum range is 14,822 m. The standard deviation of the range is 360 m. Regarding lateral drift, the average is 9 m, the minimum is –944 m, and the maximum is 946 m, with a standard deviation of 306 m. Although these elevated deviation values may appear anomalous, they are consistent with the modelling assumption of a uniform wind field - i.e., constant speed and direction throughout the entire trajectory - which does not reflect realistic atmospheric conditions. In practice, a projectile traverses multiple atmospheric layers wherein wind speed and direction vary with altitude (notably, the trajectory apex reaches approximately 2400 m when employing a small drag ring). Despite this simplification, it remains evident that, under high wind shear conditions, the aerodynamic stabilization moments generated by fin surfaces are insufficient to maintain trajectory stability. Consequently, unguided rockets exhibit substantial dispersion and are therefore unsuitable for engagement of point targets, being instead primarily employed in area saturation roles.

A second sample, generated according to the Rayleigh distribution, is expected to exhibit even less clustering around the target location. As seen in the input parameter analysis (Figure 2), the mean and median values are farther from the target compared to the normal distribution. In the Rayleigh distribution, the probability spreads more widely, resulting in a higher frequency of outliers relative to the centre. This issue is not pronounced here because the scale parameter of the Rayleigh distribution was deliberately chosen based on the threshold value (9.8 m/s), rather than another criterion—e.g., to directly compare wind speeds with a 50% probability or similar. Nevertheless, even with this parameter choice, the difference in trajectory dispersion is significant (Figure 4).

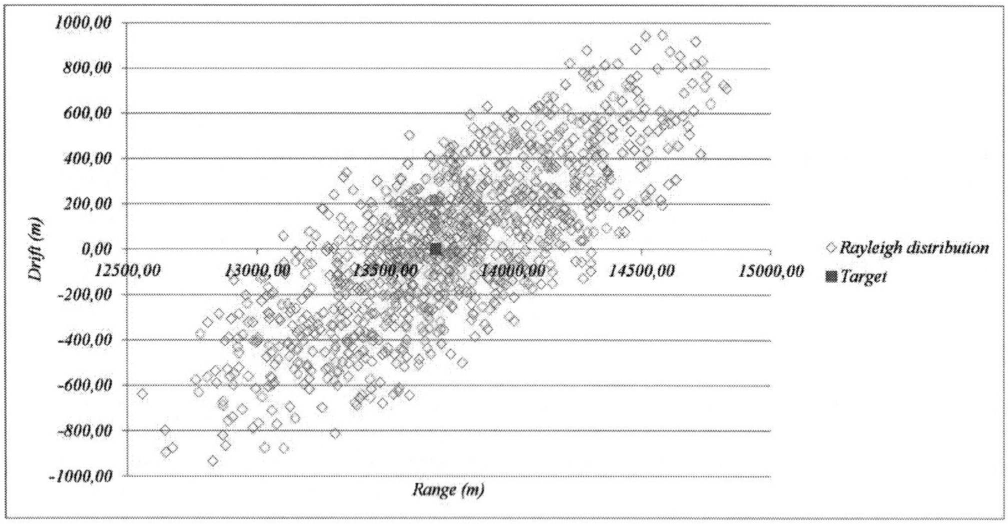

Fig. 4. Effect of Rayleigh-distributed wind on the dispersion of projectile impact points

Even with this stochastically generated wind sample, it can be observed that the dispersion pattern of impact points is rotated relative to the firing line (approximately the x-axis). The imagined trend line closely matches the trend line shown in Figure 3. However, a more important conclusion is that in Figure 4 the dispersion area is significantly less clustered around the target position and more widely spread compared to the sample generated from the normal distribution. The average range is 13,738 m, the minimum range is 12,559 m, and the maximum range is 14,829 m. The standard deviation of the range is 434 m. Regarding lateral drift, the average is 15 m, the minimum is –934 m, and the maximum is 945 m, with a standard deviation of 357 m. It is evident and analytically proven that the scatter is significantly greater. The previously stated conclusion remains valid: the large deviation values result from the imposed strong ballistic wind—assumed constant in both speed and direction along the entire trajectory—which unguided rockets are unable to compensate for through their stabilization moments.

Figure 5 presents both dispersion patterns of impact points together. The dispersion caused by normally-distributed winds is significantly narrower and more concentrated around the imagined trend line (Figure 5):

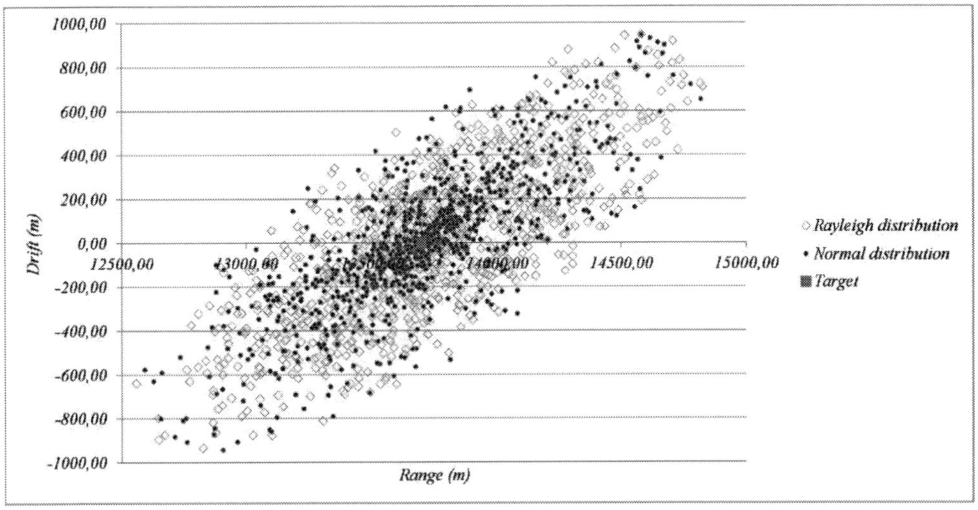

Fig. 5. Comparison of the effect of stochastic wind distributed according to two distributions (Normal vs. Rayleigh)

The displayed images excellently illustrate the value of Monte Carlo simulation. Additionally, Figure 6 shows a deterministically defined boundary line that limits the dispersion area of impact points for the case of maximum considered wind $W = 9.8\,\mathrm{m/s}$. This boundary provides valuable information and demonstrates the magnitude of deviations from the target position that can be expected when firing unguided rockets under such strong ballistic wind conditions.

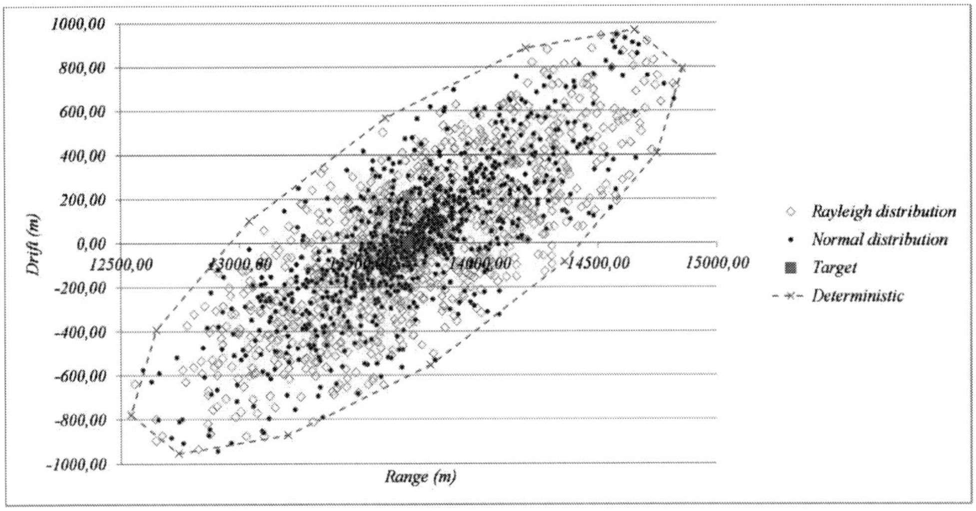

Fig. 6. Comparison of the effect of stochastic wind distributed according to two distributions, together with deterministically defined boundaries for the maximum allowable wind ($W = 9.8$ m/s)

The deterministically defined boundary clearly indicates the expected limits within which projectile impact points may fall. However, the Monte Carlo simulation provides additional information:

- It offers a statistical overview of the expected impact distribution, showing not only the outer bounds (as given by deterministic analysis) but also the internal structure of the dispersion area. Specifically, it quantifies how many projectiles are likely to impact near the target, forming a dense cluster, and how many will fall closer to the outer, deterministically defined boundary. This enables a much clearer understanding of hit probability within certain zones and supports better risk assessment and operational planning.
- It underscores the critical role of correctly selecting the probability distribution used to generate stochastic wind conditions. As shown in the comparison between normally and Rayleigh-distributed wind samples, the choice of distribution has a profound effect on the resulting dispersion pattern. A poorly chosen distribution can lead to unrealistic assumptions about rocket accuracy or safety margins, while a well-matched model leads to more reliable and actionable predictions.
- It enables probabilistic risk analysis and sensitivity studies, allowing analysts to explore how variations in environmental conditions (such as wind speed and direction) impact the overall effectiveness and precision of unguided rocket fire. This is especially important in operational scenarios where environmental uncertainty cannot be avoided but must be accounted for quantitatively.
- Finally, it bridges the gap between theoretical models and real-world variability, simulating a large number of possible outcomes that reflect the randomness of real atmospheric conditions. This makes Monte Carlo simulation a powerful complement to deterministic modelling, especially when dealing with systems—such as unguided rockets—that are highly sensitive to external perturbations like wind.

4. Conclusion

This study analysed the effect of wind on the dispersion of trajectories and impact points of unguided rockets. Two datasets, each consisting of 1,000 wind samples with stochastically defined speed (intensity) and direction, were generated. The first dataset was based on the normal distribution, while the second followed the Rayleigh distribution. The analysis of the simulation results provided answers to both research questions posed in the Introduction, within the defined scope of this investigation.

First, regarding the choice of the most appropriate probability distribution function, the authors recommend that when assessing the impact of atmospheric factors such as wind, the Weibull or Rayleigh distribution should be used. These functions yield a wider dispersion of impact points around the target, which aligns with empirical data and with the limits defined by the deterministic boundary conditions. On the other hand, for variables that are more controllable—such as manufacturing precision—the normal distribution may be more suitable, as it better reflects the expectation of smaller standard deviations and reduced uncertainty in production quality.

Second, the presented case study also illustrates the value of Monte Carlo simulation. Unlike deterministic models, Monte Carlo provides an additional layer of insight and enables more informed decision-making regarding the appropriateness of methods and acceptable dispersion limits for influencing parameters. The added value of such information has been demonstrated here in the domain of external ballistics of unguided rockets; however, the applicability of Monte Carlo simulation is equally relevant in other branches of physics and mechanics. In particular, Monte Carlo simulations prove invaluable when assessing the combined influence of multiple variable factors, especially when their higher-order interactions are unclear or difficult to model analytically.

The presented study serves as a foundation for further research that will investigate more complex atmospheric conditions, various projectile types, and target hit probability assessment. Future work will focus on extending the atmospheric model by replacing the simplified constant ballistic wind with a layered meteorological model. This enhanced model will incorporate vertical variations in wind speed and direction, as well as the influence of temperature, pressure, and relative humidity. Such an approach aims to achieve a more realistic simulation of environmental conditions affecting projectile flight throughout its trajectory. Additionally, the influence of atmospheric disturbances will be analysed across different projectile types, including variations in geometry, mass, and stabilization methods (e.g., passively stabilized, actively stabilized, and guided projectiles). This will enable a comparative evaluation of sensitivity to environmental perturbations for each configuration, and also the more accurate assessment of effectiveness and reliability of unguided rocket fire under realistic environmental conditions.

5. References

[1] Sun, D.; Liang, X.; Zhao, T.; Fang, H.; Cheng, Z.; Zheng, H.; Zhang, L.; Zhang, F.; Sun, D. & Liu, Y. (2019). Study on large scale projectile impact point positioning method based on trajectory shock wave, Proceedings of the 23rd International Congress on Acoustics, 9–13 Sept 2019, Aachen, Germany, pp. 3346–3353

[2] Hennessey, J. P. J. (1977). Some aspects of wind power statistics, J. Appl. Meteorol., Vol. 16, No. 2 (February 1977), pp. 119–128

[3] Wang, R.; Li, W. & Bagen, B. (2011). Development of wind speed forecasting model based on the Weibull probability distribution, Proceedings of 2011 International Conference on Computer Distributed Control and Intelligent Environmental Monitoring, 2011, Xi'an, China, pp. 2062–2065

[4] Rausch, C.; Nahangi, M.; Haas, C. & Liang, W. (2019). Monte Carlo simulation for tolerance analysis in prefabrication and offsite construction, Autom. Constr.

[5] Ilg, M.; Rogers, J. & Costello, M. (2011). Projectile Monte-Carlo trajectory analysis using a graphics processing unit, Proceedings of the AIAA Atmospheric Flight Mechanics Conference, 08–11 Aug 2011, Portland, Oregon, USA. DOI: 10.2514/6.2011-6266

[6] Hanson, J. M. & Beard, B. B. (2012). Applying Monte Carlo simulation to launch vehicle design and requirements verification, J. Spacecr. Rockets

[7] Trzun, Z. & Vrdoljak, M. (2020). Monte Carlo simulation of missile trajectories dispersion due to imperfectly manufactured warhead, Proceedings of the 31st DAAAM International Symposium, 21–24 Oct 2020, Mostar, Bosnia and Herzegovina, ISSN 1726-9679, ISBN 978-3-902734-29-1, Katalinic, B. (Ed.), pp. 0574–0583, DAAAM International, Vienna, DOI: 10.2507/31st.daaam.proceedings.079

[8] Kenney, W. A. (2017). Improved ballistic wind prediction using projectile tracking data, Ph.D. Dissertation, Old Dominion University, Norfolk, USA

[9] Yao, J.; Wang, X. & Li, J. (2009). Influence on rocket fall points distribution due to random wind, Proceedings of 2009 4th IEEE Conference on Industrial Electronics and Applications, 2009, Xi'an, China, pp. 1860–1863

[10] Miedziński, D.; Głębocki, R. & Jacewicz, M. (2023). Analysis of sounding rocket dispersion using Monte-Carlo simulation, In: Automation 2023: Key Challenges in Automation, Robotics and Measurement Techniques, Szewczyk, R.; Zieliński, C.; Kaliczyńska, M. & Bučinskas, V. (Ed.), pp. 39–50, Springer Nature Switzerland

[11] Chase, K. W. & Parkinson, A. R. (1991). A survey of research in the application of tolerance analysis to the design of mechanical assemblies, Research in Engineering Design

[12] Trzun, Z.; Vrdoljak, M. & Cajner, H. (2021). The effect of manufacturing quality on rocket precision, Aerospace, Vol. 8, No. 6, pp. 160, doi:10.3390/aerospace8060160

DOI: 10.2507/36th.daaam.proceedings.015

DESIGN-ORIENTED SOBOL' SENSITIVITY ANALYSIS OF KEY PARAMETERS IN PLATE HEAT EXCHANGER OPTIMISATION

Zaid Al-Dailami, Anas F. A. Elbarghthi, Vaclav Dvorak & Jaroslav Hlava

This Publication has to be referred as: Al-Dailami, Z[aid]; Elbarghthi, A[nas] [F.] [A.]; Dvorak, V[aclav] & Hlava, J[aroslav] (2025). Design-Oriented Sobol' Sensitivity Analysis of Key Parameters in Plate Heat Exchanger Optimisation, Proceedings of the 36th DAAAM International Symposium, pp.0116-0121, B. Katalinic (Ed.), Published by DAAAM International, ISBN 978-3-902734-47-1, ISSN 1726-9679, Vienna, Austria
DOI: 10.2507/36th.daaam.proceedings.015

Abstract

Plate heat exchangers (PHEs) are widely used across industries due to their compactness, high efficiency, and reliability. However, their performance can be substantially improved through targeted design optimisation. This study focuses on optimising a one-dimensional PHE model, validated through both experimental data and CFD simulations. The goal is to enhance thermal effectiveness while minimising pressure drop by adjusting key geometric parameters within feasible limits. A multi-objective optimisation strategy is employed to balance these competing performance goals. To reduce computational effort and improve efficiency, a pre-optimisation sensitivity analysis using Sobol's method identifies the most influential variables. This allows the design space to be reduced from six parameters to two without compromising accuracy. The channel gap is found to be the dominant factor affecting pressure drop, contributing over 50% to its variation. In contrast, plate thickness has the greatest impact on thermal effectiveness, accounting for more than 40% of the total and interaction effects. These insights enable more targeted design adjustments and demonstrate the value of integrating sensitivity analysis into the optimisation process. Overall, the study presents an effective and computationally efficient approach to improving PHE performance in industrial applications.

Keywords: Plate Heat Exchanger; multi-objective optimisation; Pareto front; Sobol's sensitivity analysis.

1. Introduction

PHEs have emerged as essential components in a broad range of industrial applications, including power generation, chemical processing, food manufacturing, and HVAC systems, owing to their compactness, ease of maintenance, and exceptional heat transfer efficiency [1]. Their unique design, which incorporates thin, corrugated plates to promote turbulence, results in heat transfer coefficients that far exceed those of conventional shell-and-tube exchangers [2]. This has led to widespread adoption of PHEs, driving ongoing research to further enhance their performance and reliability. The European market for PHEs, valued at $1.7 billion in 2023 and projected to reach $3.3 billion by 2032, reflects the strong demand for efficient and reliable heat exchanger solutions. Notably, over a quarter of this market (around 27.2%) cater to central heating and refrigeration applications, highlighting the increasing importance of optimisation and sensitivity analyses in PHE design [3].

In recent years, sensitivity analysis has become a valuable tool for measuring how variations in design parameters such as plate geometry, fluid properties, and fouling influence PHE performance [4]. These analyses give engineers essential insights to prioritise design changes and optimise operating conditions, thereby boosting energy efficiency and operational resilience [5]. This approach is especially important as industries face increasing demands for sustainable and cost-effective thermal management solutions [6].

Advancements in sensitivity analysis have integrated Monte Carlo simulations with global sensitivity indices, such as Sobol' indices, to quantify parameter influence effectively. To reduce computational cost, surrogate models like Polynomial Chaos Expansions and Gaussian Processes have been employed, enabling analytical or efficient estimation of sensitivity metrics [7],[8]. Additionally, machine learning–based surrogates have further enhanced sensitivity analysis by capturing complex parameter interactions and uncertainties [9]. Moreover, sensitivity analysis is often combined with multi-objective optimisation frameworks to identify critical variables and fine-tune exchanger designs for specific industrial requirements [10].

In this context, the current study demonstrates that merging sensitivity analysis with optimisation strategies, by maximising heat transfer efficiency while minimising pressure drop, leads to PHE designs with superior performance and reduced energy consumption. Specifically, the research applies a Sobol' sensitivity analysis and a pre-processing approach to identify the most influential parameters, thereby streamlining the multi-objective optimisation process and achieving faster and more efficient results.

2. Heat Exchanger Optimisation

The plate to be optimised is already available on the market and used in many refrigeration applications [11]. Figure **1** illustrates the geometrical specifications of the investigated plate, and its dimensions can be seen in Table **1**.

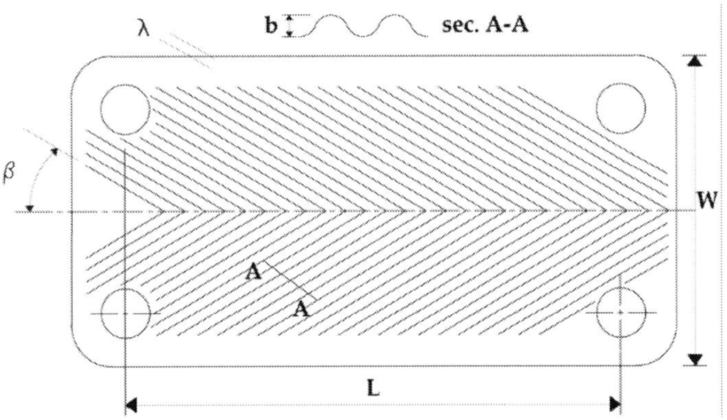

Fig. 1. Schematic illustration of the investigated PHE

Geometry parameter	Value
Effective flow length, L (mm)	250
Plate thickness, t_{pl} (mm)	0.6
Port diameter, D_{po} (mm)	30
Plate pitch, P_i (mm)	2.5
Mean channel gap, b (mm)	1.9
Plate width, W (mm)	95
Corrugation pitch, λ (mm)	6.8
Corrugation angle, β (°)	64

Table 1. Geometric characteristics of the investigated PHE.

2.1 Multi-objective optimisation Process

The main objective of this research is to identify an optimised shape that provides high thermal performance. Specifically, the aim is to maximise effectiveness ε, while reducing pressure drop ΔP across both channels of the plate heat exchanger (hot and cold). This is achieved by adjusting the geometrical parameters of the plate. Formally, the optimisation problem is defined as:

$$\max_{x \in D} \varepsilon(x) \quad , \quad \min_{x \in D} \Delta P(x) \tag{1}$$

subject to the constraints:

$$g_j(x) \leq 0, \quad j = 1, \dots, J \tag{2}$$

$$x_{lower} \leq x \leq x_{upper} \tag{3}$$

where x is the design vector of parameters in Table 2: $x = (\beta, b, \lambda, t_{pl}, D_{po}, N_{pt})$, D is the feasible design space constrained by the constraint function $g_j(x)$, x_{lower} and x_{upper} represent the lower and upper bounds of the parameters. The outcome of this multi-objective optimisation is the Pareto front, which represents the trade-off and optimal balance between the conflicting objectives. The overall optimisation procedure consists of Kriging metamodeling in the adaptive sampling step and a multi-objective genetic algorithm (NSGA-II) [12]. The NSGA-II algorithm, implemented via MATLAB's built-in gamultiobj function, is employed as the optimisation engine. It follows key steps: evaluating the objective functions for a parameter vector, selecting the best individuals based on the Pareto relationship, recombining genes to create new solutions, and applying mutations to explore additional possibilities.

2.2 Sobol' Sensitivity analysis

To identify the most influential design parameters in PHEs, Sobol' sensitivity analysis is used. This variance-based global sensitivity analysis method breaks down the output variance into contributions from individual input variables and their interactions. If the six design parameters characterised in Table **2** are within the specified optimisation range, then the performance metric of interest, such as effectiveness (ε) and pressure drop (ΔP), is denoted as cap Y. The total variance of cap Y is decomposed as[13]:

$$Var\ (Y) = \sum_{i=1}^{6} V_i + \sum_{i<j} V_{ij} + \dots + V_{1,2,\dots,6} \tag{4}$$

where V_i is the variance contribution from parameter i alone, and V_{ij} is the contribution from the interaction between parameters i and j, and so on. The first-order Sobol' index for parameter i is:

$$S_i = \frac{V_i}{Var(Y)} \tag{5}$$

This index quantifies the effect of varying parameter i alone on the output variance. The total-effect index, accounting for all interactions involving parameter i, is:

$$S_{T_i} = 1 - \frac{Var_{\sim i}(E[Y|X_{\sim i}])}{Var(Y)} \tag{6}$$

where $X_{\sim i}$ denotes all parameters except i. Applying this methodology to PHE design allows for the identification of parameters that significantly influence performance metrics.

Design Parameters	Lower Limit	Higher Limit
Number of plates N_{pt}	50	110
Plate thickness, t_{pl} (mm)	0.2	1
Port Diameter, D_{po} (mm)	20	40
Mean channel gap, b (mm)	0.9	2.9
Corrugation pitch, λ (mm)	3.8	10
Corrugation angle, β (°)	30	70

Table 2. Upper and lower limits of design parameters.

3. Results and discussion

This section presents the optimisation results both before and after incorporating the Sobol' sensitivity analysis. It demonstrates how this analysis enhances understanding of variable tendencies and their impact on system performance. By identifying and focusing on the most influential parameters, the analysis enables a reduction in the number of design variables. This leads to faster computations, reduced computational costs, and more precise optimisation results.

3.1. Sobol' sensitivity analysis

The optimisation aims to achieve two objectives: minimising the pressure drop for the two channels, hot and cold, and maximising the effectiveness. Consequently, a multi-objective optimisation approach is required. However, such an approach can be computationally expensive, particularly when multiple design variables are involved.

To address this, a preliminary advanced analysis is performed to better understand the behaviour of the design variables with respect to the objective functions. Figure **2** presents the Sobol' sensitivity analysis results for the six considered variables. The channel gap emerged as the most influential parameter for pressure drop, with first-order Sobol' indices near 52% and contributing around 70% when interactions are considered. While the corrugation angle had a minimal direct effect on effectiveness (total effect of 26.3%), it significantly influenced flow and turbulence patterns. Plate thickness was a key factor affecting effectiveness, accounting for 43% of the variance individually. Plate thickness and mean channel gap have a similar effect on effectiveness. Lastly, the corrugation pitch had a negligible impact on pressure drop but contributed 13.51% to effectiveness.

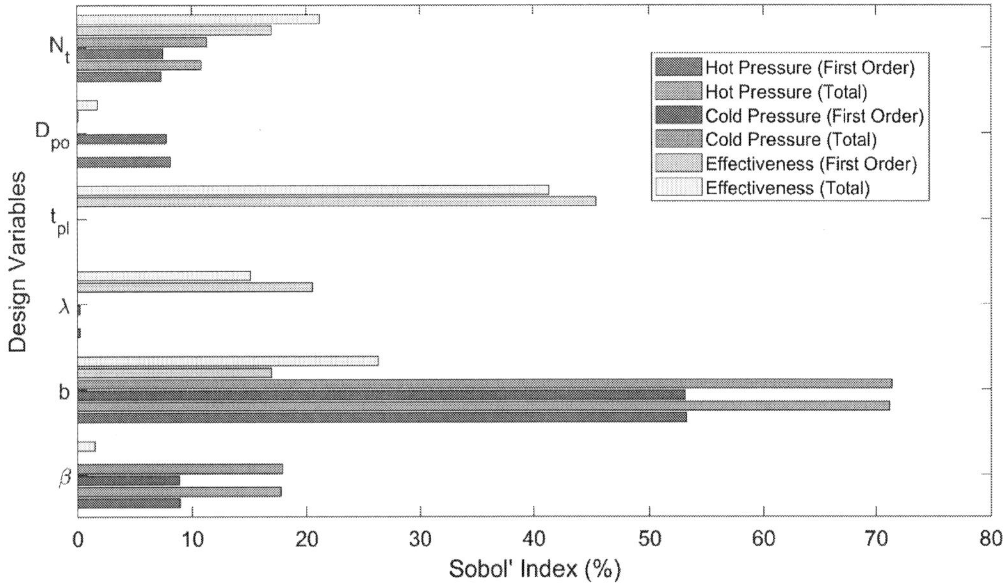

Fig. 2. Sobol' sensitivity analysis on the three objective functions for six design variables.

Figure **3** illustrates how the design variables are distributed in the best solutions along the Pareto front. From the figures, it is evident that some variables remain unchanged throughout the analysis, specifically, $\beta, \lambda, t_{pl}, N_{pt}$. As these parameters show little to no variation, they can be fixed as constants in the optimisation process. Consequently, optimisation can be focused on the two variables that exhibit variation: b and D_{po}.

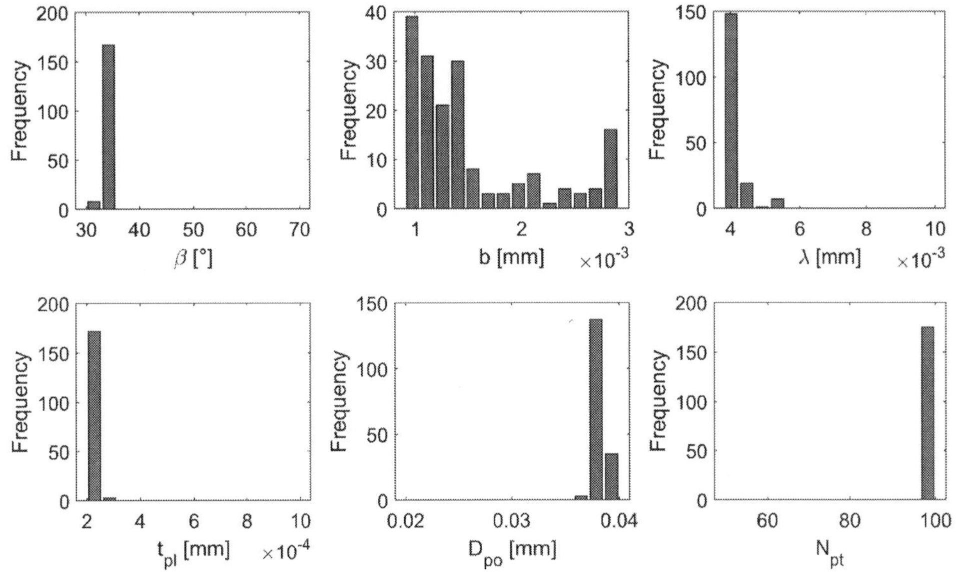

Fig. 3. Distribution of Design Variable Values in Pareto-Optimal Solutions

3.2. Results before and after the pre-analysis

The results of the six-variable optimisation and the improved two-variable optimisation are shown in Figure **4**. The results clearly indicate that concentrating on the two most adjustable variables, b and D_{po}, yields more accurate and efficient outcomes, even with fewer samples of the population and generations.

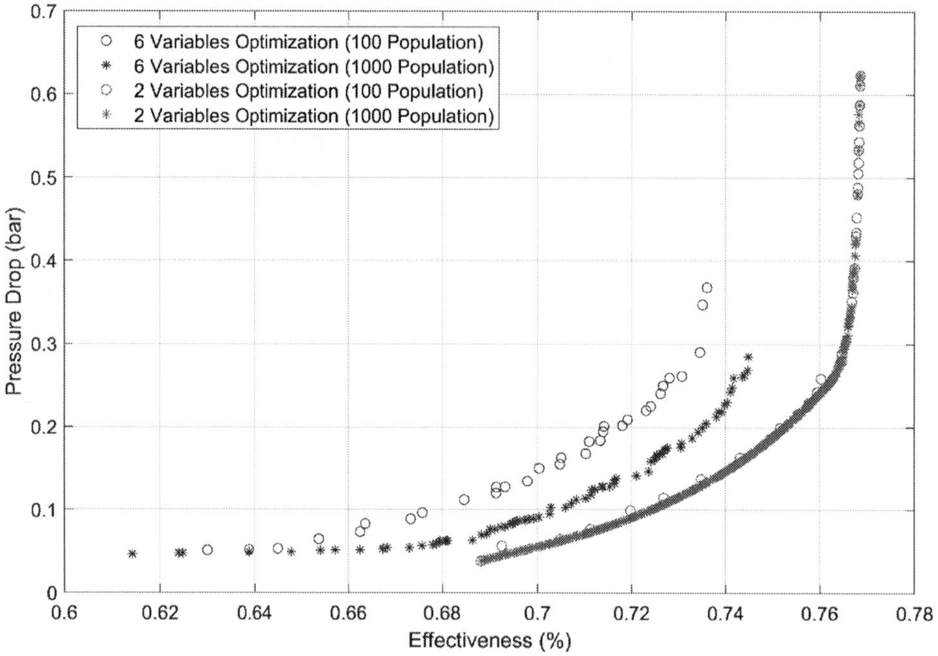

Fig. 4. Impact of Variable Reduction on Optimisation Performance.

Figure **4** shows the Pareto plots for the optimisation results. The blue plots represent optimisation with 6 design variables, while the red plots correspond to optimisation with 2 design variables. The blue circular points indicate the Pareto front obtained using six design variables, a population of 100 individuals, and 20 generations. The star points denote results with a larger population of 1000 samples, illustrating that a higher population size improves accuracy and efficiency in pressure drops and effectiveness. Conversely, the red points display the results after reducing the variables from six to two. Notably, even with fewer variables or generations, the two-variable optimisation achieves more robust and efficient outcomes. These results exemplify the core principle of simulation, which is experimenting with the purpose of better understanding and improving the system [**14**]. In terms of computational costs, simulations with a population of 100 samples took approximately one minute, while simulations with 1000 samples took around 10 minutes.

While the proposed optimisation approach combining Sobol' sensitivity analysis with multi-objective optimisation has demonstrated significant improvements in computational efficiency and accuracy, the study is subject to several limitations that should be acknowledged. First, the optimisation process was conducted within a limited operating temperature range of 20°C to 60°C, which may not reflect performance under more extreme or industrial conditions. Second, the analysis was restricted to six design variables, each constrained within specific ranges as defined in Table 2. This may limit the generalisability of the results to other heat exchanger configurations with different or additional design parameters. Furthermore, the simulation exclusively considered water as the cooling medium, meaning the influence of alternative fluids with varying thermal properties remains unexplored. In addition, the optimisation was performed using a maximum population size of 1000 and 20 generations, which, while sufficient for this case, may not capture the full solution space in more complex models or systems requiring higher-resolution optimisation. These constraints define the scope of the present study and provide direction for future research aimed at extending the applicability and robustness of the proposed method.

4. Conclusion

This study addresses a key challenge in the optimisation of plate PHEs. The challenge is the excessive computational time required when handling multiple design variables in multi-objective optimisation. To overcome this issue, a combined approach was developed that integrates Sobol' sensitivity analysis with multi-objective optimisation. This method enables the identification of the most influential variables for each objective function, significantly reducing the dimensionality of the optimisation problem.

The sensitivity analysis revealed that the channel gap is the dominant factor affecting pressure drop, accounting for more than half of the total variability, while plate thickness is the most influential variable for heat exchanger efficiency, contributing 42.6% to the combined direct and interaction effects. By isolating the parameters that have the greatest impact, the original six-variable optimisation problem was reduced to just two key variables. This reduction led to a more focused and computationally efficient optimisation process. As a result, the refined model generates accurate and high-quality Pareto fronts in a matter of minutes instead of hours without even better performance as shown in figure 4. This confirms that integrating sensitivity analysis into the optimisation workflow enhances both speed and accuracy.

Future work will explore applying this combined optimisation method to other PHE designs currently available on the market to further assess its generalisability and effectiveness. Additionally, the approach will be extended to examine the impact of alternative cooling fluids beyond water, enabling a broader evaluation of thermal performance across various operating conditions.

5. Acknowledgements

This work was supported by the Student Grant Competition of the Technical University of Liberec under project No. SGS-2023-5323

6. References

[1] Kumar, S., Singh, S. K., & Sharma, D. (2022). A comprehensive review on thermal performance enhancement of plate heat exchanger. Inter. Journal of Thermophysics, 43(7), 109.

[2] Ham, J., Lee, G., Kwon, O., Bae, K., & Cho, H. (2023). Numerical study on the flow maldistribution characteristics of a PHEs. Applied Thermal Engineering, 224, 120136.

[3] Gupta, A., & Sharma, D. (2024). Europe plate heat exchanger market – By application (oil and gas, chemical, power generation and metallurgy, marine, mechanical industry, central heating and refrigeration, food processing) and forecast, 2024–2032. Retrieved from

[4] Valencia, G., Núñez, J., & Duarte, J. (2019). Multiobjective optimisation of a plate heat exchanger in a waste heat recovery organic rankine cycle system. Entropy, 21(7), 655.

[5] Wang, Y., Yu, M., Fan, H., & Chung, J. D. (2025). Multi-Objective Topology Optimisation of Cold Plates for Enhanced Battery Thermal Management in Electric Vehicles. Case Studies in Thermal Engineering, 105794.

[6] Zhang, J., Wang, Q., Shu, C. M., Zhang, M., & Lin, J. (2021). A multi-objective optimisation method for under-the-hood thermal management of vehicles. Applied Thermal Engineering, 192, 116818.

[7] Sudret, B. (2008). Global sensitivity analysis using polynomial chaos expansions. Reliability engineering & system safety, 93(7), 964-979.

[8] Le Gratiet, L., Marelli, S., & Sudret, B. (2017). Metamodel-based sensitivity analysis: polynomial chaos expansions and Gaussian processes. In Handbook of uncertainty quantification (pp. 1289-1325). Springer, Cham.

[9] Darges, J. E., Alexanderian, A., & Gremaud, P. A. (2024). Extreme learning machines for variance-based global sensitivity analysis. International Journal for Uncertainty Quantification, 14(4).

[10] Meng, L., Liu, J., Bi, J., Özdemir, E. D., & Aksel, M. H. (2023). Multi-objective optimization of plate heat exchanger for commercial electric vehicle based on genetic algorithm. Case Studies in Thermal Engineering, 41, 102629.

[11] Tenez a.s., Welded heat exchanger accessed: 2025-02-03.

[12] Mastrippolito, F., Aubert, S., Ducros, F., Ferrand, M., & Fourmigué, J. F. (2018, June). Multi-objective shape optimisation of a plate heat exchanger: a multi-scale approach. In International Conference on Energy Engineering and Smart Grids.

[13] Le Guyadec, M., Gerbaud, L., Vinot, E., & Delinchant, B. (2019). Sensitivity analysis using Sobol indices for the thermal modelling of an electrical machine for sizing by optimization. COMPEL-The international journal for computation and mathematics in electrical and electronic engineering, 38(3), 965-976.

[14] Rakic, K. (2021). Application of agent-based modelling and simulation tools. DAAAM International Scientific Book. ISBN 978-3-902734-31-0, ISSN 1726-9687, Vienna, Austria

DOI: 10.2507/36th.daaam.proceedings.016

COMPARISON OF DIFFERENT MOTION CAPTURE TECHNOLOGIES FOR GAIT ANALYSIS AND SUGGESTIONS FOR IMPLEMENTATION IN TOURISM APPLICATIONS

Antonio Maria Coruzzolo[1]*, Chiara Forgione[2],
Elia Balugani[1,2], Rita Gamberini[1,2] & Andrea Roversi[3]

[1] Department of Sciences and Methods for Engineering, University of Modena and Reggio Emilia, Via Amendola 2, Padiglione Morselli, 42122 Reggio Emilia, Italy

[2]Department of Biomedical, Metabolic and Neural Sciences, University of Modena and Reggio Emilia, Via Giuseppe Campi 287, 41125 – Modena, Italy

[3] En&Tech Interdepartmental Center of the University of Modena and Reggio Emilia, Piazzale Europa 1, 42124 Reggio Emilia, Italy

** Corresponding author*

This Publication has to be referred as: Coruzzolo, A[ntonio] M[aria]; Forgione, C[hiara]; Balugani, E[lia]; Gamberini, R[ita] & Roversi, A[ndrea] (2025). Comparison of Different Motion Capture Technologies for Gait Analysis and Suggestions for Implementation in Tourism Applications, Proceedings of the 36th DAAAM International Symposium, pp.0122-0130, B. Katalinic (Ed.), Published by DAAAM International, ISBN 978-3-902734-47-1, ISSN 1726-9679, Vienna, Austria
DOI: 10.2507/36th.daaam.proceedings.016

Abstract

In the framework of Industry 5.0, where human well-being, ergonomics, and human–machine interaction are increasingly emphasized, gait analysis offers valuable insights to improve working conditions, prevent injuries, and enhance human-cantered design. This study compares three motion capture (MoCap) technologies namely Azure Kinect (RGB-depth camera), AiQ Synertial G6 Suit (IMU-based wearable), and Vicon Vero (marker-based optical reference system) in order to evaluate their accuracy in gait detection. Eight volunteers performed standardized walking tasks, with kinematic analysed across the three systems. Results show that the choice of device depends on the application context: Azure Kinect is more suitable for rapid and cost-effective assessments, whereas the G6 Suit enables more in-depth analyses. Beyond ergonomics, this work highlights the potential of gait analysis in tourism environments, such as optimizing visitor flow in museums and heritage sites, improving accessibility for mobility-impaired tourists, and supporting smart tourism infrastructures through adaptive systems that respond to human movement. These suggestions open the door to innovative applications of MoCap in cultural and tourism contexts, where technology can contribute both to inclusivity and to enhanced visitor experiences.

Keywords: Motion Capture Technologies; Gait Analysis; Kinect; Vicon; Tourism

1. Introduction

In the framework of Industry 5.0, where human well-being, ergonomics, and human–machine interaction are central [1] [2], motion capture technologies have become increasingly relevant to improve working conditions, prevent injuries, and foster human-centred design [3] [4], [5]. At the same time, the tourism sector has experienced a technological revolution in the study of human mobility, enabled by GPS, mobile positioning, geocoded social media data, and Bluetooth tracking [6], [7]. These advances allow the collection of high-resolution space–time data over extended periods [8], which has transformed tourism research by offering insights into flows, accessibility, and destination management [9]. Despite these progresses, location-based tracking methods primarily capture where tourists move, but not how they move [10]. This limitation is significant because travel and movement are themselves part of the tourism experience [11] and should not be overlooked.

Literature on journey experiences shows that mobility contributes to sensory, affective, and social dimensions of tourism, shaping satisfaction and sustainability outcomes [12], [13], [14]. The "experience economy" perspective [15] and mobilities geographies [16] emphasize that displacement is not merely a transition between places, but an integral process that can enrich or constrain the tourist experience. This opens space for the application of motion capture (MoCap) technologies, which can complement geolocation data with kinematic and spatiotemporal variables. Unlike conventional tracking, MoCap technologies enables the detailed assessment of gait and posture, providing new ways to understand both physical performance and experiential quality based on visitors' movement. MoCap applications in the manufacturing sector have increased during the last years but remain underexplored in other economic sectors as the tourist one [17], [18], [19].

In this study, we compare three systems: Azure Kinect (RGB-depth camera), AiQ Synertial G6 Suit (IMU-based wearable), and Vicon Vero (marker-based optical reference system) with an experiment that involved eight volunteers performed standardized walking tasks in a neutral space resembling an exposition site. The methods employed are discussed in Section 2, while the experiment and results obtained are discussed in Section 3 and Section 4 respectively. While Section 5 presents potential areas of MoCap technologies application in the tourism sector and Section 6 reports our conclusion and future research agenda.

2. Methods

The present study aims to compare different Motion Capture (MoCap) systems by analysing kinematic data. Specifically, two parameters were monitored:

1. Knee angles (right and left).
2. Hip angles (right and left).

These parameters were selected as they are representative indicators of gait analysis. The MoCap systems employed in this study are a gold standard marker-based system, used as the reference; a marker less system; and an Inertial Measurement Unit (IMU) based system.

2.1. Description of MoCap technologies

In this study, three different MoCap technologies were employed. First, the Vicon Vero v2.2, a marker-based optoelectronic system widely regarded as the gold standard in human motion analysis, was used as the reference method. The system enables highly accurate three-dimensional reconstruction of movement through multiple infrared cameras and 39 passive markers. Second, the Azure Kinect, a marker less system integrating RGB and infrared depth cameras with AI-based body tracking algorithms, was applied to capture motion without the need for physical markers. Finally, the AiQ Synertial G6 Suit, an inertial MoCap suit equipped with 19 IMU sensors, was used to record kinematic data real time. Data obtained with the Vicon system were considered the benchmark for evaluating the performance of the Kinect and the inertial suit.

2.2. Kinematic data processing

This paragraph presents an effective method for processing motion data from each device, aiming to facilitate their use in context such as scientific studies and in-depth analysis. To analyse kinematic data, it was necessary to define the body joints tracked by each technology. For the Vicon Vero system, the Plug-in Gait full-body model [20] was used, and the labelled skeleton template is shown in Figure 1. For Azure Kinect, the Body Tracking SDK [21] extracted the body joint hierarchy, as illustrated in Figure 2. For the AiQ Synertial G6 Suit, Syndash software was used [22] to record and process real-time data, generating a virtual skeleton displayed in Figure 3. Since the focus of this study was on knee and hip angles, a specific set of joints was selected for each systems to construct the vectors of interest, corresponding to the joint names in the respective system hierarchies (Table 1).

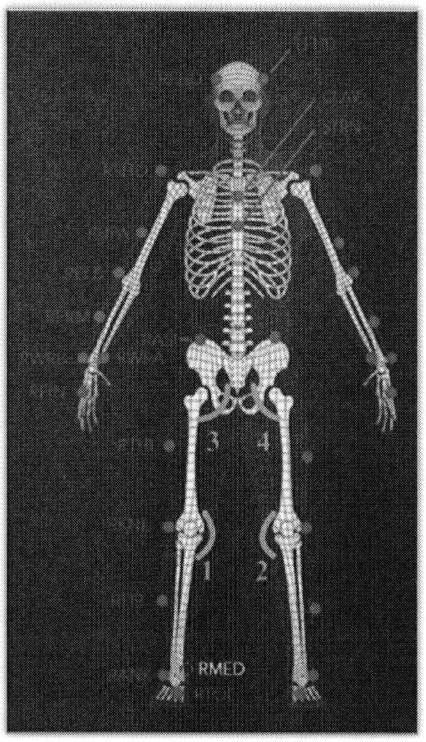

Fig. 1. Body joint hierarchy of Plug-in Gait full-body model [20].

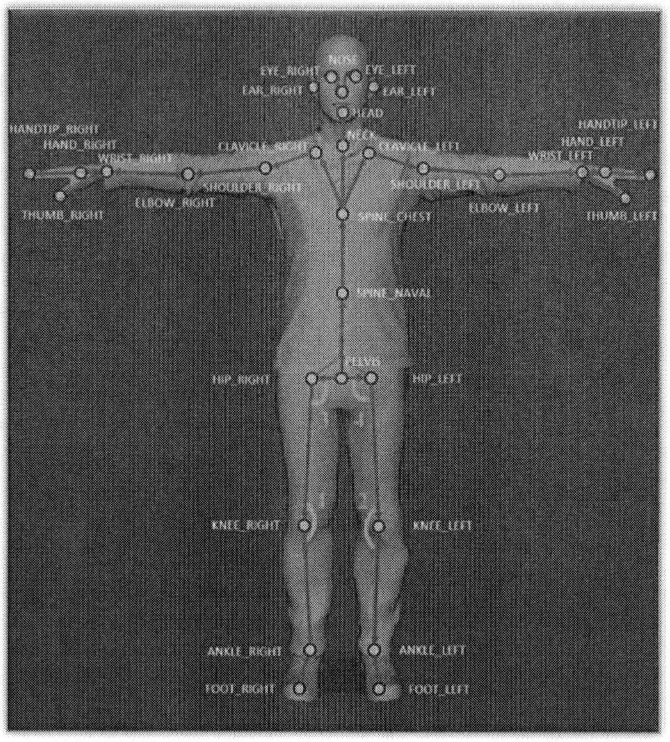

Fig. 2. Body joint hierarchy of Body Tracking SDK [21].

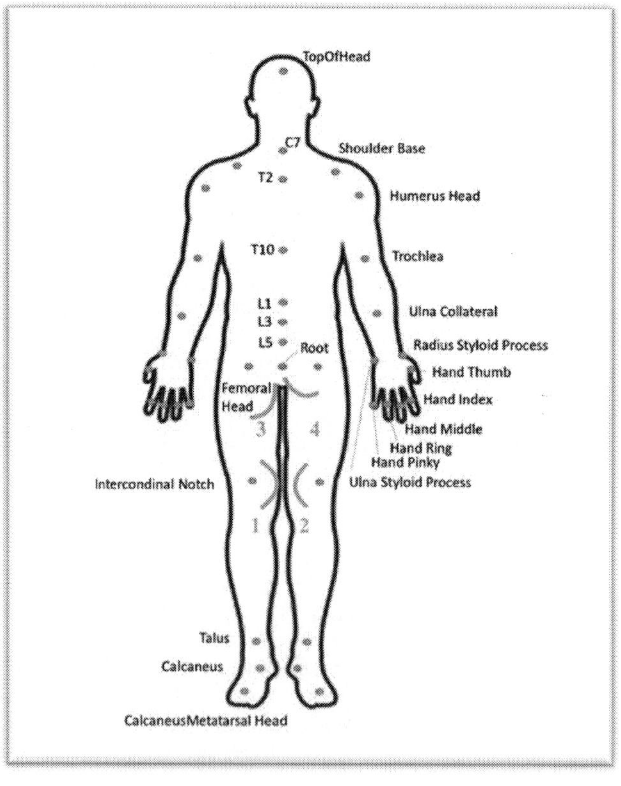

Fig. 3. Body joint hierarchy from Syndash [22].

Angle	Marker-based vectors	Markerless vector	IMU vectors
Right knee	RTHI – RKNEE and RKNEE – RTIB	HIP_RIGHT – KNEE_RIGHT and KNEE_RIGHT – ANKLE_RIGHT	Right Femoral Head – Right Intercondinal Notch and Right Intercondinal – Right Talus
Left knee	LTHI – LKNEE and LKNEE – LTIB	HIP_LEFT – KNEE_LEFT and KNEE_LEFT – ANKLE_LEFT	Left Femoral Head – Left Intercondinal Notch and Left Intercondinal – Left Talus
Right hip	RPSI – RASI and RASI – RTHIS	PELVIS – HIP_RIGHT and PELVIS – HIP_LEFT	Root – Right Femoral Head and Right Femoral Head – Right Intercondinal Notch
Left hip	LPSI – LASI and LASI – LTHIS	PELVIS – HIP_RIGHT and PELVIS – HIP_LEFT	Root – Left Femoral Head and Left Femoral Head – Left Intercondinal Notch

Table 1. Joints and vectors defined for the computation of knee and hip angles across the MoCap systems: marker-based (Vicon), markerless (Azure Kinect), and IMU-based (AiQ Synertial G6 Suit).

3. Experiments

To evaluate the effectiveness and accuracy of the Motion Capture systems under investigation, an experimental setup was established in a laboratory environment. Several walking trials were conducted with different subjects, while each system simultaneously recorded the movements. This design allowed for the assessment of each technology's functionality in a practical context, using cases that were as varied and representative as possible. The experimental setup included:

- 8 Vicon Vero v2.2 cameras and 39 body markers,
- 2 Azure Kinect cameras,
- An AiQ Synertial G6 Suit with 19 IMUs,
- 4 high-performance laptops.

Figure 4 illustrates the experimental setup.

Fig. 4. Experimental setup.

Seven participants took part in the experiment, specifically five males and two females. The participants' characteristics are reported in Table 2.

ID	Age	Weight (kg)	Height (cm)	Leg length (cm)	Knee width (cm)	Ankle width (cm)
1	22	83	183	97	10.5	7.5
2	22	76	190	102	10.5	7
3	22	81	194	107	11	7
4	27	67	171	105	11.5	7.5
5	34	69	176	104	11.5	8
6	29	92	185	106	11.5	7.5
7	35	58	160	100	10.5	6

Table 2. Demographic and anthropometric characteristics of study participants.

Each participant performed two walking trials of 120 seconds within the area shown in Figure 5, following the indicated path. The walking area and space resemble an exhibition space. At the beginning of each trial, participants performed a squat position to synchronize all systems acquisitions.

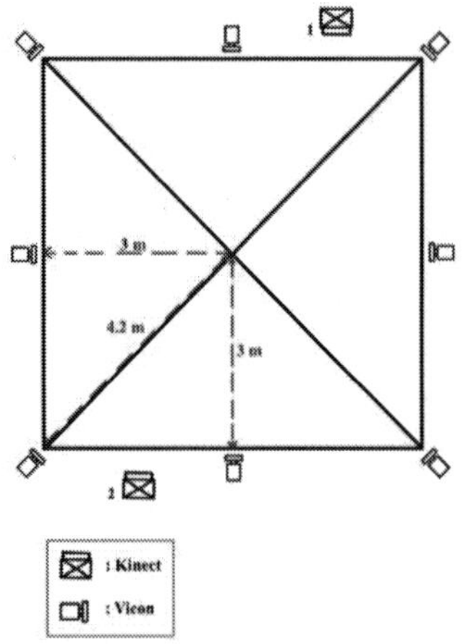

Fig. 5. Experimental walking area used in the study.

Following the preliminary analysis of the angular trend graphs, limitations were observed with the two Kinect devices. Specifically, the recording area, that was relatively wide and not a straight walkway, did not provide optimal conditions for accurate data capture with these systems. Consequently, four acquisitions were excluded from the analysis.

4. Results

During the experiment, seven subjects performed two walking trials of 120 seconds each. However, the Kinect did not provide accurate data for some acquisitions; therefore, from the total 14 recordings, only ten were analysed in this section. The results present a comparison of the three systems (Vicon marker-based, Azure Kinect markerless, and G6 Suit IMU) for the calculation of knee and hip angles during walking. Table 3 reports the mean values of the angles. The mean standard deviation is also included in the table as an indication of the oscillation range of each angle, thus reflecting specific characteristics of each subject's gait. Subsequently, data from Azure Kinect and G6 Suit were compared with the Vicon results by calculating the Root Mean Square Error (RMSE) and the Normalized Root Mean Square Error (NRMSE), which quantify the deviation of each system from the reference (Table 4).

System	Angle	Mean (°)	Mean Std. Dev. (°)
Vicon Vero	Left Knee	161.776	17.328
	Right Knee	156.242	18.113
	Left Hip	94.728	10.295
	Right Hip	99.504	10.403
Azure Kinect	Left Knee	154.593	15.407
	Right Knee	155.153	15.250
	Left Hip	99.796	7.291
	Right Hip	120.208	7.613
Synertial G6 Suit	Left Knee	153.661	23.762
	Right Knee	152.029	23.946
	Left Hip	88.066	7.482
	Right Hip	152.029	12.945

Table 3. Mean and standard deviation of joint angles across MoCap systems.

System	Angle	RMSE (°)	NMRSE
Azure Kinect	Left Knee	23.494	0.150
	Right Knee	23.787	0.152
	Left Hip	15.337	0.169
	Right Hip	20.129	0.181
Synertial G6 Suit	Left Knee	24.009	0.157
	Right Knee	23.495	0.167
	Left Hip	16.995	0.194
	Right Hip	13.872	0.144

Table 4. Root Mean Square Error (RMSE) and Normalized Mean Square Error (NRMSE) with respect to Vicon Vero.

Figure 6 presents a detailed analysis of flexion angles among the three systems. For knee flexion angles, all systems report values within the range approximately 145 to 175 degrees. The median values of both Kinects and the G6 Suit are comparable to those obtained with the Vicon Vero. Both Vicon and the Synertial G6 Suit show greater data dispersion, which can be explained by the higher number of data and the higher acquisition frequency compared to the Kinect systems. The lower amount of data acquired by the Kincts is mainly due to the larger monitored area; in practice, only about 50% of the data was used, whereas for the G6 Sut nearly 100% of the data was comparable with Vicon. For hip flexion angles, the measured range is 78 to 113 degrees, with a slight underestimation of the Kinect and G6 Suit values compared to the Vicon Vero, which appears more accurate in capturing hip movements. Nevertheless, both the Kinects and the G6 Suit demonstrated reliability in angle detection, as their overall measurements ranges were consistent with those of the Vicon Vero.

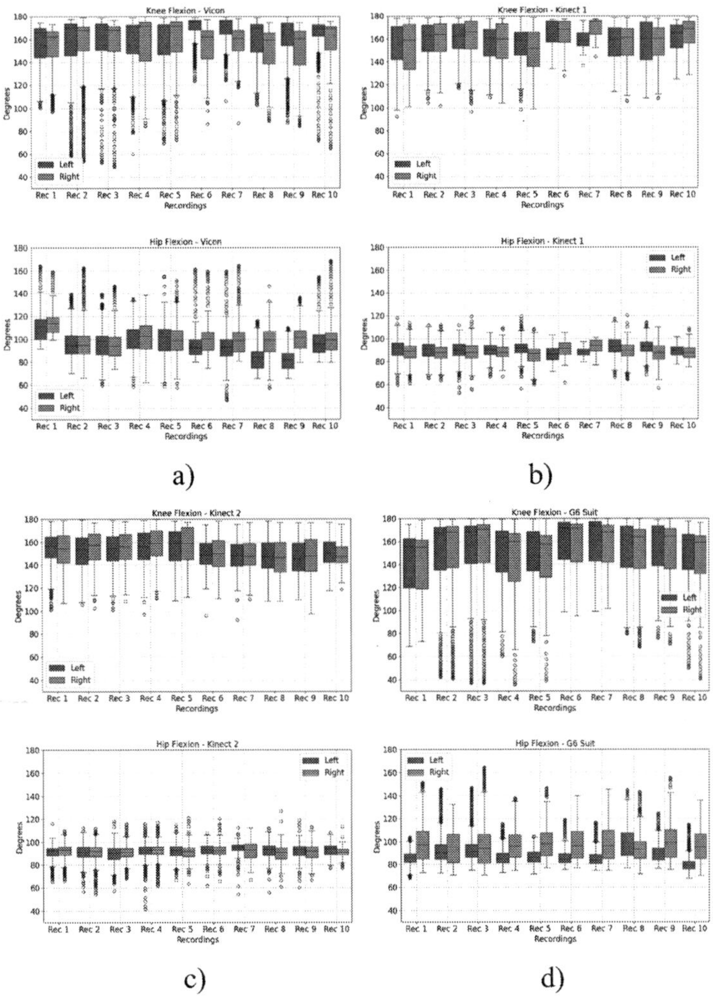

Fig. 6. Boxplots of analysed angles across 10 recordings for three technologies: a) Knee and hip flexion angles measured with Vicon Vero, b) Knee and hip flexion angles measured by the first Kinect, c) Knee and hip angles measured with the second Kinect, d) Knee and hip flexion angles measured with the Synertial G6 Suit.

5. Tourism applications suggestions

Our study focuses on the use of results collected regarding technologies experimented with in the tourism sector. The integration of these technologies enables various possible applications depending on their specific features. Motion capture technologies, for instance, can be employed either for a limited period of research or on a continuous basis with different purposes. In the case of limited use, they may support targeted studies involving a small sample, allowing precise adjustments to the structure and organization of the tourist site or in the redesign of an exhibition space. Continuous use, on the other hand, enables constant monitoring and the collection of large volumes of data, which can be leveraged to adapt and continuously improve the tourist experience. Our research, starting from the literature, has identified three main area of applications: accessibility, flow analysis, and interactivity.

5.1. Accessibility enhancement for people with impaired mobility

There has long been a growing interest in the tourism sector in promoting inclusive and accessible experiences. Investing in accessibility brings various benefits to stakeholders in tourism: it improves the image of destinations and tourism businesses; broadens the tourism market (with more than 1 billion people with disabilities worldwide [23]); contributes to a fairer and safer world for all; drives sustainable economic growth; and aligns with ESG, SDG, and ISD strategies adopted by companies and institutions [24], [25]. Accessible spaces are usually designed in advance by following conventional guidelines and hypothetical user journeys. However, real-world testing is rarely carried out, and systematic monitoring after implementation is almost never performed. The use of motion capture technology makes it possible to evaluate movement quality and monitor the real behaviour of people with reduced mobility within a given space. This would enable tourism experiences to become truly inclusive and accessible for individuals with motor disabilities or for older adults [26]. By collecting such data, it would be possible to identify personas that represent specific needs, allowing for the design of tailored pathways and support services. In this way, user journeys would be genuinely inclusive and accessible, as they would be based on real-world data rather than assumptions [27].

5.2. Attractiveness evaluation and tourists' flux design

The study of flows and the movement of people within a space makes it possible to identify areas of greatest interest. By using motion capture cameras, it is possible to measure the trajectories and walking speed of individuals or groups. These data provide indirect feedback on the level of interest in specific elements or zones, as it can be reasonably assumed that speed decreases in areas of higher interest, where visitors stop to observe, and increases in less engaging areas. For example, in museums, crowding often occurs around certain works of art, making circulation and enjoyment more difficult [28]. When aggregating this data, it becomes possible to understand the overall flow of tourists and to redesign and enhance the visitor experience by adapting pathways to better manage traffic and redistribute points of interest. In this way, the experience can be made engaging from beginning to end, while also reducing congestion and improving both quality and safety [29].

5.3. Interactive and immersive motion-based experiences

The inclusion of interactive digital elements in tourism experiences is becoming increasingly widespread, often associated with the use of touchscreens that provide information from an edutainment perspective. In the artistic domain, there are also works that interact with the public through sensors, changing and responding to external inputs processed by dedicated software, including artificial intelligence. These new opportunities enabled by creative applications of digital technologies make it possible to attract new tourism targets, such as young people and families with children, thereby broadening the audience's interest in cultural and tourism offerings [30]. In particular, the use of motion capture cameras allows the integration of visitors' movements into the experience as an integral part of the place they are visiting or of the artifacts and artworks they are observing. The reprocessing of movements captured by the cameras can be incorporated into virtual reality (VR) or augmented reality (AR) experiences. Especially in the case of AR, real elements of a tourist site can be combined with digital components that add an experiential layer through video, lights, and sounds [31]. This represents an opportunity to innovate traditional tourism and museum experiences, expanding their accessibility and appeal. At present, part of the authors of this research is engaged in an experimental project aimed at innovating a museum itinerary dedicated to the tenor Luciano Pavarotti, through the digitization of artifacts and their use within VR and AR experiences. The implementation of immersive digital experiences makes it possible to attract a new segment of younger visitors, who are not typically consumers of opera [32].

6. Conclusions

This study compared three motion capture technologies, Vicon Vero, Synertial G6 Suit, and Azure Kinect, for detecting knee and hip flexion during walking, with the aim of assessing their usability in real-world contexts and identifying suitable applications in the tourism sector. The Vicon Vero, a marker-based system, was used as the reference due to its well-established accuracy and high-quality data.

The Synertial G6 Suit, an IMU-based system, also demonstrated high accuracy; however, its applicability is limited to controlled environments, as the need to wear a full suit makes it impractical and intrusive in real-world or outdoor scenarios. The Azure Kinect, a marker less and low-cost solution, showed promising accuracy in the measurements. The main limitation, however, was the difficulty in data acquisition, likely due to the large, monitored area. Even with two Azure Kinect devices, coverage may not have been sufficient, highlighting the need to better define how many devices are required for different setups. Overall, our findings suggest that a low-cost, marker less system such as the Azure Kinect can achieve a meaningful level of accuracy without being invasive, making it particularly well-suited for application such as tourism or non-clinical motion analysis. Conversely, it remains unstable for applications requiring a high level of precision, such as medical diagnostics. Future work should therefore focus on determining the optimal number of Kinect sensors relative to the size and complexity of the monitored area. The practical implications of these findings are particularly relevant in tourism contexts. First, motion capture technologies can play a key role in enhancing accessibility for people with reduced mobility by monitoring real behaviours and identifying barriers in real time, leading to more inclusive and data-driven design of tourist spaces. Second, flow analysis based on movement trajectories can help managers understand visitor interests, reduce congestion, and optimize circulation within attractions such as museums and heritage sites. Finally, interactive and immersive applications enabled by motion capture, such as AR/VR experiences or responsive installations, offer opportunities to engage new audiences, particularly younger visitors, and expand the cultural offering of destinations.

Overall, while marker-based and wearable systems remain more suited to specialized research or clinical applications, markerless solutions like Azure Kinect provide an attractive balance of usability, affordability, and accuracy for tourism contexts. Future work should focus on identifying the optimal sensor configuration for large or complex environments and on piloting real-world applications in museums, exhibitions, and heritage sites to validate the added value of motion capture in fostering accessibility, flow management, and experiential innovation.

7. Acknowledgments

Project partially funded under the National Recovery and Resilience Plan (NRRP), Mission04 Component 2 Investment 1.5—NextGenerationEU, Call for tender n. 3277 dated 30 December2021. Award Number: 0001052 dated 23 June 2022.

8. References

[1] Lolli, F., Coruzzolo, A.M., Gonçalves Terra Neto, P., Afonso Sellitto, M., Gamberini, R. (2021). Key Enabling Technologies and Concepts for the Human-Centric Industrial Revolution. In: Human Factors and Wearable Technologies. AHFE International. https://doi.org/10.54941/ahfe1003625.

[2] Nakatova, H., Hricova, B., Badida, M. (2011). Use of Ergonomics in Design Selected Department Workspace in Engineering Company. In: Proceedings of the 22nd International DAAAM Symposium. pp. 1229–1230 (2011). https://doi.org/10.2507/22nd.daaam.proceedings.599.

[3] Bures, M. (2016). Time Demands of Virtual Ergonomic Modelling – Experimental Study. In: Proceedings of the 26th DAAAM International Symposium. pp. 0286–0291. https://doi.org/10.2507/26th.daaam.proceedings.039.

[4] Rybnikár, F., Kačerová, I., Hořejší, P., Šimon, M. (2022). Ergonomics Evaluation Using Motion Capture Technology—Literature Review. Applied Sciences. 13, 162 (2022). https://doi.org/10.3390/app13010162.

[5] Lolli, F., Coruzzolo, A.M., Forgione, C., Peron, M., Sgarbossa, F. (2024). Auto-AzKNIOSH: an automatic NIOSH evaluation with Azure Kinect coupled with task recognition. Ergonomics. 1–17. https://doi.org/10.1080/00140139.2024.2433027.

[6] Shoval, N., Isaacson, M., Chhetri, P. (2014). GPS, Smartphones, and the Future of Tourism Research. In: The Wiley Blackwell Companion to Tourism. pp. 251–261. Wiley. https://doi.org/10.1002/9781118474648.ch20.

[7] Shoval, N., Birenboim, A. (2019). Customization and augmentation of experiences through mobile technologies: A paradigm shift in the analysis of destination competitiveness. Tourism Economics. 25, 661–669. https://doi.org/10.1177/1354816618806428.

[8] Ahas, R., Miller, H.J., Witlox, F. (2014). From the Guest Editors: Mobility, Communication, and Urban Space. Journal of Urban Technology. 21, 1–7. https://doi.org/10.1080/10630732.2014.913420.

[9] Ahas, R., Aasa, A., Silm, S., Tiru, M. (2010). Daily rhythms of suburban commuters' movements in the Tallinn metropolitan area: Case study with mobile positioning data. Transp Res Part C Emerg Technol. 18, 45–54. https://doi.org/10.1016/j.trc.2009.04.011.

[10] Asakura, Y., Iryo, T. (2007). Analysis of tourist behaviour based on the tracking data collected using a mobile communication instrument. Transp Res Part A Policy Pract. 41, 684–690. https://doi.org/10.1016/j.tra.2006.07.003.

[11] Zakrisson, I., Zillinger, M. (2012). Emotions in motion: tourist experiences in time and space. Current Issues in Tourism. 15, 505–523 https://doi.org/10.1080/13683500.2011.615391.

[12] Scuttari, A. (2019). Cycling and Motorcycling Tourism. Springer International Publishing. https://doi.org/10.1007/978-3-030-17697-6.

[13] Larsen, J. (2001). Tourism Mobilities and the Travel Glance: Experiences of Being on the Move. Scandinavian Journal of Hospitality and Tourism. 1, 80–98. https://doi.org/10.1080/150222501317244010.

[14] Lohmann, G., Pearce, D.G. (2012). Tourism and Transport Relationships: The Suppliers' Perspective in Gateway Destinations in New Zealand. Asia Pacific Journal of Tourism Research. 17, 14–29. https://doi.org/10.1080/10941665.2011.613211.
[15] Pine, B.J., Gilmore, J.H. (2013). The experience economy: past, present and future. In: Handbook on the Experience Economy. Edward Elgar Publishing https://doi.org/10.4337/9781781004227.00007.
[16] Urry, John. (2010). Mobilities. Polity.
[17] Lolli, F., Maria Coruzzolo, A., Forgione, C., Gonçalves Terra Neto, P. (2023). Ergonomic Risk Reduction: A Height-Adjustable Mesh Truck for Picking Activities Evaluated With A Depth Camera. In: Ergonomics In Design. AHFE International.https://doi.org/10.54941/ahfe1003371.
[18] Lolli, F., Coruzzolo, A.M., Forgione, C., Peron, M., Sgarbossa, F. (2024). Semiautomatic rapid upper limb assessment methods: validation of AzKRULA. Int J Hum Factors Ergon. 11, 309–330 https://doi.org/10.1504/IJHFE.2024.143266.
[19] Di Natali, C., leggieri, S., Fanti, V., Forgione, C., Coruzzolo, A.M., Lolli, F., Caldwell, D.G. (2025). Pilot Test to Evaluate Ergonomic Risk of Industrial Exoskeleton with Semi-autonomous Method that Relies on AzKCLI and the Equivalent Weight. Springer Proceedings in Advanced Robotics. https://doi.org/10.1007/978-3-031-89471-8_21.
[20] Vicon Motion Systems Ltd UK: https://help.vicon.com/space/Nexus216/11607226/Full+body+modeling+with+Plug-in+Gait.
[21] Microsoft: Azure Kinect Body Tracking SDK | Microsoft Docs, https://docs.microsoft.com/it-it/azure/kinect-dk/body-sdk-download, last accessed 2021/07/30.
[22] AiQ Synertial: https://www.synertial.com/.
[23] WHO: https://www.who.int/news-room/fact-sheets/detail/disability-and-health.
[24] Kasimoglu, M. (2012). Visions for Global Tourism Industry - Creating and Sustaining Competitive Strategies. InTech. https://doi.org/10.5772/2278.
[25] UNWTO: https://tourism4sdgs.org/themes/accessible-tourism/.
[26] Yao, T., Jiang, B., Zhao, L., Chen, W., Sang, Y., Jia, Z., Wang, Z., Zhong, M. (2025). Research on Optimization of Indoor Layout of Homestay for Elderly Group Based on Gait Parameters and Spatial Risk Factors Under Background of Cultural and Tourism Integration. Buildings. 15, 2498. https://doi.org/10.3390/buildings15142498.
[27] Teixeira, P., Teixeira, L., Silva, S., Eusébio, C. (2025). Technology-based approaches for accessible tourism: Using personas to understand stakeholders' characteristics and motivations. Data Inf Manag. 100095. https://doi.org/10.1016/j.dim.2025.100095.
[28] Ferrato, A., Limongelli, C., Mezzini, M., Sansonetti, G. (2022). Using Deep Learning for Collecting Data about Museum Visitor Behavior. Applied Sciences. 12, 533. https://doi.org/10.3390/app12020533.
[29] Liu, S., Long, Y., Ge, J., Zhang, L.: Unveiling tourist movement patterns in micro-scenes: a novel framework combining computer vision and spatiotemporal clustering. Current Issues in Tourism. 1–26 (2025). https://doi.org/10.1080/13683500.2025.2526720.
[30] Zhang, L., Zou, L., Shi, M. (2024). Immersing and Perceiving in Tourism Scenarios: The Interaction Mechanism between Digital Technologies and Tourists. In: 2024 IEEE 24th International Conference on Software Quality, Reliability, and Security Companion (QRS-C). pp. 1292–1301. IEEE. https://doi.org/10.1109/QRS-C63300.2024.00168.
[31] Yu, J., Wang, Q., Chen, H. (2018). Application of kinect-based motion recognition algorithm in cultural tourism. In: ACM International Conference Proceeding Series. pp. 307–311. Association for Computing Machinery (2018). https://doi.org/10.1145/3301326.3301377.
[32] Progetto Atlante: https://www.progettoatlante.it/

DOI: 10.2507/36th.daaam.proceedings.017

A VALUE-BASED DECISION FRAMEWORK FOR EVALUATING THE INTEGRATED EFFICIENCY OF SMART GRIDS

Qian Zhao[1], Elia Balugani[1,2,*], Rita Gamberini[1,2] & Francesco Lolli[1,2]

[1]*Department of Sciences and Methods for Engineering, University of Modena and Reggio Emilia, Via Amendola 2, Padiglione Morselli, 42122, Reggio Emilia, Italy*

[2]*En&Tech Interdepartmental Centre of the University of Modena and Reggio Emilia, Piazzale Europa 1, 42124, Reggio Emilia, Italy*

* *Corresponding author*

This Publication has to be referred as: Zhao, Q[ian]; Balugani, E[lia]; Gamberini, R[ita] & Lolli, F[rancesco] (2025). A Value-Based Decision Framework for Evaluating the Integrated Efficiency of Smart Grids, Proceedings of the 36th DAAAM International Symposium, pp.0131-0140, B. Katalinic (Ed.), Published by DAAAM International, ISBN 978-3-902734-47-1, ISSN 1726-9679, Vienna, Austria
DOI: 10.2507/36th.daaam.proceedings.017

Abstract

The evaluation of Smart Grid (SG) efficiency requires a value-based perspective that incorporates decision-maker preferences alongside technical metrics. This study develops an integrated framework combining Data Envelopment Analysis (DEA) with the Flexible and Interactive Tradeoff (FITradeoff) method to assess SG efficiency. The approach begins with standardized DEA to identify an initial set of efficient DMUs, followed by an iterative preference elicitation process in which decision-makers provide pairwise comparisons of weight ratios. These preferences are translated into linear constraints that progressively contract the weight space through a robust optimization procedure with baseline weight restrictions. The algorithm systematically reduces the efficient set, terminating when a unique DMU is identified or the candidate set becomes sufficiently narrow. This framework offers a structured, transparent, and mathematically rigorous tool for identifying SG alternatives that align with strategic priorities and support preference-driven decision making.

Keywords: Smart Grids; Data Envelopment Analysis (DEA); Flexible and Interactive Tradeoff (FITradeoff)

1. Introduction

The global energy sector is undergoing a paradigm shift driven by the increasing penetration of renewable energy, digitalization, and the need for enhanced efficiency in energy management [1], [2], [3], [4]. Renewable energy sources, such as wind and solar, introduce variability and complexity that challenge traditional centralized grids, requiring sophisticated evaluation and optimization strategies to ensure reliability and sustainability [1], [2].

In this context, Smart Grids (SGs) have emerged as a cornerstone of modern energy infrastructure, integrating distributed energy resources, enabling self-healing capabilities, and optimizing large-scale grid operations [5], [6]. However, assessing the performance of these complex systems remains challenging, as it involves balancing multiple, often conflicting criteria spanning economic, technical, and environmental dimensions [7]. This multi-faceted nature creates a classic multi-criteria decision-making (MCDM) problem [8], necessitating frameworks that can holistically capture the integrated efficiency of SGs to guide strategic investment and policy decisions. To address the need for multi-dimensional benchmarking, Data Envelopment Analysis (DEA) [9] has been extensively applied in energy and environmental studies. As a non-parametric frontier analysis technique, DEA excels at evaluating the relative efficiency of homogeneous Decision-Making Units (DMUs), such as regional power grids or electricity distribution companies, without requiring a priori weight assignments. Its application in the power sector is well-established, ranging from assessing the sustainability performance of electric utilities in the United States [10] to benchmarking the operational efficiency of thermal power plants in China [11]. Nevertheless, the core strength of conventional DEA allowing each DMU to select its most favourable weights to maximize its own efficiency score, introduces a critical drawback for decision-making. This flexibility often leads to the "multiple-winner dilemma," where a large number of DMUs are identified as fully efficient [12]. The output is a set of efficient utilities or regional grids, leaving planners and investors without a definitive, justifiable answer to the fundamental question: "Which one is the best choice"? This indecisiveness severely limits DEA's utility in contexts requiring a unique selection, such as competitive funding allocation for large-scale grid modernization or strategic national infrastructure planning. Integrating preferences into DEA: The quest to incorporate decision-maker preferences into DEA to overcome its limitations has followed two primary streams, each with significant and well-documented shortcomings: (1) Direct weight restrictions, including approaches such as assurance regions [12] and cone-ratios, impose explicit bounds on the DEA weight space. By constraining the set of feasible weights, these methods can reduce the number of units classified as efficient, as overly extreme or unrealistic weight combinations are eliminated [13]. However, such restrictions have been widely criticized for their reliance on externally imposed and often subjective limits, which may not accurately capture the nuanced preferences or value system of the decision maker [14]. Consequently, while mathematically convenient, these constraints may lack a robust decision-theoretic foundation and do not necessarily produce solutions that are truly value-optimal [13]. (2) Integration with MCDM Methods. A more popular approach hybrids DEA with MCDM techniques. The most common is the AHP-DEA model, frequently applied in energy project selection, such as in the efficiency assessment of thermal power plants [15]. While decisive, this method inherits AHP's well-documented flaws: it requires many precise pairwise comparisons at the ratio scale upfront, creating a high cognitive burden [16]. Furthermore, it suffers from potential rank reversal, and its theoretical foundation is sometimes questioned for being disconnected from the axioms of multi-attribute value theory (MAVT) [17]. Other integrations, like those with TOPSIS [18], often determine a static preference structure ex-ante, failing to interactively refine the solution based on the evolving understanding of the efficient frontier.

In parallel, the field of MCDM has seen advances in more robust preference elicitation methods grounded in MAVT. The FITradeoff method [19] stands out as a state-of-the-art approach. Its core mechanism involves eliciting holistic pairwise comparisons of consequences from the DM, from which it infers linear constraints on the ratios of the scale constants in an additive utility function. Through an iterative process, it progressively reduces the feasible weight space to identify the most preferred alternative. Owing to its theoretical foundation in MAVT, which relies on more intuitive trade-off judgments rather than precise ratio-scale comparisons, a growing body of literature positions FITradeoff as a cognitively more efficient alternative to AHP, typically requiring fewer and less demanding pairwise inputs from the decision maker [20], and is theoretically sounder due to its direct foundation in MAVT axioms. Its applications have grown, including in supply chain selection and maintenance planning [21]. However, a critical analysis of the literature reveals a stark limitation: its powerful engine for iteratively constraining the weight space has been exclusively applied in traditional MCDM problems with a pre-defined, limited set of alternatives. Its potential has never been harnessed to address the fundamental weight flexibility problem in DEA, where the alternatives are the virtually infinite weight vectors within the DEA feasible region, and the goal is to rank a set of DMUs on an efficiency frontier.

The literature reveals a clear and compelling schism. DEA provides an objective efficiency benchmarking tool but fails to deliver a unique solution [12], while existing preference-integration methods are either theoretically and cognitively flawed (AHP) or have not been adapted to the specific mathematical and philosophical context of frontier-based efficiency analysis. The shortcomings of existing hybrid models highlight the need for a new synthesis that is both methodologically rigorous and practically decisive. This study bridges this gap by proposing a novel Value-Based Decision Framework that integrates DEA [9] with the FITradeoff method. Our work is designed to directly counteract the identified limitations and makes the following contributions. (1) To overcome the indecisiveness of DEA and the arbitrariness of weight restrictions [14], we develop a model where the weight space is not subjectively bounded but is systematically and rationally reduced by the DM's own value judgments, elicited through FITradeoff 's trade-off questions. This ensures the final unique solution is both Pareto-efficient on the frontier and value-aligned, directly solving the "multiple-winner dilemma" [12]. (2) To overcome the cognitive burden and theoretical issues of AHP-DEA integrations [16], [17], we replace AHP with FITradeoff. Our framework leverages FITradeoff 's cognitively efficient, holistic trade-off questions, significantly easing the DM's burden and providing a theoretically rigorous foundation for preference incorporation based on the axioms of MAVT. (3) To unlock the potential of FITradeoff for a new class of problems, we pioneer its application to frontier analysis. We create a novel, dynamic feedback loop where DEA identifies the efficient frontier and FITradeoff interactively refines the weights, converging to a single, preference-optimal SG.

This provides a defensible, decision-centric tool for energy investment and policy, effectively translating the theoretical power of trade-off-based elicitation into a practical solution for complex efficiency evaluation problems. The remainder of this paper is organized as follows. Section 2 reviews the theoretical foundations of DEA and the FITradeoff method. Section 3 presents the proposed preference-guided DEA framework integrating the FITradeoff approach. In Section 4, the applicability and effectiveness of the proposed method are demonstrated through an empirical evaluation of SG efficiency. Finally, Section 5 concludes the paper and discusses potential directions for future research.

2. Theoretical background: DEA and FITradeoff methods

This section delineates the theoretical underpinnings of the integrated evaluation framework proposed in this study. It commences with a critical exposition of DEA, establishing its role as the foundational engine for efficiency benchmarking. Subsequently, the discussion pivots to the imperatives for incorporating decision-maker preferences, thereby introducing the FITradeoff method as the mechanism for value-driven refinement. A nuanced comprehension of these constituent methodologies is paramount for appreciating their synthesis in the subsequent model development.

2.1 Data envelopment analysis

DEA is a non-parametric performance evaluation technique that utilizes linear programming to assess the relative efficiency of a set of homogeneous DMUs. Its core premise is to construct a piecewise linear efficiency frontier from the best-performing DMUs, against which all other units are benchmarked. The seminal CCR model establishes the foundational linear programming formulation under the assumption of constant returns to scale [9]. For a target DMU₀, the input-oriented model is specified as follows:

$$\text{Max } \theta_0 = \sum_{r=1}^{s} u_r y_{r0} \tag{1}$$

$$\sum_{i=1}^{m} v_i x_{i0} = 1 \tag{2}$$

$$\sum_{r=1}^{s} u_r y_{rj} - \sum_{i=1}^{m} v_i x_{ij} \leq 0, \forall j = 1,2, \dots, n \tag{3}$$

$$u_r \geq \varepsilon, v_i \geq \varepsilon, \forall\, r, i \tag{4}$$

where x_i: Input indicator i; y_r: Output indicator r; v_i, u_r: Virtual weights (multipliers) for inputs and outputs, respectively; θ_0: Efficiency score of the target DMU₀

- Equation (1) maximizes the virtual output of DMU₀.
- Equation (2) normalizes the virtual input of DMU₀ to 1, establishing a common benchmark for comparison.
- Equation (3) ensures that no DMU can achieve an efficiency score greater than 1 under the same set of weights, thereby defining the efficiency frontier.
- Equation (4) with a non-Archimedean infinitesimal (ε) prevent the model from assigning zero weights to any factor, ensuring all inputs and outputs are considered.

The efficiency score θ_0 for each DMU is obtained by solving this linear program. A DMU is considered Pareto-efficient (lying on the frontier) if $\theta_0 = 1$. A score of $\theta_0 < 1$ indicates that the DMU is inefficient, meaning there exists a combination of other DMUs that can produce at least the same level of outputs with fewer inputs. The principal strength of DEA lies in its ability to identify an objective efficiency frontier without requiring prior parametric assumptions about production technology. However, a key limitation is the potential for multiple optimal solutions, the "multiple-winner dilemma" as each DMU is evaluated using the set of weights most favourable to itself.

2.2 Flexible and Interactive Tradeoff method

The FITradeoff method is an advanced MCDM technique grounded in MAVT [19]. Its core strength lies in its adaptive preference elicitation process, which progressively refines the decision-maker's preferences through intuitive, holistic trade-off questions, significantly reducing the cognitive burden associated with methods like AHP that require precise ratio-scale comparisons. FITradeoff does not require the decision-maker to specify precise criterion weights upfront. Instead, it employs an iterative questioning procedure where the decision-maker compares pairs of consequence profiles. These preference statements are translated into linear constraints on the feasible space of criterion weights, systematically narrowing down the set of possible optimal solutions. At the heart of FITradeoff's iterative process is a linear programming model designed to test the potential optimality of an alternative a_k given the current set of preference constraints. The linear programming model for testing alternative a_k is formulated as follows:

$$\text{Max} \sum_{i=1}^{n} w_i \cdot v_i(a_{ki}) \tag{5}$$

$$\sum_{i=1}^{n} w_i \cdot v_i(a_{ki}) \geq \sum_{i=1}^{n} w_i \cdot v_i(a_{ji}), \forall j \neq k \tag{6}$$

$$w_{i+1} \leq w_i \cdot k^1 - \varepsilon, i = 1,2, \dots, n-1 \tag{7}$$

$$w_{i+1} \geq w_i \cdot k^2 + \varepsilon, i = 1,2, \dots, n-1 \tag{8}$$

$$\sum_{i=1}^{n} w_i = 1 \tag{9}$$

$$w_i > 0, i = 1,2, \dots, n-1 \tag{10}$$

where n: Number of criteria. w_i: Weight (scale constant) for criterion i. $v_i(a_k)$: Value score of alternatives a_k under criterion i (standardized). a_j: Other alternatives being compared.

- Equation (5) checks whether there exists *any* set of weights within the current feasible region that would make alternative a_k have the highest total value.
- Equation (6) ensures that under the same set of weights, the total value of a_k is at least as high as that of every other alternative.
- Equation (7) or (8) is the core of the interaction, progressively shrinking the feasible weight space to reflect the decision maker's nuanced value system.
- Equations (9) and (10) ensure the weights form a valid, positive, and normalized set.

If the LP is feasible, alternative a_k remains a candidate for the most preferred solution. If it is infeasible, a_k is eliminated. The method then intelligently selects the next most informative trade-off question based on the updated constraints. This process iterates, creating a dynamic feedback loop that rapidly converges to a single, value-optimal alternative. The power of FITradeoff lies in transforming the search for the best alternative from a static, all-information-required-a-priori judgment into a dynamic, value-driven, and highly efficient search process, making it uniquely suited for complex decision scenarios.

3. Preference-guided DEA with FITradeoff method

In this section, we propose a novel hybrid framework, the FITradeoff-DEA model, which integrates interactive preference information from MCDM. By iteratively incorporating these preferences to constrain the feasible weight space, the model converges multiple efficient DMUs to a single, preference-driven optimal solution, thereby shifting the evaluation focus from purely technical efficiency to value-based decision-making. The model operates on an iterative principle of gradually narrowing the DEA weight space based on decision-maker preferences. Initially, all DMUs are evaluated without preference constraints, potentially resulting in multiple efficient units. Through an interactive process, preference information is elicited and converted into linear constraints, progressively reducing the feasible region. The iteration continues until only one DMU achieves the maximum efficiency score under the given preferences, representing the unique preference-driven optimal solution. The following notations and definitions are introduced for clarity and consistency throughout the subsequent model formulation.

- DMUs: Consider a set of n homogeneous DMUs, denoted as DMU_j $(j = 1,2, \dots, n)$.
- Input and Output Indicators: x_{ij} $(i = 1,2, \dots, m)$ represents the amount of input i for DMU_j . y_{rj} $(r = 1,2, \dots, s)$ represent the amount of output r for DMU_j .
- Weight Variables: v_i is the weight assigned to input i; ϑ_r is the weight assigned to output r.
- Efficiency Score: For a given weight combination (v, ϑ), the efficiency of DMU_j is $\theta_j = \sum_{r=1}^{s} \vartheta_r y_{rj}$.

The following Linear Programming (LP) model is formulated for evaluating a target unit DMU_k :

$$\text{Max } \theta_k = \sum_{r=1}^{s} u_r y_{rk} \tag{11}$$

$$\sum_{i=1}^{m} v_i x_{ik} = 1 \tag{12}$$

$$\sum_{r=1}^{s} u_r y_{rj} - \sum_{i=1}^{m} v_i x_{ij} \leq 0, \forall j = 1,2, \dots, n \tag{13}$$

$$\frac{u_r}{u_{r+1}} \in [l,p], \frac{v_i}{v_{i+1}} \in [l,p] \quad (14)$$

$$u_r \geq k_1' \cdot u_{r+1} \text{ or } u_r \leq k_1'' \cdot u_{r+1} \quad (15)$$

$$v_i \geq k_2' \cdot v_{i+1} \text{ or } v_i \leq k_2'' \cdot v_{i+1} \quad (16)$$

$$u_r \geq \varepsilon, v_i \geq \varepsilon, \forall\, r, i \quad (17)$$

where (12) is the DEA normalization constraint. (13) is the DEA envelopment constraints. (14) are the weight ratio bounds. (15) and (16) are preference constraint set, a set of linear inequalities incrementally incorporated during the interactive process, C= {all preference information from decision maker}: output vs. output: $u_r \geq k^1 \cdot u_{r+1}$ implies output r is deemed more important than k_1' times output $r+1$ or $u_r \leq k_1'' \cdot u_{r+1}$ implies output r is deemed less important than k_1'' times output $r+1$; input vs. input: $v_i \geq k_2'' \cdot v_{i+1}$ implies input i is deemed more important than k_2' times input $i+1$ or $v_i \leq k_2'' \cdot v_{i+1}$ implies input i is deemed less important than k_2'' times input $i+1$. (17) is the non-negativity constraints, a minimal positive number used to enforce strict positivity of weights. The model executes through a dynamic, interactive, and iterative procedure, as shown in Algorithm 1.

Step 1: Problem initialization and data standardization

Step 1.1 Load input-output data

Inputs: x_{ij} $(i = 1,2, \dots, m)$ *represents the amount of input i for* DMU_j.

Outputs: y_{rj} $(r = 1,2, \dots, s)$ *represent the amount of output r for* DMU_j

Step 1.2 Data standardization process

For each indicator (input x_{ij}*or output* y_{rj}*), the standardized value is calculated as:*

$$z' = 0.1 + 0.9\frac{z - \min(z)}{\max(z) - \min(z)} \quad (18)$$

[0.1,1] represents the original value of any input or output variable.

Step 1.3 Set baseline weight constraints

Minimum weight: $v_i, u_r \geq \varepsilon$

Weight ratio bounds:

Input weights $\frac{v_i}{v_{i+1}} \in [L_{bound}, U_{bound}]$ for $i = 1,2 \dots, m-1$ (19)

Input weights $\frac{u_r}{u_{r+1}} \in [L_{bound}, U_{bound}]$ for $r = 1,2 \dots, s-1$ (20)

Step 2: Initial DEA efficiency calculation

Step 2.1 Solve multiplier DEA model for each DMU j by (11)-(17).

Step 2.2 Identify initial efficiency frontier.

$E^1 = \{DMU_j \mid \theta_j^* \geq 0.999\}$

Record initial efficiency scores θ_j^1 *for all DMUs on the frontier.*

Step 3: FITradeoff interactive process

While $|E^k| > 1$ ***and iteration < max iteration do:***

Step 3.1 Select weight comparison pair.

In each iteration, the algorithm selects one weight ratio from the set of candidate pairs $P = \{v_i/v_{i+1} | i = 1,2 \dots, m-1\} \cup \{u_r/u_{r+1} | r = 1,2 \dots, s-1\}$ *according to a two-level criterion:*

Primary criterion: *the pair with the smallest number of prior elicitation questions is preferred.*

Secondary criterion: *among pairs with equal numbers of prior questions, the pair with the largest current range* $(U_{bound} - L_{bound})$ *is selected, as it represents the greatest remaining uncertainty.*

Only pairs with a current range exceeding a predefined threshold (e.g., 0.05) are considered for selection.

Step 3.2 Generate comparison question.

$$\text{midpoint} = \frac{L_{bound}+U_{bound}}{2} \tag{21}$$

is the ration for the selected paired of weights.

Step 3.3 Simulate DM response.

Computing true ratio from preset weights

Response:

$$response = \begin{cases} I & if|true_{ration} - midpoint| < 0.001 \\ B & if\ true_{ration} < midpoint \\ A & if\ true_{ration} > midpoint \end{cases} \tag{22}$$

Step 3.4 Update weight boundaries

$$Case\ response = \begin{cases} I & L_{bound} = U_{bound} = midpoint \\ B & U_{bound} = midpoint \\ A & L_{bound} = midpoint \end{cases} \tag{23}$$

Step 3.5 Add linear constraints to constraint set C.

If the response is not I, add corresponding linear inequality:

For v_i/v_{i+1}*:* $v_i - midpoint \cdot v_{i+1} \geq 0$ *if (A) or* ≤ 0 *if (B)*

For u_r/u_{r+1}*:* $u_r - midpoint \cdot u_{r+1} \geq 0$ *if (A) or* ≤ 0 *if (B)*

This inequality is then incorporated into the set of preference constraints C*.*

Step 4: Constrained re-evaluation of Efficiency

Step 4.1 Solve constrained DEA for each DMU in E^k ***with constraints*** C***.***

Step 4.2 Update efficiency frontier

$$E^{k+1} = \{\text{DMU}_j \in E^k \mid \theta_j^* \geq 0.999\} \tag{24}$$

Step 4.3 Handle degenerate cases

If $| E^{k+1} | = 0$*, revert last constraint, restore previous weight boundaries, set* $E^{k+1} = E^k$

Step 5: Convergence check and termination

Step 5.1 Termination conditions

If $| E^{k+1} | = 1$*:*

Final optimal DMU = unique member of E^{k+1}

Output: Found preference-optimal solution→ terminate

Else if $| E^{k+1} | > 1$ *and maximum of iterations reached:*

Output: Multiple candidate DMUs remain

Conduct cross-efficiency analysis → terminate

Else:

$k = k + 1$*→ return to Step 3*

Step 6: Result analysis and recommendation

Step 6.1 Calculate final efficiency rankings

Compute θ_j *for all DMUs under final constraint set C*

Step 6.2 Output recommendation

If unique optimal DMU, final recommendation DMU can be considered as the optimal choice under preferences.

Algorithm 1. FITradeoff -DEA preference elicitation

4. Case study: application to power company efficiency evaluation

To empirically evaluate the effectiveness of the proposed FITradeoff -DEA model, we perform numerical analysis using a widely recognized dataset in SG efficiency studies. Specifically, we employ the input and output indicators of 15 power companies originally reported by Yu et al. [22]. This dataset is well suited to our analysis as it is constructed under the EPRI framework and sourced from the Transportation Monitoring Centre of the State Grid Corporation of China-an authoritative and reliable data provider. Building upon this established dataset allows for a direct and consistent comparison with prior research outcomes, thereby underscoring the distinctive contributions of our proposed approach.

	Input					Output		
Power company	**Power grid investment (10^4 yuan)**	**Infrastructure investment (10^4 yuan)**	**Technological investment (10^4 yuan)**	**Marketing investment (10^4 yuan)**	**Information technology (10^4 yuan)**	**Total profit (10^4 yuan)**	**Electricity sales (10^6 kwh)**	**Purchase price (Yuan / 10^3 kwh)**
Fujian	1260422	1158031	69000	153615	17139	239868	1397.80	214.16
Tianjin	620911	547966	22845	111471	12491	95404	6078.01	237.29
Hebei	912164	850828	40977	118861	13541	89700	13908.7	210.08
Jiangsu	2977447	2765963	121570	388372	17518	686197	38488.3	152.91
Shandong	2804030	2615571	107399	283868	18150	595476	32978.29	123.09
Shanghai	1079683	628907	128095	153731	29877	110202	11174.96	224.62
Shanxi	926989	807855	46466	185648	13989	88613	16295.99	126.32
Zhejiang	2290509	2134974	99994	239306	18355	588559	28026.15	217.98
Anhui	1125446	961436	107345	126241	15914	75821	1250.51	191.20
Beijing	643251	594054	28141	132705	14041	152093	7920.60	206.48
Hubei	1136556	1008999	53240	171112	17442	71546	1190.21	196.62
Hunan	891914	776652	45236	155489	16123	71723	961.89	200.39
Henan	1387564	1258219	87000	122140	18701	126149	2399.22	112.26
Jiangxi	877297	800046	22470	102954	16176	65327	730.40	252.21
Sichuan	2838232	2592478	157049	181748	19244	99523	180.40	159.64

Table 1. The dataset for SG efficiency evaluation

The FITradeoff-DEA methodology was implemented following the algorithmic procedure described in Section 3, with parameters configured to ensure both robustness and practical applicability. Key settings included: weight ratio bounds of 0.01–100 for all input (v_i/v_{i+1}) and output (u_r/u_{r+1}) comparisons, providing reasonable proportionality while allowing flexibility; a minimum weight threshold of 10^{-6} to prevent trivial solutions; a convergence threshold of 0.05 for weight ranges, excluding sufficiently determined weight pairs from further elicitation; an efficiency frontier threshold of $\theta \geq 0.999$ to account for numerical precision; and a maximum of 25 iterations to balance solution quality and computational effort. These settings were informed by preliminary sensitivity analyses and are consistent with established practices in DEA and MCDM literature. All input and output indicators were first standardized using min-max normalization to map values onto a [0.1, 1] scale, ensuring comparability across heterogeneous units. Initial DEA evaluation identified multiple efficient DMUs along the production frontier, confirming the presence of the "multiple-winner dilemma" that motivates the preference-based approach. The subsequent FITradeoff interactive preference elicitation involved six weight ratio comparisons (v_1/v_2, v_2/v_3, v_3/v_4, v_4/v_5 for inputs u_1/u_2, u_2/u_3 for outputs). Simulated decision-maker responses, derived from preset preference weights, were used iteratively to refine the feasible weight space by adding linear constraints until convergence criteria were met.

4.1 Iteration Process and Results

The FITradeoff-DEA methodology produced a systematic convergence pattern that effectively resolved the initial multiple-winner dilemma. Table 2 summarizes the iterative DMU elimination process. Initially, seven DMUs: Tianjin, Hebei, Jiangsu, Shanxi, Zhejiang, Beijing, and Jiangxi were identified on the production frontier with perfect efficiency scores ($\theta = 1.000$). As iterative weight ratio comparisons were applied, the candidate set was progressively narrowed. Early iterations exerted limited impact on the overall candidate set, whereas later iterations particularly from iteration 10 onwards resulted in a marked reduction of remaining DMUs. The majority of decision-maker responses were "B" (indicating the true ratio was below the midpoint), guiding the refinement of the feasible weight space and the elimination of less preferred DMUs. Through 16 iterations, 16 linear constraints were incorporated, progressively narrowing the candidate set and ultimately identifying Tianjin as the unique preference-optimal DMU. The corresponding output weights were $u_1 = 0.0110$, $u_2 = 0.0110$, and $u_3 = 1.1015$, and input weights were $v_1 = 0.0593$, $v_2 = 0.0593$, $v_3 = 0.0593$, $v_4 = 5.9271$, and $v_5 = 2.3018$, reflecting the relative importance of each indicator according to the decision-maker's preferences. Table 3 shows the evolution of weight ratio boundaries and the number of elicitation queries for each weight pair. The first three input ratios (v_1/v_2,v_2/v_3, v_3/v_4) converged to narrow bounds ([0.100, 1.337]) and required the maximum number of queries (three each), reflecting higher sensitivity to decision-maker preferences. In contrast, the fourth input ratio (v_4/v_5) and the two output ratios (u_1/u_2, u_2/u_3) maintained wider bounds ([1.337, 2.575] for v_4/v_5; [0.100, 2.575] for u_1/u_2 and u_2/u_3) and required fewer queries (two or fewer), indicating greater flexibility or less decisive preference information.

Iteration	Selected Weight Pair	midpoint	answer	Remaining Efficient DMUs
0 (Initial)	-	-	-	Tianjin, Hebei, Jiangsu, Shanxi, Zhejiang, Beijing, and Jiangxi
1	v_1/v_2	5.050	B	Tianjin, Hebei, Jiangsu, Shanxi, Zhejiang, Beijing, and Jiangxi
2	v_2/v_3	5.050	B	Tianjin, Hebei, Jiangsu, Shanxi, Zhejiang, Beijing, and Jiangxi
3	v_3/v_4	5.050	B	Tianjin, Hebei, Jiangsu, Shanxi, Zhejiang, Beijing, and Jiangxi
4	v_4/v_5	5.050	B	Tianjin, Hebei, Jiangsu, Shanxi, Zhejiang, Beijing, and Jiangxi
5	u_1/u_2	5.050	B	Tianjin, Hebei, Jiangsu, Shanxi, Zhejiang, Beijing, and Jiangxi
6	u_2/u_3	5.050	B	Tianjin, Hebei, Jiangsu, Zhejiang, Beijing, Jiangxi
7	v_1/v_2	2.575	B	Tianjin, Hebei, Jiangsu, Zhejiang, Beijing, Jiangxi
8	v_2/v_3	2.575	B	Tianjin, Hebei, Jiangsu, Zhejiang, Beijing, Jiangxi
9	v_3/v_4	2.575	B	Tianjin, Hebei, Jiangsu, Zhejiang, Beijing, Jiangxi
10	v_4/v_5	2.575	B	Tianjin, Hebei, Jiangsu, Zhejiang, Beijing
11	u_1/u_2	2.575	B	Tianjin, Hebei, Jiangsu, Beijing
12	u_2/u_3	2.575	B	Tianjin, Jiangsu, Beijing
13	v_1/v_2	1.337	B	Tianjin, Jiangsu
14	v_2/v_3	1.337	B	Tianjin, Jiangsu
15	v_3/v_4	1.337	B	Tianjin, Jiangsu
16	v_4/v_5	1.337	A	Tianjin

Table 2. FITradeoff iteration process and DMU elimination sequence

Weight ratio	Weight ratio bound	Question count
v_1/v_2	[0.100, 1.337]	3
v_2/v_3	[0.100, 1.337]	3
v_3/v_4	[0.100, 1.337]	3
v_4/v_5	[1.337, 2.575]	3
u_1/u_2	[0.100, 2.575]	2
u_2/u_3	[0.100, 2.575]	2

Table 3. Evolution of weight ratio bounds and question counts during iterations.

4.2 Analysis and discussion

The iterative process systematically reduced the candidate set of efficient DMUs while preserving both DEA structural constraints and decision-maker preferences. The results confirm the FITradeoff-DEA model's capability to resolve multiple-winner scenarios, yielding refined efficiency rankings and identifying a unique preference-optimal DMU, Tianjin. Notably, the first three input weight ratios exhibited pronounced convergence, reflected in narrower bounds and a higher number of elicitation queries, indicating strong sensitivity to decision-maker preferences. In contrast, the remaining input and output ratios maintained wider bounds with fewer queries, suggesting either greater flexibility or less decisive preference information. These findings demonstrate how linear constraints derived from preference elicitation can be systematically integrated into DEA, allowing simultaneous consideration of technical efficiency and subjective priorities. The differential convergence of weight ratios further highlights the algorithm's ability to prioritize under-explored or uncertain regions of the weight space, thereby reducing ambiguity among candidate DMUs and refining efficiency assessments. Overall, the results underscore the robustness and practical applicability of the FITradeoff-DEA framework in incorporating decision-maker preferences, particularly in scenarios characterized by multiple winners.

5. Conclusion

This study proposes a novel FITradeoff-DEA framework that effectively resolves the "multiple-winner dilemma" in traditional DEA by integrating decision-maker preferences through an iterative, weight-space contraction mechanism. The model successfully bridges the gap between technical efficiency analysis and value-based decision-making, converging multiple efficient DMUs to a single, preference-optimal solution, as validated in our SG case study. Notwithstanding its contributions, this study has limitations. The model's effectiveness partially relies on the consistency of decision-makers' responses during the interactive elicitation. Moreover, the current framework assumes a deterministic environment and does not account for uncertain data. Future research will focus on extending the model to handle data uncertainty through fuzzy or stochastic formulations. Exploring its application to other sectors, such as healthcare and finance, and integrating it with other advanced preference modelling techniques present promising avenues for further development. This study develops an innovative FITradeoff-DEA methodology that systematically addresses the issue of multiple efficient DMUs in conventional DEA analysis. By incorporating decision maker preferences through an interactive weight constraint mechanism, the framework progressively narrows the feasible weight space to identify the preferred efficient unit. The proposed approach effectively bridges technical efficiency measurement with value judgment, transforming the DEA frontier from a set of technically optimal DMUs to a single preference-optimal solution. The case study on Chinese power companies demonstrates the practical applicability of the method, showing how iterative preference elicitation on weight ratios leads to convergence toward a consensus solution. The algorithm's dynamic selection of comparison pairs ensures balanced preference exploration across all criteria while maintaining computational feasibility. While methodology offers significant advantages, certain limitations warrant attention. The solution quality depends on the consistency of preference statements during the interactive process. Additionally, the current implementation assumes deterministic data and may benefit from extensions to handle uncertain environments. Future research directions include developing robust versions for stochastic data, exploring applications in other regulatory contexts, and integrating machine learning techniques to reduce the cognitive burden on decision makers through preference prediction.

6. Acknowledgments

This work was co-funded by the European Regional Development Fund (ERDF) through the Regional Operational Programme of Emilia-Romagna (Italy) under Project "SACER" (CUP: J47G22000760003, POR-FESR 2021-2027) and by the Italian National Recovery and Resilience Plan (NRRP, Mission 4 Component 2 Investment 1.5) through NextGenerationEU (Tender No. 3277/30.12.2021, Award No. 0001052/23.06.2022).

7. References

[1] Blazevic, R.; Kucukalic, A.; Dzananovic, A. & Dzaferovic, E. (2022). Assessment and analysis of wind energy potential in the mountain area around Sarajevo, Proceedings of the 33rd DAAAM International Symposium, pp.0063-0070, B. Katalinic (Ed.), Published by DAAAM International, ISBN 978-3-902734-36-5, ISSN 1726-9679, Vienna, Austria, https://doi.org/10.2507/33rd.daaam.proceedings.009

[2] Maric, P. & Budimir, V. (2022). The impact of the utility scale photovoltaic power plant on the transmission grid in terms of voltage and reactive power, Proceedings of the 33rd DAAAM International Symposium, pp.0098-0104, B. Katalinic (Ed.), Published by DAAAM International, ISBN 978-3-902734-36-5, ISSN 1726-9679, Vienna, Austria, https://doi.org/10.2507/33rd.daaam.proceedings.015

[3] Midžić Kurtagić, S. & Kadrić, Dž. (2022). Improvement of airports energy efficiency, case study of airport Sarajevo, Proceedings of the 33rd DAAAM International Symposium, B. Katalinic (Ed.), Published by DAAAM International, ISBN 978-3-902734-36-5, ISSN 1726-9679, Vienna, Austria

[4] Ilko, J.; Rusko, M. & Demko, J. (2022). Non-invasive steam flow measurement as an energy medium by clamp-on ultrasonic method in context of energy management concept implementation, optimization and safety, Proceedings of the 33rd DAAAM International Symposium, pp.0238-0246, B. Katalinic (Ed.), Published by DAAAM International, ISBN 978-3-902734-36-5, ISSN 1726-9679, Vienna, Austria, https://doi.org/10.2507/33rd.daaam.proceedings.033

[5] Fang, X., Misra, S., Xue, G., & Yang, D. (2011). Smart grid-The new and improved power grid: A survey. IEEE communications surveys & tutorials, 14(4), 944-980. https://doi.org/10.1109/SURV.2011.101911.00087

[6] Gellings, C. W. (2020). The smart grid: enabling energy efficiency and demand response. River Publishers. https://doi.org/10.1201/9781003151524

[7] Zhou, Z., Zhao, F., & Wang, J. (2011). Agent-based electricity market simulation with demand response from commercial buildings. IEEE Transactions on Smart Grid, 2(4), 580-588. https://doi.org/10.1109/TSG.2011.2168244

[8] Pohekar, S. D., & Ramachandran, M. (2004). Application of multi-criteria decision making to sustainable energy planning-A review. Renewable and sustainable energy reviews, 8(4), 365-381. https://doi.org/10.1016/j.rser.2003.12.007

[9] Charnes, A., Cooper, W. W., & Rhodes, E. (1978). Measuring the efficiency of decision making units. European Journal of Operational Research, 2(6), 429-444. https://doi.org/10.1016/0377-2217(78)90138-8

[10] Sözen, A., Alp, I., & Özdemir, A. (2010). Assessment of operational and environmental performance of the thermal power plants in Turkey by using data envelopment analysis. Energy Policy, 38(10), 6194-6203. https://doi.org/10.1016/j.enpol.2010.06.005

[11] Zhang, N., Zhou, P., & Choi, Y. (2013). Energy efficiency, CO2 emission performance and technology gaps in fossil fuel electricity generation in Korea: A meta-frontier non-radial directional distance function analysis. Energy policy, 56, 653-662. https://doi.org/10.1016/j.enpol.2013.01.033

[12] Dyson, R. G., Allen, R., Camanho, A. S., Podinovski, V. V., Sarrico, C. S., & Shale, E. A. (2001). Pitfalls and protocols in DEA. European Journal of operational research, 132(2), 245-259. https://doi.org/10.1016/S0377-2217(00)00149-1

[13] Thompson, R. G., Langemeier, L. N., Lee, C. T., Lee, E., & Thrall, R. M. (1990). The role of multiplier bounds in efficiency analysis with application to Kansas farming. Journal of Econometrics, 46(1-2), 93-108. https://doi.org/10.1016/0304-4076(90)90049-Y

[14] Allen, R., Athanassopoulos, A., Dyson, R. G., & Thanassoulis, E. (1997). Weights restrictions and value judgements in data envelopment analysis: evolution, development and future directions. Annals of operations research, 73(0), 13-34. https://doi.org/10.1023/A:1018968909638

[15] Yadav, V., & Sharma, M. K. (2015). An application of hybrid data envelopment analytical hierarchy process approach for supplier selection. Journal of Enterprise Information Management, 28(2), 218-242. https://doi.org/10.1108/JEIM-04-2014-0041

[16] Luhaniwal, J., Agarwal, S., & Mathur, T. (2025). An integrated DEA-fuzzy AHP method for prioritization of renewable energy sources in India. Scientific Reports, 15(1), 838. https://doi.org/10.1038/s41598-024-84891-2

[17] Dyer, J. S. (1990). Remarks on the analytic hierarchy process. Management Science, 36(3), 249-258. https://doi.org/10.1287/mnsc.36.3.249

[18] Amini, A., Alinezhad, A., & Yazdipoor, F. (2019). A TOPSIS, VIKOR and DEA integrated evaluation method with belief structure under uncertainty to rank alternatives. International Journal of Advanced Operations Management, 11(3), 171-188. https://doi.org/10.1504/IJAOM.2019.100708

[19] de Almeida, A. T., de Almeida, J. A., Costa, A. P. C. S., & de Almeida-Filho, A. T. (2016). A new method for elicitation of criteria weights in additive models: Flexible and interactive tradeoff. European Journal of Operational Research, 250(1), 179-191. https://doi.org/10.1016/j.ejor.2015.08.058

[20] De Almeida, A. T., Cavalcante, C. A. V., Alencar, M. H., Ferreira, R. J. P., de Almeida-Filho, A. T., & Garcez, T. V. (2015). Multicriteria and multiobjective models for risk, reliability and maintenance decision analysis (Vol. 231). Cham, Switzerland: Springer International Publishing. https://doi.org/10.1007/978-3-319-17969-8

[21] Frej, E. A., Roselli, L. R. P., Araújo de Almeida, J., & de Almeida, A. T. (2017). A multicriteria decision model for supplier selection in a food industry based on FITradeoff method. Mathematical Problems in Engineering, 2017(1), 4541914. https://doi.org/10.1155/2017/4541914

[22] Xiaobao, Y., Zhongfu, T., Kangting, C., Liwei, J., & Puyu, H. (2015). Efficiency evaluation for smart grid management based on stochastic frontier model and data envelopment analysis model. Mathematical Problems in Engineering, 2015(1), 142764. https://doi.org/10.1155/2015/142764

DOI: 10.2507/36th.daaam.proceedings.018

TOWARDS SMART FACTORIES: THE DEVELOPMENT OF AUTONOMOUS MOBILE ROBOTS FOR INTRALOGISTICS 4.0

Neslihan Demir

This Publication has to be referred as: Neslihan, D[emir] (2025). Towards Smart Factories: The Development of Autonomous Mobile Robots for Intralogistics 4.0, Proceedings of the 36th DAAAM International Symposium, pp.0141-0146, B. Katalinic (Ed.), Published by DAAAM International, ISBN 978-3-902734-47-1, ISSN 1726-9679, Vienna, Austria
DOI: 10.2507/36th.daaam.proceedings.018

Abstract

The evolution of Industry 4.0 underscores intralogistics as a critical enabler of smart factories, where responsiveness and adaptability outweigh the rigid path structures of Automated Guided Vehicles (AGVs). Autonomous Mobile Robots (AMRs) have emerged as a viable alternative, yet their validation under production-like conditions remains limited. This study introduces ProdigiMover-Small, a skid steering AMR prototype designed through additive manufacturing and controlled via the Robot Operating System (ROS). The platform integrates LIDAR, wheel encoders, and inertial sensing in a compact and cost-efficient design. When validated in a workshop-like environment, the system demonstrated reliable navigation, maintaining localization accuracy within 2–4 cm in workstation targets and 4–6 cm along traversed routes, while positioning operations achieved a high overall success rate. Comparative analysis further revealed that additive manufacturing reduced development lead time by more than half and overall costs by approximately 90%. These results confirm that task proximal precision, modularity, and adaptability are decisive for intralogistics performance. The presented prototype illustrates a scalable pathway from laboratory proof-of-concept to industrial deployment. It also provides a foundation for future research on multi-robot coordination, human-aware navigation, and digital twin integration in Intralogistics 4.0.

Keywords: Industry 4.0; Intralogistics 4.0; Autonomous mobile robots; Robot Operating System; Smart factories.

1. Introduction

Manufacturing systems are undergoing a systemic reorientation with the proliferation of Industry 4.0, where digitalisation, automation, and intelligent decision-making are fundamentally reshaping traditional production processes [1], [2]. The seamless integration of cyber-physical systems, real-time data, and networked devices reinforces the necessity for redefined approaches to intralogistics [3], [4]. Conventional AGVs, constrained by fixed routes and limited adaptability to dynamic environments, prove insufficient for the requirements of modern intralogistics [3], [4], [5]. Such constraints impede factories from attaining the flexibility and responsiveness that lie at the core of smart manufacturing systems [2], [4].

Previous studies have consistently underscored intralogistics as the essential pillar of smart factories, where the timely coordination of material flows directly supports production efficiency [2], [5], [6]. In this regard, research on AMRs has primarily focused on navigation, localisation, and path planning, frequently implemented through ROS [7], [8], [9]. Although numerous prototypes and simulation models have been introduced, a persistent gap remains in translating these advances into practical intralogistics applications under real factory conditions [3], [4], [6]. To bridge this gap, the present study introduces the development of a skid steering AMR integrated with ROS, designed to facilitate intelligent material transport in manufacturing environments. The design emphasises real-time navigation capabilities and flexible control mechanisms, ensuring suitability for dynamic manufacturing environments. The proposed approach illustrates how robotics and intralogistics can be systematically integrated in alignment with Industry 4.0 principles.

This work contributes by illustrating how ROS enabled AMRs extend beyond the conventional capabilities of AGVs, delivering modularity, adaptability, and interoperability in intralogistics. As a proof-of-concept, the system demonstrates the integration of robotics and material handling functions, advancing the operational foundations of smart factories. Subsequent research will advance toward multi-robot collaboration, the incorporation of advanced perception systems, and leveraging digital twins to enable predictive logistics management. These research directions are expected to foster the evolution of Intralogistics 4.0, consolidating the role of AMRs as key enablers of fully automated and intelligent manufacturing ecosystems.

2. Literature Review

The transformation toward Industry 4.0 has shifted manufacturing from rigid, centralised production lines to smart factories that operate as cyber-physical systems, emphasizing reconfigurability, real-time data integration, and human-machine collaboration. Within this transformation, intralogistics has become a decisive driver of industrial competitiveness, as the precise synchronization of material delivery with production demands directly determines the agility and resilience of manufacturing networks [5]. At the core of this transition stand AMRs, surpassing the rigid path dependency of AGVs. By uncoupling intralogistics from static infrastructure, AMRs enable decentralised, adaptive, and human-aware operations across dynamic manufacturing environments [7]. Foundational studies have introduced the concept of autonomous intralogistics systems, proposing autonomy levels across transport, storage, and order-handling tasks and highlighting the shift from centralised AGV style control toward networked, adaptive decision-making; in this regard AMRs equipped with onboard perception and navigation systems provide greater responsiveness to environmental changes, thereby enabling decentralised and reconfigurable logistics operations [8], [10].

The adoption of AMRs in intralogistics brings clear advantages, such as enhanced productivity, improved safety, and reduced manual workload [11]. However, challenges remain, including high integration costs, energy constraints, and limited human-robot collaboration in semi-structured environments [12], [13]. Recent studies further show that AMR enabled networks can outperform traditional conveyor (or AGV) based systems when cost and shift conditions are favourable [5], while reviews continue to highlight persistent gaps in safety, interoperability, and scalability that limit widespread deployment.

At the motion level, AMR performance hinges on robust localisation, mapping, and planning capabilities. Surveys classify approaches ranging from graph-based global routing to local obstacle avoidance under sensing and computation limits, underscoring trade-offs among optimality, smoothness, and real-time feasibility [8], [9], [14]. Comparative analyses further highlight when sampling-based planners (e.g., RRT variants), search-based planners (A* derivatives), or optimisation-based model predictive control (MPC) planners are best suited under industrial constraints such as narrow aisles, mixed traffic, and safety zones [15]. Beyond technical motion planning, human-aware navigation introduces additional requirements (predicting worker trajectories, respecting proxemics, and signalling intent) so that robots can coexist safely without excessive conservatism [16]. Industrial contributions complement this by integrating explicit environmental cues, such as safety line detection, to simplify perception and enforce compliance with plant safety policies [17]. Locomotion strategy likewise shapes intralogistics performance. Kinematic comparisons of differential, mecanum, and omnidirectional wheelbases clarify how controllability and manoeuvrability affect throughput in tight layouts, positioning operations, and pallet pickup [15]. Within this spectrum, differential or skid steering platforms remain the most widely adopted due to their structural simplicity, robustness, and ability to handle higher payloads at lower cost [7], [15]. While omnidirectional bases offer superior manoeuvrability, their sensitivity to floor conditions and higher integration cost often limit their scalability in industrial contexts. This positions skid-steering AMRs as a pragmatic solution for dynamic shop floors, where reliability and cost efficiency are prioritized over unrestricted mobility. Although their inability to perform true lateral motion imposes kinematic constraints in confined or mixed traffic environments, recent studies suggest that advanced navigation algorithms, hybrid path planners, and adaptive control strategies can mitigate these limitations thereby extending the applicability of skid steering robots beyond traditional domains. Against this backdrop, the present research contributes by developing a skid steering AMR specifically tailored for intralogistics operations, aiming to reconcile industrial practicality with the flexibility demands of Intralogistics 4.0.

From a systems perspective, the challenge extends beyond driving robots to orchestrating them within complex workflows. Recent navigation architectures adopt modular, platform-agnostic layers with semantic mapping, profile switching, and distributed control to improve robustness across heterogeneous fleets and changing layouts [10]. Parallel work on process-level orchestration emphasises intuitive, visual programming interfaces that allow non-experts to configure missions, parameterise behaviours, and monitor fleets, thus facilitating adoption in brownfield sites [12].

Application-level evidence demonstrates both benefits and pitfalls of AMR integration. In Smart Factory contexts, MiR100 platforms integrated with MES/WMS and wireless I/O achieved millimetre-scale repeatable positioning accuracy using visual markers, while also exposing failure modes in narrow spaces [18]. Case studies in remanufacturing show that AMR-driven material delivery increases throughput and reduces nonvalue adding time when paired with lean practices such as 5S and Kaizen [13]. Multi-case warehouse research further identifies adoption drivers (layout decoupling, ERP/WMS integration) and barriers (legacy safety zoning, change management), highlighting how organisational readiness influences total cost of ownership [11].

Despite these advances, several gaps persist. Current approaches to production logistics integration often remain offline or myopic, lacking predictive scheduling and receding horizon fleet control [5]. Safety-aware navigation requires verifiable behaviours that degrade gracefully under sensor faults, layout changes, and mixed traffic [19]. Human-in-the-loop orchestration remains underdeveloped, yet essential for acceptance, requiring explainable intent, ergonomic task design, and clear collaboration protocols [20]. Finally, learning-augmented forecasting, particularly trajectory prediction of workers and forklifts, offers a promising direction to reduce idle time and congestion by coupling human motion prediction with task allocation and path planning [16].

Collectively, the literature positions AMRs as critical enablers of Intralogistics 4.0, but only when navigation, fleet control, and process orchestration are co-designed with production scheduling, safety governance, and human workflows. These converging insights provide a rationale for research into new robotic platforms such as skid steering AMRs tailored for dynamic manufacturing environments aiming to bridge the persistent gap between theoretical advances and practical deployment in smart factories.

3. Prototype Development for Intralogistics Applications

Although recent research in intralogistics has made significant progress in path planning, localisation, and fleet coordination for AMRs, much of this work remains confined to simulations or idealized laboratory tests. To ensure applicability to manufacturing practice, AMRs must be evaluated in production-like environments that capture the complexity of real material handling workflows. Simulation studies often fail to reflect practical constraints such as narrow aisles, positioning operations, or interactions with human workers. To address this gap, the present study introduces a 3D printed skid steering prototype conceived as a proof-of-concept platform for intralogistics applications. Skid steering was selected due to its mechanical simplicity and robustness under varying payload conditions, despite its limited lateral mobility, which was considered acceptable for intralogistics environments dominated by linear and aisle-based flows.

The prototype was developed to be compact, cost efficient, and sufficiently robust for repeated laboratory testing. PLA filament was selected for its accessibility, low cost, and suitability for rapid prototyping. While adequate for laboratory testing, its limited durability under high-load or high-temperature conditions suggests that future iterations could adopt reinforced polymers or composite materials to meet industrial robustness requirements. The platform integrates four wheels in a skid steering configuration, a locomotion principle widely adopted in industrial applications due to its mechanical simplicity and manoeuvrability. Differential motion is achieved through the coordinated actuation of the left and right motor pairs. To ensure realistic kinematic behaviour at reduced scale, proportional downscaling was applied, harmonizing chassis dimensions, wheel diameter, and motor torque. In this way, the compact platform-maintained consistency with larger AMR concepts while being more accessible for experimental validation. This scaling approach ensured that turning radii, wheel slip characteristics, and manoeuvrability constraints remained representative of industrial-scale AMRs, thereby preserving the validity of experimental observations despite the reduced size. Key specifications of the prototype are summarized in Table 1, highlighting the compact yet functional design of the robot.

Component	Specification
Chassis	3D-printed frame, compact skid-steering design
Wheels	4 wheels, 85 mm diameter × 38 mm width, rubber with nylon-reinforced rims
Motors	DC motors with proportional downscaling, integrated incremental encoders
Sensors	2D LiDAR (YDLidar G2), wheel encoders, IMU
Control Unit	Onboard computing with Ubuntu + ROS Noetic
Power Supply	Rechargeable battery pack (compact, lightweight)

Table 1. Technical Specifications of the Prototype

The physical prototype developed for this research is presented in Fig. 1. The compact skid steering platform integrates 3D printed structural components, encoder-equipped motors, and a ROS-based control unit. Its lightweight and portable design enables efficient testing of navigation and mapping functionalities, serving as a proof-of-concept platform for intralogistics research.

Fig. 1. 3D Printed Skid-Steering Robot Prototype for Intralogistics Applications

The compact prototype, named ProdigiMover-Small, was developed using additive manufacturing to replicate the essential features of larger autonomous mobile robots. The platform integrates four wheels in a skid steering configuration, encoder-equipped DC motors, and a ROS-based control unit. Its lightweight and modular structure provides a cost-effective solution for proof-of-concept experiments in production-like laboratory environments. Validation was conducted in a production-like laboratory environment designed to resemble a small-scale manufacturing workshop. The facility is equipped with conventional machine tools such as turning and milling machines, a KUKA robotic manipulator for handling tasks, and buffer areas arranged with palletised storage zones. These elements provided realistic obstacles and spatial constraints that reflect actual shop floor conditions. The layout of the laboratory allowed the prototype to be tested in scenarios involving manoeuvring through narrow pathways, positioning in front of workstations, and navigating around static industrial equipment. Such an environment ensured that the prototype was not only tested under controlled conditions but also challenged by the types of constraints typical of intralogistics operations.

The platform employs four rubber wheels (85 mm diameter and 38 mm width) with nylon-reinforced rims, selected to provide sufficient grip on laboratory floor surfaces. Motion is actuated by DC motors with integrated incremental encoders, controlled via a compact motor driver. Power is supplied by an 11.1 V 3S Li-Po battery, balancing autonomy with portability. For perception and localisation, the robot integrates a YD LIDAR sensor and an MPU-9250 IMU, complemented by encoder feedback for odometry. Low-level actuation and sensor interfacing are managed by an ESP32 microcontroller, while high-level navigation and decision-making tasks are executed on a Raspberry Pi. Beyond raw sensor fusion, the combined odometry estimates were subjected to filtering within the ROS framework, which attenuated noise and compensated for drift. This additional layer of processing yielded smoother trajectories and reinforced localisation stability, particularly during extended navigation tasks.

The control system was designed as a layered architecture in which low-level actuation and sensor interfacing were managed by the ESP32 microcontroller, while high-level decision-making and navigation modules were executed on the Raspberry Pi. This hardware division enabled reliable real-time control of motors and sensors, while ensuring sufficient computational capacity for ROS-based tasks such as localisation, mapping, and path planning. The overall architecture was implemented within the ROS framework, with nodes for odometry calculation, LIDAR based environment scanning, navigation via the ROS Navigation Stack, and teleoperation. Standard ROS message types (e.g., nav_msgs/Odometry, geometry_msgs/Twist) facilitated communication among modules. By relying on widely available open-source ROS packages, the system demonstrated both modularity and portability, ensuring that the prototype's architecture can be directly transferred to larger AMR platforms.

4. Results and Discussion

The developed skid-steering AMR prototype (ProdigiMover-Small) was validated in a production-like lab setup. The 3D printed chassis showed adequate rigidity for repeated trials; ROS-based modules (odometry, LIDAR perception, local/global planners) communicated reliably and supported both teleoperation and autonomous runs. For the small-scale platform, localization accuracy remained within 2–4 cm in target-proximal positioning operations and within 4–6 cm along the traversed routes, which is consistent with the literature's emphasis that fine, localized accuracy, rather than uniform global precision, determines success in station approach and precise positioning [14], [21]. Path planning experiments confirmed that the standard ROS Navigation Stack generated feasible routes in semi-structured layouts; using Hybrid A* improved path smoothness and reduced in-place turning compared to grid-based A*. Positioning trials in front of workstation targets reached a 92% success rate, demonstrating practical applicability despite the kinematic limits of skid-steering noted in prior work [15]. These observations are consistent with recent results that recommend coarse-to-fine localization and task-phase-aware planning to bridge global navigation with high precision, target-zone maneuvers [21]. From a prototyping perspective, additive manufacturing proved decisive for iteration speed. In our case, 3D printing reduced chassis production lead-time by more than 50% relative to conventional fabrication, enabling rapid design test loops. This aligns with current manufacturing science that highlights AM as a lever for fast iteration and shorter time to market in robotics development [22].

A comparative bill of materials assessment indicated that, relative to typical industrial scale AMRs reported in the literature, our scaled down prototype achieved approximately 90% lower material and component cost while reproducing core behaviors. The primary substitutions were smaller wheels and lower torque motors (consistent with reduced wheel diameter from geometric down-scaling) and a PLA-based 3D printed chassis; sensing and compute subsystems remained the same as LIDAR and IMU, ESP32, Raspberry Pi, and power distribution. This pattern underscores a modular architecture where scale primarily affects mechanical elements, not perception or control, a finding that dovetails with industry reports advocating modular, reconfigurable intralogistics solutions [18], [23]. Modularity and transferability also emerged on the software side. Although our implementation used ROS Noetic, the broader ecosystem's evolution (e.g., Navigation2 on ROS 2) emphasizes reliability, modularity, and practical operational feasibility in facilities, qualities that make it straightforward to port the presented stack from small proof-of-concept platforms to larger AMRs as requirements scale [24]. These simulation results reinforced the experimental findings by showing similar accuracy trends in localization and navigation performance. Experiments conducted in Gazebo mirrored real world performance trends unsurprising given that Gazebo is reported as the most commonly used ROS-compatible simulator for Industry 4.0/5.0 contexts, making it a sensible testbed before shop-floor piloting [25].

Although this study focused on a single prototype, preliminary insights at the fleet level suggest broader implications for multi-robot systems. At the fleet-level, our single robot results point to expected system level bottlenecks, chiefly deadlocks in narrow intersections and contention around shared operational stations. Recent agent-based evaluations show that proactive deadlock/collision prevention policies can improve throughput by about 39%, reinforcing the value of integrating positioning-aware planning with multi-robot coordination in follow up work [26]. It should be emphasized that this study does not claim a larger platform; rather, it demonstrates that a smaller than typical prototype can faithfully reproduce key behaviors of industrial AMRs at a fraction of the cost and time, while preserving architectural modularity. This supports prior DAAAM findings on pragmatic intralogistics deployment and human-robot workflow integration [18], [21], and aligns with the growing consensus that high-accuracy positioning near targets, not omnipresent precision, is the practical lever for robust intralogistics [14], [21].

5. Conclusion

This paper presented the development and validation of a compact skid-steering AMR prototype, ProdigiMover-Small, designed as a proof-of-concept platform for intralogistics applications under Industry 4.0 principles. The study confirmed that a scaled-down, 3D printed platform can reproduce the essential behaviours of industrial AMRs such as navigation, localisation, and positioning while significantly reducing cost and development time. ROS-based implementation proved modular and transferable, demonstrating how widely available open-source tools can accelerate prototyping and facilitate the transition from laboratory validation to industrial deployment. The findings emphasize that precision near workstations, modularity of design, and adaptability to dynamic layouts are more decisive for intralogistics than universal high-accuracy positioning. Additive manufacturing further supported rapid iteration, reinforcing its role as an enabler of agile development in robotics research. Future work will extend the presented prototype towards multi-robot coordination, advanced human-aware navigation, and integration with digital twin frameworks, thereby advancing the realization of Intralogistics 4.0 and resilient smart factory ecosystems.

6. Acknowledgments

The author gratefully acknowledges the support of Bogrekci Robotics & AI Lab, where the prototype development and experiments were conducted. Special thanks are extended to Prof. Dr. Pınar Demircioglu and Prof. Dr. Ismail Bogrekci for their valuable guidance and encouragement throughout this research. The author also acknowledges DAAAM International for fostering an academic platform that enables the dissemination and exchange of research on smart manufacturing and intralogistics.

7. References

[1] Smit, J.; Kreutzer, S.; Moeller, C. & Carlberg, M. (2016). Industry 4.0, European Parliament, Directorate General for Internal Policies, Policy Department A: Economic and Scientific Policy, Brussels.

[2] Albrecht, T.; Baier, M.-S.; Gimpel, H.; Meierhöfer, S.; Röglinger, M.; Schlüchtermann, J. & Will, L. (2023). Leveraging Digital Technologies in Logistics 4.0: Insights on Affordances from Intralogistics Processes, Information Systems Frontiers, 26, pp. 755–774, DOI: 10.1007/s10796-023-10394-6

[3] Yan, J.; Zhang, M. & Fu, Z. (2019). An intralogistics-oriented Cyber-Physical System for workshop in the context of Industry 4.0, Computers in Industry, 115, 103162, pp. 1178–1190, DOI: 10.1016/j.compind.2019.103162

[4] Aron, C.; Sgarbossa, F.; Ballot, E. & Ivanov, D. (2024). Cloud Material Handling Systems: A Cyber-Physical System to Enable Dynamic Resource Allocation and Digital Interoperability, Journal of Intelligent Manufacturing, 35, pp. 3815–3836, DOI: 10.1007/s10845-023-02262-6

[5] Fragapane, G.; Ivanov, D.; Peron, M.; Sgarbossa, F. & Strandhagen, J. O. (2020). Increasing flexibility and productivity in Industry 4.0 production networks with autonomous mobile robots and smart intralogistics, Annals of Operations Research, 308(1–2), pp. 125–143, DOI: 10.1007/s10479-020-03526-7

[6] Fragapane, G.; Ivanov, D.; Peron, M.; Sgarbossa, F. & Strandhagen, J. O. (2021). Autonomous mobile robots in intralogistics: Potentials and challenges for the smart factory of the future, Annals of Operations Research, 314(1), pp. 1–29, DOI: 10.1007/s10479-021-04445-0

[7] Fottner, J.; Mascolo, J. E. & Freund, T. (2021). Autonomous systems in intralogistics: State of the art and future research challenges, Logistics Research, 14(1), pp. 1–17, DOI: 10.23773/2021_11

[8] Bernardo, R.; Sousa, J. M. C. & Gonçalves, P. J. S. (2022). Survey on robotic systems for internal logistics, Journal of Manufacturing Systems, 62, pp. 738–762, DOI: 10.1016/j.jmsy.2021.12.013

[9] Demir, N.; Demircioglu, P. & Bogrekci, I. (2024). Advancing Industry 4.0 with ROS: A Case Study on Autonomous Mobile Robot Technological Advancements, International Journal of 3D Printing Technologies and Digital Industry, Vol. 8, No. 1, pp. 130–142. DOI: 10.46519/ij3dptdi.1366132.

[10] Taranta, D.; Marques, F.; Lourenço, A.; Prates, P. A.; Souto, A.; Pinto, E. & Barata, J. (2021). An autonomous mobile robot navigation architecture for dynamic intralogistics, Proceedings of the 2021 IEEE 19th International Conference on Industrial Informatics (INDIN), pp. 1–6, ISSN 2156-1676, ISBN 978-1-6654-4131-5, DOI: 10.1109/INDIN45582.2021.9557531

[11] Grover, R. & Ashraf, M. H. H. (2024). Leveraging autonomous mobile robots for Industry 4.0 warehouses: A multiple case study analysis, The International Journal of Logistics Management, 35, No. 3, pp. 188–209, ISSN 0957-4093, DOI: 10.1108/IJLM-09-2022-0362

[12] Niermann, H.; Schake, L.; Gómez Palm, D.; Grimm, S.; Lukas, S.; Schomburg, G. & ten Hompel, M. (2023). Intuitive and flexible process control for autonomous mobile robots: A case study in a large logistics enterprise, Procedia CIRP, 119, pp. 123–128, ISSN 2212-8271, DOI: 10.1016/j.procir.2023.08.028

[13] Pizoń, J.; Wójcik, Ł.; Gola, A.; Kański, Ł. & Nielsen, I. (2024). Autonomous mobile robots in automotive remanufacturing: A case study for intra-logistics support, Advances in Science and Technology Research Journal, 18, No. 1, pp. 213–230, ISSN 2299-8624, DOI: 10.12913/22998624/177398

[14] Sánchez-Ibáñez, J. R.; Pérez-del-Pulgar, C. J. & García-Cerezo, A. (2021). Path planning for autonomous mobile robots: A review, Sensors, 21, No. 23, 7898, ISSN 1424-8220, DOI: 10.3390/s21237898

[15] Tagliavini, L.; Colucci, G.; Botta, A.; Cavallone, P.; Baglieri, L. & Quaglia, G. (2022). Wheeled mobile robots: State of the art overview and kinematic comparison among three omnidirectional locomotion strategies, Journal of Intelligent & Robotic Systems, 106, No. 3, 57, ISSN 0921-0296, DOI: 10.1007/s10846-022-01745-7

[16] Löcklin, A.; Dettinger, F.; Artelt, M.; Jazdi, N. & Weyrich, M. (2022). Trajectory prediction of workers to improve AGV and AMR operation based on the manufacturing schedule, Procedia CIRP, 107, pp. 283–288, ISSN 2212-8271, DOI: 10.1016/j.procir.2022.04.046

[17] Karamanos, X.; Bechtsis, D.; Vlachos, D. & Mastos, T. (2021). Safety lines detection as a means of navigation in industrial facilities, Proceedings of the 32nd DAAAM International Symposium, pp. 0256–0261, Katalinic, B. (Ed.), DAAAM International, Vienna, Austria, ISBN 978-3-902734-33-4, ISSN 1726-9679, DOI: 10.2507/32nd.daaam.proceedings.036

[18] Herčík, F.; Zavřel, J.; Heriban, P.; Štefánik, J. & Kočí, V. (2022). Implementation of an autonomous mobile robot in a SmartFactory: Case study with MiR100, Applied Sciences, 12, No. 17, 8912, ISSN 2076-3417, DOI: 10.3390/app12178912

[19] Niloy, M. A. K.; Shama, A.; Chakrabortty, R. K.; Ryan, M. J. & Badal, F. R. (2021). Critical design and control issues of indoor autonomous mobile robots: A review, IEEE Access, 9, pp. 35338–35370, ISSN 2169-3536, DOI: 10.1109/ACCESS.2021.3062227

[20] Sidiropoulos, A.; Sidiropoulos, V.; Bechtsis, D. & Vlachos, D. (2021). An Industry 4.0 tool to enhance human–robot collaboration, Proceedings of the 32nd DAAAM International Symposium, pp. 0617–0622, Katalinic, B. (Ed.), DAAAM International, Vienna, Austria, ISBN 978-3-902734-33-4, ISSN 1726-9679, DOI: 10.2507/32nd.daaam.proceedings.087

[21] Yilmaz, A. & Temeltas, H. (2024). A multi-stage localisation framework for accurate and precise docking of autonomous mobile robots (AMRs). Robotica, Published online. ISSN 0263-5747, DOI: 10.1017/S0263574724000602.

[22] Klenam, J. S.; Amegah, K. E. & Asante, J. P. (2025). Additive manufacturing of non-metallic materials: A review and outlook. Applications in Engineering Science, 40, 100224, ISSN 2666-4968, DOI: 10.1016/j.apples.2025.100224.

[23] Raamets, T.; Majak, J.; Karjust, K.; Mahmood, K. & Hermaste, A. (2024). Autonomous mobile robots for production logistics: A process optimisation model modification. Proceedings of the Estonian Academy of Sciences, 73, No. 2, pp. 134–141, ISSN 1736-6046, DOI: 10.3176/proc.2024.2.06.

[24] Horelican, V.; Cveticanin, D.; Ali, S. & Dobi, L. (2022). Utilizability of Navigation2/ROS 2 stack in industrial facilities and its reliability. IFAC-PapersOnLine, 55, No. 10, pp. 261–266, ISSN 2405-8963, DOI: 10.1016/j.ifacol.2022.06.018.

[25] Flores-Gonzalez, F.; Coronado, P. D. U. & Yamanobe, N. (2025). ROS-compatible robotics simulators: A review in the context of Industry 4.0 and Industry 5.0. Applied Sciences, 15, 8637, ISSN 2076-3417, DOI: 10.3390/app15158637.

[26] Turhanlar, E.; Ren, J.; Wen, J.; O'Brien, P. & Cormican, K. (2024). Autonomous mobile robot travel under deadlock and collision prevention algorithms by agent-based modelling in warehouses. International Journal of Logistics Research and Applications, 27, No. 8, pp. 1351–1376, ISSN 1367-5567, DOI: 10.1080/13675567.2022.2138290.

DOI: 10.2507/36th.daaam.proceedings.019

LEARNING DESIGN OF EXPERIMENTS THROUGH PROJECT-BASED LEARNING

Edwing Isaac Rosales Molina

This Publication has to be referred as: Rosales Molina, E[dwing] I[saac] (2025). Learning Design of Experiments through Project-Based Learning, Proceedings of the 36th DAAAM International Symposium, pp.0147-0153, B. Katalinic (Ed.), Published by DAAAM International, ISBN 978-3-902734-47-1, ISSN 1726-9679, Vienna, Austria
DOI: 10.2507/36th.daaam.proceedings.019

Abstract

Design of Experiments (DOE) is a very versatile tool with broad applicability across disciplines. Fields such as research and engineering can greatly benefit from its systematic approach. For this reason, learning DOE is fundamental for engineering students. Due to its relevance, the use of project-based approaches to learning DOE was chosen for evaluation. Students chose from various project alternatives, designing and conducting a factorial experiment and subsequently analysing the results. Learning outcomes were assessed by reviewing the academic papers submitted as a final project deliverable. Overall, the submitted deliverables demonstrate a satisfactory understanding of DOE principles. However, several areas for improvement were identified. There is a need to reinforce the use of prior research to contextualise the specific focus of each project. Furthermore, as visual resources are commonplace in technical communication, many reports could have been substantially improved by integrating photographs and schematics to support the findings. In addition, the interpretation of numerical results was not consistently clear in terms of the response variables, and the connection to decision-making was sometimes weak. These aspects merit further attention in future implementations.

Keywords: design of experiments; learning; project-based learning; DOE.

1. Introduction

1.1. Design of Experiments in Engineering Education

Design of experiments (DOE) "is a body of knowledge and techniques that enables an investigator to conduct better experiments, analyse data efficiently, and make the connections between the conclusions from the analysis and the original objectives of the investigation" [1]. The experiments referred to, as well as the act of investigation, should be understood in a broad sense: experimentation may take place either in the laboratory or on the shop floor, and investigation refers to the pursuit of a deeper understanding of a phenomenon, system, or process. This set of tools has a wide range of applications, from design[2] to quality assurance [3]. Given its relevance to industrial settings, learning DOE is essential for engineering students. The Accreditation Board for Engineering and Technology (ABET) includes, as a student outcome in its Criteria for Accrediting Engineering Programs, "an ability to develop and conduct appropriate experimentation, analyse and interpret data, and use engineering judgment to draw conclusions" [4].

Moreover, in the context of professional practice, DOE is recognised in the United States as part of the knowledge base required for obtaining Professional Engineer certification in Industrial and Systems Engineering[5], as well as for the Fundamentals of Engineering examination in Chemical Engineering[6] and Industrial and Systems Engineering [7]. Although the reference is to the United States, it is intended merely as an illustrative example of how DOE is integrated into professional certification frameworks. In view of DOE's relevance to engineering education and practice, it is essential to identify appropriate tools to ensure its effective learning by engineering students. Current pedagogical approaches advocate for student-centred methods that promote skill development, respond to contextual needs, and foster cognitive, procedural, and attitudinal dimensions. One method that has emerged as a viable option to meet these demands is project-based learning (PBL).

1.2. Project-Based Learning

Work with projects in education is often traced back to William Kilpatrick's 1918 article *The Project Method* [8]; however, Knoll suggests that the use of projects in educational contexts began "in the early eighteenth century" [9]. According to Knoll, the project method has its origin in the professionalization... of architecture" and later "it did not long remain the privilege of architects". Since then, the idea of "learning by projects" has evolved and become a key concept in education. *Project-based learning* (PBL), as a specific term, emerged much later. In the Web of Science (WOS) database, this term first appeared in a 1976 article [10]. Nevertheless, it is not possible to say that it was not being used earlier or referred to by an equivalent expression.

Given this historical trajectory, a unified definition of the term is quite difficult to achieve, though it is possible to examine several proposals. In [11], the author synthesises the concept of PBL drawing on [12] and [13], defining it as "an inquiry-based instructional method that engages learners in knowledge construction by having them accomplish meaningful projects and develop real-world products". Kokotsaki et al. [14] reference Cocco [15], who characterises project-based learning as a student-centred methodology and highlights that it "is based on three constructivist principles: learning is context-specific, learners are involved actively in the learning process, and they achieve their goals through social interactions and the sharing of knowledge and understanding". On the other hand, [16] emphasises "the development of higher-order thinking skills such as analysis, synthesis, and evaluation. Just as definitions differ, it must be acknowledged that implementations of PBL are highly varied, "depending on the context, resources, and curriculum design" [17].

The extensive historical trajectory of the concept does not preclude its continued relevance; throughout its development, it has been adapted to various fields and has acquired distinctive nuances. Nevertheless, certain elements remain central: the involvement and centrality of the learner, meaningful learning, interaction, and context-specific learning, among others. Accordingly, it has been recognised as an appropriate pedagogical strategy for facilitating learning in DOE environments.

1.3. Project-Based Learning and Design of Experiments

Previous studies have reported successful implementations of PBL in the teaching of DOE [18], [19]. An important key to success in using a tool lies in its suitability to the task at hand, in [19], the authors declared that PBL methodology was selected because it "enables learners to apply theoretical concepts to a controlled real-world environment and to make decisions based on practical experience". In contrast, reference [18] seeks to test the hypothesis that "a structured PBL curriculum with targeted learning outcomes provides an efficient and engaging learning strategy that addresses the interface between engineering analysis and practice". Nevertheless, both cases are aligned with the core elements of PBL previously discussed.

It is appropriate to highlight some general conclusions drawn from both studies. According to [19], PBL has enabled the acquisition of theoretical knowledge in a practical manner and has facilitated the development of a set of transversal competences, several of which fall within the domain of what are commonly referred to as soft skills. This study also highlights the potential of PBL as a tool for interdisciplinary education, since the project is reportedly developed within the framework of all the semester's subjects. The authors of [18] conclude that the transition from seminar-based course to a project-base course improved the students' engineering judgement "and demonstrated to students the value of engineering analysis and mathematical models in practical engineering design". The studies mentioned, which were identified for their explicit reference to "project-based learning" and to "design of experiments", reinforce what was previously stated: that PBL can be a highly useful tool for learning DOE.

1.4. Statement of Purpose

In line with the foregoing, this study aims to evaluate the application of PBL in the learning of DOE by identifying opportunities for improvement in various elements of DOE. In this case, the elements of DOE are considered to be reflected in the article submitted by students as one of the deliverables of their project. This objective guides the subsequent analysis, which seeks to assess how effectively PBL supports the learning of DOE through the examination of the student-submitted article.

2. Methodology

The project was proposed as part of the module Design and Analysis of Experiments. Students, organised into teams, defined projects within the thematic areas provided: urban and home gardening, domestic space cooling, by-product utilisation, manufacturing processes, food processing, and physical activity. The provision of multiple topics fulfilled two functions: (1) to provide a basis for idea generation and (2) to stimulate engagement in areas in which students had a greater interest. Each project was required to use a factorial experimental design comprising at least two factors at two levels (2^2), with four replicates per treatment ($n = 4$); resulting in 16 runs. Other designs could also be used, provided that the minimum number of runs was met. During the term, formative and summative assessments were conducted to encourage progress and provide feedback. However, this study focuses solely on the final deliverable, namely a paper, as it marks the end of the process and remains a highly valued format within academia. The typical IMRaD structure (Introduction, Methods, Results and Discussion) was proposed for the article. Key aspects to be evaluated in each section were outlined and are presented in Table 1. These aspects were used in this study to identify opportunities for improvement through a qualitative evaluation of the student submissions. The key aspects mentioned in Table 1 align with the recommended guidelines for article writing.

Paper Section	Key aspects	Description
Introduction	Context	Contextual framework is provided, and it is appropriate to the specific project being developed.
	Key concepts	Relevant concepts are presented, and they are appropriate to the project's experimental approach.
	Previous research	Relevant prior work is cited.
	Aim	Aim is explicit and pertinent to the specific project being developed.
Methods	Equipment and materials	The information on the materials and equipment used is presented.
	Experimental procedure	The experimental procedure is described in such a way that replication would be possible.
	Experimental design	The experimental design provides evidence of both the understanding and application of DOE concepts.
	Variables	The factors and the response variable(s) are identified and described clearly, if necessary.
	Measuring	The method used to measure the response variable is clearly described.
Results and Discussion	Visual aids	Appropriate visual aids (such as graphs, diagrams, images, etc.) are used.
	Qualitative analysis	A qualitative analysis of the results is presented.
	Data analysis	A complete analysis of the collected data is presented, including assumption checking and mean comparison, if appropriate.
	Interpretation	The numerical results are interpreted in a way that informs decision-making.

Table 1. Key elements of each section

The aspects listed do not exhaust the possibilities for evaluating the application of PBL in learning DOE; however, they are considered to provide valuable information for implementing improvements. The evaluation is not limited to DOE understood as mere calculation but rather seeks to highlight its use as a problem-solving tool, as expected in the application of PBL.

3. Results and Discussion

The results of the evaluation are presented below, grouped according to the key aspects mentioned previously. Sixteen projects were evaluated.

3.1 Context

Most of the projects were contextualised as responses to specific needs; however, in one case, the contextualisation was rather superficial. In another, no context was provided for the experiment. Although an experiment was conducted in this instance, which led to data collection and subsequent analysis, it must be noted that DOE is intended to address problems that arise within a specific context. The opportunity for improvement lies in emphasising that projects involving DOE respond to concrete realities and must be properly situated. It should be noted that, in most cases, PBL has facilitated context-specific learning.

3.2 Key Concepts

The identification of key concepts is an aspect that could be improved in the vast majority of the projects. Given that these are experimental processes, the use of terminology is expected to go beyond general vocabulary. In most cases, it would be necessary to provide definitions that demonstrate a deeper engagement with the subject matter. The recommendation for this aspect is that key concepts relevant to the problem being addressed should be identified, either independently or with the support of experts.

3.3 Previous Research

In nearly all projects, previous research was consulted. Improvements that could be expected in future work include an increase in the number of sources reviewed and a deeper engagement with what those sources express, applied to the case under study as concretely as possible. One case was found in which no references to research articles were made, and another in which citations were not indicated in the text. From a teaching perspective, emphasis should be placed on the importance of contextualising the work in relation to existing research on the topic.

3.4 Aim

Only one case did not explicitly present an aim. Although most projects met expectations in this regard, given the centrality of this element, it is recommended that it be highlighted in the project guidelines, preferably as a separate component. It should also be noted that, in article format, the aim is typically presented at the end of the introduction.

3.5 Equipment and Materials

This aspect can, in general terms, be considered satisfactory. A key consideration is the need to provide sufficient procedural detail to allow replication of the conditions under which the experiment was conducted.

3.6 Experimental Procedure

As with the previous aspect, the observed area for improvement concerns highlighting the importance of replicating experimental conditions. There were projects in which insufficient information was provided regarding the experimental procedure, or where the inclusion of images or diagrams could have enhanced the presentation.

3.7 Experimental Design

The descriptions of the experimental designs are acceptable when evaluated against the minimum requirements established for the project, as outlined in the methodology. Randomisation, a key element in DOE, is mentioned and its implementation is explained in several deliverables. It is noteworthy that students employed designs that went beyond the minimum requirements. Fractional factorial designs and factorial designs with blocks were implemented. This highlights the development of higher-order thinking skills as described in [16]: "analysis, synthesis, and evaluation". The student, within a specific context [15] — or what [19] refers to as a "controlled real-world environment" — engages in analysis, evaluation, and decision-making [19]. Based on observations, two levels of performance emerged spontaneously: those that meet expectations and those that are outstanding. The evidence associated with each performance level is presented in Table 2.

Level	Design used
Meet expectations	- Full factorial design o Factors: 2 o Levels: 2
Outstanding	- Full factorial design o Factors: 3 o Levels: 2 - Full factorial design o Factors:3 o Levels:2 o Blocks: 1 - Fractional factorial design o Factors: 5 o Fraction: 1/2 o Levels: 2 o Blocks: 1

Table 2. Levels of performance in design selection

3.8 Measuring

The complexity of measuring the response variable varies depending on the project. Measurement instruments range from a simple measuring tape to specialised laboratory equipment. A concrete area for improvement would be to ensure greater support for projects requiring more complex measurement procedures. Additionally, this again highlights the importance of describing experimental conditions in sufficient detail to allow replication. One practical way to achieve this is using illustrations, as noted in the previous aspect.

3.9 Visual Aids

Graphical outputs specific to DOE are the standard visual resource used to accompany results. However, images, photographs, diagrams, and similar resources can enhance the interpretation of findings. In many cases, visual analysis is a valuable tool for explaining what the data reveal. This is particularly important when emphasising that numerical results—such as those presented in an ANOVA—require interpretation and explanation. One recommended improvement is to draw on examples in which visual elements form part of the evaluative process used to assess the quality of a given item. Although this aspect was not explicitly included as an evaluation criterion in the methodology, its importance in supporting explanation and replicability was previously acknowledged. Moreover, it was already anticipated that this aspect is related to the qualitative analysis.

3.10 Qualitative Analysis

There is virtually no evidence of any qualitative analysis of the results produced by the experiments. That is, no observations were documented—for example, through photographs—that would allow for additional assessments beyond those provided by the response variable measurements. Given that conducting experiments involves the use of resources (whether financial or otherwise), it is important to make use of any additional information that their execution may yield. Furthermore, as previously noted, the observation of qualitative aspects can provide valuable input for explaining numerical results (Table 3).

Level	Examples
Defects detection	- Slag in plasma cutting - Delamination in 3D printed part. - Insufficient bonding within the briquette material
Cause identification	- Pest occurring in a home garden - Poor bed adhesion in 3D printing
Abnormality detection	- Inappropriate microbial growth in sourdough starter

Table 3. Examples of the uses of qualitative analysis

3.11 Data Analysis

The minimum expectations for a basic analysis of factorial design results—namely, ANOVA and assumption checking—have been met. However, the examination of effects was not addressed in all projects. It would be advisable to reinforce what is meant by "effect" and how effects are to be interpreted.

Level	Elements
Meet expectations	- ANOVA - Model adequacy checking: graphical tools o Normal probability plot o Plot of residuals in time sequence o Plot of residuals versus fitted values
Outstanding	- ANOVA - Model adequacy checking: graphical tools o Normal probability plot o Plot of residuals in time sequence o Plot of residuals versus fitted values - Model adequacy checking: analytical tests o Normality: Shapiro-Wilk test o Homoscedasticity: Levene test

Table 4. Levels of performance in model adequacy checking

Similarly, mean comparisons were not consistently included. With regard to data analysis, it should also be noted that several students went beyond the elements that were merely introduced in class, such as the use of non-parametric tests or analytical procedures for verifying assumptions. This was facilitated by access to data analysis software and training in its use. In terms of software, students employed various tools according to their preferences, namely RStudio, Excel, and Minitab. Similarly to the experimental design (3.7), two levels of performance can be identified here: meets expectations and outstanding. The evidence associated with each performance level is presented in Table 4. These practices are consistent with recommendations found in recent literature, such as [20], where DOE has been applied to evaluate a full-scale energy consumption project. This example illustrates the broader applicability of DOE in engineering contexts and reinforces the relevance of the skills developed through project-based learning.

3.12 Interpretation

Numerical results are interpreted in all cases; however, in some instances, the interpretation lacks sufficient clarity and assertiveness, particularly with regard to decision-making orientation. This final point may be the most appropriate to reinforce: the results of a DOE are intended to inform decision-making. One specific issue that stood out is the need to appropriately interpret when the analysis indicates that a source of variation is not significant. In this regard, it is important to recall the fundamental premise of factorial models: the comparison of treatments—that is, determining whether a treatment differs significantly or not.

3.13 Additional Remarks

In several projects, the limitations of the study were explicitly stated; this is a feature that could be revisited in future deliverables. Furthermore, in one of the papers, the development of the methodology was explained with reference to other authors. This idea may be integrated as a recommendation in future implementations of the PBL methodology. In some cases, the conclusions are not entirely clear in terms of the response variables; this is also a point worth emphasising, as it connects directly with the decision-making dimension. Using graphs to present the raw data before performing an ANOVA could help students to become familiar with the data and identify possible patterns. However, there is a risk that they will accept these trends as valid without considering the numerical analysis.

4. Conclusions

A qualitative evaluation of the implementation of project-based learning (PBL) in the teaching of Design of Experiments (DOE) has been conducted, based on a deliverable in the form of an academic paper containing thirteen key aspects expected in this format. The conclusions reached are presented below. First of all, the principles underlying PBL appear to be well suited to the learning of DOE—particularly when it is understood that DOE refers not only to data analysis but also to experimental design. The principles underlying PBL appear well suited to learning DOE, particularly when DOE is understood to encompass not only data analysis but also experimental design. This was observed in the qualitative evaluation, as most deliverables met the key aspects assessed. Several areas for improvement were identified. There is a need to reinforce the use of prior research to contextualise the specific focus of each project. Furthermore, as visual resources are commonplace in technical communication, many reports could have been substantially improved by integrating photographs and schematics to support the findings. In addition, the interpretation of numerical results was not consistently clear in terms of the response variables, and the connection to decision-making was sometimes weak. These aspects merit further attention in future implementations. Addressing these areas for improvement would require data analysis that not only describes the observations but also moves closer to explaining the phenomena analysed and guides decision-making, in a manner consistent with several aspirations of the PBL methodology.

5. References

[1] Wu, C. F. J. & Hamada, M. S. (2021) Experiments - Planning, Analysis and Optimization, 3rd ed.

[2] Chiang, Y., (2023) Fundamentals of Design of Experiments for Automotive Engineering, vol. 1. SAE International

[3] Benbow, D. W. (2021) Mistakes in Quality Statistics and How to Fix Them. American Society for Quality (ASQ)

[4] https://www.abet.org/accreditation/accreditation-criteria/criteria-for-accrediting-engineering-programs-2025-2026/ (2025) "Criteria for Accrediting Engineering Programs, 2025 - 2026 - ABET." Accessed: Sep. 23, 2025.

[5] https://ncees.org/wp-content/uploads/2024/10/PE-Ind-Oct-2020_updated-2024_no-RDS.pdf (2020) "Industrial and Systems CBT Exam Specifications.", Accessed: Sep. 23, 2025

[6] https://ncees.org/wp-content/uploads/FE-Che-CBT-specs.pdf, (2020) "Fundamentals of Engineering (FE) Chemical CBT Exam Specifications Effective Beginning with the July 2020 Examinations.", Accessed: Sep. 23, 2025

[7] https://ncees.org/wp-content/uploads/FE-Industrial-and-Systems-CBT-specs.pdf, (2020), "Fundamentals of Engineering (FE) Industrial and Systems CBT Exam Specifications Effective Beginning with the July 2020 Examinations.", Accessed: Sep. 23, 2025

[8] Kilpatrick, W. H. (1918) "The Project Method," Teachers College Record: The Voice of Scholarship in Education, vol. 19, no. 4, pp. 1–5, Sep. 1918, doi: 10.1177/016146811801900404.

[9] Knoll, M. (2012) "'I Had Made a Mistake': William H. Kilpatrick and the Project Method," Teachers College Record: The Voice of Scholarship in Education, vol. 114, no. 2, pp. 1-45, https://doi.org/10.1177/016146811211400202

[10] Morgan, A. (1976) "The Development of Project-based Learning in the Open University," Programmed Learning and Educational Technology, vol. 13, no. 4, pp. 55–59, Oct. 1976, doi: 10.1080/1355800760130407.

[11] Guo, P.; Saab, N.; Post, L. S.; & Admiraal, W. (2020) "A review of project-based learning in higher education: Student outcomes and measures," Int J Educ Res, vol. 102, p. 101586, 2020, doi: 10.1016/j.ijer.2020.101586.

[12] Brundiers, K. & Wiek, A. (2013) "Do We Teach What We Preach? An International Comparison of Problem- and Project-Based Learning Courses in Sustainability," Sustainability, vol. 5, no. 4, pp. 1725–1746, Apr. 2013, doi: 10.3390/su5041725.

[13] Krajcik, J. S. & Shin N. (2014) "Project-Based Learning," in The Cambridge Handbook of the Learning Sciences, Cambridge University Press, 2014, pp. 275–297. doi: 10.1017/CBO9781139519526.018.

[14] Kokotsaki, D.; Menzies V., & Wiggins, A. (2016) "Project-based learning: A review of the literature," Improving Schools, vol. 19, no. 3, pp. 267–277, Nov. 2016, doi: 10.1177/1365480216659733.

[15] Cocco, S. (2006) "Student leadership development: The contribution of project-based learning", Royal Roads University, Victoria, BC, Canada.

[16] Ramírez de Dampierre, M.; Gaya-López, M. C. & Lara-Bercial, P. J. (2024) "Evaluation of the Implementation of Project-Based-Learning in Engineering Programs: A Review of the Literature," Educ Sci (Basel), vol. 14, no. 10, p. 1107, Oct. 2024, doi: 10.3390/educsci14101107.

[17] Lavado-Anguera, S.; Velasco-Quintana, P.J. & Terrón-López, M.J. (2024) "Project-Based Learning (PBL) as an Experiential Pedagogical Methodology in Engineering Education: A Review of the Literature," Educ Sci (Basel), vol. 14, no. 6, p. 617, Jun. 2024, doi: 10.3390/educsci14060617.

[18] Herber, D.; Deshmukh, A.; Mitchell, M. & Allison, J. (2016) "Project-Based Curriculum for Teaching Analytical Design to Freshman Engineering Students via Reconfigurable Trebuchets," Educ Sci (Basel), vol. 6, no. 1, p. 7, Feb. 2016, doi: 10.3390/educsci6010007.

[19] Unzueta, G. & Eguren, J. A., "Implementation of project-based learning for design of experiments using 3D printing," (2023) Journal of Industrial Engineering and Management, vol. 16, no. 2, p. 263, May 2023, doi: 10.3926/jiem.5254.

[20] Kadric, D.; Blazevic, R.; Zukic, E. & Kadric, E. (2023) "Applying the Design of Experiments to Model the Energy Consumption of Typical Single-Family House in Bosnia and Herzegovina," 2023, pp. 0219–0228. doi: 10.2507/34th.daaam.proceedings.028.

DOI: 10.2507/36th.daaam.proceedings.020

EVALUATION OF FIRING RELIABILITY AND PENETRATION DEPTH OF 3D PRINTED PLA PROJECTILES FOR 9 X 19 MM AMMUNITION

Muhamed Bisić, Adi Pandžić, Nikola Jokić, Mustafa Bevrnja & Predrag Elek

This Publication has to be referred as: Bisic, M[uhamed]; Pandzic, A[di]; Jokic, N[ikola]; Bevrnja, M[ustafa] & Elek, P[redrag] (2025). Evaluation of Firing Reliability and Penetration Depth of 3D Printed PLA Projectiles for 9 X 19 mm Ammunition, Proceedings of the 36th DAAAM International Symposium, pp.0154-0162, B. Katalinic (Ed.), Published by DAAAM International, ISBN 978-3-902734-47-1, ISSN 1726-9679, Vienna, Austria
DOI: 10.2507/36th.daaam.proceedings.020

Abstract

Recent advances in 3D printing technologies and polymer materials have opened new possibilities in various engineering fields, including the defense industry. Among these developments, the concept of 3D printed ammunition has attracted growing attention due to its low cost, rapid prototyping potential, and material flexibility. However, this trend is also raising concerns regarding the reliability, safety, and possible misuse of such ammunition, especially given the limited research available in this domain. In this study, several series of bullet projectiles were produced using an FDM 3D printer with PLA filament, designed according to NATO standards and integrated into live 9×19 mm ammunition. An experiment was conducted to evaluate firing reliability, weapon function, and target penetration behavior using a wooden board as a simplified terminal medium. All shots were successfully fired and analyzed in terms of operational consistency and penetration depth. Based on the experimental results, a preliminary analytical model was developed to support future research and enhance the predictive understanding of such projectiles. The results indicate promising potential for further development of 3D printed ammunition, with practical implications for both research and controlled applications.

Keywords: FDM technology; Ammunition; PLA projectiles; Penetration; Reliability; Ballistics.

1. Introduction

Conventional small-caliber ammunition, which is still the most widely used today, has been manufactured following established technological procedures that have remained almost unchanged for more than a century [1]. Its production does not require highly sophisticated infrastructure, as the machines used to produce projectiles and cases are similar to those used in other metalworking processes. The absence of significant technological barriers for the production of small-caliber ammunition has enabled the development of manufacturing capacities all over the world [1]. Production facilities are capable of mass-producing all components of a cartridge. However, the rapid modernization of the defense industry, the need for mobile production units, the demand for specialized projectiles, and various emergency scenarios have encouraged the adoption of new, more advanced technologies for the manufacturing of ammunition and even weapons. One of the most notable modern methods is additive manufacturing (AM), particularly 3D printing.

FDM (Fused Deposition Modeling) is the most common 3D printing technique, where thermoplastic material is melted and deposited layer by layer to form a three-dimensional object. Its simplicity, low cost, and wide availability of materials make it suitable for producing various components. In the defense industry, this technology is increasingly used for the rapid fabrication of spare parts, functional prototypes for testing ballistic and mechanical properties, sighting systems, electronics housings, and unmanned aerial vehicle components. FDM printing allows military units to produce critical components directly in the field, reducing logistics and procurement time [2], [3], [4].

According to one of the simplest definitions, a projectile is a body that is launched by an external force and continues its motion due to inertia while being influenced by gravity and air resistance [5]. Therefore, any object that is propelled through the air (or another medium) with an initial velocity caused by an external force—most commonly a weapon system—can be considered a projectile [5]. The primary purpose of a projectile is to damage or destroy a target, making it a key element of ammunition and modern weapon systems [6]. The development of projectiles has a long history, starting with stone, iron, and later lead as materials. Their shapes and materials changed slowly, adapting to various types of weapons such as bows, catapults, cannons, and smoothbore firearms, with significant advancements beginning in the mid-19th century. The most notable progress during this period occurred in the field of small-caliber ammunition and the improvement of the projectiles it uses [7], [8], [9], [10], [11].

In small-caliber ammunition, all elements are often combined into a single unit, which sometimes causes confusion in terminology, particularly with the term "bullet." For example, a 9×19 mm "bullet" refers to the entire cartridge assembly, which consists of a primer, case, propellant, and projectile. The projectile is only the part that leaves the barrel, often called the "bullet" or "slug." Since this study deals with designing a polymer projectile for a 9×19 mm cartridge and testing it, it is important to clarify how the system operates. The functioning of a 9×19 mm round involves a precise sequence of events that ensure reliable ignition and energy transfer to the target. After pulling the trigger, the firing pin strikes the primer at the base of the case. The primer ignites the propellant inside the case (typically nitrocellulose in modern ammunition), generating high-pressure gases that push the projectile through the barrel. The projectile then travels toward the target, transferring kinetic energy upon impact.

The most common 9×19 mm projectile is the FMJ (Full Metal Jacket) type, consisting of a lead-antimony core with a copper alloy (tombac) jacket. The material selection is optimized to ensure controlled deformation when hitting hard surfaces, allowing for maximum energy transfer and destructive effect. Since this type of ammunition is primarily designed for use against human targets, the goal is to cause bone fractures (including comminuted fractures that produce multiple fragments, creating dangerous secondary injuries), temporary and permanent cavities in soft tissue, and severe blood loss. If the target is wearing body armor, the projectile still transfers significant energy, causing a strong blunt force impact [12], [13], [14], [15]. The branch of physics that studies the effect of projectiles on targets is called terminal ballistics. Small-caliber projectiles generally do not contain secondary high explosives. Exceptions include pyrotechnic compositions used in tracer rounds.

While conventional ammunition has proven reliable, the ongoing modernization of defense technologies calls for exploring alternatives. Additive manufacturing offers several advantages for ammunition development: fast prototyping, low-cost design iterations, and the potential for lightweight or customized projectiles. It is therefore reasonable to analyze whether 3D-printed projectiles, produced on standard desktop machines using polymers like PLA, can meet the functional requirements of small-caliber ammunition. However, ammunition imposes much stricter demands on reliability, consistency, and safety compared to most other parts, which makes careful testing essential [16].

At the same time, the widespread availability of desktop FDM printers, cheap PLA filament, and free CAD models has significantly lowered the barrier to unregulated or home-made ammunition. This raises security and forensic concerns because such projectiles could work in standard firearms, including "ghost guns," and may still acquire rifling marks that make—or complicate—identification in forensic labs. Recent forensic studies have confirmed that 3D-printed bullets can be functional and pick up class and individual characteristics, but these markings are sometimes harder to visualize due to layer artifacts [17], [18].

There is also growing interest in finding environmentally safer ammunition, especially lead-free designs, which has led researchers to explore polymer and composite alternatives. Moreover, the unpredictability and risks of so-called "less-lethal" projectiles (e.g., rubber bullets) highlight the need for carefully engineered, predictable solutions that could be manufactured under controlled conditions. In this area, 3D printing offers precise control over geometry and energy transfer [19], [20]. Despite this growing interest, there is still a lack of hard experimental data on the firing reliability, weapon performance, and penetration depth of fully 3D-printed polymer projectiles such as PLA. Without such data, it is difficult to build analytical models, define safety margins, or establish guidelines for controlled testing. This work aims to fill that gap by experimentally evaluating the firing reliability and penetration behavior of FDM-printed PLA projectiles for 9×19 mm ammunition, and by proposing a simple analytical model as a foundation for future studies.

1.1. Literature review

Research on AM and firearms mostly focuses on 3D printed parts, entire weapons or protective structures, while 3D printed ammunition, especially polymeric bullets and projectiles for small calibers, has very limited scientific data. Regarding the polymer casings, company "TrueVelocity" (Garland, Texas, USA) offers 5.56 x 45 mm bullets with polymer-based casings. The manufacturer states that this production approach leads to increased projectile accuracy, less barrel wear and minimal muzzle deflection in fast projectiles. [21], [22], [23], [24].

A report from 2015 on light weapons and 3D printing stated that there was no proof of functional 3D printed ammunition at that time, with studies focusing only on parts and accessories [25]. Forensic tests on 3D printed firearms, like the single-shot Liberator pistol, showed that they can fire but usually fail after one or two shots. Bullets fired from these weapons can pick up rifling marks, but the marks are often hard to see because of the layer lines from printing [26], [27]. A more recent study from 2025 tested 3D printed PLA bullets on plywood, drywall, glass, and even soft body armor. The bullets penetrated the targets and left marks that could be traced back to the printer's build plate [28]. Research on metal 3D printing also shows that stainless steel (316L) projectiles made by selective laser melting (SLM) have high precision and strength, but this is a completely different process compared to polymer FDM printing [29].

There is also a previously mentioned trend towards lead-free ammunition for environmental and health reasons, and tests on copper bullets and other alternatives show they often have lower muzzle energy and different penetration because of lower density compared to lead [30]. Unconventional designs like the Ruger/PolyCase ARX polymer-copper bullets, made with injection molding, show that alternative materials can be shaped for specific terminal effects [31]. In terms of customization, design and 3D print of non-standard bullet shapes can increase ammunition efficiency, as described in [32], where authors produced a fully lethal projectile of 12.7 mm calibre, but non-standard shape. Custom shaped bullets are also used in [33] for producing bio-degradable projectiles containing animal medicine. Authors in [34] 3D printed arrow tips from the ancient period and tested their penetration capabilities.

There is no peer-reviewed study that has tested fully 3D-printed PLA projectiles loaded into live 9×19 mm cartridges. Existing data on 3D-printed bullets is still very limited, with almost no hard measurements of firing reliability, weapon performance, or penetration depth. Most examples of 3D-printed small-caliber ammunition found online come from amateur experiments, blog posts, or social media videos, which often lack proper testing methodology, reproducibility, or detailed data. These sources are therefore unreliable and cannot be considered a solid technical reference.

1.2. Aim of the study

There is still very little reliable data on the firing reliability, overall weapon performance, and penetration depth of fully 3D-printed polymer projectiles such as PLA. Most information available today comes from amateur tests, often shared as short videos or posts on social media, without any clear methodology or reproducible measurements. Such sources cannot provide the technical detail needed to understand the real potential or risks of these projectiles, which is why controlled experimental research is necessary. This study examines whether a 3D-printed PLA projectile can meet the basic requirements of internal, transitional, external, and terminal ballistics when used in standard 9×19 mm cartridges. The central assumption is that a carefully designed PLA bullet should be able to fire safely from a conventional handgun, exit the barrel without damaging the weapon, and deliver measurable penetration to a target medium.

To test this, PLA projectiles were produced using FDM 3D printing and evaluated for firing reliability, weapon integrity, and penetration depth against wooden targets. Besides penetration, the experiments also observed how the bullets deform after impact, how the firearm cycles during shooting, and whether the printed projectiles maintain structural stability throughout the process. The broader aim of this research is to explore the feasibility of manufacturing specialized small-caliber ammunition using desktop 3D printing. The collected data can serve as a foundation for future analytical models, the development of predictable non-lethal ammunition, or even on-site production in conditions where traditional manufacturing is not practical.

2. Materials and methods

2.1. Design

This study represents the first step in a broader research effort on 3D-printed ammunition and its potential applications. It is therefore intentionally based on simple principles that require only a minimal level of engineering knowledge.

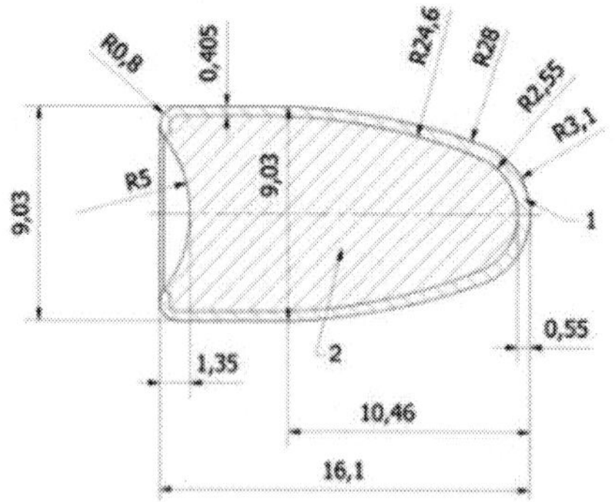

Fig. 1. Dimensions of a standard 9 x 19 mm projectile [36]

One of the goals of this research is also to examine the risks of misuse, considering that FDM 3D printers are widely accessible and allow the production of bullets with very little technical expertise. In line with this, no custom 3D model was designed for the purpose of this work. Instead, a ready-made model was downloaded for free and fully legally from one of the major online repositories of files intended for 3D printing [35]. The chosen model is a projectile for 9×19 mm ammunition, commonly known as „Parabellum". Its dimensions are shown in Figure 1. The 9×19 mm cartridge is NATO-standardized (STANAG 4090 / AEP-97), which guarantees interchangeability and well-defined performance limits across allied weapons. It is also the most commonly studied handgun caliber in modern forensic validation papers, which makes comparison with prior work straightforward and the results broadly relevant [37], [38].

2.2. 3D Printing

The polymer projectiles were manufactured using a desktop FDM 3D printer. FDM is currently the most accessible and widely used 3D printing technology due to its low cost, simplicity, and broad availability to both hobbyists and engineers [39], [40]. Due to the simplicity of the projectile model, the choice of FDM 3D printer is largely irrelevant—whether it's a miniature, homemade, or high-end industrial machine, the only differences may be in dimensional tolerances, which can easily be adjusted for in this case. This further highlights how feasible and accessible potential misuse of 3D printing for producing such ammunition really is [41]. PLA (polylactic acid) was selected as the printing material. PLA is a biodegradable thermoplastic commonly used in desktop 3D printing because it is easy to process, has low warping, and does not require a heated enclosure [42]. It is also cheaper than other polymers, easy to find, simple to use, and doesn't need any special storage, which makes it a good choice for this study based on the goals explained earlier. In order to ensure consistency, black PLA was selected for all samples, based on findings that filament color can impact the physical behavior of PLA prints [43]. The models were printed using a 0.4 mm nozzle and main parameters based on filament manufacturer guidelines presented in Table 1.

Parameter	Value
Layer height	0.2 mm
Infill density	100 %
Nozzle temperature	205 °C
Build plate temperature	60 °C
Print speed	200/300 mm/s

Table 1. Main printing parameters

A 100% infill density was used to ensure maximum strength. Projectiles were printed in a vertical orientation, with their axis aligned to the Z-axis of the printer. This alignment was chosen to minimize interlayer weaknesses in the direction of expected firing loads [44]. No raft or support structures were used, as the geometry did not require them. Model printing time was 5 minutes per projectile (without preparation), with total mass approximately 1 g. All printed projectiles were visually inspected for defects such as delamination, voids, or warping. To simulate conditions of homemade ammunition production without specialized tools, a kinetic bullet puller was used to disassemble a factory (Figure 2a) -made 9×19 mm cartridge. The original projectile was removed, while the primer, casing, and powder charge were preserved. The 3D-printed PLA projectile was then manually pressed into the original case, (Figure 2b).

a) b)

Fig. 2. a) Disassembled bullet, b) Bullet with a 3D printed projectile and Factory made bullet

2.3. Experimental testing

For the first phase of the experiment, the powder charge was varied from 0.065 g to 0.13 g to ensure safety during testing, both for the shooter and the structural integrity of the weapon. Once functionality and accuracy were confirmed, the second phase used a powder charge of 0.26 g, which is slightly lower than the standard charge used for metal projectiles. The appearance of the test-ready cartridges is shown in Figure 3.

Fig. 3. 9 x 19 mm ammunition with polymer projectiles

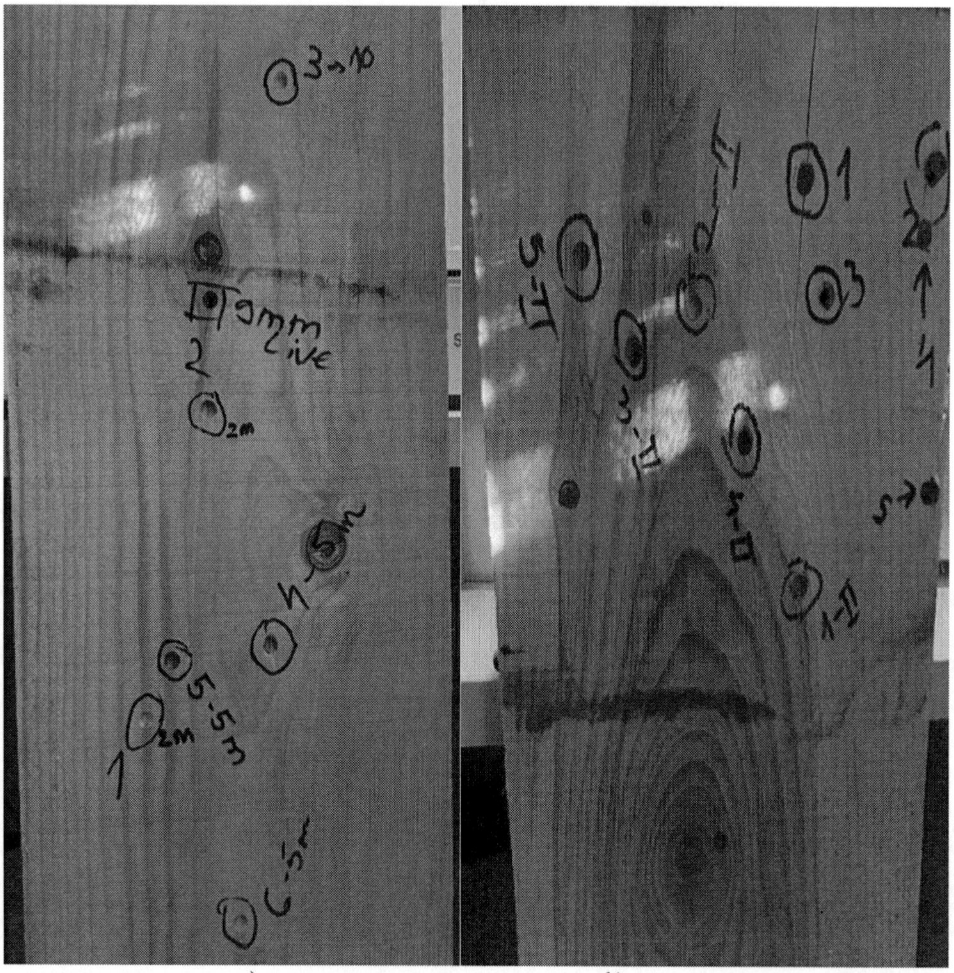

a) b)

Fig. 4. Target appearance after testing, a) precision and reliability testing, b) penetration testing

At this stage of testing, due to the lack of existing data on the functionality of this type of ammunition, it was decided to conduct the experiment in two previously mentioned phases, on an open-range testing ground using a real firearm. The decision to use a real weapon instead of a reference barrel was based solely on cost considerations, since damaging a reference barrel would be significantly more expensive than damaging a standard pistol barrel. The firearm used for testing was a Grand Power K100 pistol with a barrel length of 108 mm. As mentioned earlier, the first phase of testing involved firing projectiles with reduced powder charges, with the main goal of analyzing the internal ballistic cycle. The target was a fir board with a thickness of 25.4 mm (1"), which is commonly used in studies such as [32], [45] as a standard for evaluating projectile lethality. This is one of the historically used methods to replace the often expensive and less accessible ballistic gel. The first phase did not include projectile velocity measurements, only visual inspection of the weapon's barrel and the test board. A total of 6 projectiles were fired from various distances, and it was confirmed that this projectile type is safe to use and successfully completes all phases of the ballistic cycle. After that, using a powder charge of 0.26 grams and a constant distance of 3 meters between the muzzle and the target, 10 projectiles were fired (5 test samples and 5 reference) at different points on the target to evaluate penetration capability. The average muzzle velocity was 250 m/s, with a standard deviation of 20 m/s. Projectile speed was measured using a ProChrono DLX chronograph. The appearance of the targets after testing is shown in Figure 4.

An interesting fact is that none of the 16 projectiles deformed during the initial movement through the barrel, despite high pressure and temperature, nor upon high-velocity impact with the target, even though PLA is often considered as a material originally intended for low-load applications [46]. The appearance of the projectiles after firing and impact is shown in Figure 5. Marks from the barrel rifling are also visible. Additional observations include that projectiles with reduced powder charges did not result in cartridge ejection, likely due to insufficient energy. In the first phase, the projectiles showed a deviation of less than 1 cm between the laser sight point and the actual point of impact. Angular deviation (projectile yaw during flight) was recorded in only two cases and may be attributed to human error.

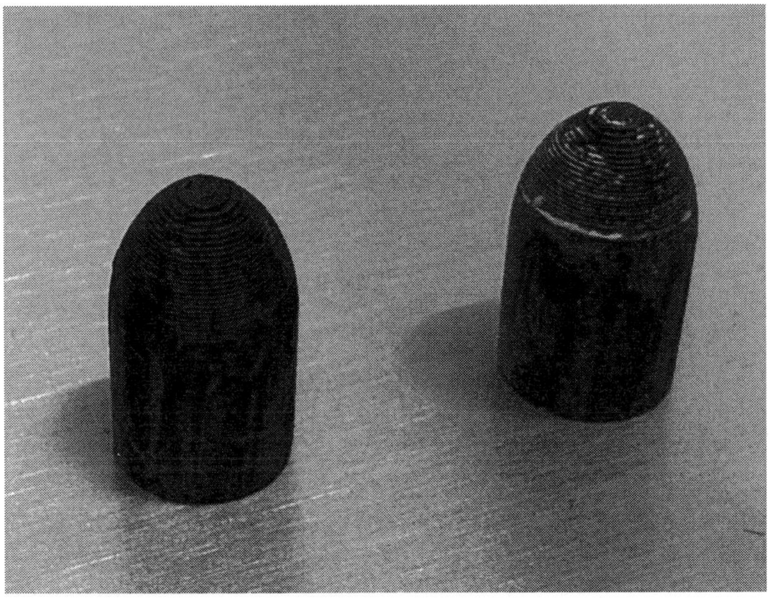

Fig. 5. Samples of projectiles after the experiment

3. Results and discussion

Analysis of the impact holes on both wooden targets revealed significant structural damage. Even on the target used in the first phase, penetration depths reached up to 1 cm in certain areas. Regarding the second phase, Table 2 presents the observed muzzle velocities and corresponding penetration depths.

Shot #	Velocity (m/s)	Penetration (mm)	Kinetic Energy (J)
1	231.3	16.0	26.74
2	237.1	12.8	28.11
3	278.9	25.4 (full)	38.91
4	287.1	20.0	41.23
5	238.7	15.8	28.49

Table 2. Observed values after the experiment

With an average penetration depth of 18 mm, the results suggest that 3D printed PLA projectiles weighing 1 g are not lethal under typical conditions (despite one recorded case of full penetration, which is insufficient according to the V50 standard), but they are capable of causing serious injuries. However, the 12.6 mm variation in penetration depth between the best and worst-performing projectile highlights the unreliability of such ammunition. It also illustrates why wooden targets are generally avoided in ballistic testing due to their material inhomogeneity and the strong correlation between impact location and penetration depth.

Based on the collected data, the average velocity was 254.6 m/s, with a spread of 55.8 m/s and a standard deviation of 26.2 m/s. Compared to a standard 9×19 mm cartridge, these results would typically be classified as representing aged, poorly stored, or inconsistent ammunition. Given that the tests were conducted outdoors and the projectile had extremely low mass, it is realistic to assume that each shot encountered different air resistance, even at a distance of just 3 meters, resulting in data dispersion. Another contributing factor to result dispersion is the inherent inconsistency in 3D printing. Although the projectiles may appear identical, minor variations in print quality, such as porosity, inconsistent mass, microcracks, and imperfect centering, can significantly impact kinetic energy and flight stability. These imperfections lead to unstable trajectories and varying impact angles, where even slight deviations can alter penetration depth. Furthermore, variations in launch velocity suggest that no two projectiles were identical, resulting in inconsistent pressure build-up and barrel friction. It is also noteworthy that the projectile with the highest muzzle velocity did not achieve the greatest penetration. Taking into account the initial kinetic energy and aerodynamic drag, it is possible to approximate the resistance of the wooden target to frontal impact in this specific experiment. The calculated average resistance is around 28 MPa, with a variation of approximately 8 MPa, representing a 30% difference. This again confirms that wooden targets are suboptimal for evaluating penetration performance.

Nevertheless, projectile #3 demonstrated that such a round can indeed be effective and potentially fatal against human targets. The results suggest that further research and development of 3D printed small-caliber projectiles is justified. Future studies should aim to standardize projectile fabrication, use homogeneous target materials, and maintain controlled testing conditions. Additionally, increasing the projectile mass, either by using denser materials or incorporating a mass insert into the projectile's center of gravity could significantly enhance performance. One of the most important findings of this study is that a polymer with relatively low thermal resistance and moderate mechanical properties can still survive the launch phase and travel through a firearm barrel, a process characterized by extreme heat, pressure, and mechanical stress [47]. This observation further supports the feasibility of fully functional polymer firearms without any metallic components, which is a central focus of research in this domain. There is little doubt that in the coming years, interest in such studies will grow as modern manufacturing methods are increasingly applied to the development of special-purpose ammunition and weaponry. Although regression analysis could not be reliably performed on this data set due to the material inhomogeneity of the wooden targets, the experiment provided valuable input data that can support future analytical modeling and the development of predictive models for projectile behavior.

4. Conclusion

The conducted experiments confirm that polymer-based 3D printed projectiles are not only functional but can also be dangerous and, in some cases, potentially lethal. Despite the simplicity of the setup, the rounds demonstrated the ability to survive firing, maintain structural integrity in flight, and achieve significant penetration into soft targets. These findings reinforce the need for more rigorous ballistic testing using standardized materials such as calibrated ballistic gel instead of wood, in order to reduce variability and obtain more reliable penetration data. Future studies must focus on improving the consistency of projectile manufacturing to reduce dispersion in muzzle velocity and kinetic energy. This includes refining print quality, exploring different infill structures, and testing variations in projectile tip geometry, such as spiral features, vents, or other shape optimizations that could affect flight stability and penetration behavior. Moreover, traditional forensic methods for analyzing rifling marks should be applied to 3D printed bullets to evaluate their traceability and potential forensic implications. The current study has shown that even low-cost, entry-level FDM printing techniques are sufficient to produce functional ammunition without requiring any advanced knowledge or specialized equipment—a significant observation in itself. The results serve as a successful proof-of-concept for a low-barrier, fully printable round and confirm that further development is both technically feasible and justified. In future iterations, the research will involve controlled environments, standardized targets, more advanced measuring systems, and greater sample sizes. There is also a realistic possibility of developing a completely polymer-based cartridge, possibly incorporating electronic initiation systems. Subsequent research will aim to establish controlled analytical and numerical models that can reliably predict projectile behavior, while also ensuring safety in use and exploring the full potential—and limits—of such additive-manufactured ammunition.

5. References

[1] Bevan, J. & Pézard, S. (2006). Targeting Ammunition: A Primer – The Production of Ammunition for Small Arms and Light Weapons. Small Arms Survey, Geneva.

[2] Ngo, T.D.; Kashani, A.; Imbalzano, G.; Nguyen, K.T.Q. & Hui, D. (2018). Additive manufacturing (3D printing): A review of materials, methods, applications and challenges. Composites Part B: Engineering, Vol. 143, pp. 172–196. https://doi.org/10.1016/j.compositesb.2018.02.012

[3] Berman, B. (2012). 3D printing: The new industrial revolution. Business Horizons, Vol. 55, No. 2, pp. 155–162.
[4] Lough, K.G.; Cox, D.C.; Lough, R. & Kreger, S.T. (2021). 3D printing in military applications: Opportunities and challenges. Defense Manufacturing Conference (DMC), U.S. Department of Defense.
[5] U.S. Army Materiel Command. (1963). Engineering Design Handbook: Guns Series—Ballistics.
[6] Carlucci, D.E. & Jacobson, S.S. (2013). Ballistics: Theory and Design of Guns and Ammunition, 2nd Edition. CRC Press.
[7] Schubert, F.N. & Krause, T.L. (Eds.). (1995). The Whirlwind War: The United States Army in Operations Desert Shield and Desert Storm. Center of Military History, U.S. Army.
[8] Headrick, D.R. (1981). The Tools of Empire: Technology and European Imperialism in the Nineteenth Century. Oxford University Press.
[9] McNeill, W.H. (1983). The Pursuit of Power: Technology, Armed Force, and Society since A.D. 1000. Oxford: Basil Blackwell.
[10] Fackler, M.L. (1986). Ballistic injury. Annals of Emergency Medicine, Vol. 15, No. 12, pp. 1451–1455.
[11] Bevan, J. & Pézard, S. (2006). Targeting Ammunition: A Primer – Characteristics of Ammunition: From Handguns to MANPADS. Small Arms Survey, Geneva.
[12] Hayward, D.; Baum, G.R.; Baum, J. & MacKay, B. (2023). Gunshot Wounds: Ballistics, Pathology, and Treatment with a Focus on Retained Bullets.
[13] Kieser, D. (2013). Gunshot Induced Indirect Femoral Fracture: Mechanism of Injury and Fracture Morphology. Journal of the Royal Army Medical Corps.
[14] da Cunha Neto, J.; Pereira dos Santos, L.M. & Issa, J.P.M. (2024). Terminal ballistics performance of 9×19 mm cartridges in 10% ballistic gelatin: FBI Protocol based analysis of Brazilian-made law enforcement ammunition. Forensic Sci Med Pathol. https://doi.org/10.1007/s12024-024-00870-z
[15] Dodd, M.J. (2005). Terminal Ballistics: A Text and Atlas of Gunshot Wounds, 1st Edition. CRC Press. https://doi.org/10.1201/9781420037463
[16] Colorado, H.A.; Cardenas, C.A.; Gutierrez-Velazquez, E.I.; Escobedo, J.P. & Monteiro, S.N. (2023). Additive manufacturing in armor and military applications: research, materials, processing technologies, perspectives, and challenges. Journal of Materials Research and Technology, Vol. 27, pp. 3900–3913. https://doi.org/10.1016/j.jmrt.2023.11.030
[17] Hays, G. & Ivan, T. with Jenzen-Jones, N.R. (2020). Desktop Firearms: Emergent Small Arms Craft Production Technologies. ARES Research Report No.8.
[18] Vandenburgh, J.R. (2025). Examining the effects of 3D printed projectiles on firearm identification. Journal of Forensic Sciences. https://doi.org/10.1111/1556-4029.70122
[19] Kanstrup, N.; Swift, J.; Stroud, D.A. & Lewis, M. (2018). Hunting with lead ammunition is not sustainable: European perspectives. Ambio, Vol. 47, No. 8, pp. 846–857. https://doi.org/10.1007/s13280-018-1042-y
[20] Haar, R.J.; Iacopino, V.; Ranadive, N. et al. (2017). Death, injury and disability from kinetic impact projectiles in crowd-control settings: a systematic review. BMJ Open, Vol. 7, e018154. https://doi.org/10.1136/bmjopen-2017-018154
[21] Available at: https://www.tvammo.com/, Accessed on: 2025-07-30
[22] Bisić, M. & Pandžić, A. (2024). Advances in Additive Manufacturing Application in Military Industry. In: Mitrovic, N.; Mladenovic, G. & Mitrovic, A. (Eds.), New Trends in Engineering Research, Lecture Notes in Networks and Systems, Vol. 792. Springer, Cham. https://doi.org/10.1007/978-3-031-46432-4_14
[23] Marciniak, M. (2023). The 3D Printing in Military Applications: FDM Technology, Materials, and Implications. Advances in Military Technology, Vol. 18, No. 2, pp. 241–257. https://doi.org/10.3849/aimt.01846
[24] Bisić, M.; Pandžić, A.; Jusufbegović, M.; Ćerimović, M. & Elek, P. (2025). Influence of Infill Pattern on Ballistic Resistance Capabilities of 3D-Printed Polymeric Structures. Polymers, Vol. 17, 1854. https://doi.org/10.3390/polym17131854
[25] Jenzen-Jones, N. R. (2015). Small Arms and Additive Manufacturing: An Assessment of 3D-Printed Firearms, Components, and Accessories. Small Arms Survey.
[26] Szwed, A., Schaufelbühl, S., Gallusser, A., Werner, D., & Delémont, O. (2023). Was a 3D-printed firearm discharged? Study of traces produced by the use of six fully 3D-printed firearms. Forensic Science International, 348, 111736. https://doi.org/10.1016/j.forsciint.2023.111736
[27] Falardeau, M. S., Mireault, C., Daoust, B., & Muehlethaler, C. (2024). Chemical analysis of polymers used for 3D printing of firearms. Forensic Science International, 357, 111999. https://doi.org/10.1016/j.forsciint.2024.111999
[28] Vandenburgh, J. R. (2025). Examining the effects of 3D printed projectiles on firearm identification. Journal of Forensic Sciences, 00, 1–12. https://doi.org/10.1111/1556-4029.70122
[29] Xue, H.; Wang, T.; Cui, X.; Wang, Y. & Huang, G. (2024). Ballistic performance of additive manufacturing 316L stainless steel projectiles based on topology optimization method. Defence Technology, Vol. 35, pp. 1–17. https://doi.org/10.1016/j.dt.2023.06.010
[30] Courtney, E., Courtney, A., Andrusiv, L., & Courtney, M. (2016). Terminal performance of lead-free pistol bullets in ballistic gelatin using retarding force analysis from high speed video. arXiv preprint, arXiv:1604.01000. https://doi.org/10.48550/arXiv.1604.01000
[31] Vandenburgh, J.R. (2022). The forensic implications of polymer-copper ARX bullets. AFTE Journal.

[32] Bisić, M.; Razić, F.; Pandžić, A. et al. (2023). Penetration testing of 3D printed projectiles made of various types of polymers. J Mech Sci Technol, Vol. 37, pp. 5535–5539. https://doi.org/10.1007/s12206-023-2304-6
[33] Long, J.; Nand, A.V.; Ray, S.; Mayhew, S.; White, D.; Bunt, C.R. & Seyfoddin, A. (2018). Development of customised 3D printed biodegradable projectile for administrating extended-release contraceptive to wildlife. International Journal of Pharmaceutics, Vol. 548, Issue 1, pp. 349–356. https://doi.org/10.1016/j.ijpharm.2018.07.002
[34] Grady, J.H. & Churchill, S.E. (2023). Projectile point morphology and penetration performance. Journal of Archaeological Science: Reports, Vol. 48, Article No. 103893. https://doi.org/10.1016/j.jasrep.2023.103893
[35] Available at: https://www.thingiverse.com/thing:3144261
[36] Wiśniewski, A. & Gmitrzuk, M. (2013). Validation of numerical model of the Twaron CT709 ballistic fabric. Proceedings - 27th International Symposium on Ballistics, Vol. 2, pp. 1535–1544.
[37] Available at: https://edstar.eda.europa.eu/Standards/Details/037daad4-96cf-4f7b-ac28-0acfb582c830
[38] Mattijssen, E.J.A.T. (2020). Interpol review of forensic firearm examination 2016–2019. Forensic Sci Int Synerg, Vol. 2, pp. 389–403. https://doi.org/10.1016/j.fsisyn.2020.01.008
[39] Bevrnja, M.; Imamovic, Z. & Bisić, M. (2022). Additive Manufacturing Application for the Inclusion of the Vision Impaired Population in Public Transport. Proceedings of the 33rd DAAAM International Symposium, pp. 0519–0525. https://doi.org/10.2507/33rd.daaam.proceedings.072
[40] Pandžić, A.; Hodžić, D.; Hajro, I. & Tasić, P. (2020). Strength Properties of PLA Material Obtained by Different Models of FDM 3D Printer. Proceedings of the 31st DAAAM International Symposium, pp. 0315–0322. https://doi.org/10.2507/31st.daaam.proceedings.044
[41] Torrado, A.R. & Roberson, D.A. (2016). Failure analysis and anisotropy evaluation of 3D-printed tensile test specimens of different geometries and print raster patterns. Journal of Failure Analysis and Prevention, Vol. 16, pp. 154–164. https://doi.org/10.1007/s11668-016-0067-y
[42] Pandžić, A.; Hodžić, D. & Milovanović, A. (2019). Influence of Material Colour on Mechanical Properties of PLA Material in FDM Technology. Proceedings of the 30th DAAAM International Symposium, pp. 0555–0561. https://doi.org/10.2507/30th.daaam.proceedings.075
[43] Pandžić, A.; Kadrić, E. & Kolesar, S. (2023). Investigating the Influence of Printing Orientation and Filament Drying on Tensile and Flexural Strength of FDM-Printed Carbon Fiber Reinforced Polyamide Composites. Proceedings of the 34th DAAAM International Symposium, pp. 0146–0153. https://doi.org/10.2507/34th.daaam.proceedings.020
[44] Cheaper Than Dirt. Bullet Penetration in Home Environments. Available at: https://blog.cheaperthandirt.com/bullet-penetration-home-environments . Accessed on: 2025-07-30
[45] Letcher, T. & Waytashek, M. (2014). Material Property Testing of 3D-Printed Specimen in PLA on an Entry-Level 3D Printer. ASME International Mechanical Engineering Congress and Exposition, Vol. 3A: Design and Manufacturing. https://doi.org/10.1115/IMECE2014-39379
[46] Kim, H.; Jeong, H.; Lee, S.; Kim, S. & Lee, Y. (2022). Ballistic performance of 316L stainless steel projectiles made by SLM. Additive Manufacturing Letters. https://doi.org/10.1016/j.addlet.2022.100069
[47] Perego, G. & Cella, G.D. (2010). Mechanical Properties. In: Poly(Lactic Acid): Synthesis, Structures, Properties, Processing, and Applications, pp. 141–153. https://doi.org/10.1002/9780470649848.ch11

DOI: 10.2507/36th.daaam.proceedings.021

IMPACT OF TRAP-NET FISHERY ON MARINE POLLUTION AND SOLUTIONS

Astrida Rijkure, Kristine Cerbule & Janis Megnis

This Publication has to be referred as: Rijkure, A[strida]; Cerbule, K[ristine] & Megnis, J[anis] (2025). Impact of Trap-Net Fishery on Marine Pollution and Solutions, Proceedings of the 36th DAAAM International Symposium, pp.0163-0168, B. Katalinic (Ed.), Published by DAAAM International, ISBN 978-3-902734-47-1, ISSN 1726-9679, Vienna, Austria

DOI: 10.2507/36th.daaam.proceedings.021

Abstract

Current data on the impact of "ghost nets" on the biological diversity of the Baltic Sea indicate that no comprehensive inventory has yet been conducted of their proportion in the region, nor have methods been developed to assess it. The results of the study showed that net fishing is one of the main causes of gear loss, as nets are often lost due to storms and strong currents. These derelict fishing gears remain active for long periods of time, accidentally catching fish, marine mammals and other organisms. In addition, the nets of some materials gradually break down, forming microplastics that enter the food chain and negatively affect both the marine ecosystem and human health. In order to effectively reduce the negative impact of riser fishing, their effectiveness largely depends on the participation of fishermen and public awareness. In the future, it is essential to develop new sustainable fishing gear and improve international cooperation to ensure responsible and environmentally friendly fishing in the Baltic Sea.

Keywords: blue economy; fisheries; ghost nets; sustainable; trap-net

1. Introduction

The ecological challenges facing the Baltic Sea, together with the problem of overfishing, pose significant threats to fish stocks, marine ecosystems, livelihoods, and coastal communities. In Latvia's coastal regions, it is essential to foster economic development while simultaneously safeguarding natural and cultural heritage. To achieve this objective, community-led local development strategies have been implemented along the entire coastline. The overarching aim of these strategies is to ensure balanced and sustainable development. They support the growth of fishing and aquaculture communities while preserving the integrity and resilience of the marine ecosystem. The principal aim of this study is to evaluate the impact of longline fishing on the Baltic Sea ecosystem, with particular emphasis on lost and abandoned fishing gear, commonly referred to as "ghost nets". Such nets continue to capture fish and other marine organisms indiscriminately, thereby generating both ecological and environmental hazards. The study further examines potential solutions and management strategies to mitigate this problem and to foster sustainable fishing practices in the Baltic Sea. The research methodology is grounded in a critical review of scientific literature, an assessment of technological capabilities, and an exploration of innovative approaches that could enhance the marine environment and support the development of the blue economy. The theoretical and methodological foundation of this study draws upon the work of Raid T., Miethe T., Kotta J., and Moll D.

2. Materials and methods

The Baltic Sea is confronted with considerable ecological challenges, primarily driven by eutrophication, climate change, and overfishing. These stressors have resulted in widespread degradation of marine ecosystems, posing serious risks to fish stocks and the livelihoods of coastal communities [9], [10], [11], [21], [26]. In particular, the overexploitation of key species such as cod has precipitated the collapse of several fish stocks, with recovery remaining insufficient despite the implementation of reduced fishing pressures [7], [18]. Herring is the most abundant and extensively targeted species along the Latvian coastline. The peak fishing season occurs between April and June, when herring migrate towards shallow coastal waters to spawn before returning to deeper areas. To safeguard the population, a prohibition on targeted fishing is introduced once catch limits are reached.

The ecosystems of the Baltic Sea are under increasing pressure from a complex range of interacting anthropogenic stressors, which frequently lead to biodiversity loss and habitat degradation. These intensifying and diversifying human impacts threaten the long-term sustainability of marine ecosystems and the essential services they provide. Among the most prevalent human activities in the Baltic Sea region are marine transportation and fishing [3]. In the Gulf of Riga, which forms part of the Baltic Sea, the principal methods used to harvest herring are ship-based trawl fishing and coastal riser fishing. To safeguard the herring population within the gulf, additional conservation measures—such as restrictions on vessel capacity—have been implemented. Trawl fishing is permitted throughout the year, with the exception of a 30-day prohibition during May and June, coinciding with the peak of the herring spawning season.

The management of Baltic herring stocks is governed by international agreements that establish Total Allowable Catch (TAC) limits designed to prevent overfishing [24], [25]. The herring fishery utilises several techniques, including trawl and pound-net fishing. Pound-net fisheries are particularly effective at capturing older age groups with higher mean body weights and avoiding the by-catch of juvenile fish, thereby contributing to the reduction of fishing mortality [24]. Standpipe fishing typically commences in April along the coast and continues until the end of June. During May, when coastal weather remains cool and water temperatures are relatively low, fishing remains productive. Although herring is not the most economically valuable species, it accounts for approximately 90% of the total coastal catch in the Gulf of Riga. There are two distinct herring populations in Latvian waters: bay herring and open-sea herring, each managed under separate catch quotas. Bay herring, one of the smallest herring types in the Baltic Sea, predominantly inhabits and reproduces within the Gulf of Riga. Following spawning, a small proportion of the older bay herring migrate during the summer and autumn months to the Irbe Strait and the central Baltic Sea, but ultimately return to the gulf.

Herring from the neighbouring central Baltic Sea stock also migrate into the gulf, which results in both Latvian and Estonian fishers occasionally catching small quantities of open-sea herring while operating in the Gulf of Riga. Due to the gulf's relatively shallow waters, additional conservation measures—such as restrictions on vessel engine power—have been introduced to protect the bay herring population. Trawl fishing is permitted year-round, except for a 30-day closure in May and June, coinciding with the period of peak spawning migrations. Fishing practices differ considerably between Latvia and Estonia: in Latvia, the largest catches are obtained using trawl gear, with 15% of the national quota allocated to coastal fisheries, whereas in Estonia, 40% of the national quota is reserved for coastal fishing.

Herring populations in the Baltic Sea, including the Western Baltic spring-spawning herring (WBSS), undertake pronounced seasonal migrations to shallow coastal areas for spawning [5], [8], [23]. Environmental stressors such as eutrophication and coastal modifications have altered spawning habitats, potentially influencing the efficiency and preference for different fishing methods [8], [15], [16], [19]. In recent years, there has been a marked decline in the volume of illegally caught or misreported fish. This improvement is largely attributed to healthier herring stock levels and enhanced fleet management, including a significant reduction in the number of active vessels. In 2023, fishers operating in the Gulf of Riga were authorised to use 115 standpipe risers and 610 herring nets. However, according to official logbook data, only 70 of the permitted 115 risers were actually employed along the gulf's coastline. The highest coastal catches were recorded along its southern and western shores.

The herring stock in the Gulf of Riga remains in a stable and healthy condition, and it is regarded by scientists as one of the best-managed fisheries in the Baltic Sea. Since the late twentieth century, herring populations have remained within biologically safe limits, supported by favourable environmental conditions for spawning and juvenile development. Typically, warm winters lead to the formation of strong and productive year classes, whereas colder winters result in weaker cohorts. In recent years, a succession of mild winters has created optimal breeding conditions for bay herring, thereby contributing to the continued resilience of the stock. A range of effective management measures has been implemented, including restrictions on vessel engine power and fishing gear size. These regulatory interventions have played a crucial role in maintaining stock levels within biologically safe boundaries and ensuring that fishing activities do not exert a detrimental impact on the herring population [4], [8], [25]. Scientific advice recommends a total allowable catch of 41,635 tonnes of herring in the Gulf of Riga for 2025. This figure includes both the permissible catch within the gulf and the projected catch of Baltic Sea herring in the same area. This represents a 9.7% increase compared with the previous year. The proposal by the European Commission aligns with the scientific advice provided by ICES and similarly advocates a 9.7% rise in catch opportunities [17], [20].

Coastal riser fishing, like other fishing practices, can also give rise to ecological challenges. One particularly significant issue is the loss or abandonment of fishing gear, commonly referred to as 'ghost nets'. These nets continue to capture fish and other marine organisms indiscriminately, resulting in unnecessary by-catch, marine pollution, and detrimental impacts on the wider marine ecosystem.

Ghost nets are especially harmful as they persist in the environment for extended periods, continuing to entrap and kill marine species, including non-target organisms. This contributes to increased by-catch and exacerbates biodiversity loss [6], [12], [14]. Moreover, such nets can entangle and damage benthic organisms, including branching corals, which are essential for providing habitat and breeding grounds for a variety of marine species [22].

3. Results and discussion

To address this issue effectively, it is essential to implement a range of targeted initiatives, including:

- Regular net retrieval expeditions – systematic searches and recovery operations for abandoned fishing gear at sea.
- Tagging of fishing gear – enabling the identification of lost nets and the fisheries responsible.
- Environmentally friendly materials – testing and deploying biodegradable nets that degrade naturally, thereby reducing long-term marine pollution.
- Fishing waste collection programmes – encouraging fishers to return used nets to shore for appropriate recycling and disposal.

In September 2024, the authors of this study identified a network of lost riser nets in the Gulf of the Baltic Sea, near the Port of Engure, and subsequently organised an expedition to recover it. The net was located approximately 1.6 nautical miles offshore at a depth of only 6 metres, posing a serious hazard to small vessels. The riser net, measuring approximately 290 metres in length, had remained submerged for around 25 years, during which it had become heavily overgrown and had entrapped multiple fish. Divers detached the net from its moorings; however, a larger vessel equipped with a winch was required to lift it from the seabed. This operation was carried out in collaboration with professionals from the State Environmental Service, who assisted in documenting and highlighting the presence of these derelict nets.

Fig. 1. Clean-up of a shipwreck near the port of Engure, Baltic Sea (Latvian territorial waters)

The recovered fishing gear weighed more than two tonnes, although its precise dimensions could not be determined underwater. Constructed from capron, a durable plastic material, the net exhibited minimal signs of degradation despite having remained submerged for an extended period. This underscores its potential to persist in the marine environment for many more years. Several sections of the net were torn, indicating that it had likely come into contact with watercraft at some stage.

Fig. 2. Ghost nets, Baltic Sea (Latvian territorial waters)

At present, no technology exists that can accurately determine the dimensions of lost fishing gear underwater; current estimates rely exclusively on visual inspection. The development of technologies capable of detecting and measuring such gear, as well as other submerged objects, would provide a valuable preventive tool. [2] Such innovations could significantly enhance the monitoring of underwater debris, allowing for more precise assessments of its size and volume. The authors of this study have identified the coordinates of several ghost nets along the coast of the Gulf of Riga. These sites should be systematically mapped and prioritised for future clean-up operations. Riser nets, in particular, must not be left in the water, as they are authorised for use only during a two-month fishing season. As highlighted previously, these nets should be properly retrieved and not abandoned.

Currently, no comprehensive inventory or standardised methodology exists for assessing the extent or proportion of ghost nets in the Baltic Sea. This observation is consistent with previous findings, which indicate that such an inventory has yet to be undertaken [13]. It is evident that the Baltic Sea, which ranks among the most polluted seas in the world, provides a highly degraded habitat for its marine life. Numerous shipwrecks are scattered across the Baltic Sea, including in Latvian territorial waters, the majority of which have not yet been fully identified or surveyed. Surveys conducted to date have revealed that a considerable amount of lost fishing gear—commonly referred to as 'ghost nets'—has accumulated on all examined wrecks. These nets continue to function as passive fishing gear, causing extensive damage to the marine environment, including aquaculture areas, and posing technical hazards to vessels. Due to their synthetic composition, they degrade very slowly, if at all, and can persist in the marine environment for decades.

At present, wrecks and submerged objects are inspected exclusively through direct visual surveys by divers. This method is constrained by limited air supply and the maximum safe time that can be spent at depth, thereby restricting the scope and effectiveness of such assessments [1]. Sustainability and the conservation of fish stocks are fundamental to the long-term viability of any fishery. Overfishing can lead to the collapse of fish populations, with severe consequences for marine ecosystems and for the livelihoods of communities that depend on fishing as a primary source of income. To ensure the sustainable management of fish stocks, fisheries must adopt responsible and environmentally sound practices, including:

1. Fishing within sustainable limits – Harvesting only what the stock can replenish, and avoiding the capture of juvenile or undersized fish to allow populations to reproduce and thrive.
2. Using selective fishing gear – Employing gear that minimizes bycatch and helps avoid the capture of non-target species, thus reducing unnecessary ecological harm.
3. Protecting critical habitats – Safeguarding vital areas such as spawning grounds and nursery habitats from damage or destruction is essential for supporting healthy fish populations.
4. Promoting responsible behavior – Encouraging fishers to dispose of gear properly, reduce marine litter, and minimize the use of single-use plastics contributes to a healthier marine environment.

4. Conclusion

The presence of 'ghost nets' in the Baltic Sea poses a serious threat to marine biodiversity. However, a comprehensive inventory assessing their prevalence and ecological impact has yet to be conducted. These abandoned fishing nets—often lost during storms or as a result of strong currents—continue to capture fish, marine mammals, and other marine organisms indiscriminately over extended periods. Many of these nets are composed of synthetic materials that gradually fragment into microplastics, thereby polluting the marine environment, entering the food chain, and threatening both ecosystem health and human well-being. This study underscores the urgent need for collaborative solutions involving both the fishing industry and the wider public. Raising awareness and encouraging the active engagement of fishers are essential to effectively addressing this issue. Future strategies should prioritise the development and adoption of sustainable fishing gear, the improvement of waste management systems, and the strengthening of international cooperation to ensure environmentally responsible fishing practices across the region. In the Gulf of Riga—where herring constitutes the principal target species—fishing primarily takes place from April to June, with seasonal restrictions introduced to protect stocks during spawning. The area supports two distinct herring populations: bay herring and open-sea herring, each managed through specific catch quotas. Conservation measures such as restrictions on vessel capacity and temporary fishing bans have contributed to the long-term stability of these stocks. In Latvia, trawl fishing accounts for the majority of herring catches, whereas Estonia allocates a larger proportion of its quota to coastal fishing activities. Despite the generally healthy status of herring stocks, the persistent problem of lost or abandoned nets remains a significant ecological concern. To mitigate these impacts, several initiatives are recommended. These include organising regular net retrieval expeditions, promoting the use of selective fishing gear to minimise by-catch and avoid the capture of juvenile fish, and protecting critical habitats such as spawning and nursery grounds. Fishers should also be encouraged to adopt responsible practices, including the proper disposal of used gear, the reduction of single-use plastics, and participation in recycling programmes. Tagging fishing equipment would facilitate the tracing of lost nets and the identification of responsible parties. Furthermore, investment in the research and development of biodegradable fishing gear represents a promising long-term solution, as such materials degrade naturally and reduce environmental harm. Overfishing continues to pose a persistent threat to marine ecosystems and to the coastal communities that depend on them.

Ensuring the long-term sustainability of the Baltic Sea's fisheries requires a sustained commitment to responsible fishing practices, robust conservation measures, and the implementation of innovative, environmentally conscious solutions. The fisheries sector is crucial for global food security and the livelihoods of millions worldwide. Ensuring the sustainability of this industry requires effective management, responsible fishing practices, and the conservation of fish stocks. By embracing sustainable approaches, we can protect the long-term viability of fisheries and the health of our oceans. To minimize the environmental impact of practices such as longline fishing, stronger collaboration is needed between fishermen, scientists, and regulatory authorities. While some regulations are already in place, their success depends heavily on the active engagement of the fishing community and increased public awareness. Looking ahead, the development of innovative, sustainable fishing gear and the strengthening of international cooperation will be essential. These efforts will support responsible and environmentally friendly fishing in the Baltic Sea and contribute to the broader goal of preserving marine ecosystems for future generations.

5. Acknowledgments

Supported by research and development grant No. RTU-PA-2024/1-0084 under the EU RRF project No 5.2.1.1.i.0/2/24/I/CFLA/003 grant "Innovative underwater drone equipment with manipulators for cleaning the sea from ghost nets (UDRON)"(No. C4835.ZPD.PI.0084.P1)

6. References

[1] Beneli T.M.,.Pereira P.H.C ,.Nunes J.A.C.C ,.Barros F, "Ghost fishing impacts on hydrocorals and associated reef fish assemblages" (2020), Marine Environmental Research, Volume 161, 2020, 105129, ISSN 0141-1136, https://doi.org/10.1016/j.marenvres.2020.105129 . [Accessed: March. 1, 2025]

[2] Chernyshev V., Ksenzenko A., Nechaev A., Petrakov M., Pryanichenikov V., Prysev E., The Development of Service and Underwater Robots using Network and 3D vision Technologiest (2019), 31 in DAAAM International Scientific Book 2019, pp.359-366, B. Katalinic (Ed.), Published by DAAAM International, ISBN 978-3-902734-24-2, ISSN 1726-9687, Vienna, Austria, DOI: 10.2507/daaam.scibook.2019.31 https://www.daaam.info/Downloads/Pdfs/science_books_pdfs/2019/Sc_Book_2019-031.pdf [Accessed: October. 18, 2025]

[3] Dabuleviciene T., Jucevicius D., Zolubas T., Vaiciute D., Nika N., "The Effect of Short-Term Upwelling Events on Fish Assemblages at the South-Eastern Coast of the Baltic Sea". Water 2023, 15, pp.452. https://doi.org/10.3390/w15030452 [Accessed: March. 1, 2025]

[4] Filaretov V., Konoplin A., Zuev A. & Krasavin N., 2020 "System of High-precision Movements Control of Underwater Manipulator" Proceedings of the 31st International DAAAM Virtual Symposium "Intelligent Manufacturing & Automation" Volume 31, No.1, ISSN 1726-9679, ISBN 978-3-902734-29-7 Ed. B. Katalinic / Published by DAAAM International, Vienna, Austria, EU, 2020 DOI:10.2507/31st.daaam.proceedings.104; https://daaam.info/31st-proceedings-2020 [Accessed: October. 18, 2025]

[5] Hammer C., Dorrien C., Zimmerman C., "A Review on Herring, Clupea Harengus (Actinopterygii: Clupeiformes: Clupeidae) Recruitment and Early Life Stage Ecology in the Western Baltic Sea" (2013) Acta Ichthyologica et Piscatoria 43(3):pp.169-182, DOI:10.3750/AIP2013.43.3.01 [Accessed: March. 3, 2025]

[6] Kraufvelin P., Pekcan-Hekim Z., .Bergström U,, Britt F.A., Lehikoinen A., Mattila J., Arula T., Briekmane L., Brown E.J., Celmer Z., Dainys J., Jokinen H., Kääriä P., Kallasvuo M., Lappalainen A, L.Lozys, P.Möller, A.Orio, M.Rohtla, L.Saks, M.Snickars, J.Støttrup, G.Sundblad, I.Taal, D.Ustups, A.Verliin, M.Vetemaa, H.Winkler, A.Wozniczka, J.Olsson, "Essential coastal habitats for fish in the Baltic Sea", Estuarine, Coastal and Shelf Science,Volume 204, 2018, pp.14-30, ISSN 0272-7714, https://doi.org/10.1016/j.ecss.2018.02.014 [Accessed: February 23, 2025]

[7] Lehtonen K.K., Schiedek D., "Monitoring biological effects of pollution in the Baltic Sea: Neglected—but still wanted?", Marine Pollution Bulletin, Volume 53, Issues 8–9, 2006, pp.377-386, ISSN 0025-326X, https://doi.org/10.1016/j.marpolbul.2005.11.016 [Accessed: February 22, 2025]

[8] Márcio L.V., Barbosa F., Seminara C.I., Castro Tavares D., Siciliano S., Hauser-Davis R.A., Silva Mourão J., "Artisanal fisher perceptions on ghost nets in a tropical South Atlantic marine biodiversity hotspot: Challenges to traditional fishing culture and implications for conservation strategies", Ocean & Coastal Management, Volume 192, 2020, 105189, ISSN 0964- https://doi.org/10.1016/j.ocecoaman.2020.105189. [Accessed: February 13, 2025]

[9] Lively J.A, Good T.P., '"Ghost fishing" World Seas: An Environmental Evaluation Volume III: Ecological Issues and Environmental Impacts., pp.183–196

[10] Margonski P., Hansson S., Tomczak M.T., Grzebielec R., "Climate influence on Baltic cod, sprat, and herring stock-recruitment relationships"(2010) Progress in Oceanography, Volume 87, Issue 1-4, pp.277–288. Progress in Oceanography,ISSN 0079-6611, https://doi.org/10.1016/j.pocean.2010.08.003 . [Accessed: February 23, 2025]

[11] Miethe T., Gröhsler T., Böttcher U., Dorrien C., "The effects of periodic marine inflow into the Baltic Sea on the migration patterns of Western Baltic spring-spawning herring", ICES Journal of Marine Science, Volume 71, Issue 3, March/April 2014, pp.519–527, https://doi.org/10.1093/icesjms/fst166 [Accessed: February 20, 2025]

[12] Moll D., Kotterba P., Nordheim L., "Storm-Induced Atlantic Herring (Clupea harengus) Egg Mortality in Baltic Sea Inshore Spawning Areas". Estuaries and Coasts 41, pp.1–12 (2018). https://doi.org/10.1007/s12237-017-0259-5 [Accessed: February 23, 2025]

[13] Moyano M., Illing B., Akimova A., "Caught in the middle: bottom-up and top-down processes impacting recruitment in a small pelagic fish". Rev Fish Biol Fisheries 33, pp.55–84 (2023). https://doi.org/10.1007/s11160-022-09739-2 [Accessed: February 23, 2025]

[14] Peltonen H., Weigel B., "Responses of coastal fishery resources to rapid environmental changes". Journal of Fish Biology, 101(3),(2013) pp.686–698. https://doi.org/10.1111/jfb.15138698 [Accessed: February 23, 2025]

[15] Rahikainen M., Hoviniemi K.M., Mäntyniemi S., Kuikka S, "Impacts of eutrophication and oil spills on the Gulf of Finland herring stock" (2017) Canadian Journal of Fisheries and Aquatic Sciences, Volume 74, Issue 8, pp.1218 – 1232 Canadian Journal of Fisheries and Aquatic Sciences 10 January 2017, https://doi.org/10.1139/cjfas-2016-0108

[16] Raid T., Sepp E., Vetemaa M., "Use of autonomous research vehicles in Baltic fisheries acoustic surveys: Potential benefits and pitfalls?" (2022) Trends in Maritime Technology and Engineering, pp.8 DOI: 10.1201/9781003320289-66

[17] Raid T., Shpilev H., Järv L., Järvik A., "Towards sustainable Baltic herring fishery: trawls vs. pound nets" (2011) Conference: IMAM 2011, Sustantainable Maritime Transportation and Exploitation of the Sea Resources, Genova, Italy, 13.-16. September; pp.1055 – 1062, Volume: 2

[18] Rijkure A., Megnis J., "Technical Methods of Cleaning Shipwrecks form Ghost Nets", ETR, vol. 3, pp. 253–256, Jun. 2024, https://doi.org/10.17770/etr2024vol3.8160 [Accessed: March. 1, 2025]

[19] Rosciszewski-Dodgson M.J., Cirella G.T., "Environmental drivers affecting the status of top commercial fish stocks in the Baltic Sea: review" (2024) Marine Fisheries, Aquaculture and Living Resources, Volume 11 - 2024 | https://doi.org/10.3389/fmars.2024.1399707 [Accessed: February 20, 2025]

[20] Ruskule A., Kotta J., Saha C.R., Arndt P., Ustups D., Strāķe S., Bergström L., "Testing the concept of green infrastructure at the Baltic Sea scale to support an ecosystem-based approach to management of marine areas" (2023) Marine Policy, Volume 147, 2023, 105374, ISSN 0308-597X, https://doi.org/10.1016/j.marpol.2022.105374. [Accessed: February 23, 2025]

[21] 19.Schiewer U., Schernewski G., "Self-Purification Capacity and Management of Baltic Coastal Ecosystems." Journal of Coastal Conservation, vol. 10, no. 1/2, 2004, pp. 25–32. JSTOR,DOI: http://www.jstor.org/stable/25098434. [Accessed: February 20, 2025]

[22] Sepp E., Vetemaa M., Raid T., Arula T., "Trawler Engine Size Had No Effect on Baltic Herring Size/Age Structure: An Experimental Study in the Gulf of Riga, Baltic Sea.", (2022), Fishes 7(46):pp.2-11, DOI:10.3390/fishes7010046 [Accessed: February 23, 2025]

[23] Vetemaa M., Eschbaum R., Saat T., "The transition from the Soviet system to a market economy as a cause of instability in the Estonian coastal fisheries sector", (2006), Marine Policy, Volume 30, Issue 6, 2006, pp.635-640, ISSN 0308-597X, https://doi.org/10.1016/j.marpol.2005.08.001 [Accessed: February 25, 2025]

[24] Vozbinaite G.K., "Recycled landscapes: Beyond the horizon of the Baltic Sea," 2016 Techno-Ocean (Techno-Ocean), Kobe, Japan, 2016, pp. 1-6, DOI:10.1109/Techno-Ocean.2016.7890745. [Accessed: March. 2, 2025]

[25] William C., Long P., Cummiskey A.J., Munk E., "Effects of ghost fishing on the population of red king crab (Paralithodes camtschaticus) in Womens Bay, Kodiak Island, Alaska" (2014) Fishery Bulletin, 112(2-3):pp.101-111, DOI: 10.7755/FB.112.2-3.1 [Accessed: February 19, 2025]

[26] Zimmermann C., "Fish stocks and fisheries in the Baltic Sea" (2022) Geographische RundschauVolume 74, Issue 11, pp.22–26.

DOI: 10.2507/36th.daaam.proceedings.022

TRANSFORMATION OF HR AREAS AS A RESULT OF DIGITIZATION AND AUTOMATION: ANALYSIS AND FINDINGS

Natália Vraňaková & Zdenka Gyurák Babeľová

This Publication has to be referred as: Vranakova, N[atalia] & Gyurak Babelova, Z[denka] (2025). Transformation of Hr Areas as a Result of Digitization and Automation: Analysis and Findings, Proceedings of the 36th DAAAM International Symposium, pp.0169-0174, B. Katalinic (Ed.), Published by DAAAM International, ISBN 978-3-902734-47-1, ISSN 1726-9679, Vienna, Austria
DOI: 10.2507/36th.daaam.proceedings.022

Abstract

Digitalization and automation are currently ones of the main trends transforming the functioning of industrial enterprises. These changes significantly affect not only production processes, but increasingly affect also the areas of human resource management. Technologies such as electronic recruitment tools, e-learning platforms, performance evaluation software, and automated administrative systems are changing the way organizations work with employees throughout the entire work life cycle. The aim of the article is to present the results of a quantitative survey conducted in industrial enterprises in Slovakia, which analyses the extent to which individual HR (human resource) areas have been affected by digitalization and automation. The article presents the results of a questionnaire survey, in which participated 177 respondents from large and medium-sized industrial enterprises in Slovakia. The results of the survey also revealed areas in which employees at various levels of organizational hierarchy perceive the greatest potential for further use of modern technologies within human resource management. Areas of human resource management where the implemented digitalization has contributed to their improvement were also identified.

Keywords: automation; digitization; employees; HR; industrial enterprises

1. Introduction and Theoretical Background

Digitalization and automation in industrial enterprises encompass a complex interplay of digital transformation, technology adoption, and the challenges we face during this transition. Understanding these topics is crucial for enterprises that want to navigate effectively in the evolving industrial environment. Digital transformation is an evolutionary process that optimizes organizational capabilities to leverage digital technologies, which has a significant impact on manufacturing sectors. The rise of smart manufacturing technologies is revolutionizing the manufacturing environment and increasing interest in digital transformation [1]. Companies are increasingly focusing on automating industrial processes and creating intelligent, connected manufacturing systems [2]. Today's society is strongly influenced by the development and implementation of modern information and communication technologies [3]. Organizations face significant challenges in acquiring the necessary knowledge and technical skills to effectively implement new digital technologies.

Despite the challenges, digitalization presents opportunities for radical and incremental process innovations that increase competitiveness [4]. From an innovation perspective, the essence of digitalization and automation lies in the development of advanced technologies aimed at achieving savings in processes, products or services [5]. Digital transformation in human resources (HR) is essential to maintain competitiveness, especially in the wake of the pandemic, which has accelerated the adoption of digital HR technologies [6]. The integration of e-HRM (electronic human resource management) facilitates better service delivery to employees and managers and streamlines human resource operations [7] [8]. Organizations are being urged to modernize their HR functions to optimize processes and improve the overall employee experience [9]. The pandemic has accelerated the transformation of human resource management, highlighting the importance of digital tools to remain competitive in a rapidly changing environment. The digitalization of human resources is seen as a revolutionary opportunity to change corporate strategies and culture, emphasizing the importance of continuous learning and adaptation [10]. Digital HR processes involve implementing selection, training, and evaluation functions using advanced digital technologies, thereby increasing efficiency and effectiveness [11].

In summary, digitalization and automation are significantly impacting the HR area by transforming HR processes, influencing workforce dynamics, and evolving the competencies required of HR professionals. These changes require strategic adjustments to ensure that organizations can effectively manage the challenges and opportunities presented by the digital age.

2. Materials and Methods

The main aim of the article is to present the results of a quantitative survey conducted in industrial enterprises in Slovakia, which analysed the extent to which individual HR areas have been affected by digitalization and automation. For the purpose of data collection, a questionnaire was designed and filled out by respondents in various job positions in medium-sized and large enterprises in Slovakia. Large and medium-sized enterprises in particular face significant challenges in the field of digitalization due to barriers to technology adoption and a lack of necessary skills. Data collection took place from February 11, 2025, to May 26, 2025. The questionnaire also identified areas in which respondents in various job positions perceive the greatest potential for further use of modern technologies in human resource management. To meet the main objective of the article, three research questions were set:

- **Research question 1**: Which areas of human resource management have been affected by automation and digitalization and the leveraging of new technologies?
- **Research question 2**: In which areas of human resource management do respondents see the greatest scope for using automation and digitalization?
- **Research question 3**: How have been improved processes of human resource management if digitalization of these processes was implemented?

The following Fig. 1 and Table 1 contain the basic characteristics of the respondents of the questionnaire survey.

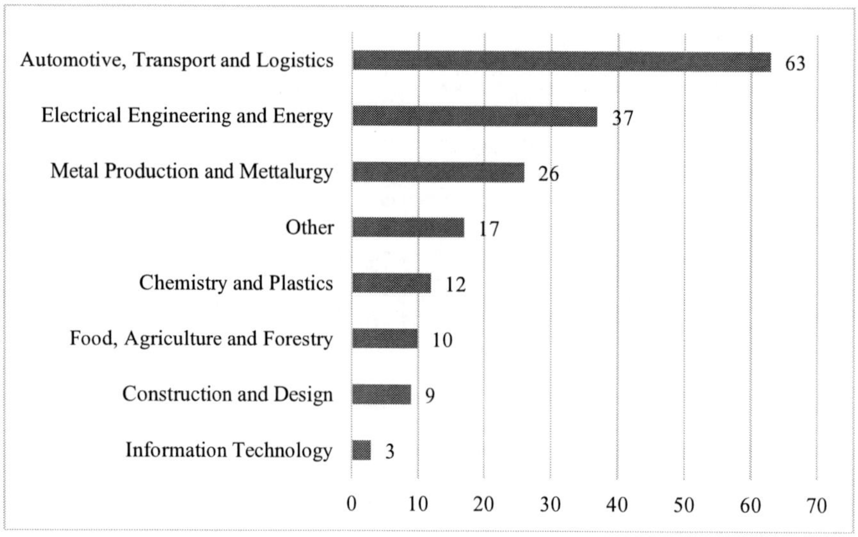

Fig. 1. Characteristics of respondents from the perspective of sector affiliation (own elaboration, 2025)

Fig. 1 shows the characteristics of respondents by industry sector in which they operate. The largest share of respondents comes from the automotive, transport and logistics sector, which represents 63 responses, making it the largest group. This is followed by electrical engineering and energy with 37 respondents and metallurgy and metallurgy with 26 responses. Other sectors are represented to a lesser extent.

Respondents	Automotive, Transport and Logistics	Electrical Engineering and Energy	Chemistry and Plastics	Information Technology	Food, Agriculture and Forestry	Construction and Design	Metal Production and Metallurgy	Other
Gender								
Male	37	28	7	0	5	3	17	8
Female	26	9	5	3	5	6	9	9
Job position								
Production position	14	4	6	0	2	0	2	3
Administrative position	14	11	4	0	3	6	9	8
Employee specialist	22	14	1	3	1	2	8	3
Managerial position	13	8	1	0	4	1	7	3
Education level								
Secondary school without diploma	3	1	2	0	1	0	3	2
Secondary school with diploma	26	15	5	1	3	2	7	6
University level 1	15	8	3	1	0	1	6	6
University level 2	18	12	2	1	6	4	9	2
University level 3	1	1	0	0	0	2	1	1
Age								
21 - 30	18	12	5	3	3	1	4	6
31 - 40	19	13	0	0	1	4	8	3
41 - 50	20	10	2	0	2	2	9	4
51 - 60	6	0	5	0	3	2	4	4
> 61	0	2	0	0	1	0	1	0

Table 1. Characteristics of respondents in terms of gender, job position, education level and age (own elaboration, 2025)

Table 1 provides a detailed overview of respondents by gender, job title, education and age in individual sectors. Majority of respondents were men, especially in the automotive and energy sectors. Specialists and administrative employees are most frequently represented job positions, with a significant share also being managers. In terms of education, secondary education with a high school diploma and a second-level university education prevail. Most numerous age group is respondents aged 31 - 50, which indicates an active working population in industry in middle age.

3. Results and Discussion

This section of the article deals with the analysis of the three research questions that were set out in the previous section. The research questions were analysed using IBM SPSS software. The results are presented using descriptive statistics, tables and graphs in absolute frequencies.

Descriptive Statistics					
	N	Minimum	Maximum	Mean	Std. Deviation
Recruitment and selection of employees	177	0	5	1.92	1.494
Employee adaptation and onboarding	177	0	5	1.95	1.538
Employee training and development	177	0	5	3.01	1.394
Employee performance evaluation	177	0	5	2.66	1.515
Career planning and talent management	177	0	5	2.18	1.485
Administration (employment contracts, attendance records, ...)	177	0	5	3.32	1.481
Internal communication (within the organization)	177	0	5	3.36	1.459
External communication (e.g. with customers)	177	0	5	3.05	1.609
Employee motivation	177	0	5	2.19	1.429
Presence at the workplace (e.g. home office)	177	0	5	2.82	1.599
Work-life balance	177	0	5	2.16	1.441
Valid N (listwise)	177				

Table 2. Result of RQ1 analysis (own elaboration, 2025)

Research question 1: Which areas of human resource management have been affected by automation and digitalization and the leveraging of new technologies? The aim of the research question (RQ1) is to find out which specific HR activities or processes (e.g. recruitment, adaptation, training, administration, communication, etc.) have experienced a change or transformation due to the introduction of modern technologies. The result of analysis can be seen in Table 2. Table 2 summarizes the results of a questionnaire survey that examined the extent to which different areas of human resource management in organizations have been affected by digitalization, automation, and other new technologies. The sample consisted of 177 respondents who rated the impact on a scale of: 0 (cannot say), 1 (no impact), 2 (minimal impact), 3 (medium impact), 4 (significant impact), 5 (very significant impact). The highest average impact was recorded in the areas of internal communication (M = 3.36; SD = 1.459) and administration (M = 3.32; SD = 1.481), indicating that technologies have significantly affected primarily operational and communication processes. Employee training and development (M = 3.01; SD = 1.394) and external communication (M = 3.05; SD = 1.609) also show a significant impact, with higher variance in external communication indicating greater diversity of experiences across organizations. Performance appraisal (M = 2.66; SD = 1.515) and workplace presence (M = 2.82; SD = 1.599) show a moderate impact of technology, likely related to the use of performance measurement systems and the range of remote work.

Lower impact was noted in areas such as career planning (M = 2.18; SD = 1.485), motivation (M = 2.19; SD = 1.429) and work-life balance (M = 2.16; SD = 1.441), with mean standard deviations indicating moderate variability in responses. The lowest mean values occurred in recruitment and selection (M = 1.92; SD = 1.494) and employee adaptation (M = 1.95; SD = 1.538), suggesting that these areas are less transformed by technology so far, and while some organizations may be using digital tools, many remain traditional in their approach. In general, the standard deviations range between 1.39 and 1.61, indicating moderate to higher variability in the responses, i.e. that the degree of technology impact varies between organizations – likely depending on their size, sector and level of digitalization.

Research question 2: In which areas of human resource management do respondents see the greatest scope for using automation and digitalization? The second research question (RQ2) aims to identify which HR areas respondents believe have the greatest potential for further implementation of digital and automated solutions. The specific results are presented in Table 3 below.

Descriptive Statistics					
	N	Minimum	Maximum	Mean	Std. Deviation
Simplifying the recruitment and selection process	177	0	5	2.58	1.562
Automation of administrative tasks (e.g. attendance)	177	0	5	3.68	1.387
More effective employee training and education through e-learning	177	0	5	3.72	1.317
Automated performance evaluation	177	0	5	3.02	1.524
Expansion and improvement of internal communication tools	177	0	5	3.07	1.495
Faster and more accurate career planning and talent management	177	0	5	2.45	1.515
Improving and simplifying processes related to remuneration and benefits	177	0	5	2.77	1.479
Personalizing work plans and schedules using digital tools	177	0	5	2.97	1.502
Optimization of payroll processing and administration processes	177	0	5	3.08	1.609
Improving the management and protection of employee data through digital solutions	177	0	5	3.03	1.593
Valid N (listwise)	177				

Table 3. Result of RQ2 analysis (own elaboration, 2025)

Table 3 captures the views of 177 respondents on where they see the greatest scope for the use of automation and digitalization in human resource management. The assessment was carried out on a scale: 0 (cannot say), 1 (no scope for use), 2 (small scope for use), 3 (medium scope for use), 4 (large scope for use), 5 (large scope for use). The highest average was recorded in more effective training and education through e-learning (M = 3.72; SD = 1.317), which indicates a strong consensus among respondents on the potential of digital platforms for employee development. The low standard deviation also indicates a relatively uniform perception of this area. The second highest value was obtained by automating administrative tasks such as attendance records (M = 3.68; SD = 1.387), which confirms that digitalization of routine activities is very attractive and feasible for organizations. Optimization of payroll management (M = 3.08; SD = 1.609), improvement of internal communication (M = 3.07; SD = 1.495) and improvement of management and protection of employee data (M = 3.03; SD = 1.593) also achieved average values above 3, which indicates a medium-sized scope for digitalization and at the same time considerable variability in responses – probably due to different levels of digitalization between organizations.

Automated performance evaluation (M = 3.02; SD = 1.524) and personalization of work plans and schedules (M = 2.97; SD = 1.502) are also perceived as areas with potential, although with a slightly higher dispersion of responses, which may reflect different experiences with the use of advanced digital tools in HR practice. Lower averages were recorded for compensation and benefits (M = 2.77; SD = 1.479), recruitment processes (M = 2.58; SD = 1.562), and lowest for career planning and talent management (M = 2.45; SD = 1.515). These areas may be perceived as more complex, less standardisable, or more demanding to implement technologies. Overall, the standard deviations range from 1.3 to 1.6, indicating moderate to high variability in respondents' opinions and indicating that many organizations are at different levels of readiness or digital maturity. The data confirms that respondents see the greatest digitalization potential in the HR sphere in the areas of education and administration, while more strategic or individually tailored processes are so far less digitalized.

Research question 3: How have been improved processes of human resource management if digitalization of these processes was implemented? The aim of the third research question (RQ3) is to evaluate the effectiveness and benefits of digitalization in specific areas of human resource management from the perspective of the respondents. The analysis of the given research question is presented graphically in the following Fig. 2.

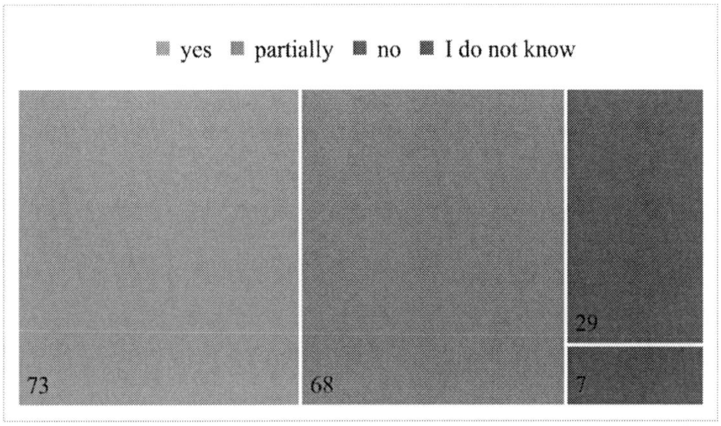

Fig. 2. Result of RQ3 analysis (own elaboration, 2025)

Based on the data in Fig. 2, it can be stated that 73 respondents (the largest share) answered “yes”, which means that they perceive digitalization as a clear benefit in terms of improving HR processes. Subsequently, 68 respondents chose the option “partially”, which indicates a perceived positive impact, but with certain reservations or only in some areas. A total of 29 respondents indicated that they “cannot answer”, which may indicate either a lack of information or the absence of direct contact with digitalized processes. Only 7 respondents answered “no”, which indicates that only a small percentage did not notice any improvement. Fig. 2 shows that the majority of respondents perceive digitalization positively, having completely or at least partially improved HR processes in organizations.

Integrating digital technologies into HR is essential to improve efficiency and compliance with standards and regulations [12]. Organizations must navigate the complexities of digital transformation, which is not just an IT project but a comprehensive business change initiative [13]. HR departments often face challenges related to the fear of job loss among employees, which can create resistance to the implementation of automation technologies [14]. High levels of technostress can hinder job satisfaction and are exacerbated by techno-uncertainty, which negatively impacts the perceived benefits of HR technologies [15]. Lack of management expertise and support can hinder the successful execution of HR automation initiatives, highlighting the need for leadership engagement [16] and [17]. Training opportunities and organizational support are essential to address the human aspects of integrating automation into HR [18].

4. Conclusion

Based on the survey, several key conclusions can be drawn. Digitalization and automation have most significantly affected the areas of administration and internal communication, i.e. mainly operational and routine HR processes. Respondents also identified the greatest potential for further use of technologies in the areas of education and e-learning, as well as in the automation of attendance and payroll management. More strategic areas, such as career planning, recruitment or motivation, are so far less affected by technologies. The results also indicate that the degree of digitalization varies between organizations depending on the size, industry and level of technological maturity. Most respondents perceive digitalization as a positive contribution to making HR processes more efficient, although some of them report only partial improvements. Overall, technologies already play an important role in HR today and at the same time there is wide scope for their further use, especially in areas with a higher degree of standardization.

The main limitations of the research include focusing only on medium and large industrial enterprises, which limits the possibility of generalizing the results to other types of organizations (e.g. small businesses or the public sector). Another limitation is the subjectivity of the respondents' answers, which may affect the objectivity of the findings. In addition, the research focused only on quantitative data, which lacks a deeper qualitative view of the reasons, obstacles or practical experiences with digitalization in HR. Future research could focus on analysing the factors influencing the rate of digitalization in different types of organizations, as well as exploring the potential of technologies in strategic HR areas such as recruitment, motivation, and career planning. It would also be beneficial to evaluate the effectiveness of specific digital tools in improving the performance of HR processes.

5. Acknowledgments

Funded by the EU NextGenerationEU through the Recovery and Resilience Plan for Slovakia under the project No. 09I03-03-V05-00005.

6. References

[1] Wu, C.-H., Chou, C.-W., Chien, C.-F., & Lin, Y.-S. (2024). Digital transformation in manufacturing industries: Effects of firm size, product innovation, and production type. Technological Forecasting and Social Change, 207, 123624.

[2] Bzhalava, L., Hassan, S. S., Kaivo-oja, J., Köping Olsson, B., & Imran, J. (2022). Mapping the Wave of Industry Digitalization by Co-Word Analysis: An Exploration of Four Disruptive Industries. International Journal of Innovation and Technology Management, 19(02), 2250001.

[3] Budic, H., Horvat, D., & Ruzic, I. (2022). ICT in Education Development. In B. Katalinic (Ed.), DAAAM Proceedings (1st ed., Vol. 1, pp. 0017–0023). DAAAM International Vienna.

[4] Knapcikova, L., Sukic, E., Behunova, A., Tauberova, R., Ljajic, S., & Dedic, N. (2024). Digitalization and Technology as Main Tools to Assure the Company's Performance. International Journal for Quality RESEARCH, 18(4), 953–964.

[5] Tesic, S., Bajic, B., Cosic, I., & Rikalovic, A. (2021). Frugal innovations in Industry 4.0: The Identification of Key Indicatiors. In B. Katalinic (Ed.), DAAAM Proceedings (1st ed., Vol. 1, pp. 0262–0269). DAAAM International Vienna.

[6] Flechsig, C., Anslinger, F., & Lasch, R. (2022). Robotic Process Automation in purchasing and supply management: A multiple case study on potentials, barriers, and implementation. Journal of Purchasing and Supply Management, 28(1), 100718.

[7] Balasundaram, S., & Venkatagiri, S. (2020). A structured approach to implementing Robotic Process Automation in HR. In A. George, A. S. Pandian, B. Ayshwarya, & R. Stephen (Eds.), Third National Conference on Computational Intelligence (NCCI 2019) (Vol. 1427, p. 012008). IoP Publishing Ltd.

[8] Bindra, S., Bhattacharya, S., & Bhattacharya, S. (2025). Traditional to digital: Human resource management transformation. Journal of Work-Applied Management.

[9] Marsikova, K., & Mazurchenko, A. (2019). Digitalization: Transforming the nature of HRM processes and HR professionals' competencies. In K. Antlova & T. Semeradova (Eds.), Liberec Economic Forum 2019 (pp. 291–301). Technical University Liberec.

[10] Munsamy, M., & Telukdarie, A. (2019). Digital HRM Model for Process Optimization by Adoption of Industry 4.0 Technologies. 2019 IEEE International Conference on Industrial Engineering and Engineering Management (IEEM), 374–378.

[11] Zhang, J., & Chen, Z. (2024). Exploring Human Resource Management Digital Transformation in the Digital Age. Journal of the Knowledge Economy, 15(1), 1482–1498.

[12] Mohamed, S. A., Mahmoud, M. A., Mahdi, M. N., & Mostafa, S. A. (2022). Improving Efficiency and Effectiveness of Robotic Process Automation in Human Resource Management. Sustainability, 14(7), Article 7.

[13] Sobczak, A. (2021). Robotic Process Automation Implementation, Deployment Approaches and Success Factors—An Empirical Study. Entrepreneurship and Sustainability Issues, 8(4), 122–147.

[14] Arslan, A., Cooper, C., Khan, Z., Golgeci, I., & Ali, I. (2021). Artificial intelligence and human workers interaction at team level: A conceptual assessment of the challenges and potential HRM strategies. International Journal of Manpower, 43(1), 75–88.

[15] Florkowski, G. W. (2019). HR technologies and HR-staff technostress: An unavoidable or combatable effect? Employee Relations: The International Journal, 41(5), 1120–1144.

[16] Krutikova, V. M., Isavnin, A. G., Eremina, I. I., Lysanov, D. M., & Ishmuradova, I. I. (2019). Solutions for Reduction of Time and Expenses of Business Management Via 1c: Payroll & Hr Management Software. International Transaction Journal of Engineering Management & Applied Sciences & Technologies, 10(16), 10A16I.

[17] Shani, A., & Tesone, D. V. (2010). Have human resource information systems evolved into internal e-commerce? Worldwide Hospitality and Tourism Themes, 2(1), 30–48.

[18] Yadav, D. S., Yadav, A., & Chhabria, S. G. (2024). Challenges in implementation of robotic process automation (RPA) in human resource management (HRM) and its effectiveness. Journal of Information & Optimization Sciences, 45(7), 1949–1961.

DOI: 10.2507/36th.daaam.proceedings.023

IMPACT OF REMEDIAL INTERVENTIONS ON RISING DAMP: INSIGHTS FROM A SINGLE-CASE STUDY

Patrik Šťastný

This Publication has to be referred as: Stastny, P[atrik] (2025). Impact of Remedial Interventions on Rising Damp: Insights From a Single-Case Study, Proceedings of the 36th DAAAM International Symposium, pp.0175-0179, B. Katalinic (Ed.), Published by DAAAM International, ISBN 978-3-902734-47-1, ISSN 1726-9679, Vienna, Austria
DOI: 10.2507/36th.daaam.proceedings.023

Abstract

This paper addresses the problem of rising damp and its negative impact on residential buildings. The focus is a specific case study of a family house, where remedial measures were applied and in situ monitoring of their effects was carried out. The main objective is to evaluate the effectiveness of the chosen technology and its contribution to improving the indoor environment as well as extending the lifespan of the construction. The results demonstrate that the intervention significantly contributed to drying the masonry and stabilizing building elements, thereby improving their technical condition. At the same time, a positive influence on the indoor microclimate was confirmed, particularly in reducing moisture-related issues and secondary defects such as mold growth or damage to surface finishes. The case study highlights the practical importance of appropriately selected remediation technologies and their application in typical residential construction. The conclusions provide insights applicable to the protection and maintenance of similar buildings, emphasizing the importance of early diagnosis and targeted implementation of measures.

Keywords: Rising damp; Moisture remediation; Remedial interventions; Undercutting; Case study

1. Introduction

Moisture problems in masonry structures, particularly rising damp driven by the capillary transport of groundwater, are recognized as a prevalent and destructive form of building pathology. This phenomenon leads to significant material degradation, reduced thermal performance, and a decline in indoor air quality, posing a major challenge for the preservation of older buildings and the sustainability of residential buildings. In Europe, this issue is particularly critical due to a large and aging building stock, where old construction methods and materials are highly susceptible to moisture ingress. Today, however, by combining modern technologies and materials, we are able to effectively address the problem of rising damp [1]. Extensive research has focused on remedial strategies for damp-affected masonry, including mechanical undercutting, chemical injections, and the application of specialized remedial plasters. While various remediation technologies have demonstrated varied levels of effectiveness, their performance is often context-specific, dependent on factors such as building material composition, environmental conditions, and specific construction details [2], [3], [4]. A notable gap in the current literature is the scarcity of comprehensive in situ case studies that systematically evaluate the long-term performance of combined remedial interventions, particularly in the specific climatic and material conditions [5].

This paper addresses this research gap by presenting a detailed in situ case study of a family house in Western Slovakia, severely affected by rising damp. The primary objective is to evaluate the effectiveness of a combined remedial approach—specifically, mechanical undercutting complemented by the application of remedial plasters. We aim to quantify the improvements in both the structural integrity of the masonry and the indoor environmental quality, thereby providing empirical evidence that contributes to a more nuanced understanding of these rehabilitation methods.

2. Methodology

The research methodology consisted of several sequential steps (Figure 1). The investigation was carried out primarily in situ, as continuous measurements were required. The first step involved an on-site inspection of the building, the preparation of necessary documentation, and an assessment of the current condition of the construction, during which specific measurement points were identified. At these locations, the moisture content of the investigated construction was monitored throughout the entire study.

In the second step, initial measurements were conducted at the designated points across the construction. The collected data were systematically recorded in tables, which included the identification of the measurement point, the measurement height, and the date of data collection. Based on these values, the degree of dampness was subsequently diagnosed.
The following step focused on the selection of an appropriate remediation technology. This required the consideration of several factors, including the extent of moisture penetration and the spatial accessibility of the construction for the application of the chosen technology. After implementing the selected remedial intervention, additional measurements were performed at regular intervals, typically every one to two months, to continuously monitor the development of moisture content in the construction. On the basis of these successive measurements, it was possible to observe the temporal progression of moisture levels within the construction and to evaluate the effectiveness of the applied damp-proofing intervention.

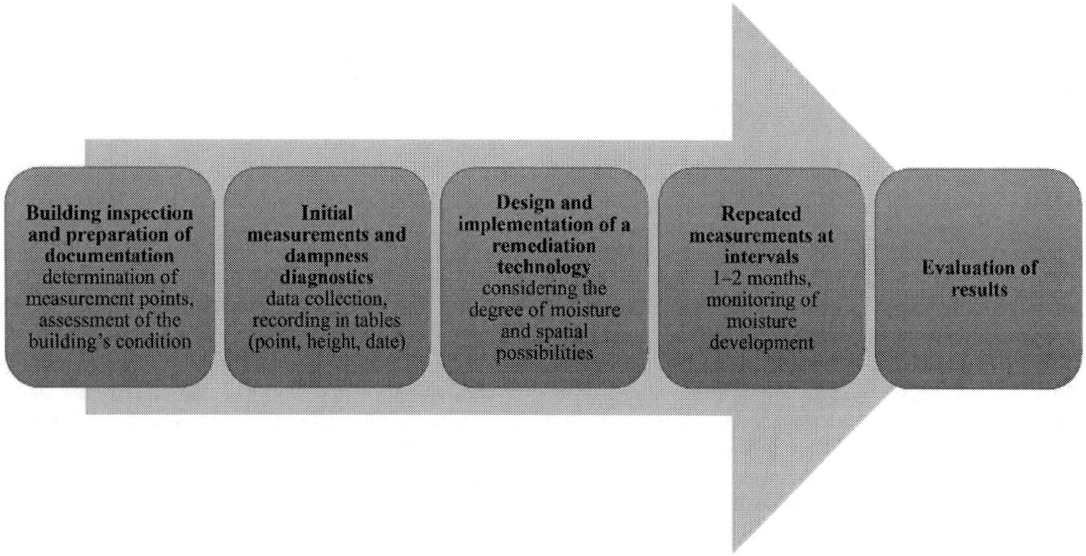

Fig. 1. Flowchart of methodological steps in building dehumidification (author)

3. Description of the Applied Remedial Technology

This article focuses on the technology applied to the investigated building, specifically the undercutting method. The fundamental principle of this approach is the creation of an additional impermeable barrier within the masonry, designed to serve as supplementary waterproofing against capillary rise and groundwater infiltration. In practice, this intervention aims to restore the structural integrity of the building envelope by interrupting the upward movement of moisture through the porous material of the wall. From a methodological perspective, the formation of additional impermeable layers represents a well-established approach within the field of building rehabilitation. As noted by Makýš [1], undercutting masonry is classified among the most effective methods in this category, particularly when dealing with historic or severely damaged structures. The process itself is executed in carefully sequenced stages to ensure stability, precision, and durability of the intervention. The undercutting procedure typically employs a diamond-wire saw, which, in combination with guiding rollers, enables controlled cutting across the masonry. This creates a narrow cavity or slot suitable for the insertion of a waterproofing barrier. Following the cut, the slot is meticulously cleaned to remove debris and dust, after which a waterproofing strip—usually composed of durable synthetic materials—is inserted. The strip is subsequently wedged to prevent settlement and ensure continuous contact with the surrounding masonry.

Work then progresses incrementally to the next section of the wall. The length of each section depends on multiple factors, including the type of masonry, the thickness of the wall, and the structural condition of the building; nonetheless, a working length of approximately 0.5 m per stage is considered standard practice. To secure the inserted barrier, wedges are applied systematically along the joints, preventing vertical displacement and maintaining even distribution of forces within the structure. Once a complete segment is stabilized, the process is repeated for the adjacent section, thereby advancing progressively across the wall. The final stage of the intervention involves sealing the joint, commonly using mortar injection or mechanical mortar application, which consolidates the barrier and re-establishes the wall's structural continuity. To complete the rehabilitation process, a finishing plaster is applied, serving both protective and aesthetic functions. Overall, the undercutting method represents a precise and labor-intensive technology that, when properly implemented, provides a highly effective and durable solution for the remediation of moisture-damaged masonry. Accordingly, it can be evaluated as effective [6], yet it imposes considerable economic costs [7].

4. Research Results

The building under investigation can be characterized as a residential structure with a floor area of approximately 120 m², which is permanently inhabited. It is situated on flat terrain with sufficient distance from surrounding developments. From a construction perspective, the building is predominantly made of brick, although in certain areas, mixed masonry was used. During the initial measurements, the building exhibited high moisture levels, which, according to ČSN P 73 0610 [8], can be classified as waterlogging. This standard divide moisture content into five categories (Table 1).

	Degree of humidity	Moisture (uM) [%]
1	Very low moisture	< 3.0
2	Low moisture	3.0 – 5.0
3	Increased moisture	5.0 – 7.5
4	High moisture	7.5 – 10
5	Very high moisture	> 10

Table 1. Degree of moisture of constructions, ČSN P 73 0610 [8]

As mentioned in Chapter 2, measurements were carried out at designated locations and subsequently recorded in tables. These data were then evaluated in accordance with the above-mentioned standard. Average moisture values before the implementation of remedial measures, taken at a height of approximately 30 cm above the floor, exceeded 10%. This indicated a high level of dampness in the building and the necessity for remediation. Following the implementation of remedial measures in September 2023, these values gradually decreased. The measured values may have been slightly influenced by prevailing weather conditions at the time of measurement; however, as shown in Figure 2, it is evident that moisture receded from the structure over time, and the masonry above the newly installed waterproofing layer gradually dried out. The use of heating during the colder season may also have contributed to this accelerated drying process.

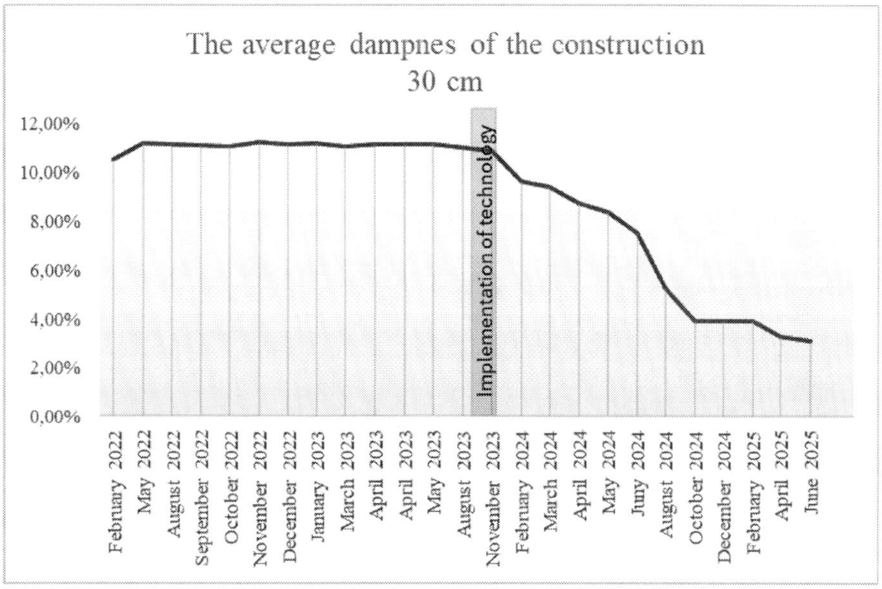

Fig. 2. Graph illustrating the average moisture development in the construction measured at 30 cm height, presented for the periods before, during, and after the application of damp-proofing technologies. (author)

Measurements were also conducted at a height of approximately 150 cm above the floor. From the very beginning, these values remained consistent, averaging around 1.30%, which according to the standard indicates that the masonry at this level was dry. These findings also excluded the possibility of condensation within the building and pointed directly to the influence of rising damp on the construction. Since the measured values at these points were consistently low, it is not necessary to present their course in this contribution, as they have no direct impact on the evaluation of the remediation technology's effectiveness and would only provide a supplementary overview of moisture levels at this height.

5. Discussion

The results of the research clearly demonstrated that the selected remedial technology—undercutting in combination with the application of remedial plasters—represents an effective solution for eliminating rising damp in masonry constructions. Prior to the intervention, measurements taken at 30 cm above floor level exceeded 10%, which, according to ČSN P 73 0610, corresponds to the category of "very high moisture." Following the intervention, however, the measured values gradually declined and stabilized within a few months at levels that can be considered technically acceptable. This trend highlights the ability of the applied technology not only to stop capillary rise but also to promote the progressive drying of the construction.

These findings may serve as a basis for further improvements through the application of active thermal protection (ATP), which could enhance indoor environmental quality and improve the efficiency of heating systems, even with reduced thicknesses of conventional thermal insulation, as indicated in publication [9], [10]. Active thermal protection represents a controlled dynamic process characteristic of building constructions with integrated energy-active components that provide one or more energy functions under different operational regimes of heating and energy supply systems. Such an approach could significantly contribute to the long-term sustainability of older buildings, many of which fail to meet today's energy performance standards. Nevertheless, the practical implementation of ATP remains the subject of ongoing research, which is expected to directly build upon the findings presented in this study. Preliminary assessments suggest that, for the walls investigated in this case study, the integration of ATP could enable a reduction in insulation thickness from 210 mm to approximately 50 mm, representing a substantial decrease while maintaining adequate performance.

6. Conclusion

The findings of this study confirm the positive impact of undercutting technology in mitigating rising damp. The applied intervention has proven its effectiveness not only in previous research but also within the framework of this case study. However, long-term monitoring remains essential in order to fully verify the durability and reliability of the system, as well as to establish realistic service-life expectations. Certain risks of reduced performance do exist, particularly when the intervention is carried out improperly or when low-quality workmanship compromises critical construction details. For this reason, strict adherence to professional execution and established technological procedures is indispensable. The study also emphasizes the importance of systematic in situ measurements as an objective basis for evaluating the effectiveness of remedial measures. Beyond confirming the suitability of the applied technology under the specific conditions of Western Slovakia, the research provides valuable empirical data that may serve as a reference for future studies and as decision-making support for the application of similar interventions in comparable buildings.

7. Acknowledgments

Funded by the EU NextGenerationEU through the Recovery and Resilience Plan for Slovakia under the project No. 23-01-03-B.

8. References

[1] Makýš, O. (2023). Technológia Obnovy Budov, Ochrana a Oprava Spodných a Obalových Konštrukcií [Technology of Building Rehabilitation: Protection and Repair of Substructure and Building Envelope Systems], SPEKTRUM STU, [Online], ISBN: 978-80-227-4848-3, Bratislava, Slovakia, Available at: https://www.vydavatelstvo.stuba.sk/obchod/technologia-obnovy-budov-ochrana-a-oprava-spodnych-a-obalovych-konstrukcii

[2] Møller, E. B. & Olsen, B. (2011). Rising damp, a reoccurring problem in basements: The 9th Nordic Symposium on Building Physics (NSB 2011), 9th Nord. Symp. Build. Phys., Vol. 2, p. 765–772

[3] Wójcik, R., Panuś, A., Tunkiewicz, M. & Hamdy, M. (2017). Influence of chemical damp proof cream on the capillary action and microstructure of mortars, Energy Procedia, vol. 132, p. 670–675, doi: 10.1016/j.egypro.2017.10.004.

[4] Šťastný, P., Gašparík, J. & Makýš, O. (2021). Analysis of moisture and salinity of historical constructions before and after the application of REMEDIATIONS, J. Build. Eng., vol. 41, s. 102785, sep. 2021, doi: 10.1016/j.jobe.2021.102785.

[5] Franzoni, E. (2018). State-of-the-art on methods for reducing rising damp in masonry, J. Cult. Herit., vol. 31, p. S3–S9, doi: 10.1016/j.culher.2018.04.001.

[6] Schmidt, H. (1999). Historische Bauwerksabdichtungen - Traditionelle und neuzeitliche Maßnahmen zum Schutz gegen Bodenfeuchtigkeit und zur Trockenlegung feuchter Wände [Historical Waterproofing of Buildings: Traditional and Modern Measures for Protection Against Ground Moisture and Drying of Damp Walls]. Bautechnik, Vol. 76, No. 7, p. 581–598, DOI: 10.1002/bate.199903790.

[7] Kiesow, G. (2012). Památková péče v Německu - kniha [Heritage Conservation in Germany], Barrister & Principal, ISBN: 978-80-87474-54-9, Brno, Czech Republic

[8] ČSN P 73 0610 (2020). Hydroizolace staveb - Sanace vlhkého zdiva - Základní ustanovení [Waterproofing of buildings - The rehabilitation of damp masonry and additional pro-tection of buildings against ground moisture and against atmospheric water - Basic provision]. [Online]. Available at: https://csnonline.agentura-cas.cz/Detailnormy.aspx?k=58539

[9] Stastny, P. (2024). Application of Remedial Measures in Combination with Active Thermal Protection for Optimizing the Indoor Environment of Older Buildings, Proceedings of the 35th DAAAM International Symposium, pp.0368-0372, B. Katalinic (Ed.), Published by DAAAM International, ISBN 978- 3-902734-44-0, ISSN 1726-9679, Vienna, Austria DOI: 10.2507/35th.daaam.proceedings.050

[10] Kalús, D. et al. (2025). Energy Sustainability, Resilience, and Climate Adaptability of Modular and Panelized Buildings with a Lightweight Envelope Integrating Active Thermal Protection: Part 2—Design and Implementation of an Experimental Prototype of a Building Module for Modular Buildings", Coatings, vol. 15, issue 7, p. 781, doi: 10.3390/coatings15070781.

DOI: 10.2507/36th.daaam.proceedings.024

SYSTEM DYNAMICS AND DIGITAL TWINS: A BIBLIOMETRIC INSIGHT INTO REAL-TIME DATA INTEGRATION

Lea Masnec & Zlatko Stapić

This Publication has to be referred as: Masnec, L[ea] & Stapic, Z[latko] (2025). System Dynamics and Digital Twins: A Bibliometric Insight into Real-Time Data Integration, Proceedings of the 36th DAAAM International Symposium, pp.0180-0185, B. Katalinic (Ed.), Published by DAAAM International, ISBN 978-3-902734-47-1, ISSN 1726-9679, Vienna, Austria
DOI: 10.2507/36th.daaam.proceedings.024

Abstract

With the increasing complexity of business systems, there is a growing need for models that support real-time decision-making based on up-to-date data and dynamic system behaviour. One direction is the integration of system dynamics, a methodology for modeling complex systems based on feedback loops, with digital twins that enable real-time data integration. Although interest in combining system dynamics with other methods is increasing, the methodological connection between these two approaches is still insufficiently explored. This paper presents a bibliometric analysis of existing research in area of digital twins and system dynamics, using the Scopus, Web of Science and IEEE scientific literature databases. The analysis identifies leading researchers, journals, conferences, research groups, and scientific-professional associations active in this field. The database query resulted in 236 publications after 79 duplicates were removed. Based on the analysis and defined criteria, three researchers, three journals, three conferences, and scientific-professional sources were selected. The presented results reveal publication trends and collaboration networks, as well as existing research gaps that hinder practical integration. The research findings will contribute to concrete practical application in future phases, aiming to connect system dynamics with real-time data.

Keywords: system dynamics; digital twins; real-time data; bibliometric analysis; simulation modeling

1. Introduction

Understanding and modeling complex systems is important for decision-making under conditions of uncertainty [1], [2]. In this context, there is an increasing need for the integration of models with up-to-date data to ensure adaptability and accuracy of predictions in dynamic environments [3]. The application of computer models enables the simulation of different scenarios and the prediction of possible consequences of decisions before they are implemented in the real system. Such models are used in a wide range of domains – from business management and production to ecological systems [4], [1]. There are numerous methodologies for modeling complex systems, but for the purpose of this research, system dynamics was chosen. System dynamics was chosen due to the ability to view the system as a whole and to observe feedback loops, flows in the system, as well as stock accumulation, which is especially relevant nowadays when the need for circularity is growing, and in that context, it is increasingly important to monitor where resources go in enterprises.

Originally developed for understanding long-term changes in complex systems, system dynamics today increasingly require extensions in order to respond to the challenges of an ever-changing environment [5], [6], [2]. Its application is widely spread in the field of business decision-making, sustainable development, resource management, and process optimization [7], [5], [2]. Classical system dynamics models primarily rely on historical data, which can limit their ability to predict in dynamic systems [8]. One of the potential solutions to this problem is the connection of system dynamics with the digital twin methodology, which enables the integration of real-time data into simulation models. Digital twins enable dynamic modeling and process optimization based on real-time data [9], [10], [11]. The use of real-time data, as a concept of digital twins, allows for the adjustment of models and decision-making based on current conditions, thereby increasing the accuracy and efficiency of the process [12], [2]. However, there still exists a research gap in how digital twins can relate to system dynamics to improve the precision of predictions and adaptability of the models. The motivation for studying the topic of this paper arose from participation in the project focused on modeling and circular economy, and from the motivation due to the identified gap in current research.

The aim of this paper is to conduct a bibliometric analysis of relevant literature to explore the current state of research in the field of connecting system dynamics and digital twins. The following are the research questions that will be answered in this paper, and it is noted that the questions are related to the field of system dynamics and digital twins: Who are the leading researchers in the research field and their research groups?; What are the two to three strongest scientific journals in the field?; What are the three strongest international conference in this field?; What international scientific-professional associations operate in this field? In the paper, after the introduction, materials and methods are presented for the purpose of literature search. This is followed by the results chapter where the results obtained by searching databases are presented. Then, a discussion describes the review of the results obtained, with a conclusion at the end.

2. Methods

The paper uses a bibliometric analysis closely related to system dynamics as a methodology in combination with the methodology of digital twins in the form of real-time data. In order to reach relevant publications on the mentioned topic, the Scopus, WoS and IEEE databases were used. It was necessary to create a query that would cover a satisfactory spectrum of publications in order to conduct research and analysis. The search strategy was focused on the areas of system modeling and digital twins. Considering that system modeling is a broad term, it was decided to immediately concentrate on the area of system dynamics. Likewise, when speaking of digital twins in this context, they are often also referred to as real-time data, the concept of which is indeed part of the scope of this research. The keywords were selected for the purpose of accurately searching for relevant information on the possibility of connecting the methodology of system dynamics and digital twins, as real-time data. It was decided that in all databases the query would be used:

TITLE-ABS-KEY (("system dynamics" OR "system modeling" OR "dynamics system" OR "simulation modeling") AND ("decision-making" OR "sustainability" OR "business process*" OR "resource management") AND ("real-time data" OR "digital twin" OR "data-driven simulation" OR "predictive analytics" OR "Internet of Things" OR "IoT")) AND PUBYEAR > 2001 AND PUBYEAR < 2026*

The stated query was originally created in the Scopus database, but the same query was also used in other databases, Web of Science and IEEE, with only the names of the search fields being different. The search field in Scopus was "title, abstract, keywords", in Web of Science it was "Topic" (which also covers titles of papers, abstracts, and keywords), while in IEEE there is no similar way to select the search field, therefore the "Abstract" field was selected. The year set as a limit in the query, 2002, is the year when the concept of digital twins was first mentioned. In the paper, as the research questions state, the most significant researchers, their research groups, scientific journals, scientific conferences, and international scientific-professional associations are determined according to precisely defined criteria. Researchers were included in the selection if they met at least three of the following four criteria: if there is at least one publication relevant to the field of system dynamics and digital twins, if there is citation of relevant publications in the field of system dynamics and digital twins, editorial membership in at least one scientific journal, and the total citation count of relevant works related to the topic of system dynamics and digital twins must be greater than 40. Authors whose works are not cited, authors whose research is not connected with the research area related to system dynamics and digital twins, and those who have not published in scientific journals were excluded from the analysis. For each author, it is also a condition that they must be active in at least one research group. The initial set of journals is identified by searching in relevant databases using the predefined query, with all journals that publish works related to digital twins and system dynamics initially included. In order to keep the analysis focused on the most relevant sources, journals that primarily cover medical topics were excluded from the set, except if they directly deal with system modeling. By reviewing the Scimago Journal Country Rank, journals with a small number of citations and those not belonging to ranked categories were also eliminated. For the ranking of the remaining journals, criteria of impact in the academic community were used according to inclusion in the Q1 or Q2 quartile according to SJR, as well as the citation of relevant works published in a particular journal by reviewing Research Gate. Priority was given to journals in the Q1 quartile with a higher H-index and greater citation count. For the analysis of relevant scientific conferences, clear selection criteria were defined, which include inclusion and exclusion criteria.

Selected conferences must be thematically related to system dynamics and digital twins and have international recognition. Priority is given to conferences whose papers are indexed in the Scopus, WoS, or IEEE databases and those with a long-standing tradition. Scientific conferences excluded from the initial set are those conferences held at irregular intervals or one-time conferences. The selection criterion for scientific-professional associations was the sponsorship of scientific conferences identified as the most prominent. For additional review of the content of identified scientific-professional associations, the Google search engine was used.

3. Results

The aforementioned search resulted in a total of 236 publications; the total number was obtained after eliminating duplicates, of which there were 79 duplicates in total. Table 1 shows the most relevant authors of the published publications related to the research area, as well as their research groups. Additionally, the table shows whether the author is an editor of a scientific journal and which ones, as well as the number of papers related to the research area and the citation count of the papers. Table 1 was created after the application of the criteria described in the chapter "Materials and Methods", and after applying the criteria, only three authors were obtained.

Author	Research groups[1]	Journal editorial[1]	Number of papers related to system dynamics and digital twins	Citation count of selected papers[2]
Hu, Xiaolin	SIMS Lab (Georgia State University) - USA	International Journal of Modeling, Simulation, and Scientific Computing (IJMSSC), and International Journal of Agent Technologies and Systems (IJATS)	6	49
Ding, Zhikun	TRIZ and patent laboratory (TP Lab - Shenzhen University), Guangdong Laboratory of Artificial Intelligence and Digital Economy(SZ) - China	/	3	509
Onggo, Bhakti Stephan	Centre for Operational Research, Management Science and Information Systems (CORMSIS) - UK	Journal of Simulation	4	177
[1] Source: Websites of institutions and researcher profiles (Scopus, ORCID profile, ResearchGate profile)				
[2] Source: Scopus				

Table 1. Overview of the leading researchers

Table 2 shows the most relevant journals of published publications related to the research area through a presentation of the total citation count of selected papers, Q and H-index of the scientific journal, and additionally the publications where the potential for further research on the research topic was recognized (references).

Journal	Total citation counts of papers on ResearchGate	Q1[1]	H-index[1]	References
System Dynamics Review	SIMS Lab (Georgia State University) - USA	International Journal of Modeling, Simulation, and Scientific Computing (IJMSSC), and International Journal of Agent Technologies and Systems (IJATS)	65	[13], [14]
IEEE Access	TRIZ and patent laboratory (TP Lab - Shenzhen University), Guangdong Laboratory of Artificial Intelligence and Digital Economy(SZ) - China	/	243	[15], [16]
Sustainability	Centre for Operational Research, Management Science and Information Systems (CORMSIS) - UK	Journal of Simulation	169	[17], [18], [19], [4]
[1] Source: Scimago Journal & Country Rank				

Table 2. Overview of the leading journals

Table 3 provides an overview of the most relevant conferences based on the number of publications retrieved through database queries, the publisher, and the H-index of each conference. A review of relevant scientific conferences is also important due to their thematic focus, which provides insight into current research and technological trends. As stated, the focus of this paper is on the topic of digital twins and system dynamics, as interconnected methodologies for the simulation, analysis, and optimization of complex systems. A clear example of thematic alignment can be seen in the DAAAM conference: For instance, [20] demonstrate how the digital twin methodology is implemented in the context of intelligent automation, with particular emphasis on the integration of physical and virtual systems, creating opportunities for dynamic modeling of system behavior over time [20]. [21] further highlight the challenges in developing digital twins for technologically complex environments, in which key elements—such as feedback loops, accumulations, and time delays—are precisely those addressed by system dynamics. [22] explore the visualization and dynamic analysis of spatial manipulators with closed structure, providing a concrete example of the use of mathematical modeling and simulation within virtual environments, which is highly relevant to digital twin development [22].

Conference	Number of publications[1]	Publisher[2]	H-index[3]	Sponsor (Scientific-Professional Associations)[4]
Winter Simulation Conference	5	ACM, IEEE	64	ASA
DAAAM (Danube Adria Association For Automation & Manufacturing)	3		22	DAAAM International Vienna
International Academic Mindtrek Conference	2	ACM	12	ACM

[1] Number of publications obtained by querying the Scopus, WoS and IEEE databases
[2] Source: Official websites of scientific conferences
[3] Source: Scimago Journal & Country Rank
[4] Source: Official websites of the conference and proceedings

Table 3. Overview of the leading conferences

Recognized scientific-professional associations include the Association for Computing Machinery (ACM), DAAAM International Vienna and the American Statistical Association (ASA).

4. Discussion

The use of bibliometric analysis enabled the identification of the most prominent researchers, scientific journals, and conferences in the field of system dynamics and digital twins. This approach is based on objective data from relevant scientific databases (Scopus, IEEE, WoS), ensuring an unbiased mapping of the research field. The identification of researchers Hu X., Zhikun D., and Onggo B. S. confirms that the analyzed publications are focused on advanced research in systems modeling and digital twins and provides an overview of researchers based on predefined criteria. The inclusion of research groups contributes to understanding the organizational context of the research and the influence of specific institutions on the development of the field.

The established inclusion and exclusion criteria enabled the formation of the final results. The focus on indexed papers contributed to ensuring scientific quality, but at the same time, there is a possibility of excluding smaller, yet innovative studies that are not part of major academic databases. Additionally, the elimination of journals and conferences with a low impact factor meant that sources such as Sustainability, IEEE Access, and System Dynamics Review were recognized as leading in the field, but niche publications with specialized audiences may have been omitted. The selection of keywords was fundamental to the relevance of the search. By using terms such as system dynamics, digital twin, and real-time data, precise filtering of publications was enabled, but publications using alternative terminology may have been overlooked. The results of the journal and conference analysis confirm that the identified sources are key to the scientific community in this field. The high H-index of journals such as IEEE Access (243) and System Dynamics Review (65) indicates their scientific importance and impact. The journals presented in Table 2 cover a wide range of research topics.

With the selected journals, coverage of the research themes is ensured, whereby Sustainability and System Dynamics Review cover sustainability and system dynamics, while IEEE Access addresses the technical aspects of digital twins and simulation models. Conferences such as the Winter Simulation Conference, DAAAM, and International Academic Mindtrek Conference have also proven to be relevant for the exchange of the latest research. In terms of research limitations, it is possible that the inclusion of additional databases would allow a broader understanding of the interconnections within the field.

5. Conclusion

The research provided a bibliometric analysis of literature in the field of integrating system dynamics and digital twins. Using bibliometric analysis, the most prominent researchers, leading scientific journals, conferences, as well as relevant research institutions and associations were identified. The results show that in recent years there has been a significant increase in the number of publications in this field, especially in 2024. The advantages of this research include quantitative analysis of trends, as well as a foundation for future research. The study also has certain limitations. These include the exclusive focus on publications in selected databases (WoS, Scopus, and IEEE), which may result in missing relevant papers from other sources. The literature review identified a research gap, namely that there is still no established methodology for linking digital twins and system dynamics. The lack of such a methodology, as identified through this quantitative research, lays the groundwork for new practical research and further investigation toward the development of a methodology. Importantly, to date, no bibliometric study has specifically addressed the integration of system dynamics and digital twins. This makes the presented analysis valuable foundation for future research. By mapping the current scientific state, this paper lays the groundwork for a more structured approach to methodological development in this emerging field. The results show that, despite increasing interest, there is still no standardized or widely accepted framework for the integration of real-time data from digital twins into system dynamics models. This research gap reveals a need for interdisciplinary efforts. Future studies can expand on this work by exploring concrete case studies, developing domain-specific integration frameworks.

6. References

[1] Hu, X. (2023). Dynamic data-driven simulation: Real-time data for dynamic system analysis and prediction. World Scientific. https://doi.org/10.1142/13166

[2] Gejo-García, J;, Reschke, J., Gallego-García, S., & García-García, M. (2022). Development of a system dynamics simulation for assessing manufacturing systems based on the digital twin concept. Applied Sciences, 12(4), 2095. https://doi.org/10.3390/app12042095

[3] Coelho, F.; Relvas, S., & Barbosa-Póvoa, A. P. (2021). Simulation-based decision support tool for in-house logistics: The basis for a digital twin. Computers & Industrial Engineering, 153, 107094.

[4] Korder, B.; Maheut, J., & Konle, M. (2024). Simulation methods and digital strategies for supply chains facing disruptions: Insights from a systematic literature review. Sustainability, 16(14), 5957. https://doi.org/10.3390/su16145957

[5] Hussain, H.; & De Vries, M. (2025). System dynamics applied in enterprise engineering – A systematic literature review. Journal of Modelling in Management, 20(1), 1–29.

[6] Secci, D.; Saysel, A. K., Uygur, İ.; Yoloğlu, O. C., Zanini, A., & Copty, N. K. (2024). Modeling for sustainable groundwater management: Interdependence and potential complementarity of process-based, data-driven and system dynamics approaches. Science of The Total Environment, 951, 175491. https://doi.org/10.1016/j.scitotenv.2024.175491

[7] Bestle, D., & Bielitz, T. (2024). Real-time models for systems with costly or unknown dynamics. Proceedings in Applied Mathematics and Mechanics, 24(2), e202400008.

[8] Ganguli, R., & Adhikari, S. (2020). The digital twin of discrete dynamic systems: Initial approaches and future challenges. Applied Mathematical Modelling, 77, 1110–1128.

[9] Onggo, B. S., & Currie, C. S. M. (2024). Extending simulation modeling methodology for digital twin applications. In 2024 Winter Simulation Conference (WSC) (pp. 3058–3069). IEEE.

[10] Chakraborti, A.; Heininen, A.; Koskinen, K. T., & Lämsä, V. (2020). Digital twin: Multi-dimensional model reduction method for performance optimization of the virtual entity. Procedia CIRP, 93, 240–245.

[11] Relich, M. (2023). Predictive and prescriptive analytics in identifying opportunities for improving sustainable manufacturing. Sustainability, 15(9), 7667.

[12] Choi, J.; Moon, S., & Min, S. (2023). Digital twin simulation modeling process with system dynamics: An application to naval ship operation. International Journal of Robust and Nonlinear Control, 33(16), 10136–10150.

[13] Pluchinotta, I., Zhou, K., & Zimmermann, N. (2024). Dealing with soft variables and data scarcity: Lessons learnt from quantification in a participatory system dynamics modelling process. System Dynamics Review, 40(4), e1770. https://doi.org/10.1002/sdr.1770

[14] Chen, W.; Hong, W.; Zhang, H.; Yang, P., & Tang, K. (2023). Multi-fidelity simulation modeling for discrete event simulation: An optimization perspective. IEEE Transactions on Automation Science and Engineering, 20(2), 1156–1169.

[15] Gorelova, A.; Meliá, S. & Gadzhimusieva, D. (2024). A discrete event simulation of patient flow in an assisted reproduction clinic with the integration of a smart health monitoring system. IEEE Access, 12, 46304–46318. https://doi.org/10.1109/ACCESS.2024.3380021

[16] Kumar, R.; Ather, D.; Sagar, P.; Yusupov, I.; Duggal, S., & Jain, V. (2024). Integrating predictive models and remote sensing for sustainable agriculture, energy, and environmental management. In 2024 4th International Conference on Technological Advancements in Computational Sciences (ICTACS) (pp. 1930–1934). IEEE.

[17] Rocca, R.; Rosa, P.; Sassanelli, C.; Fumagalli, L., & Terzi, S. (2020). Integrating virtual reality and digital twin in circular economy practices: A laboratory application case. Sustainability, 12(6), 2286.

[18] Chidozie, B.; Ramos, A.; Vasconcelos, J;, Ferreira, L. P., & Gomes, R. Highlighting Sustainability Criteria in Residual Biomass Supply Chains: A Dynamic Simulation Approach. Sustainability, vol. 16, no. 22, p. 9709, 2024. doi: 10.3390/su16229709.

[19] Ding, Z.; Gong, W.; Li, S., & Wu, Z. System Dynamics versus Agent-Based Modeling: A Review of Complexity Simulation in Construction Waste Management. Sustainability, vol. 10, no. 7, p. 2484, 2018. doi: 10.3390/su10072484.

[20] Janda, P.; Hájíček, Z.; & Bernardin, P. (2019). Implementation of the digital twin methodology. In Proceedings of the 30th International DAAAM Symposium "Intelligent Manufacturing & Automation". DAAAM International. https://doi.org/10.2507/30th.daaam.proceedings.072

[21] Ponomarev, K.; Kudryashov, N.; & Popelnukha, N. (2017). Main principals and issues of digital twin development for complex technological processes. In Proceedings of the 28th International DAAAM Symposium 2017. DAAAM International. https://doi.org/10.2507/28th.daaam.proceedings.074

[22] Damić, V., & Cohodar Husić, M. (2016). Dynamic analysis and visualization of spatial manipulators with closed structure. In Proceedings of the 26th International DAAAM Symposium 2016. DAAAM International. https://doi.org/10.2507/26th.daaam.proceedings.016

DOI: 10.2507/36th.daaam.proceedings.025

DESIGN AND EVALUATION OF A PIEZO-DRIVEN TURNING TOOL FOR MACHINING NON-CIRCULAR GEOMETRIES

Maximilian Krause, Friedrich Bleicher & Christoph Habersohn

This Publication has to be referred as: Krause, M[aximilian]; Bleicher, F[riedrich] & Habersohn, C[hristoph] (2025). Design and Evaluation of a Piezo-Driven Turning Tool for Machining Non-Circular Geometries, Proceedings of the 36th DAAAM International Symposium, pp.0186-0190, B. Katalinic (Ed.), Published by DAAAM International, ISBN 978-3-902734-47-1, ISSN 1726-9679, Vienna, Austria
DOI: 10.2507/36th.daaam.proceedings.025

Abstract

Conventional turning processes are inherently limited to the generation of rotationally symmetric geometries due to the fixed kinematics between the cutting tool and the rotating workpiece. To overcome this limitation and enable the fabrication of non-circular profiles, a superimposed radial oscillation of the cutting tool can be introduced. This additional degree of freedom allows for the controlled deviation from circular paths, thereby enabling the machining of a broad spectrum of arbitrary geometries, constrained primarily by the actuator's stroke amplitude. Precise synchronisation between the oscillatory motion of the tool and the angular position of the workpiece is essential to ensure contour fidelity and surface quality. In this context, a piezo-actuated turning tool was developed and experimentally evaluated. The system is capable of generating a sinusoidal radial displacement with a maximum amplitude of 85 μm. Experimental results demonstrate the tool's ability to consistently and accurately reproduce non-circular profiles, confirming the feasibility and repeatability of piezo-assisted turning for advanced shape generation. The approach offers a promising solution for applications requiring high-precision non-rotationally symmetric components.

Keywords: Non-circular turning; Piezoelectric actuation; Shape control

1. Introduction

Various methods exist to produce non-circular profiles in turning operations. One of these approaches is polygon or non-circular-rotary-turning, where a defined oscillatory motion is superimposed on the cutting tool while the workpiece rotates. This technique enables the direct machining of geometries like polygons, lobes or cam shapes on conventional lathes [1]. Another method of manufacturing non-round geometries on a lathe is equipping the tool with a feed in one or multiple axes. These systems are called Fast Tool Servos (FTS) and can be realised with a voice coil drive, a linear synchronous motor, a reluctance-based drive [2] or a piezoelectric actuator [3].

A practical application of non-circular profile machining is the production of engine pistons. During combustion, the pistons are subjected to thermal and mechanical loads that cause deformation resulting in reduced performance and efficiency. This issue can be mitigated by designing the piston geometry with a middle-convex, varying elliptical (MCVE) profile to better accommodate distortions and maintain optimal guidance [4].

Fig. 1. shows the characteristic geometry of a MCVE piston: the cross-section of the piston is an ellipse with a varying degree of ellipticity along the axial direction ("varying ellipse") while the axial profile has a cam-like, "middle-convex" shape [5].

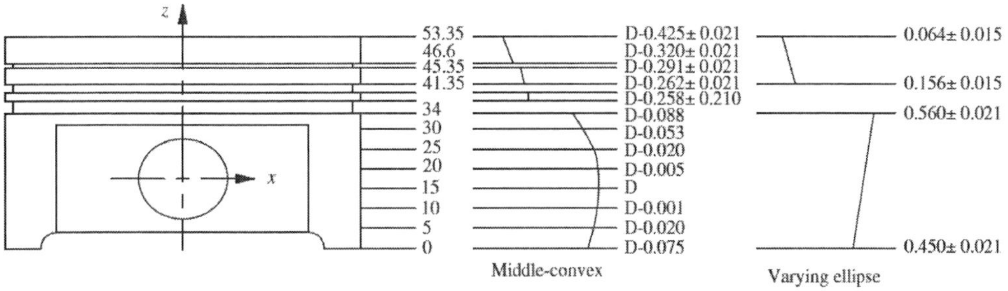

Fig. 1. Geometry of a MCVE piston [5]

Although surface roughness is typically regarded as detrimental, particularly due to its negative influence on friction and the functional performance of sliding interfaces [6], intentionally engineered surface features can provide significant advantages in specific tribological applications. A practical application of piezo-assisted machining is surface texturing, as tribological research has demonstrated its effectiveness in reducing friction in mechanical components [7]. Kurniawan et al. [8] used a piezoelectric actuator to drive a tool holder on a standard CNC lathe, allowing the tool to create surface textures directly during the machining process through controlled vibration. Ali et al. [9] examined the tribological properties of cylindrical surfaces textured with micro-dimples under starved lubrication. Comparing the resulting Stribeck curves of the textured surfaces to their non-textured control group, the authors found that the textured surfaces have a lower coefficient of friction in regions of high contact pressure.

This study presents the development and design of a piezo-actuated turning tool intended for machining defined non-circular geometries. The primary objectives are to investigate the tool's capability to accurately reproduce specified geometrical features and to systematically assess its performance in terms of precision, repeatability and surface quality. In addition, the influence of actuation parameters on machining outcomes is evaluated to verify the suitability of the approach for advanced manufacturing applications.

2. System design

The respective piezo turning tool is a prototype setup designed for producing non-circular geometries in the micrometer range. The schematic structure is depicted in Fig. 2. Two piezoelectric actuators excite the tool holder, which is mounted in the housing by two membranes. A return spring counteracts the force of the actuators and pushes the tool back into its initial position. The moving part is also equipped with a linear encoder that measures the stroke of the tool. The piezo-actuators can be powered with a voltage of up to 200 V and a maximum frequency of 300 Hz. The geometry generated during the turning process is influenced by the applied actuation frequency, enabling the production of varying profile shapes depending on the excitation conditions.

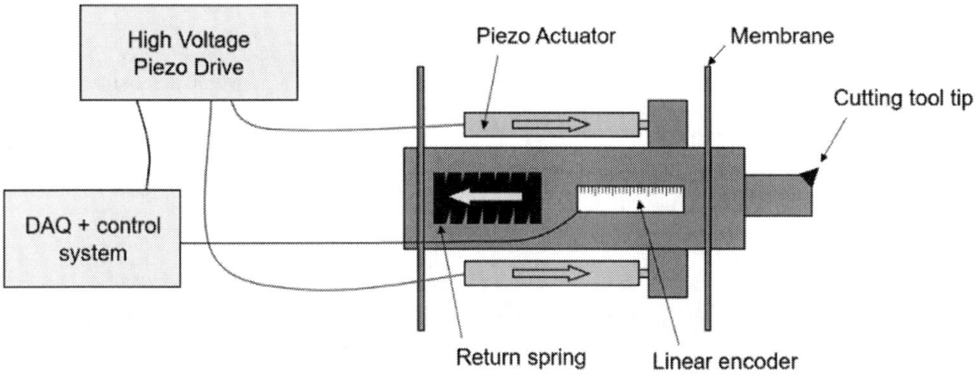

Fig. 2. Schematic of the piezo-driven tool

The piezo drive is a PD200X4 from Scientific Instruments with a maximum output voltage of 150 V and a peak current of 2 A per channel. The DAQ and control unit is realised with a National Instruments USB-6343 X-Series featuring analog and digital I/O ports.

3. Experimental Evaluation

3.1. Determination of the characteristic map of motion

To characterise the dynamic behaviour of the tool, experiments were performed to determine the amplitude response as a function of the excitation voltage and frequency. The objective was to construct a characteristic map representing the peak-to-peak values of the stroke (in micrometers) as a function of the applied sinusoidal voltage and its corresponding frequency. The piezoelectric actuators were driven by a sinusoidal input signal ranging from 10 V to 150 V peak-to-peak in increments of 10 V, centered on a DC offset of 75 V. Excitation frequencies varied from 10 Hz to 300 Hz in steps of 10 Hz. Each test configuration was held for five seconds, followed by a two-second idle period to prevent thermal drift and actuator saturation. Following completion of the full parameter sweep, the displacement data recorded via a high-resolution linear encoder was post-processed using fast Fourier transformation (FFT). This frequency-domain analysis enabled automated extraction of the dominant frequency component and its corresponding amplitude, facilitating the accurate quantification of peak-to-peak displacement for each excitation condition.

The resulting characteristic curve is presented in Fig. 3. The actuator displacement exhibits a predominantly linear dependence on the applied voltage amplitude. Frequency-dependent effects on stroke amplitude were observed to be minimal, although a slight increase at higher frequencies was detected. Beyond a critical voltage-frequency threshold, the actuator power supply entered an overload protection mode, limiting the system's operational envelope and preventing further displacement amplification.

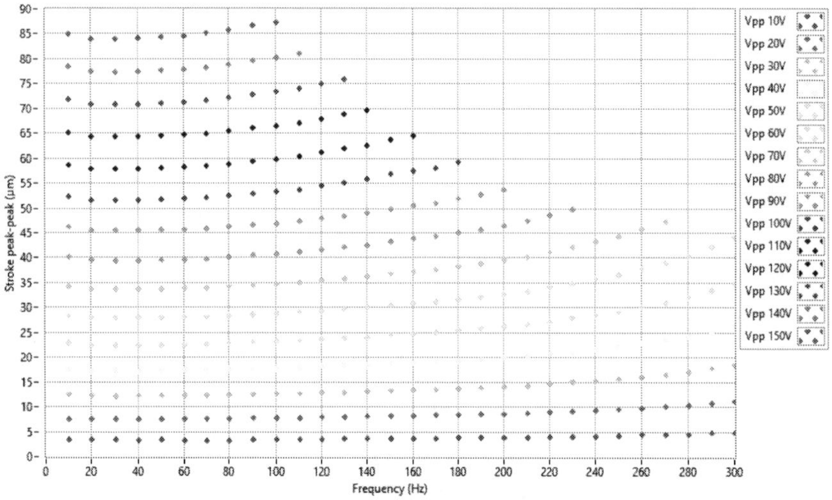

Fig. 3. Characteristic map of the tool

3.2. Machining tests

For the cutting experiments, the piezo-actuated tool was mounted on the machine table of a CNC machining center as shown in Fig. 4. The specimen is an aluminium rod with a raw diameter of $D = 25$ mm, which is mounted in the machine's tool holder. An inductive proximity switch is mounted next to the tool holder and generates a pulse once per spindle revolution, which is subsequently used as a trigger signal for the control of the actuating system.

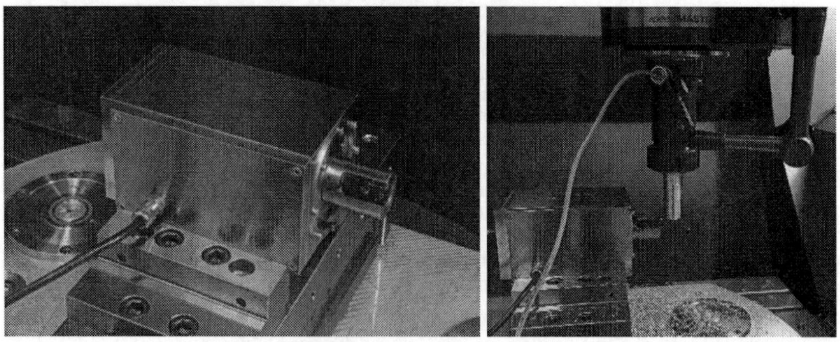

Fig. 4. Experimental setup

Table 1 lists the machining parameters corresponding to the various specimens depicted in Fig. 5. The rotational speed of the spindle was set to 1500 rpm which results in a cutting speed of v_c = 118 m/min for the given workpiece diameter of 25 mm, suitable for turning aluminum alloys.

Test run	Waveform	n (s^{-1})	Freq (Hz)	V_{pp} (V)	Offset (V)	Stroke peak-peak (µm)
a)	Sine	25	50	140	80	72
b)	Sine	25	100	140	80	76
c)	Sine	25	100	80	80	40
d)	Sine	25	75,2	140	80	74
e)	Triangle	25	75,2	140	80	74
f)	Sawtooth	25	75,2	140	80	98

Table 1. Parameters and results of the cutting tests

Following the machining experiments conducted under these systematically varied process parameters, the resulting workpiece geometries were characterised using a high-precision coordinate measuring machine (CMM) to quantify dimensional accuracy and assess surface characteristics. The experimental results are depicted in Fig. 5.

A comparative analysis between four-fold sinusoidal oscillations per spindle revolution at excitation levels of 140 V_{pp} (b) and 80 V_{pp} (c) reveals an approximately linear correlation between the peak-to-peak stroke amplitude and the magnitude of the applied voltage. Specifically, at an excitation amplitude of 140 V_{pp}, two oscillations per revolution (a) result in a stroke of 72 µm, while three (d) and four oscillations (b) yield strokes of 74 µm and 76 µm, respectively. This moderate stroke amplification with increasing frequency is consistent with the previously discussed actuator response behaviour shown in the characteristic map in Fig. 3. Furthermore, an investigation of the influence of waveform shape on tool displacement - while maintaining constant frequency and amplitude - demonstrates that sinusoidal (d) and triangular (e) excitation signals produce nearly identical stroke amplitudes. In contrast, a sawtooth waveform (f) induces a higher peak displacement. However, the surface profile generated by a sawtooth input exhibits superimposed oscillations along the ramp section of the geometry. These undesired oscillations are absent in the triangular waveform, which yields a smooth and continuous contour. The disturbance observed in the sawtooth case is attributed to the abrupt phase discontinuity inherent in the waveform, which induces dynamic perturbations in the actuator's mechanical support system, comprising a compliant suspension mechanism formed by a spring and membrane. This excitation of structural resonances compromises the kinematic fidelity of the actuator’s motion, thus adversely affecting surface integrity.

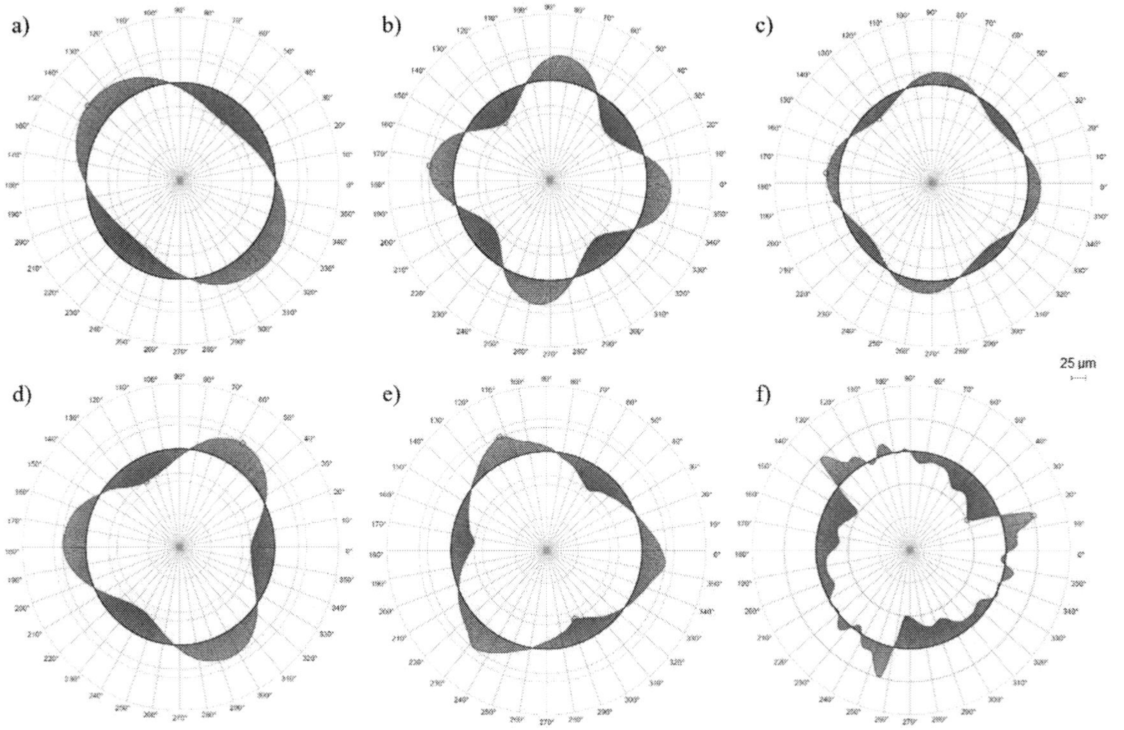

Fig. 5. Roundness measurement of the specimens

4. Conclusion and Outlook

This study has demonstrated the feasibility of using a piezo-actuated turning tool to machine non-circular geometries by superimposing controlled tool path modulations onto the cutting path. One key area for future investigation is the influence of waveform steepness on dynamic stability. Preliminary results indicate that excessively steep ramp transitions, such as those present in sawtooth waveforms, can induce unwanted oscillations due to the rapid retraction of the actuator and the resulting excitation of structural resonances. A more detailed analysis is needed to quantify the critical slope thresholds beyond such effects occur. Furthermore, the system's ability to produce controlled, periodic tool motions opens up opportunities for automated chip-breaking strategies where continuous chips are undesirable. Overall, the integration of programmable piezo actuation into turning processes represents a promising direction for achieving high-precision, flexible and functionally tailored surface geometries.

5. References

[1] Arndt, T., & Schulze, V. (2024). Non-Circular-Rotary-Turning process for manufacturing parts with non-circular contours. CIRP Annals, 73(1), 61-64.

[2] Krall, S., Caudr, H., Prießnitz, M., Maier, S., Baumann, C., & Bleicher, F. (2024). Reluctance Based Actuator for Discrete Machine Hammer Peening. Procedia CIRP, 123, 517–522

[3] Bleicher, F., Biermann, D., Drossel, W.-G., Moehring, H.-C., & Altintas, Y. (2023). Sensor and actuator integrated tooling systems. CIRP ANNALS-MANUFACTURING TECHNOLOGY, 72(2), 673–696

[4] Zhang, Y., Huang, Y., Shao, W., & Ming, W. (2017). Research on MCVE piston machining and process parameter optimization. The International Journal of Advanced Manufacturing Technology, 93, 3955-3966.

[5] Wang, H., & Yang, S. (2013). Design and control of a fast tool servo used in noncircular piston turning process. Mechanical Systems and Signal Processing, 36(1), 87-94.

[6] Qehaja, N., Jakupi, K., Bunjaku, A., Bruçi, M., & Osmani, H. (2015). Effect of machining parameters and machining time on surface roughness in dry turning process. 25th DAAAM International Symposium on Intelligent Manufacturing and Automation, Procedia Engineering, 100, 135-140.

[7] Ali, S., Kurniawan, R., Chul, P. G., & Ko, T. J. (2023). Tribological properties of hierarchical micro-dimples produced on a cylindrical surface by dual-frequency texturing. Friction, 11(2), 246-258.

[8] Kurniawan, R., & Ko, T. J. (2013). A study of surface texturing using piezoelectric tool holder actuator on conventional CNC turning. International Journal of Precision Engineering and Manufacturing, 14, 199-206.

[9] Ali, S., Kurniawan, R., Chul, P. G., & Ko, T. J. (2023). Tribological properties of hierarchical micro-dimples produced on a cylindrical surface by dual-frequency texturing. Friction, 11(2), 246-258.

DOI: 10.2507/36th.daaam.proceedings.026

EFFICIENT COMPUTATION OF 2D SELF-MOTION MANIFOLDS: A METHOD COMPARISON IN A 5-DOF ROBOT

Marc Fabregat-Jaén, Adrián Peidró, Luis Payá & Óscar Reinoso

This Publication has to be referred as: Fabregat-Jaén, M[arc]; Peidro, A[drian]; Paya, L[uis] & Reinoso, O[scar] (2025). Efficient Computation of 2D Self-Motion Manifolds: A Method Comparison in a 5-DoF Robot, Proceedings of the 36th DAAAM International Symposium, pp.0191-0200, B. Katalinic (Ed.), Published by DAAAM International, ISBN 978-3-902734-47-1, ISSN 1726-9679, Vienna, Austria
DOI: 10.2507/36th.daaam.proceedings.026

Abstract

This paper presents a comparative evaluation of methods for computing self-motion manifolds in redundant robotic manipulators. Redundant robots have more degrees of freedom than required to perform their task. We consider a 5-degree-of-freedom robot tasked with reaching three-dimensional positions, which results in two degrees of redundancy. This implies that the inverse kinematic problem of the robot has infinitely many solutions that lie on two-dimensional self-motion manifolds. Computing these manifolds is of paramount importance for planning feasible and globally optimal motions. In this paper, we compare different methods for computing these manifolds: cellular automata, continuation and sampling-based methods. Our results show that the established techniques like continuation and sampling provide better performance. We further discuss the individual strengths and limitations of each approach. By offering this balanced evaluation, our study aims to guide researchers and users in selecting appropriate tools for analysing redundancy in industrial robotic systems.

Keywords: Redundant robots; Self-motion manifolds; Redundancy resolution; Robotics;

1. Introduction

Kinematic redundancy arises when a robotic manipulator possesses more degrees of freedom (DOF) than are strictly required to accomplish a given task. This property, common in many industrial and collaborative robots, enables the system to exploit its extra DOF to achieve secondary objectives such as obstacle avoidance, singularity evasion, or energy minimization [1]. However, this flexibility comes at the cost of increased computational complexity: the inverse kinematics problem (IKP) for redundant manipulators is underdetermined, admitting infinitely many solutions for a given end-effector position. Redundancy resolution methods are typically classified into velocity-level and position-level approaches. Velocity-level methods, such as those based on the pseudoinversion of the Jacobian [2] or optimization-based schemes [3], provide local solutions by integrating joint velocities to follow a desired task trajectory. While effective for real-time control, these methods are inherently local and may fail to capture the global structure of the solution space, often resulting in suboptimal or non-cyclic motions [4]. In contrast, position-level methods aim to characterize the entire set of solutions for a given task. Self-motion manifolds [5] and Feasibility Maps [6] are examples of approaches that describe the infinite set of solutions.

Among these, self-motion manifolds (SMMs) are of particular interest: they are surfaces in the joint space along which the end-effector position remains constant. The computation and analysis of SMMs are crucial for applications requiring global optimality, cyclic motions, or exhaustive feasibility analysis: capabilities that are especially valuable in industrial contexts such as welding, painting, or assembly, where repeatability and reliability are paramount. Several methods have been proposed for computing self-motion manifolds. Continuation methods systematically trace the solution set from an initial configuration. The original continuation approach, employed by [7], incrementally constructs the manifold by following the Jacobian's null space, which is tangent to the manifold at each point, to predict subsequent points. These points are then corrected using the Newton-Raphson method, ensuring convergence to the manifold. This technique allows for the computation of one-dimensional self-motion curves for robots with a single degree of redundancy. Later, [8] generalized the method to higher-dimensional manifolds by constructing an atlas of local parameterizations (charts) that collectively cover the manifold.

Sampling-based methods typically identify the self-motion manifold in two stages: first, they sample a set of valid joint configurations to generate a point cloud that approximates the manifolds, and then cluster these points to identify disjoint manifolds. Ref. [9] proposed sampling the point cloud by randomly selecting joint configurations and checking whether they map (via forward kinematics) sufficiently close to the desired task value. [10] presented a more systematic approach by sweeping every possible combination of redundant joint variables over their feasible range and solving the remaining joint variables to satisfy the task constraint. This method avoids computing points that would later be discarded, as the resulting points are guaranteed to belong to the manifold. Other strategies for SMM computation have also been explored. For instance, [11] proposed a cellular automata-based method that discretizes the joint space and iteratively updates the state of each cell. [12] introduced a swarm intelligence approach, specifically the artificial bee colony algorithm, to sequentially trace the SMMs, similarly to continuation methods, without requiring the Jacobian matrix. Multi-objective optimization has also been proposed [13] to compute the manifolds, and a branch-and-bound method [14] systematically shrinks the search space to exclude infeasible regions.

Despite the variety of available techniques, comprehensive comparisons of their performance and suitability for practical industrial problems remain limited in the literature. This paper addresses this gap by evaluating and comparing several representative methods for computing two-dimensional self-motion manifolds in a 5-DOF robot. By analysing their strengths and limitations, we aim to provide guidance for researchers and practitioners in selecting appropriate tools for redundancy analysis and exploitation in industrial robotic systems. The remainder of the paper is organized as follows: Section 2 presents the inverse kinematics of the studied 5-DOF robot; Section 3 describes the method presented in [10] for computing self-motion manifolds by sweeping two redundant joint angles; Section 4 presents the higher-dimensional continuation method [8]; Section 5 presents a discussion of the results obtained from the different methods; and Section 6 concludes the paper and outlines future work.

2. Inverse Kinematics of the Studied 5-DOF Robot

The robot studied in this paper is a 5-DOF serial robot defined by the Denavit-Hartenberg (DH) parameters defined in Table 1. The robot and DH frames are shown in Fig. 1. This robot is taken from an example presented in [11].

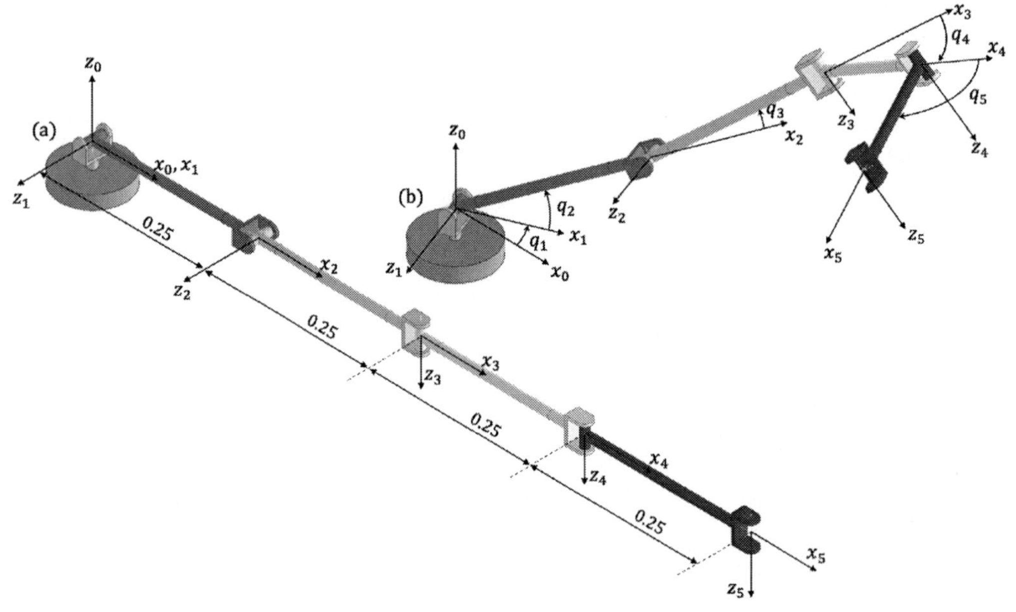

Fig. 1. (a) Robot in the reference pose. (b) Robot in a more arbitrary pose defined by: $q_1 = 20°$, $q_2 = 30°$, $q_3 = 10°$, $q_4 = 30°$, and $q_5 = 45°$.

Transformation between frames $i-1 \rightarrow i$	θ_i (rad)	d_i	a_i (m)	α_i (rad)
$0 \rightarrow 1$	q_1	0	0	$\pi/2$
$1 \rightarrow 2$	q_2	0	0.25	0
$2 \rightarrow 3$	q_3	0	0.25	$\pi/2$
$3 \rightarrow 4$	q_4	0	0.25	0
$4 \rightarrow 5$	q_5	0	0.25	0

Table 1. Denavit-Hartenberg (DH) parameters of the studied 5-DOF robot.

By multiplying the DH matrices from the fixed base of the robot to its end-effector, we obtain the homogeneous transformation matrix that represents the position and orientation of the end-effector with respect to the fixed base frame:

$$\mathbf{T} = {}^{0}\mathbf{T}_1 \, {}^{1}\mathbf{T}_2 \, {}^{2}\mathbf{T}_3 \, {}^{3}\mathbf{T}_4 \, {}^{4}\mathbf{T}_5, \tag{1}$$

where the DH matrix between consecutive frames $i-1$ and i is:

$${}^{i-1}\mathbf{T}_i = \begin{bmatrix} \cos\theta_i & -\cos\alpha_i \sin\theta_i & \sin\alpha_i \sin\theta_i & a_i \cos\theta_i \\ \sin\theta_i & \cos\alpha_i \cos\theta_i & -\sin\alpha_i \cos\theta_i & a_i \sin\theta_i \\ 0 & \sin\alpha_i & \cos\alpha_i & d_i \\ 0 & 0 & 0 & 1 \end{bmatrix} \tag{2}$$

In this paper, we are interested only in the position coordinates of the end-effector, neglecting its orientation. The position of the end-effector is taken from the last column of $\mathbf{T}$ as computed in (1), which has the following symbolic expression:

$$\begin{aligned} x &= \tfrac{1}{4}[\cos q_1 \cdot (a \cdot \cos(q_2+q_3) + \cos q_2) + b \cdot \sin q_1] \\ y &= \tfrac{1}{4}[\sin q_1 \cdot (a \cdot \cos(q_2+q_3) + \cos q_2) - b \cdot \cos q_1] \\ z &= \tfrac{1}{4}[a \cdot \sin(q_2+q_3) + \sin q_2], \end{aligned} \tag{4}$$

where $a = 1 + \cos q_4 + \cos(q_4+q_5)$ and $b = \sin q_4 + \sin(q_4+q_5)$.

The inverse kinematic problem consists in determining the values of the joint coordinates $\mathbf{q} = [q_1, \dots, q_5]^T$ to achieve a desired position $\mathbf{p} = [x, y, z]$. Since we are using five angles to control the position of only three coordinates, the robot is kinematically redundant, that is, there are two redundant degrees of freedom to control the position of the end-effector. The inverse kinematic problem in that case is underdetermined: there are three equations from which to solve five unknowns. For a given $\mathbf{p}$, the solution set of these equations in the space $(q_1, \dots, q_5)$ will generically be a surface, or a set of disjoint surfaces. These surfaces are the self-motion manifolds (SMMs): moving the angles $[q_1, \dots, q_5]$ along these surfaces does not modify the position $\mathbf{p}$ of the end-effector. Since we cannot solve all joints $\mathbf{q} = [q_1, \dots, q_5]^T$ from (3) for a given $\mathbf{p} = [x, y, z]$, we will solve $[q_1, q_2, q_3]$ in terms of $[q_4, q_5]$, which will remain as free parameters in the following. Note that, for given values of $[q_4, q_5]$, the values of $[a, b]$ in (3) are known.

To solve this system of equations assuming that $[q_4, q_5]$ are known, first we note that the first two equations are linear in $\cos q_1$ and $\sin q_1$:

$$\begin{bmatrix} a \cdot \cos(q_2+q_3) + \cos q_2 & b \\ -b & a \cdot \cos(q_2+q_3) + \cos q_2 \end{bmatrix} \begin{bmatrix} \cos q_1 \\ \sin q_1 \end{bmatrix} = \begin{bmatrix} x \\ y \end{bmatrix} 4, \tag{4}$$

from which we solve the sine and cosine of q_1:

$$\begin{bmatrix} \cos q_1 \\ \sin q_1 \end{bmatrix} = \begin{bmatrix} a \cdot \cos(q_2+q_3) + \cos q_2 & b \\ -b & a \cdot \cos(q_2+q_3) + \cos q_2 \end{bmatrix}^{-1} \begin{bmatrix} x \\ y \end{bmatrix} 4 \tag{5}$$

$$\begin{bmatrix} \cos q_1 \\ \sin q_1 \end{bmatrix} = \frac{1}{D} \underbrace{\begin{bmatrix} a \cdot \cos(q_2+q_3) + \cos q_2 & -b \\ b & a \cdot \cos(q_2+q_3) + \cos q_2 \end{bmatrix}}_{\mathbf{P}} \begin{bmatrix} x \\ y \end{bmatrix} 4, \tag{6}$$

with $D = \det(\mathbf{P}) = (a \cdot \cos(q_2+q_3) + \cos q_2)^2 + b^2$. We can eliminate the unknown q_1 by imposing that $\cos^2 q_1 + \sin^2 q_1 = 1$, which yields:

$$\cos^2 q_1 + \sin^2 q_1 = \begin{bmatrix}\cos q_1\\ \sin q_1\end{bmatrix}^T \begin{bmatrix}\cos q_1\\ \sin q_1\end{bmatrix} = \frac{16}{D^2}\begin{bmatrix}x\\ y\end{bmatrix}^T \underbrace{\mathbf{P}^T\mathbf{P}}_{\begin{bmatrix}D & 0\\ 0 & D\end{bmatrix}} \begin{bmatrix}x\\ y\end{bmatrix} = \frac{16}{D}(x^2+y^2) = 1 \tag{7}$$

That is:

$$16(x^2+y^2) = D = (a\cdot\cos(q_2+q_3)+\cos q_2)^2 + b^2 \tag{8}$$
$$16(x^2+y^2) - b^2 = (a\cdot\cos(q_2+q_3)+\cos q_2)^2 \tag{9}$$
$$\pm 4\sqrt{(x^2+y^2)-b^2} = a\cdot\cos(q_2+q_3)+\cos q_2 \tag{10}$$

Now, (5) can be written as:

$$4z = a\cdot\sin(q_2+q_3)+\sin q_2 \tag{11}$$

Equations (10) and (11) constitute a system of two equations and two unknowns. More importantly: the form of these equations is identical to those of a 2R planar serial arm, as demonstrated in Fig. 2. Thus, q_2 and q_3 can be solved by solving the inverse kinematic problem of such equivalent 2R arm, which is a straightforward solution that can be found in any textbook on robot kinematics. The inverse kinematics of the 2R arm is summarized in Fig. 2, which yields two possible solutions for a given position of its end-effector $[x', y']$. Since there are two possible values of x' due to the "$\pm$" as noted in Fig. 2, the system formed by (10) and (11) has up to four possible solutions for (q_2, q_3). For each solution, a unique value of $\sin q_1$ and $\cos q_1$ is obtained from (6), which yields a unique corresponding value of $q_1 = \text{atan2}(\sin q_1, \cos q_1)$, completing the resolution of the inverse kinematic problem of the 5-DOF robot of Fig. 1 for a given pair $[q_4, q_5]$ and a desired position $\mathbf{p} = [x, y, z]$.

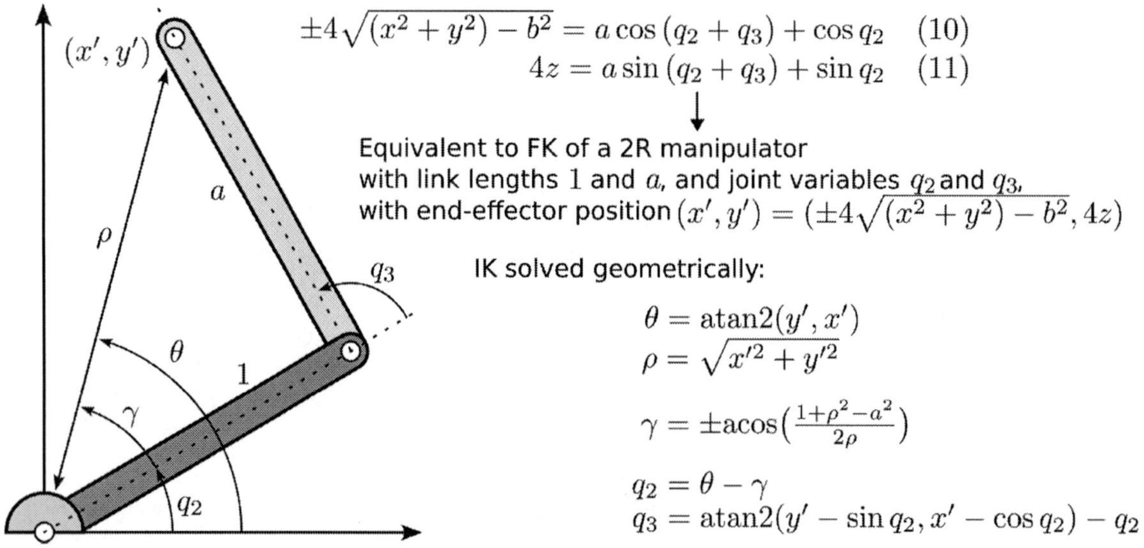

Fig. 2. Resulting system of equations, equivalent to that of a 2R planar arm.

3. Computing self-motion manifolds by sweeping two redundant joint angles

The sweeping method for computing self-motion manifolds, introduced by [10], systematically explores the feasible ranges of r redundant joint variables, where $r = n - m$ is the degree of redundancy, n is the number of DOFs, and m is the dimension of the task. For each discretized values of these variables, the remaining $n - r$ joint variables are solved to satisfy the task and constraints. After sweeping through the entire range, the resulting point cloud is clustered in order to identify disjoint self-motion manifolds. The algorithm to compute SMMs via sweeping redundant joints proceeds as follows, and is illustrated in Fig. 3, which corresponds to the SMMs of the 2R arm of Fig. 2 for task value $y' = 1$ m, and an arbitrary task constraint.

1. Discretize each of the r redundant joints into N points within their feasible ranges, forming a grid of N^r combinations.
2. For each grid point, solve the inverse kinematics for the remaining $n - r$ joints to satisfy the task constraint.
3. Collect all valid solutions to construct a point cloud representing the self-motion manifolds.
4. Cluster the point cloud to identify disjoint self-motion manifolds.

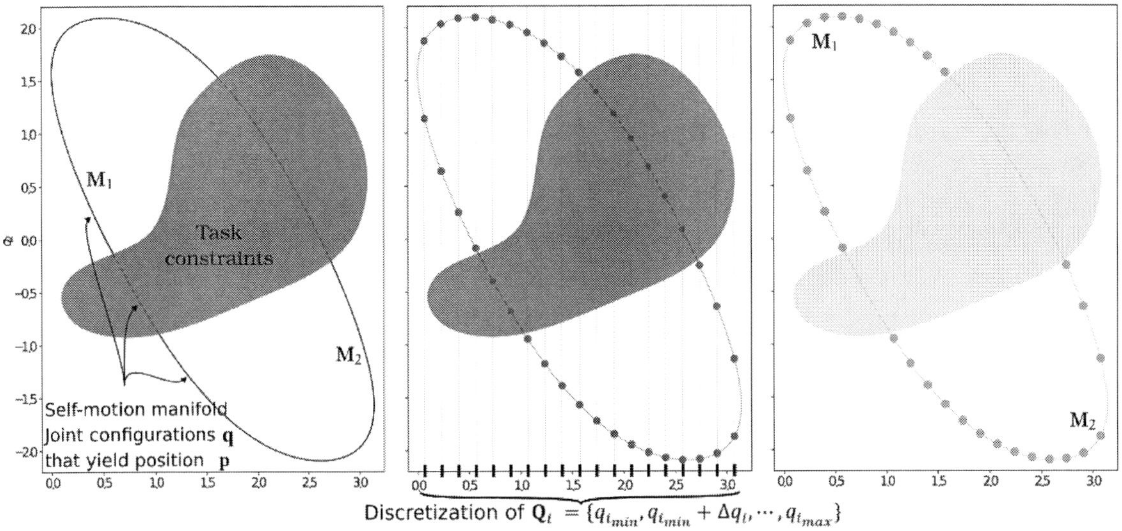

Fig. 3. Illustration of the sweeping method for computing self-motion manifolds of the 2R arm with $y' = 1$ m.

Let us consider the 5-DOF robot in Fig. 1 to further exemplify the algorithm. It has $n = 5$ joints and $m = 3$ task dimensions (3D end-effector position), yielding $r = n - m = 2$ redundant joints. We select the last two joints, (q_4, q_5), to sweep over their feasible ranges. Each joint i is discretized into a set $\mathbf{Q}_i = \{q_{i_{min}}, q_{i_{min}} + \Delta q_i, \cdots, q_{i_{max}}\}$, where $\Delta q_i = (q_{i_{max}} - q_{i_{min}}) / (N - 1)$, and N is the number of discretization points per joint. The Cartesian product $\mathbf{Q}_r = \mathbf{Q}_4 \times \mathbf{Q}_5$ yields a grid of N^2 pairs (q_4, q_5). For each pair (q_4, q_5), the inverse kinematics are solved for all possible solutions of the remaining joints (q_1, q_2, q_3), as detailed in Section 2. Solutions are checked for validity (e.g., joint limits, collision avoidance) and retained if feasible. The resulting point cloud accurately represents the self-motion manifolds at the specified end-effector position. Since the point cloud may contain points from multiple disjoint self-motion manifolds, a clustering algorithm is applied to distinguish them. Specifically, we use the DBSCAN (Density-Based Spatial Clustering of Applications with Noise) algorithm [15], which is well-suited for identifying clusters of arbitrary shape without requiring the number of clusters *a priori*. DBSCAN groups closely packed points and returns the identified clusters as disjoint self-motion manifolds, as Fig. 3(c) illustrates.

4. Computing self-motion manifolds by continuation

4.1. One-dimensional self-motion manifolds (curves)

The basis for the continuation method for computing self-motion manifolds was first introduced by [7]. Starting from an initial robot configuration, this approach generates a one-dimensional self-motion curve by iteratively advancing along the null space of the Jacobian, which is tangent to the manifold at each point. Fig. 4 illustrates this process.

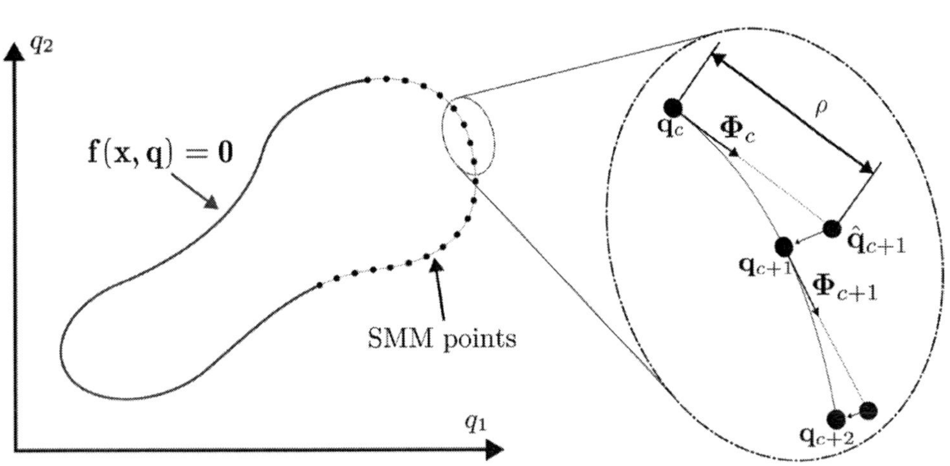

Fig. 4. Illustration of the continuation method for computing 1D self-motion manifolds.

The logic of the algorithm can be summarized as follows:
1. Compute a seed configuration $\mathbf{q}_0$ that lies on the SMM, if not already provided, and initialize $\mathbf{q}_c \leftarrow \mathbf{q}_0$.
2. Determine an orthonormal basis $\mathbf{\Phi}_c$ for the null space of the Jacobian matrix $\mathbf{J}$ at $\mathbf{q}_c$.
3. Take a step of size ρ along $\mathbf{\Phi}_c$, which is tangent to the SMM at $\mathbf{q}_c$, to obtain the candidate point $\widehat{\mathbf{q}}_{c+1}$.
4. Project $\widehat{\mathbf{q}}_{c+1}$ back onto the SMM by means of Newton's method, yielding the corrected point $\mathbf{q}_{c+1}$.
5. Repeat steps 2-4, updating the current configuration $\mathbf{q}_c \leftarrow \mathbf{q}_{c+1}$, until the entire manifold is traced.

The Jacobian matrix $\mathbf{J}(\mathbf{q})$ is the matrix of partial derivatives of the task function $\mathbf{f}(\mathbf{x}, \mathbf{q})$ with respect to the joint variables $\mathbf{q}$:

$$\mathbf{J}(\mathbf{q}) = \begin{bmatrix} \frac{\partial f_1}{\partial q_1} & \cdots & \frac{\partial f_1}{\partial q_n} \\ \vdots & \ddots & \vdots \\ \frac{\partial f_m}{\partial q_1} & \cdots & \frac{\partial f_m}{\partial q_n} \end{bmatrix} \tag{12}$$

To obtain an orthonormal basis $\mathbf{\Phi}_c$ for the null space, the Singular Value Decomposition (SVD) of the Jacobian is used:

$$\mathbf{J}(\mathbf{q}) = \mathbf{USV}^T, \tag{13}$$

where $\mathbf{U}$ and $\mathbf{V}$ are orthogonal matrices and $\mathbf{S}$ is a diagonal matrix of singular values. The null space is spanned by the $n-m$ last columns of $\mathbf{V}$.

After taking a step, the candidate configuration $\widehat{\mathbf{q}}_{c+1}$ must be corrected back onto the SMM defined by $\mathbf{f}(\mathbf{x}, \mathbf{q}) = \mathbf{0}$. This is accomplished by using the Newton-Raphson method, which iteratively updates $\widehat{\mathbf{q}}_{c+1}$ as follows:

$$\widehat{\mathbf{q}}_{c+1} \leftarrow \widehat{\mathbf{q}}_{c+1} - \mathbf{J}(\widehat{\mathbf{q}}_{c+1})^{\dagger}\, \mathbf{f}(\mathbf{x}, \widehat{\mathbf{q}}_{c+1}), \tag{14}$$

where $\mathbf{J}(\widehat{\mathbf{q}}_{c+1})^{\dagger}$ denotes the Moore-Penrose pseudoinverse of the Jacobian. This process is repeated until convergence, i.e., until $\|\mathbf{J}(\widehat{\mathbf{q}}_{c+1})^{\dagger}\, \mathbf{f}(\mathbf{x}, \widehat{\mathbf{q}}_{c+1})\|$ is below a specified threshold ε. At convergence, $\widehat{\mathbf{q}}_{c+1}$ lies on the manifold, and $\mathbf{q}_{c+1}$ is set to this value.

4.2. Higher-dimensional self-motion manifolds

The continuation method can be generalized to compute higher-dimensional self-motion manifolds, as described by [8]. We have based our approach in the interpretation by [16]. The central idea is to build an atlas of local parameterizations (charts) that collectively cover the manifold. The main concepts and steps are outlined below, and illustrated in Fig. 5.

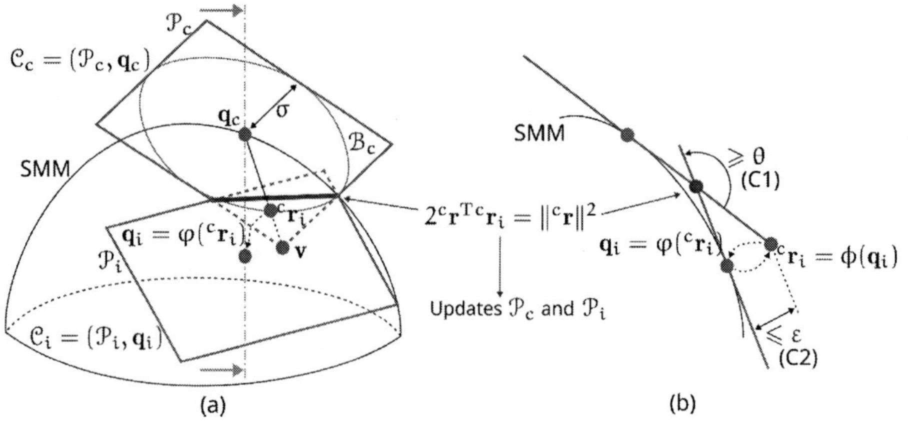

Fig. 5. Illustration of the continuation method for computing 2D self-motion manifolds. (b) is a cross-section defined by the plane represented in (a) in dash-and-dotted orange.

4.2.1. Charts

Given a point $\mathbf{q}_c$ on the manifold, a chart C_c provides a local parametrization in the neighbourhood of $\mathbf{q}_c$. This is achieved via a mapping $\varphi: \mathbb{R}^r \rightarrow \mathbb{R}^n$, which maps local r-dimensional coordinates in the chart to n-dimensional joint coordinates, with $\varphi(\mathbf{0}) = \mathbf{q}_c$.

Following [7] and in analogy to Section 4.1, the mapping $\mathbf{q}_i = \varphi_c(^c\mathbf{r}_i)$ is constructed in two steps:

1. Given a local coordinate vector $^c\mathbf{r}_i = [\rho_1, \dots, \rho_r]$ (expressed relative to chart C_c, denoted by superscript ^{c}), a candidate point $\widehat{\mathbf{q}}_i$ in the joint space is computed by stepping in the tangent space:

$$\widehat{\mathbf{q}}_i = \mathbf{q}_c + \mathbf{\Phi}_c\, ^c\mathbf{r}_i, \tag{15}$$

   where $\mathbf{\Phi}_c$ is an orthonormal basis for the tangent space of the manifold at $\mathbf{q}_c$.

2. The candidate $\widehat{\mathbf{q}}_i$ is then projected onto the manifold by solving

$$\mathbf{f}(\mathbf{x}, \widehat{\mathbf{q}}_i) = \mathbf{0} \tag{16}$$

   using the Newton-Raphson method as described in Section 4.1, yielding at convergence the corrected point $\mathbf{q}_i$.

Note that our approach differs slightly from the original method, which solves the following system of equations for correcting $\widehat{\mathbf{q}}_i$:

$$\begin{cases} \mathbf{f}(\mathbf{x}, \widehat{\mathbf{q}}_i) = \mathbf{0} \\ \mathbf{\Phi}_c^T(\mathbf{q}_i - \widehat{\mathbf{q}}_i) = 0 \end{cases} \tag{17}$$

This formulation ensures that the correction step projects orthogonally to the tangent space $\mathbf{\Phi}_c$ of the manifold at $\mathbf{q}_c$. In contrast, our approach (by employing the Moore-Penrose pseudoinverse, which minimizes the change in $\widehat{\mathbf{q}}_i$), updates the candidate configuration by moving it towards the manifold along a direction perpendicular to the manifold (as it is the least-norm solution) at the convergence point $\mathbf{q}_i$, i.e., perpendicular to $\mathbf{\Phi}_i$.

The inverse mapping $\phi_c: \mathbb{R}^n \to \mathbb{R}^r$, which recovers local chart coordinates from a joint configuration, is given by:

$$^c\mathbf{r}_i = \phi_c(\mathbf{q}_i) = \mathbf{\Phi}_c^T(\mathbf{q}_i - \mathbf{q}_c), \tag{18}$$

which projects the joint configuration $\mathbf{q}_i$ onto the tangent space $\mathbf{\Phi}_c$.

4.2.2. *Atlas of charts*

A atlas is a coordinated collection of charts that together cover the entire manifold. An atlas A is defined as a set of charts $A = \{C_1, C_2, \dots\}$, where each chart C_c is centred at a point $\mathbf{q}_c$ (i.e., $\varphi_c(\mathbf{0}) = \mathbf{q}_c$), as described in Section 3.2.1.

In [8]'s algorithm, each chart C_c is associated with an r-dimensional polytope $\mathcal{P}_c$ that defines its region of validity. This polytope is initially a hypercube enclosing an r-dimensional hyperball $\mathcal{B}_c$ of radius σ centered at $\mathbf{q}_c$, as illustrated in Fig. 5.

A chart is considered *bounded* only when every vertex of its polytope $\mathcal{P}_c$ lies within the hyperball $\mathcal{B}_c$. To bound the chart C_c, the algorithm iteratively selects a vertex $\mathbf{v}$ of $\mathcal{P}_c$ that is outside $\mathcal{B}_c$, and generates a local coordinate $^c\mathbf{r}_i$ in chart C_c as:

$$^c\mathbf{r}_i = \alpha\sigma\frac{\mathbf{v}}{\|\mathbf{v}\|}, \tag{19}$$

where α is a scaling factor (initially 1) ensuring $^c\mathbf{r}_i$ lies within $\mathcal{B}_c$. This local coordinate is mapped to joint space using φ_c, yielding $\mathbf{q}_i = \varphi_c(^c\mathbf{r}_i)$. The validity of $\mathbf{q}_i$ is checked by two conditions:

$$\|\mathbf{q}_i - \mathbf{q}_c + \mathbf{\Phi}_c\, ^c\mathbf{r}_i\| \le \varepsilon \qquad \text{(C1)} \tag{20}$$

$$\|\mathbf{\Phi}_c^T\mathbf{\Phi}_i\| \ge \cos\theta \qquad \text{(C2)} \tag{21}$$

Condition (C1) controls the deviation from the tangent space, while (C2) ensures the angle between the tangent spaces at C_c and C_i is not too large, which would indicate the need for more charts in regions of high curvature. If both conditions are satisfied, $\mathbf{q}_i$ is used to initialize a new chart C_i with its own local parameterization φ_c and tangent space $\mathbf{\Phi}_i$. If not, α is reduced by a user-defined factor β (e.g., $\beta = 0.8$), and the process repeats until a valid $\mathbf{q}_i$ is found or α reaches a minimum threshold. Starting with a larger α (i.e., 1) allows the algorithm to attempt larger charts before subdividing, improving coverage efficiency. When a new chart C_i is created, the algorithm coordinates neighbouring charts C_c and C_i by adding the following inequality constraint to $\mathcal{P}_c$:

$$2\,{}^c\mathbf{r}\,{}^c\mathbf{r}_i = \|\,{}^c\mathbf{r}_i\|^2 \tag{22}$$

This constraint restricts the validity region of C_c by cropping a half-space defined by the hyperplane orthogonal to ${}^c\mathbf{r}_i$ at ${}^c\mathbf{r}_i$. Fig. 5 shows the new chart C_i in red and the restricted region in blue. Importantly, this boundary constraint is applied to every chart in the atlas A for which $\mathbf{q}_i$ is a valid point, i.e., for every chart C_j where ${}^j\mathbf{r}_i = \phi_j(\mathbf{q}_i)$ is within $\mathcal{P}_j$. This iterative process continues until all vertices of every polytope $\mathcal{P}_c$ are contained within their respective hyperballs $\mathcal{B}_c$, at which point all charts are considered bounded and the atlas fully covers the self-motion manifold. Additional constraints, such as joint limits or obstacle avoidance, can be incorporated by introducing further conditions ((C3), (C4), etc.) in the same manner as (C1) and (C2), ensuring that all generated points $\mathbf{q}_i$ remain within the feasible region defined by these constraints.

5. Discussion

In this section, we compare the performance of the three methods for computing self-motion manifolds: the sweeping method, the continuation method, and the cellular automata method. Sweeping and continuation methods have been implemented in Python, on a machine with an Intel Core i7-9700F CPU and 32 GB of RAM. The cellular automata method has not been implemented by us, but we have used the results provided in [11] for comparison. The cellular automata (CA) method for computing self-motion manifolds, as proposed in [11], models the robot's configuration space (C-space) as a grid of high-dimensional elements. Instead of exhaustively sampling or clustering the entire C-space, the method uses a grid-based evolution search strategy inspired by cellular automata. Each grid element represents a continuous region of joint space, and its state evolves according to simple transition rules based on whether it contains part of the self-motion manifold. The computation begins by identifying a set of seed elements that intersect the manifold. These elements then propagate the search to their neighbours, iteratively expanding the set of elements identified as containing the manifold. This process continues until no new elements are found. For consistency and fairness in our comparison, we sought to visually match the resolution used in [11], as the original work did not specify the discretization parameters. In our implementations of both the sweeping method and the higher-dimensional continuation method, we carefully selected comparable resolutions. Specifically, the sweeping method was implemented by discretizing each axis of the redundant joints (q_4, q_5) into $N = 300$ points. The continuation method was executed using the following parameters: $\theta = \frac{\pi}{6}$, $\varepsilon = 0.1$, $\sigma = 0.3$. Fig. 6 illustrates the two-dimensional self-motion manifold computed by each method for the studied 5-DOF robot when the end-effector is at the position $\mathbf{x} = [0.2, 0, 0.3]$ m. Fig. 6(a) and Fig. 6(c) show the results of our implementations of the sweeping and continuation methods, respectively; while Fig. 6(d) shows the result of the CA method, directly taken from [11]. Note how the manifolds produced by the sweeping method in Fig. 6(a) present regions of discontinuities in the areas where the manifold is tangent to the horizontal lines defined by fixed values of the redundant joints (q_4, q_5). A solution to this problem is to increase the resolution, which produces the results presented in Fig. 6(b) when $N = 6000$.

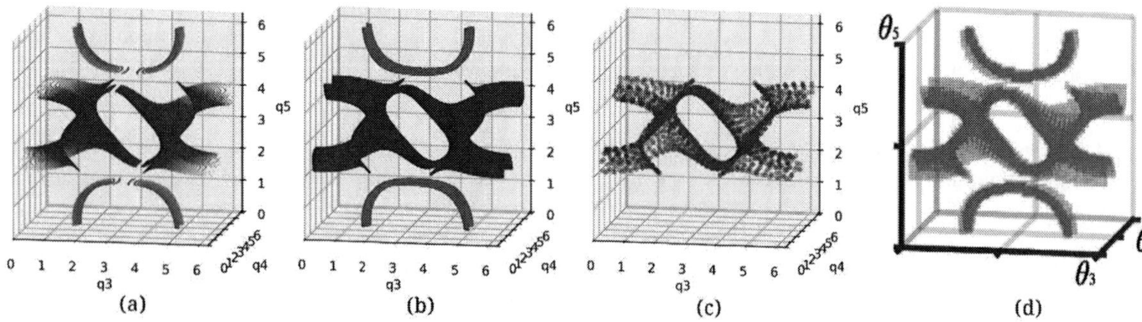

Fig. 6. Results of the different methods for computing SMMs: sweeping (a-b), continuation (c), cellular automata (d).

Table 2 shows the computation times for each method, where the CA results are directly taken from [11], and the sweeping and continuation times are averaged over 50 runs. The results of our comparative study reveal distinct strengths and limitations for each method of computing self-motion manifolds in the 5-DOF robot.

Method	Runtime (s)	Number of points
Sweeping [10]	0.46	14,4648
Sweeping [10] ($N = 6000$)	165.46	5.7×10^7
Higher-dimensional continuation [8]	11.9	2,227
Cellular automata [11]	11347	N/A

Table 2. Performance of the different methods for computing SMMs.

The sweeping method stands out as the fastest approach by a significant margin, with computation times orders of magnitude lower than the other methods. Moreover, it is able to identify all disjoint self-motion manifolds present in the configuration space. The primary limitation of the sweeping method is that it does not densely sample regions where the manifold is tangent to the horizontal lines defined by fixed values of the redundant joints (q_4, q_5). This can result in sparse point clouds in certain areas of the solution set. However, this drawback can be mitigated by increasing the resolution of the sweep or by employing redensification strategies, which we propose as a direction for future work. Another drawback is the requirement of deriving an analytical solution of the inverse kinematics, as done in Section 2.

The higher-dimensional continuation method offers a more uniform and dense coverage of the self-motion manifold. Unlike the sweeping method, it does not require an analytical solution to the inverse kinematics; it only relies on the Jacobian, which can be computed numerically if necessary. This flexibility comes at the cost of increased computational time, although it remains substantially faster than the cellular automata approach. A notable limitation of the continuation method is that it only explores the manifold connected to the initial seed configuration, and cannot identify all disjoint manifolds without multiple initializations.

The cellular automata (CA) method is the slowest by a wide margin, with computation times exceeding 10,000 seconds. While it is capable of identifying all disjoint self-motion manifolds, the computational cost is prohibitive for practical applications. The method's exhaustive grid-based search ensures completeness, but the trade-off in efficiency is substantial. We recommend its use when the identification of all disjoint manifolds is paramount, but no analytical solution to the inverse kinematics problem is available and computation speed is not critical.

6. Conclusion and Future Work

This paper has presented a comparative evaluation of three representative methods for computing two-dimensional self-motion manifolds in a 5-DOF redundant robot: the sweeping method, the higher-dimensional continuation method, and the cellular automata approach. Our results demonstrate that the sweeping method is the most efficient, providing rapid identification of all disjoint manifolds with minimal computational cost. The continuation method, while slower, offers dense coverage of the manifold and does not require an analytical inverse kinematic solution, making it suitable for complex robots where only the Jacobian is available. The cellular automata method, although exhaustive and complete, requires prohibitive computational times when compared to the other methods. Future work will focus on addressing the limitations identified in each method. For the sweeping approach, we plan to investigate redensification strategies to improve sampling density in regions where the manifold is tangent to the discretized horizontal lines. Additionally, we will explore new methods for computing self-motion manifolds that combine the strengths of the studied methods, while addressing their issues. Overall, our study provides guidance for researchers and practitioners in selecting appropriate tools for redundancy analysis and exploitation in industrial robotic systems, and lays the groundwork for further advances in efficient and comprehensive computation of self-motion manifolds.

7. Acknowledgments

Work supported by grant PRE2021-099226, funded by MCIN/AEI/10.13039/501100011033 and the ESF+, and by project PID2024-159765OA-I00, funded by the State Research Agency of the Spanish Government.

8. References

[1] Diek, D., Stuja, K., & Aburaia, M. (2022). Design and development of a flexible, cable-controlled and hyper redundant robot with Strength-Adjustable ball joints. In Proceedings of the 33rd DAAAM International Symposium (pp. 0024-0028).

[2] Flacco, F., De Luca, A., & Khatib, O. (2015). Control of redundant robots under hard joint constraints: Saturation in the null space. IEEE Transactions on Robotics, 31(3), 637-654.

[3] Faroni, M., Beschi, M., Pedrocchi, N., & Visioli, A. (2018). Predictive inverse kinematics for redundant manipulators with task scaling and kinematic constraints. IEEE Transactions on Robotics, 35(1), 278-285.

[4] Albu-Schäffer, A., & Sachtler, A. (2023). Redundancy resolution at position level. IEEE Transactions on Robotics, 39(6), 4240-4261.

[5] Fabregat-Jaén, M., Peidró, A., Colombo, M., Rocco, P., & Reinoso, Ó. (2025). Topological and spatial analysis of self-motion manifolds for global redundancy resolution in kinematically redundant robots. Mechanism and Machine Theory, 210, 106020.

[6] Wenger, P., Chedmail, P., & Reynier, F. (1993, May). A global analysis of following trajectories by redundant manipulators in the presence of obstacles. In [1993] Proceedings IEEE International Conference on Robotics and Automation (pp. 901-906). IEEE.

[7] Burdick, J. W. (1989, May). On the inverse kinematics of redundant manipulators: Characterization of the self-motion manifolds. In Advanced Robotics: 1989: Proceedings of the 4th International Conference on Advanced Robotics Columbus, Ohio, June 13–15, 1989 (pp. 25-34). Berlin, Heidelberg: Springer Berlin Heidelberg.

[8] Henderson, M. E. (2002). Multiple parameter continuation: Computing implicitly defined k-manifolds. International Journal of Bifurcation and Chaos, 12(03), 451-476.

[9] DeMers, D., & Kreutz-Delgado, K. (1994, May). Canonically parameterized families of inverse kinematic functions for redundant manipulators. In Proceedings of the 1994 IEEE International Conference on Robotics and Automation (pp. 1881-1886). IEEE.

[10] Peidro, A., Reinoso, O., Gil, A., Marín, J. M., & Paya, L. (2018). A method based on the vanishing of self-motion manifolds to determine the collision-free workspace of redundant robots. Mechanism and Machine Theory, 128, 84-109.

[11] Wu, T., Zhao, J., & Xie, B. (2023). A novel method for computing self-motion manifolds. Mechanism and Machine Theory, 179, 105121.

[12] Zhou, Z., Zhao, J., Zhang, Z., & Li, X. (2023). Motion planning method of redundant dual-chain manipulator with multiple constraints. Journal of Intelligent & Robotic Systems, 108(4), 69.

[13] Banfield, I., & Rodríguez, H. (2021, August). Generation of the self-motion manifolds of a functionally redundant robot using multi-objective optimization. In Climbing and Walking Robots Conference (pp. 438-452). Cham: Springer International Publishing.

[14] Yang, Y., Wu, Y., & Pan, J. (2021). An Interval Branch-and-Bound-Based Inverse Kinemetics Algorithm Towards Global Optimal Redundancy Resolution. arXiv preprint arXiv:2104.12183.

[15] Ester, M., Kriegel, H. P., Sander, J., & Xu, X. (1996, August). A density-based algorithm for discovering clusters in large spatial databases with noise. In kdd (Vol. 96, No. 34, pp. 226-231).

[16] Jaillet, L., & Porta, J. M. (2012). Path planning under kinematic constraints by rapidly exploring manifolds. IEEE Transactions on Robotics, 29(1), 105-117.

DOI: 10.2507/36th.daaam.proceedings.027

DESIGN OF A HANGING PARALLEL ROBOT FOR PLASTERING BRICK WALLS IN CONSTRUCTION

Jorge Francisco Aznar-Ortega, Adrián Peidró, Luis Payá & Óscar Reinoso

This Publication has to be referred as: Aznar Ortega, J[orge] F[rancisco]; Peidro, A[drian]; Paya, L[uis] & Reinoso, O[scar] (2025). Design of a Hanging Parallel Robot for Plastering Brick Walls in Construction, Proceedings of the 36th DAAAM International Symposium, pp.0201-0209, B. Katalinic (Ed.), Published by DAAAM International, ISBN 978-3-902734-47-1, ISSN 1726-9679, Vienna, Austria
DOI: 10.2507/36th.daaam.proceedings.027

Abstract

In the construction industry, plastering brick walls is necessary for structural and aesthetic reasons, as it provides smooth and even surfaces for decoration, protects against moisture and weather damage, and enhances the durability of the walls. Traditionally, plastering has been executed by human workers. When the wall is so tall that it spans several stories, it is necessary to use scaffolds, gondolas or ropes from which humans hang while they manually perform the plastering, which endangers their lives due to the risk of falling. To avoid this and automate the process, some plastering robots and machines have been proposed in the last years, but they have limitations such as limited vertical reach or the necessity of resting on horizontal surfaces at the bottom of the wall. To overcome these limitations, we propose a new robot consisting of a planar parallel mechanism that hangs from cables, increasing vertical reach and avoiding the need of resting horizontally. In this paper, we demonstrate the kinematic design, and the workspace and force analysis to guarantee that sufficient pressure is applied stably to the wall during plastering while cables remain taut.

Keywords: Wall plastering; parallel robot; cable robot; workspace; force analysis.

1. Introduction

Construction is still a sector with low levels of automation and is not free of hazards for operators [1], [2]. Many of the tasks in this area are labour-intensive, require precision and skill and consist of repetitive movements. In addition, the demanding conditions and unclean working environment have led to a decreasing number of skilled workers willing to engage in this field. Therefore, robots need to be developed to alleviate the human workload and compensate for the shortage of personnel in construction [3], [4], [5], [6]. One of the tasks commonly performed in construction is the plastering of vertical walls. This type of work is still mostly done manually. For small buildings, plastering can be done from the ground or small heights and does not pose a significant hazard for the labourers. However, for larger buildings with high walls, it is necessary for workers to hang from the facade or work from cranes or scaffolding, which entails a risk of falling. There are currently robots for plastering, both for research and commercial use. Some of them make use of technologies such as convolutional neural networks (CNN) to detect defects in plasterwork and correct them [7], [8]. However, most existing robots consist of a trolley with a retractable mast that extends vertically up the wall to perform the plastering [9], [10], [11].

This type of robot model has limitations that make it unsuitable for some cases, such as when the height of the surface to be plastered is greater than the range achievable by the robot's mast. Likewise, these robots need horizontal support to hold up the trolley, which is a problem when the lower part of the wall is not horizontal, but another adjacent roof or other sloping surface. In summary, despite the existence of some robots that can perform the plastering task, their use is limited to certain workplace conditions. For this reason, the design and creation of a new robot that can be used in a wider range of situations is beneficial to help alleviate the lack of workers and to avoid dangerous situations for operators and occupational accidents in the construction sector. The solution proposed in this article, which will be described in more detail in Section 2, consists of a hanging robot with a parallel robot on its gondola and an adhesion arm. This configuration overcomes the limitations of previous robots, such as limited vertical reach and the need to rest on a horizontal surface, among others. The remainder of this paper is organized as follows: Section 2 describes in detail the configuration of the robot, Section 3 presents its kinematic and workspace analysis. Next, Section 4 focuses on force analysis. Later, Section 5 explains the singularity analysis. Following, Section 6 analyses the results of these analyses and, finally, Section 7 draws conclusions and suggests future lines of work.

2. Description of the proposed robot for plastering

The wall plastering robot presented in this paper is shown in Figure 1. It consists of a gondola suspended by cables (C_1 and C_2) that moves vertically along the walls of buildings. The cable-driven system allows flexible positioning at different heights. Mounted on the gondola there is a parallel robotic mechanism (Figure 2) with three degrees of freedom (P_1, P_2 and P_3), responsible for adjusting both the location and orientation of the scraper (triangle formed by points A_1, A_2 and A_3). The use of a parallel structure offers high rigidity, which is important for achieving uniform plastering across the entire surface of the wall.

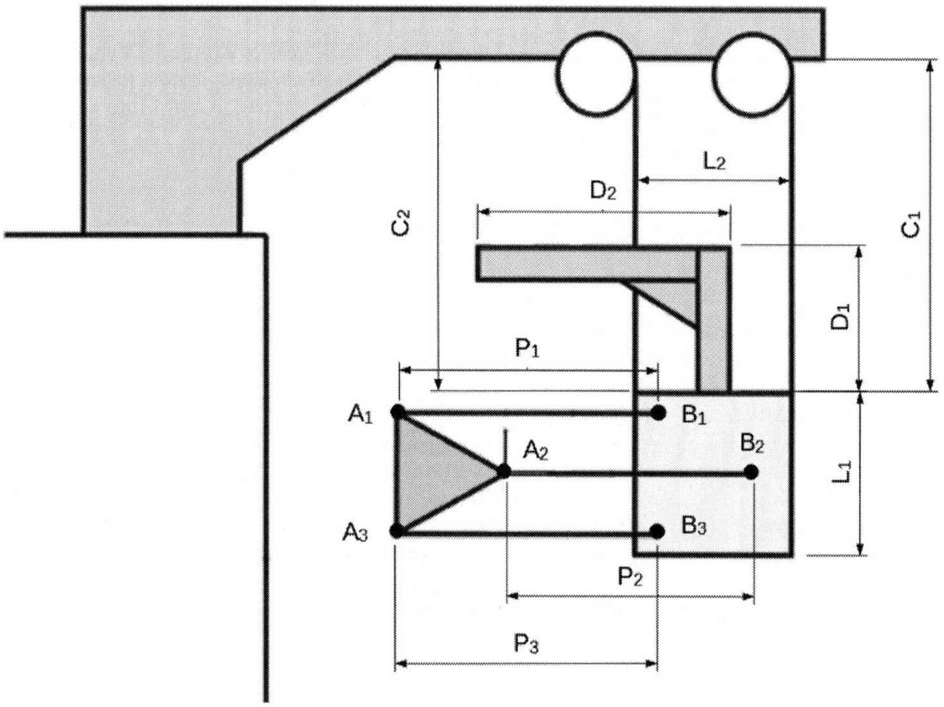

Fig. 1. Representation of the plastering robot

Since the support cables only control vertical movement and do not exert any force towards the wall, the reaction force generated during the application of the mortar could push the robot away from the surface. To solve this, the robot incorporates a wall adhesion mechanism, preferably by suction, which ensures that the robot remains in contact with the wall while operating (bars of length D_1 and D_2, where D_2 is extensive/retractile). This mechanism will maintain constant pressure during application, preventing the robot from malfunctioning. The robot's operating sequence begins with its placement at the desired height. Once there, the adhesion system is activated to secure the robot to the wall. When stable contact is established, the mortar applicator adjusts to the required position and orientation to begin the plastering process. The robot then covers the entire accessible surface from that height. Once the plastering in that area is complete, the adhesion mechanism disengages from the wall, and the robot repositions itself via the cables to a new area to continue the task. This sequence is repeated until the entire target surface is plastered.

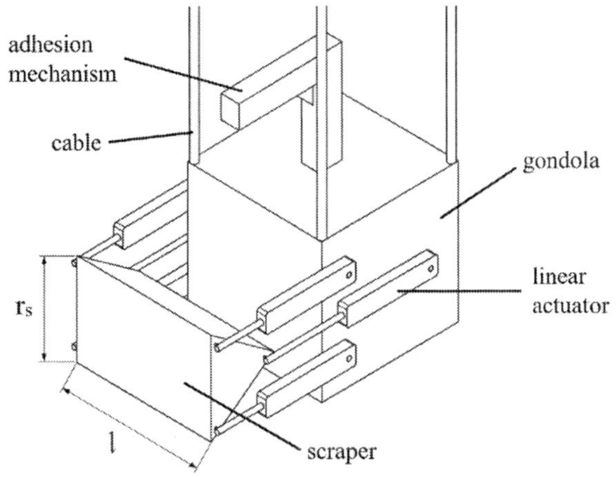

Fig. 2. Location of the robot's parallel mechanism

3. Kinematics and workspace

When defining the location of the robot's elements, three different reference systems have been used, as shown in Figure 3. The first of these is the fixed inertial system, which is defined by the X and Y axes. The origin of the second system is located on the gondola, with its axes being X' and Y'. Finally, the third reference system is fixed on the scraper, with axes X'' and Y'' and its orientation φ.

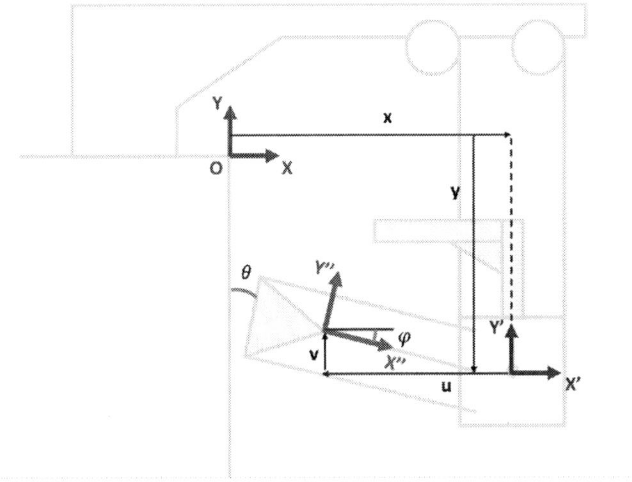

Fig. 3. Location of reference systems

For the robot presented in this paper, the kinematic input variables are the lengths of the prismatic actuators: P_1, P_2 and P_3. Likewise, the output variables (or platform pose) are the coordinates of the platform reference point (u, v) and its orientation relative to the carriage (φ). There are two types of kinematic calculations. First, the direct kinematics of the robot is based on determining the position and orientation of the end effector (scraper) from the known values of the actuator lengths. The solution is obtained by solving a sixth-degree polynomial, as explained in [12]. However, the calculation of direct kinematics will not be necessary for this work, so it will not be developed here. Secondly, inverse kinematics consists of calculating the prismatic lengths of the actuators based on a desired pose (position and orientation) of the end effector. This calculation is quite simple and is more relevant because it is the most useful in practice. To calculate inverse kinematics, the following data is used as a starting point:

- Fixed points on the gondola:

$$B_i = \begin{bmatrix} B_{i_x} \\ B_{i_y} \end{bmatrix} \quad i = 1, 2, 3$$

- Points of the end effector, expressed in its local system:

$$A_{i\_local} = \begin{bmatrix} A_{i_{x\_local}} \\ A_{i_{y\_local}} \end{bmatrix} \quad i = 1,2,3$$

- Platform pose:

$$p = \begin{bmatrix} u \\ v \end{bmatrix}, \qquad angle: \varphi$$

- Rotation matrix:

$$R(\theta) = \begin{bmatrix} \cos(\varphi) & -\sin(\varphi) \\ \sin(\varphi) & \cos(\varphi) \end{bmatrix}$$

Thus, the end effector points represented in the reference system of the gondola (X' Y') are obtained using the following expression:

$$A_{i\_global} = \begin{bmatrix} \cos(\varphi) & -\sin(\varphi) & u \\ \sin(\varphi) & \cos(\varphi) & v \\ 0 & 0 & 1 \end{bmatrix} \begin{bmatrix} A_{i_{x\_local}} \\ A_{i_{y\_local}} \\ 1 \end{bmatrix} \quad i = 1,2,3 \tag{1}$$

Given the start and end points of the prismatic actuators in gondola coordinates, the inverse kinematics is solved using this formula:

$$P_i = \| A_i - B_i \| \quad i = 1,2,3 \tag{2}$$

Finally, the calculation of the robot's range of motion is discussed. To do this, the constant orientation workspace is defined, which consists of the positions (x, y) that the scraper can reach for a fixed orientation and given minimum (P_{min}) and maximum (P_{max}) piston lengths ($P_{min} <= P_i <= P_{max}$). To perform this analysis, it is necessary to determine the range of orientations compatible with the plastering task, which requires incorporating the force analysis presented in the following section.

4. Static force analysis

To begin solving the static force analysis, the formulation described in [13] is adopted. Figure 4 illustrates the variables that will be used in (3), (4) and (5). First, the angle formed by the scraper and the wall is called θ. Likewise, the support force of the scraper on the mortar is represented by F. On the other hand, the compaction force of the mortar is indicated as N. Finally, the variables f_l and f_w are the friction force of the scraper on the mortar and of the wall on the mortar, respectively.

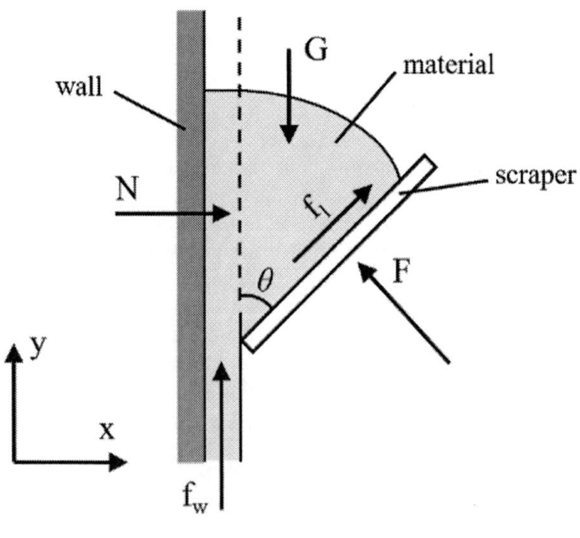

Fig. 4. Physical model for scraper plastering. Adapted from [13]

Therefore, the equations used to obtain the valid range of orientations for the plastering task are as follows:

$$G = \frac{1}{4} \cdot r_s^2 \cdot l \cdot \rho \cdot g \cdot sin(2\theta) \quad (3)$$

$$\left.\begin{aligned} F \cdot \cos(\theta) &= N + f_l \cdot \sin(\theta) \\ F \cdot sin(\theta) + f_w + f_l \cdot cos(\theta) &= G \end{aligned}\right\} \quad (4)$$

Since $f_w = \mu_w \cdot N$, $f_l = \mu_l \cdot F$, it can be derived:

$$N = \frac{G}{(1 - \mu_l \cdot \mu_w) \cdot sin(\theta) + (\mu_l + \mu_w) \cdot cos(\theta)} \cdot (cos(\theta) - \mu_l \cdot sin(\theta)) \quad (5)$$

The constants are the density of the mortar (ρ), the force of gravity (g), the coefficient of friction between the wall and the mortar (μ_w) and the coefficient of friction between the scraper and the mortar (μ_l). Also, the width (r_s) and length (l) of the scraper (Figure 2). If it is considered that the compaction force, N, must be greater than 100 N, an inequality that depends on the angle θ is obtained. By solving this inequality, the valid angles (θ) for the plastering task (Figure 5), the values of the forces N and F, as well as the friction forces f_l and f_w are obtained. It should be noted that the range of valid angles obtained is small (from 16° to 37°), but comparable in amplitude to that of [13], which was from 12° to 37°. This range is reasonable because it must be considered that the inclination of the scraper is limited by the conditions of the task, since it cannot be too horizontal or too vertical to perform the plastering.

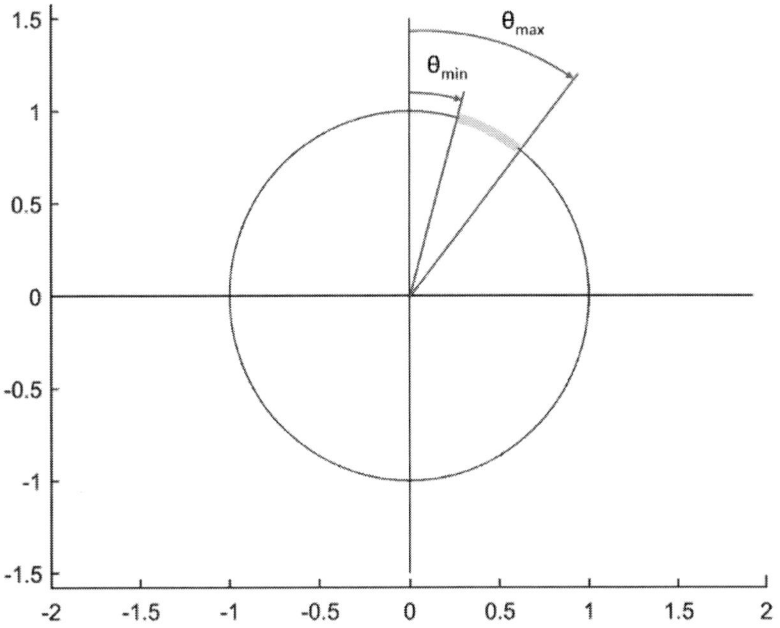

Fig. 5. Range of valid values for theta angle (θ)

Next, based on the data represented in Figure 6, the equations for the static equilibrium of the robot (6), (7) and (8) are established, considering all forces and moments. For the calculation of the moments, the reference point is indicated by the letter A. Known values include the adhesion force ($F_{adhesion}$), some distances and the weight of the gondola, which have been reasonably estimated. The rest of the variables involved (F_x, F_y, d_{Fx}, and d_{Fy}) depend on the position and orientation of the scraper, so their values will change according to these variables. Finally, the unknown variables are the reaction force ($F_{reaction}$) and the cable tensions (T_1 and T_2).

$$F_{reaction} = F_{adhesion} - F_x \quad (6)$$

$$P = T_1 + T_2 + F_y \quad (7)$$

$$M_1 + F_x \cdot d_{Fx} + T_2 \cdot d_{T2} = F_y \cdot d_{Fy} + T_1 \cdot d_{T1} \quad (8)$$

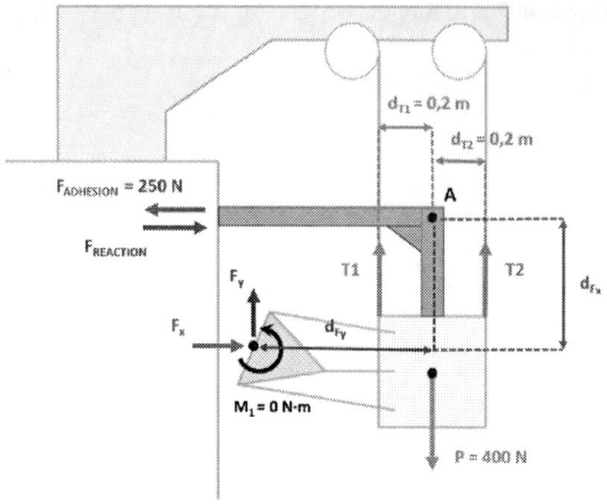

Fig. 6. Diagram of forces and moments on the robot

Solving the unknown variables in the above system of equations yields a table of results (one result for each angle θ) that meets the constraints. The values obtained must then be checked, as only those greater than zero are valid. This is because the cables can only withstand tensile forces and the reaction force can never be negative, as this would mean that it is exerting a force against the wall.

5. Singularity analysis

A singularity is a geometric configuration of a robot in which control over its movements or the forces it can transmit is lost. In parallel robots, these singularities occur when the geometry of the mechanism causes the end effector to move in a direction not controlled by the actuators, or when the actuators are unable to generate certain movements required of the end effector. Mathematically, singularities are detected when the Jacobian matrices that relate velocities/displacements between actuators and end effectors lose rank or, in other words, their determinant is zero or almost zero. To analyse the singularities of this robot, we will first determine the expression of its Jacobian matrix and then calculate the determinant for each of the valid points in the workspace. The Jacobian matrix that will be considered is formed by the derivatives of the right-hand side of (2) with respect to (u, v, φ). If there were sign changes between the points in the workspace, this would mean that those points would be singular. Otherwise, the robot would have no singularities and could be controlled at all points in the workspace.

6. Results

Based on the values of angle θ obtained in the previous section, an algorithm has been programmed in MATLAB that performs a sweep of the positions (x, y) of the end effector with a step of 0.01 between u = [-2, -0.5] and v = [-1.6, 1.6]. This range of values has been chosen because it covers the space between the wall and the robot, ensuring that the scraper does not come into contact with other objects. The representation of the point mesh that will be traversed by point A_2 of the scraper is shown in Figure 7.

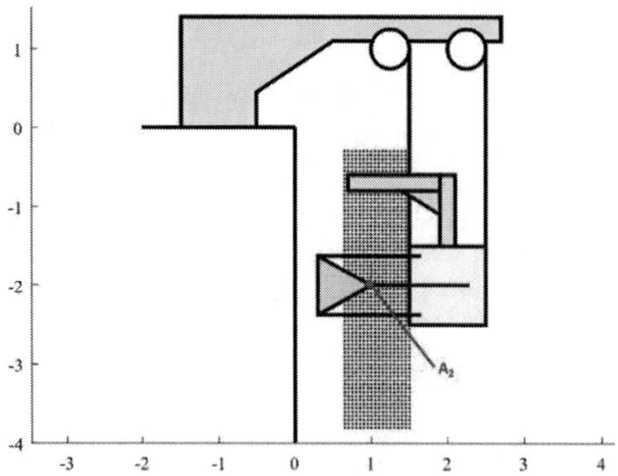

Fig. 7. Grid of points

For each point (x, y), the inverse kinematics is solved, as explained in Section 3, obtaining the lengths that the pistons should have in each case. For the initial design of the robot, commercial pistons with the following characteristics will be chosen: stroke of 20 cm, minimum length of 40 cm and maximum of 60 cm. Therefore, it must be verified that the lengths obtained through inverse kinematics are within these values. If the lengths of the three pistons are in the specified range, before qualifying the point as valid, a static check must be performed to determine whether the cable forces and the reaction force are positive. If all the above requirements are met, the point in the workspace is accepted as valid. The variation in the robot's workspace for some of the valid values of angle θ is shown in Figure 9, which spans the current page and the following one. These points have been obtained based on the midpoint of segment A_1 - A_3 of the scraper because it is where the force F is applied (Figure 4), so it simplifies the calculation of static equilibrium. To have a reference for the workspace with respect to the robot, Figure 8 represents the same workspace as in Figure 9a, but with the robot superimposed.

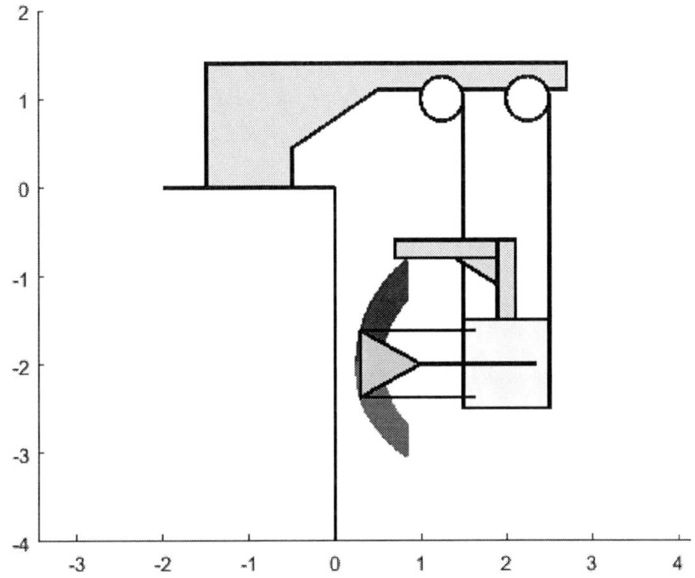

Fig. 8. Workspace and the plastering robot

Of all the points in the initial mesh (Figure 7), only those shown in Figure 9 are within the range of lengths of both actuators. Furthermore, of those points, only the blue and green ones also comply with the static equilibrium constraints. It can be seen in the sequence of workspaces that, as the angle θ increases, the number of reachable points decreases and, among those that are reachable, there are increasingly those that do not meet the conditions of static equilibrium: points at which the tension forces in the cables or the reaction in the attachment arm become negative (red points). Once the workspaces have been displayed, the force values obtained in the static equilibrium analysis ($F_{reaction}$, T_1 and T_2) will now be discussed. Since the value of these three variables is known at each point in the workspace, the intensity of the magnitude of these variables can be represented on the points of the workspace. Nevertheless, the values for the reaction force and tension T_1 will not be shown because their minimum values are 60 N and 270 N respectively, so they do not represent a stability problem for this robot. On the other hand, the T_2 tension does have values close to 0, being the most critical force. In fact, at the boundary between the points that do and do not satisfy the static equilibrium of forces (the boundary between the red and blue points), the tension values in the right cable are practically zero, indicating that it is this variable that prevents there being more valid points in the workspace. The magnitude, in Newtons, of the tension T_2 is expressed as a vertical-coloured bar in each of the subfigures in Figure 9.

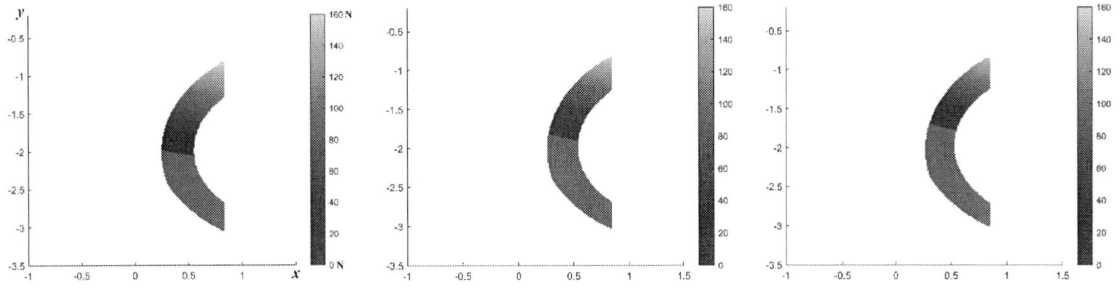

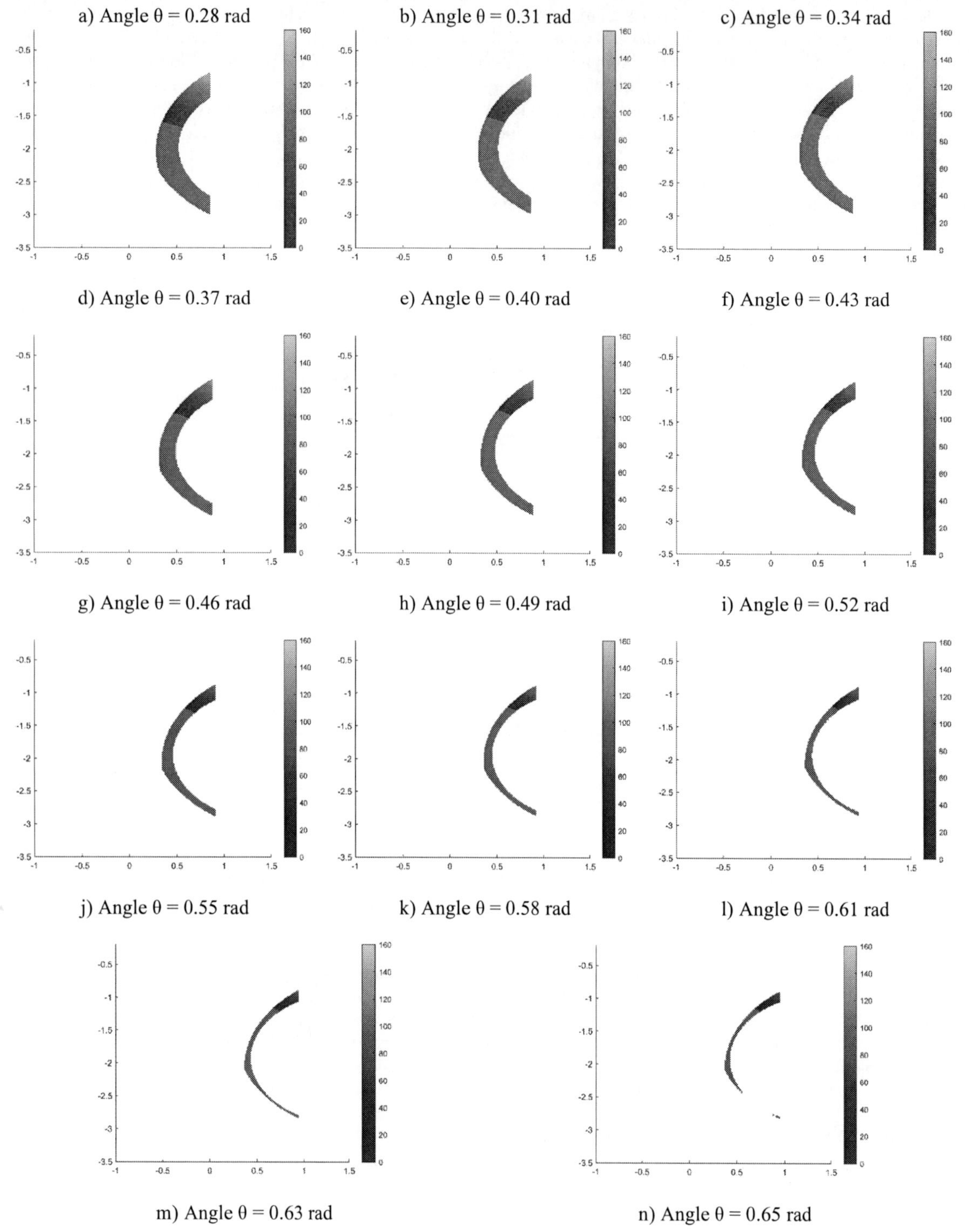

Fig. 9. Variation of tension T_2 on the workspace

Finally, the results of the singularity analysis are discussed. The magnitude of the value of the determinant of the robot's Jacobian matrix at all points in the workspace is always of the same sign (negative), which implies that there are no singular points for the studied orientations. This indicates that the design presented in this work avoids the appearance of singularities and, therefore, the robot will be able to move throughout its workspace without having to evade any area in which control may be lost due to singularities.

7. Conclusion

This paper presented the initial design and analysis of a new robot for plastering walls in the construction sector, an activity which, despite its apparent simplicity, involves significant risks for operators and presents technical challenges that have not yet been satisfactorily resolved by existing commercial solutions. The structural approach of this robot is different from that used by most current machines, which will allow it to be used effectively in environments where existing systems were not practical: walls too tall and walls without stable horizontal support at their bottom. The analyses carried out in the previous sections have validated the feasibility of the proposed design for the plastering task and provide information that will be used to improve the current design of the robot. The results have shown that the robot can operate within an adequate workspace, without any singular point in it, and guarantee positive tension in the cables, which are essential conditions for its safe operation. Future work will focus on refining the design and validating its performance in real conditions. More specifically, the next steps include designing an algorithm to calculate the optimal sequence of robot movements, improving the robot's design by optimizing the arrangement of joints, resizing the end effector, and modifying the geometry of certain parts, as well as constructing a prototype to carry out the necessary tests and verify the robot's performance in a controlled environment to ensure that it meets the desired specifications and performs the task correctly.

8. Acknowledgments

This work was supported by project PID2024-159765OA-I00, funded by State Research Agency of the Spanish Government.

9. References

[1] Aghimien, D. O., Aigbavboa, C. O., Oke, A. E., & Thwala, W. D. (2020). Mapping out research focus for robotics and automation research in construction-related studies: A bibliometric approach. Journal of Engineering, Design and Technology, 18(5), 1063–1079. https://doi.org/10.1108/JEDT-09-2019-0237

[2] Gharbia, M., Chang-Richards, A., Lu, Y., Zhong, R. Y., & Li, H. (2020). Robotic technologies for on-site building construction: A systematic review. Journal of Building Engineering, 32, 101584. https://doi.org/https://doi.org/10.1016/j.jobe.2020.101584

[3] Akinlolu, M., Haupt, T. C., Edwards, D. J., & Simpeh, F. (2022). A bibliometric review of the status and emerging research trends in construction safety management technologies. International Journal of Construction Management, 22(14), 2699–2711. https://doi.org/10.1080/15623599.2020.1819584

[4] Liu, Y., A.H., A., Haron, N. A., N.A., B., & Wang, H. (2024). Robotics in the Construction Sector: Trends, Advances, and Challenges. Journal of Intelligent & Robotic Systems, 110(2), 72. https://doi.org/10.1007/s10846-024-02104-4

[5] Kim, M. J., Chi, H. L., Wang, X., & Ding, L. (2015). Automation and Robotics in Construction and Civil Engineering. Journal of Intelligent & Robotic Systems, 79(3-4), 347. https://doi.org/10.1007/s10846-015-0252-9

[6] Pajaziti, A., Stuja, K., Gjelaj, A., & Calvin, L. (2023). Design and Optimization of the Welding Robot System for Aluminium Frames using Simulation Software, Proceedings of the 34th DAAAM International Symposium, pp.0040-0049, B. Katalinic (Ed.), Published by DAAAM International. ISBN 978-3-902734-41-9, ISSN 1726-9679, Vienna, Austria. DOI: 10.2507/34th.daaam.proceedings.006

[7] Liu, Z., Chen, D., Eldosoky, M., Ye, Z., Jiang, X., Liu, Y., & Ge, S. (2024). Puttybot: A sensorized robot for autonomous putty plastering. Journal of Field Robotics, 41, 1744–1764. https://doi.org/10.1002/rob.22351

[8] Bard, J., Bidgoli, A., & Chi, W. W. (2019). Image Classification for Robotic Plastering with Convolutional Neural Network. In J. Willmann, P. Block, M. Hutter, K. Byrne, & T. Schork (Eds.), Robotic Fabrication in Architecture, Art and Design 2018 (pp. 3–15). Springer International Publishing.

[9] Forsberg, J., Graff, D., & Wernersson, Å. (1995). An Autonomous Plastering Robot for Walls and Ceilings. IFAC Proceedings Volumes, 28(11), 301–306. https://doi.org/https://doi.org/10.1016/S1474-6670(17)46989-8

[10] Liu, G., Li, Z., Zhang, J., & Bao, H. (2024). A mortar scraping robot for a whole wall in construction. 2024 IEEE International Conference on Mechatronics and Automation (ICMA), 1688–1693.

[11] Yadav, B., Mandal, N., Thakur, A., Jaiswal, B., Yadav, S., Pandit, S., Yadav, R., Shrestha, A., & Risal, S. (2020). International Journal of Advanced Engineering Modelling and Fabrication of Automatic Wall Plastering Machine.

[12] Gosselin, C. M., & Sefrioui, J. (1991, June). Polynomial solutions for the direct kinematic problem of planar three-degree-of-freedom parallel manipulators. In Fifth International Conference on Advanced Robotics' Robots in Unstructured Environments (pp. 1124-1129). IEEE.

[13] Wang, Y., Xie, L., Chen, J., Chen, M., Hu, T., Liao, H., Sun, S., & Chen, J. (2024). Mortar spraying and plastering integrated robot for wall construction. Automation in Construction, 165, 105533.

DOI: 10.2507/36th.daaam.proceedings.028

FROM UNBALANCE TO PRECISION: DEVELOPMENT OF A SENSOR-BASED BALANCING CONCEPT FOR GRINDING TECHNOLOGY

Markus Schneider[1], Manfred Bobertag[2], Jan Glück[2],
Viktoria Hendricks[1] & Justus Kreißl[1]

[1] University of Applied Sciences Ruhr West
[2] PMB – Präzisionsmaschinenbau Bobertag GmbH

This Publication has to be referred as: Schneider, M[arkus]; Bobertag, M[anfred]; Glück, J[an]; Hendricks, V[iktoria] & Kreißl, J[ustus] (2025). From Unbalance to Precision: Development of a Sensor-Based Balancing Concept for Grinding Technology, Proceedings of the 36th DAAAM International Symposium, pp.0210-0216, B. Katalinic (Ed.), Published by DAAAM International, ISBN 978-3-902734-47-1, ISSN 1726-9679, Vienna, Austria
DOI: 10.2507/36th.daaam.proceedings.028

Abstract

This report examines the development and testing of a measuring system for analysing the imbalance of grinding wheels. The system is based on a microcontroller in combination with MEMS-based inertial technology in form of acceleration sensors. The sensor technology is integrated into a custom-designed testing device whose geometry corresponds to that of a conventional grinding wheel. Due to the identical geometric dimensions, it can be mounted on the grinding wheel flange of a standard grinding machine, thus avoiding the need for external balancing outside the machine. The experimental investigations show that vibrations can be recorded reproducibly and reliably. By applying the Fast Fourier Transform for frequency analysis, changes in vibration behaviour as a result of targeted relocation of balancing weights can be clearly identified. Challenges arose in particular in the form of signal losses and limited measuring ranges of the sensors used during data acquisition. Solutions to these problems are being designed and discussed as part of the investigations. Based on the results obtained, an extended system concept for process monitoring can be developed.

Keywords: surface grinding; condition monitoring; data-driven process optimization; grinding wheel balancing; vibration-based process monitoring.

1. Introduction

Grinding is a key manufacturing process in production and is used in numerous industrial sectors. It is used to machine workpieces in the semiconductor industry, aerospace and the automotive industry for example and is generally positioned as the final process stage in machining production. At this point, the workpiece already contains a considerable amount of the added value previously generated by processes such as turning or milling.

Errors in this final process stage can therefore lead to significant economic losses. For this reason, grinding technology plays a key role in ensuring the dimensional accuracy and final quality of the product. The high precision and the demand for a consistently high surface quality of the workpiece mean that process stability and repeatability play a key role. At the same time, the increasing use of extremely hard machining materials in grinding technology and the stress on the tools require continuous monitoring of the tool. Disturbance variables in the form of externally or internally stimulated machine oscillations have a negative impact on the grinding process and thus on the grinding result. In particular, imbalances in the grinding wheel caused by externally stimulated oscillations result in increased waviness and roughness of the ground workpiece and should be minimised as far as possible. These imbalances are caused during the production of the grinding wheel by unavoidable irregularities in the structure or by uneven wear of the grinding coating during the machining process. [1],[2]

In addition, imbalances also contribute to additional stress on individual machine components such as the spindle bearings and the tool itself, which ultimately results in increased wear on the entire machining centre and reduced tool life. These effects lead to an inefficient grinding process. In addition, the oscillations caused by imbalances also generate increased noise emissions, placing additional strain on the machine operator. [3] For the reasons mentioned above, precise balancing of the grinding wheel before process and in the event of high grinding wheel wear is essential and is considered an integral part of process reliability in grinding operations. This article discusses the development of a measuring system for analysing imbalance in grinding technology. The aim of the measuring system is to continuously record and store sensor data during the rotation process. Based on this data, the extent to which the developed system is suitable for reliable detection of imbalances is investigated.

2. Principles of balancing

A basic distinction is made between manual and automatic balancing: while manual balancing is carried out by manually moving balancing weights inside or outside the machine on external peripherals, automatic balancing is based on the machine-side, automatic positioning of weights or liquids to compensate for the imbalance. Balancing is also divided into static or dynamic balancing principles, depending on the type of imbalance. While static imbalance, which only occurs in one plane, can be measured at standstill or during rotation, dynamic imbalance can only be measured and corrected during rotation due to its occurrence in two planes. If a manual, static balancing principle is used on a balancing stand for balancing tasks with lower precision, the process conditions mean that a considerable amount of time and the associated machine downtime as well as errors due to manual operation must be taken into account. Overall, this leads to lower surface quality of the workpiece, increased tool deformation and thus reduced tool life. As a result of the process conditions involved in balancing, automatic dynamic balancing (during rotation) offers high accuracy and reproducibility as well as low time expenditure compared to manual static balancing (at standstill). The industry standard for precise grinding processes at high speeds is therefore dynamic balancing of the grinding wheels. However, high investment and retrofitting costs must be taken into account. The grinding process is therefore massively influenced by the selected balancing principle. [4], [5], [6]. Industrial manufacturing currently faces the challenge of meeting the growing demand for high-precision machining processes while facing an increasing shortage of skilled workers. The integration of intelligent, automated systems that do not require operating personnel is therefore becoming increasingly important in this context. The balancing concept presented in this publication creates the conditions for such a system.

3. State of research on process monitoring

Targeted process monitoring enables early detection of imbalances by identifying oscillations and can promote adaptive process control by linking it to grinding forces, grinding performance and other relevant process data. The first step in implementing such a modern process monitoring system in grinding technology is to integrate suitable sensor technology.

3.1. Use of sensors for wear detection

Monitoring the general condition of the grinding wheel is crucial for process stability and workpiece quality. Sensor principles for acoustic emissions (AE) and oscillations are often used for this purpose. Researchers at Cardiff University used AE sensors attached to the workpiece surface to monitor the condition of a grinding pin. The aim was to develop an indirect, retrofittable tool monitoring system that could be used in particular for machining high-performance materials such as ceramics or glass. In the study, four grinding tests were carried out with identical process parameters, in each case using a new tool that was used until it was completely worn out. Optical wear measurements were taken after each grinding pass. The results show the development of a typical three-phase characteristic with surface craters, which lead to significant increases in concentricity and form deviation. It was found that the amplitudes of the AE signal show high sensitivity to irregularities in the grinding contact and can therefore be used as an indicator of tool wear. The indicator correlates strongly with the optically measured tool errors. AE kurtosis is therefore considered a useful measurement parameter for evaluating tool wear and grinding quality. [6] Another method of monitoring the wear of grinding tools was investigated by Hassui et al. In addition to measuring the AE signals, an acceleration sensor was also used to detect vibrations.

To evaluate the condition of the grinding wheel, the surface roughness, roundness and removed volume of the workpiece are recorded in addition to the sensor signals. The results show that roughness increases continuously as the grinding wheel wears. The main cause of this is the loss of grain sharpness, which leads to higher friction and cutting forces and, in turn, manifests itself in increased vibration. The vibration curves recorded by the measurement technology confirm a correlation between vibration intensity and the volume of material removed. It should be noted that the use of an absolute vibration limit value alone can only be used to a limited extent to determine the state of wear due to the influence of threshold value selection and feed rate. Vibration data should therefore only be used in combination with other measured variables to evaluate the state of wear of the grinding wheel. [7] In order to obtain a comprehensive overview of the grinding wheel condition and the grinding process, the measured values recorded by the AE sensor can be supplemented with data from an additional force measurement. The data provides further information about the grinding wheel wear and the surface quality of the workpiece. [8]

3.2. Use of sensors for oscillation measurement

Jiménez et al. developed a method for measuring vibrations on machines in which MEMS acceleration sensors were installed inside a hollow rotor to detect oscillations in the rotating reference system. The study shows that these system-internal sensors can function as a virtual displacement sensor mounted on the stator by means of appropriate signal processing. The internal integration of the sensor technology enabled active vibration control, which traditional stator-mounted sensor arrangements could not offer. [9] Further investigations into the behaviour of rotor-mounted acceleration sensors were carried out by Feng et al. Condition monitoring was performed using an acceleration sensor mounted on the flywheel of a piston compressor. The development of a mathematical model showed how gravitational acceleration, which has a disruptive effect on the measurement signals, can be effectively eliminated from the measured values. As a result, defects such as leaks in outlet valves could be identified and classified. The special feature of this research is the number of sensors required. While statically mounted sensor concepts require several sensors to obtain and interpret a meaningful signal, rotor-mounted concepts require only a single acceleration sensor. [10]

4. Systematic development of a balancing concept for rotating grinding tools

Acceleration is a key physical parameter for assessing oscillation intensity. Due to the periodically generated oscillations, repetitive frequencies are also analysed. This allows conclusions to be drawn about the magnitude and position of the imbalance. The phase angle between the measured oscillation and the rotational movement of the grinding wheel provides further crucial information for the precise localisation of the imbalance. The interpretation of the measured data is supported by the speed information of the grinding spindle, which is a crucial measurement parameter due to its direct correlation with the oscillation frequency.

4.1. Implementation of sensor technology and system design

In contrast to conventional methods for measuring unbalance, the acceleration sensors in this test setup are mounted directly on the flange of the grinding wheel. This allows the sensors and the grinding wheel to rotate together in a rotating reference system. This approach makes it possible to measure oscillations and accelerations directly at the source without them being distorted by transmission effects via bearings or machine structures. The sensors are not only exposed to the oscillations due to unbalances, but also to the permanently acting force of gravity, which generates an oscillating component in the measurement due to the rotation.

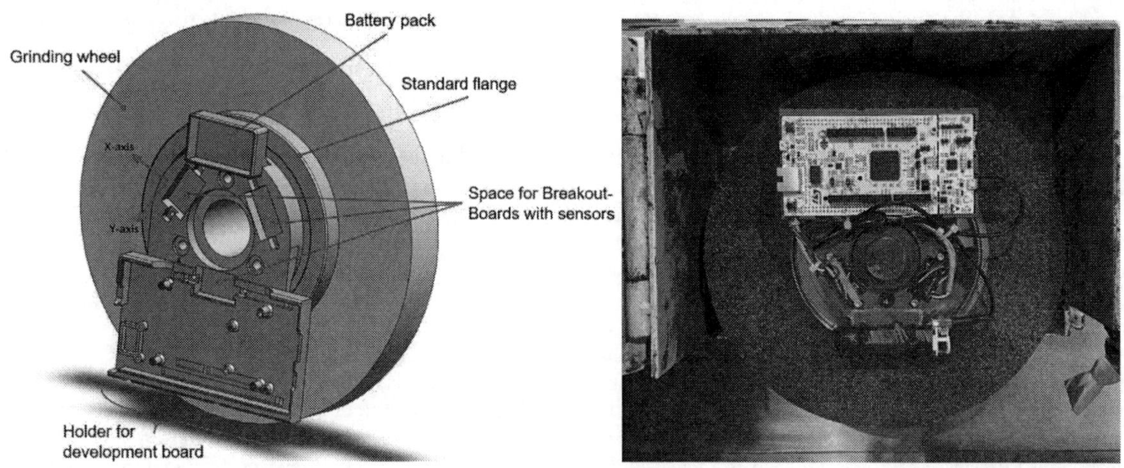

Fig. 1. Left: Base unit with sensor pockets and hardware attachments, right: testing model clamped in place

As the sensor moves continuously around the axis of rotation, there is a constant change in the magnitude of the measured gravitational force, resulting in a sinusoidal gravitational signal. This must be taken into account during data processing. The effect described is particularly relevant at low speeds, where the influence of gravity in relation to rotational acceleration is not negligible. In the developed test model (Fig. 1, left), three sensors are therefore used to calculate and compensate for the influence of gravity if necessary. The testing model consists of an additively manufactured base unit, breakout boards including the sensors mounted on them for measuring data acquisition, and suitable hardware, which is attached to the base body together with the cabling. The hardware includes a development board and a battery for powering the electronics. To ensure precise positioning of the three acceleration sensors, three mounting pockets are attached to the surface of the testing model at intervals of 120° in a circular pattern. These pockets ensure precise and secure attachment of the sensors. Furthermore, a mounting option is provided for the breakout boards of the sensors, which are also enclosed in a pocket and additionally secured with screws. The axis alignment can be seen in Figure 1, left. Here, the Y-axis represents the tangential plane and the X-axis represents the radial plane. The components breakout boards with sensors, development board and power supply could then be attached to the base of the laboratory sample (Fig. 1, right).

4.2. Data flow and data security

With regard to data acquisition, transmission and security, relevant requirements are placed on the system, which are considered below. Data management must be carried out via a real-time operating system to ensure continuous monitoring in real time. This enables immediate identification of deviations and anomalies in the process and in the machine status, thus ensuring an efficient machining process in compliance with the applicable safety and quality requirements. The data collected by the sensors is stored on an SD card with a sampling rate of 7.68 kHz per sensor and axis and a digital resolution of 16 bits. Simultaneous data acquisition from all sensors must be taken into account in order to rule out possible synchronisation problems. The time difference between the measurements of the individual sensors must not exceed a few microseconds. To ensure the simultaneous connection of all sensors, at least three parallel SPI or I2C interfaces are required. Optionally, interfaces for direct data transfer to the PC via USB and for outputting debug information via UART (Universal Asynchronous Receiver/Transmitter) should be available. The power supply should be stabilised and allow charging via conventional connectors such as USB. During the grinding process, very high temperatures may arise. Since the temperature and its changes have a direct influence on the function, service life and safety of individual machine components and the entire machine system, it is essential that the components are designed to be temperature-resistant. The developed balancing system must therefore be designed for an operating temperature range of at least -20 °C to +85 °C. In addition, resistance to cooling lubricants is absolutely essential. The breakout board in particular must be designed to withstand the environmental influences of cooling lubricants and contamination from grinding particles.

4.3. Experimental results on the effectiveness of the balancing concept

To analyse the effectiveness of the developed balancing concept, the test sample was tested on an FS 840 Z CB surface grinding machine from Geibel & Hotz GmbH. Due to the maximum measuring range of the sensor of ±16 g, radial acceleration can only be recorded up to approx. 600 rpm. The radial acceleration a_r corresponds to the centripetal acceleration generated towards the centre of rotation (1). The sensors move at an angular velocity ω on a circular path with radius r.

$$a_r = -r * \omega^2 \tag{1}$$

Due to the grinding wheel mounting, the distance between the sensors and the axis of rotation cannot be reduced any further. However, in order to avoid inaccuracies, the tests are not carried out at the maximum measuring range of the sensors, but at a speed of 500 rpm. In order to be able to measure acceleration even at higher speeds, sensors with a measuring range of ± 320 g are to be used in further prototypes of the test model. In order to visualise the influence of gravity, a manually generated rotation of the system was first recorded. Data acquisition begins when the power supply is switched on and is started by activating the user button on the development board. The measurement is stopped after one or two revolutions. The acceleration values in the radial and tangential directions are shown in Figure 2. This shows the respective angular position of the sensor axes and the sinusoidal oscillation caused by gravity during rotation. The relationship between angular position and acceleration during manual rotation, where centripetal acceleration is negligible, is described by equation (2), where θ describes the phase angle of the rotor and τ describes the angular offset of the individual sensors [11]. In this case, the angular offset is 0°, 120° or 240°, depending on the sensor position.

$$\begin{bmatrix} a_x \\ a_y \end{bmatrix} = \begin{bmatrix} g * \cos(\theta + \tau) \\ g * \sin(\theta + \tau) \end{bmatrix} (m/s^2) \tag{2}$$

For a further series of tests, the system is first balanced manually on a balancing machine. To compensate for the mass of the development board, a brass plate is attached to the rotor. Balancing is carried out using the sliding weights on the flange. After balancing, the system is mounted on the grinding machine.

The circumferential speed is set so that a spindle speed of 500 rpm is achieved. Two measurements are then taken at the same speed for 20 seconds each. During the second measurement, the brass plate is removed in order to deliberately create the imbalance. The acceleration values are then evaluated and transferred to the frequency range using a Fast Fourier Transform (FFT) (Fig. 3).

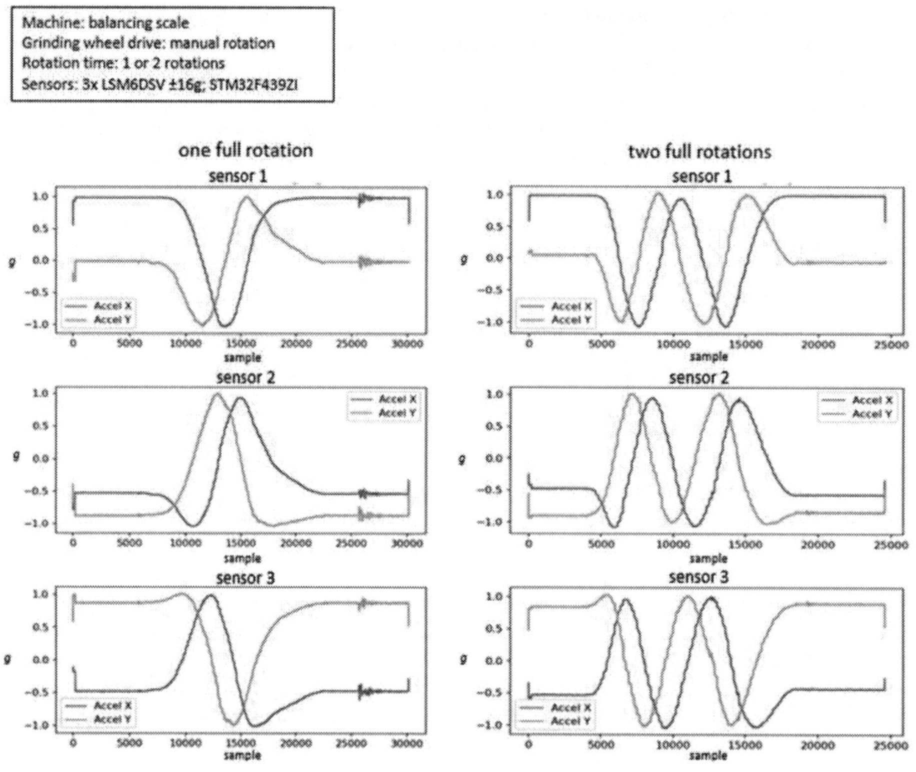

Fig. 2. Acceleration values during manual rotation

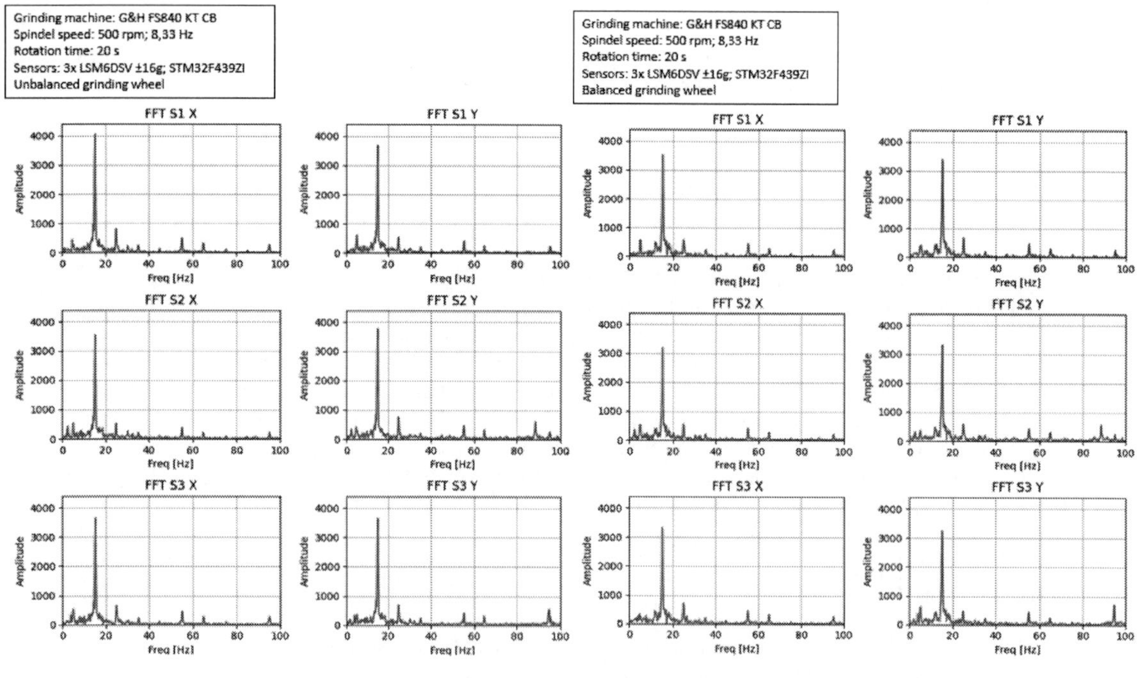

Fig. 3. Left: minimal imbalance, right: maximum imbalance

The speed of 500 rpm results in an angular frequency of 8.33 Hz. The base frequency of the system should be set to this value. On the one hand, the frequency could not be mapped from the measurement data, which is why it can be assumed that it was lost in the system noise. On the other hand, amplitude deflections can be found at multiples of the nominal speed, occurring at approximately 16 Hz and 24 Hz. These values are twice and three times the fundamental frequency of 8.33 Hz. Signal filtering should be tested to identify the fundamental frequency of the system. The deviation in the amplitude of the three sensors within a measurement also indicates that the sensors are not calibrated, resulting in an offset that should be taken into account in future. At a frequency of 16 Hz, clear differences in amplitude can be seen between balanced and unbalanced states. In the X and Y directions, the amplitudes of the unbalanced system are approximately 16% higher than those of the balanced rotor. This verifies the basic functionality of the developed balancing system for detecting imbalance.

5. Research contribution in the context of industrial requirements and trends

The relevance of a precise balancing system in grinding technology has already been mentioned in this article. The balancing system developed in this work was therefore designed to be highly flexible, enabling industrial application in the context of modern, sensor-based production and measurement systems. The high degree of flexibility in integrating the system into conventional grinding machines makes the concept an innovative solution for grinding technology. In addition to identifying the magnitude and position of the imbalance based on a prior definition of limit values, the recorded signals can also be used to draw conclusions about the wear condition of the grinding wheel. In combination with Internet of Things (IoT)-enabled data management, it is possible to document the machining history and condition of individual grinding wheels. Overall, this not only opens up the potential for reducing non-productive times, but also for permanent process monitoring. The primary aim of the measurement system developed is to improve process results and increase operational reliability. With a share of around 43% of machining processes in industry, grinding is the largest area alongside turning, milling and drilling, which account for around 23%, 20% and 14% respectively [12]. Innovations in the field of grinding technology therefore have a significant influence and relevance in industry.

6. Conclusion

The system described in this article shows that dynamic measurement of grinding wheel imbalance is possible using the innovative flange system. The tests carried out prove that differences in oscillation – especially in the tangential direction – can be reliably recorded and analysed. FFT analysis can be used to clearly identify changes in oscillation behaviour. However, various challenges arose during the course of the experiment. In particular, plug connections in the signal line led to interruptions in data transmission, which is why greater focus should be placed on soldered and shielded connections in future. In addition, it was found that the radial accelerations occurring during operation reach the limits of the maximum possible sensor measuring range. This means that future prototypes will need to be equipped with sensors with a larger measuring range capacity or with an optimised installation position for the sensor technology in order to reduce the distance between the sensor and the axis of rotation. This article builds on existing research results and applies the concept for the first time to a practical application directly on the grinding wheel under real machine conditions. The successful integration of MEMS sensor technology into the rotating structure demonstrated the basic feasibility of such a system. This lays an essential foundation for further developments towards a series-ready, integrated process monitoring system. The system developed for measuring imbalance can be expanded by integrating additional sensors, including those for measuring force and power. Overall, this enables the development of a comprehensive process monitoring system with adaptive process control, which allows early fault detection in real time, ensuring high machining quality and process stability.

7. Acknowledgment

This work is a research result within the Central Innovation Programme for SMEs (ZIM) project "IBG Intelligent-Balancing-Grinding" in cooperation between PMB - Präzisionsmaschinenbau Bobertag GmbH and the Institute of Mechanical Engineering at University of Applied Sciences Ruhr West.

8. References

[1] F. Klocke (2018), Zerspanung mit geometrisch unbestimmter Schneide,engl. Machining with geometrically undefined cutting edge, 6th ed. Berlin: Springer Vieweg.

[2] M. Ahrens, M. Dagen, B. Denkena & T. Ortmaier (2016), "An Active Damping Method for Chatter Vibration in Plunge Grinding Using Electromagnetic Actuators," Procedia CIRP, vol. 46, pp. 197–200, doi: 10.1016/j.procir.2016.03.200.

[3] PMB - Präzisionsmaschinenbau Bobertag GmbH (2025), Spindeln auswuchten für Drehzahlsteigerung, Lebensdauer und Genauigkeit, engl. Balancing spindles for increased speed, service life and accuracy. [Online]. Available: https://pmb-bobertag.de/produkte/auswuchten-nach-anwendung/spindeln-auswuchten/ (accessed: Sep. 10 2025).

[4] PMB - Präzisionsmaschinenbau Bobertag GmbH (2025), Die statische Unwucht, engl. The static imbalance. [Online]. Available: https://pmb-bobertag.de/unwucht/statische-unwucht/ (accessed: Sep. 11 2025).
[5] PMB - Präzisionsmaschinenbau Bobertag GmbH (2025), Die dynamische Unwucht, engl. The dynamic imbalance. [Online]. Available: https://pmb-bobertag.de/unwucht/dynamische-unwucht/ (accessed: Sep. 11 2025).
[6] T. Jessel, C. Byrne, M. J. Eaton & R. Pullin (2024), "The capability of Acoustic Emission features to monitor diamond-coated burr grinding wear and effectiveness," eJNDT, vol. 29, no. 10, doi: 10.58286/30231.
[7] A. Hassui, A. E. Diniz, J. Oliveira, J. Felipe & J. Gomes (1998), "Experimental evaluation on grinding wheel wear through vibration and acoustic emission," Wear, vol. 217, no. 1, pp. 7–14, doi: 10.1016/S0043-1648(98)00166-5.
[8] J. Farsky, T. Baksa & M. Zetek (2018), "Influence of Grinding Parameters on Forces when Grinding Maraging Steel X3Nicomo 18 9 5 with SiC Grinding Wheels," in DAAAM Proceedings, Proceedings of the 29th International DAAAM Symposium 2018, B. Katalinic, Ed.: DAAAM International Vienna, pp. 767–771.
[9] S. Jiménez, M. O. Cole & P. S. Keogh (2016), "Vibration sensing in smart machine rotors using internal MEMS accelerometers," Journal of Sound and Vibration, vol. 377, pp. 58–75, doi: 10.1016/j.jsv.2016.05.014.
[10] G. Feng, N. Hu, Z. Mones, F. Gu & A. D. Ball (2016), "An investigation of the orthogonal outputs from an on-rotor MEMS accelerometer for reciprocating compressor condition monitoring," Mechanical Systems and Signal Processing, 76-77, pp. 228–241, doi: 10.1016/j.ymssp.2015.12.033.
[11] D. Janssens, J. Croes, C. Claeys & F. Naets (2025), "Rotor states of motion estimation using on-shaft MEMS accelerometers," Mechanical Systems and Signal Processing, vol. 232, p. 112658, doi: 10.1016/j.ymssp.2025.112658.
[12] H. K. Tönshoff, B. Karpuschewski, T. Mandrysch & I. Inasaki (1998), "Grinding Process Achievements and their Consequences on Machine Tools Challenges and Opportunities," CIRP Annals, vol. 47, no. 2, pp. 651–668, doi: 10.1016/S0007-8506(07)63247-8.

DOI: 10.2507/36th.daaam.proceedings.029

AI-DRIVEN APPROACHES TO GENETIC RISK STRATIFICATION AND EARLY DETECTION OF BREAST CANCER

Elzana Dupljak & Ervin Domazet

This Publication has to be referred as: Dupljak, E[lzana] & Domazet, E[rvin] (2025). AI-Driven Approaches to Genetic Risk Stratification and Early Detection of Breast Cancer, Proceedings of the 36th DAAAM International Symposium, pp.0217-0224, B. Katalinic (Ed.), Published by DAAAM International, ISBN 978-3-902734-47-1, ISSN 1726-9679, Vienna, Austria
DOI: 10.2507/36th.daaam.proceedings.029

Abstract

Breast cancer is one of the leading causes of mortality among women worldwide. One of the crucial factors when determining breast cancer predisposition is genetics. Traditional statistical models for risk assessment often have difficulty in detecting complex interaction among genetics variants, environmental factors, and clinical variables. Artificial Intelligence has developed as a transformative approach for analysing large scale genomic data, enabling more accurate genetic risk stratification and early detection of breast cancer. This paper focuses on review of AI-driven methods applied to breast cancer genetics. We examine machine learning and deep learning techniques for polygenic risk score improvement, single nucleotide polymorphism interaction modelling, and chromosomal-scale genomic analysis. Additionally, we discuss the integration of genomics with multi-omics data and imaging for precision diagnostics. In this paper, current challenges, including population bias and variant interpretation, are explored. Our findings underscore the potential of Artificial Intelligence to advance screening and prevention strategies in breast cancer care.

Keywords: Artificial Intelligence; Breast Cancer; Genetic Risk Stratification; Machine Learning; Precision Oncology.

1. Introduction

Breast cancer is known as the most prominent malignancy among women globally and it is a leading cause of cancer-related mortality [1]. Genetic factors play a crucial role in determining personal vulnerability, with highly penetrant genetic variants such as BRCA1 and BRCA2, which represent a substantial proportion of genetic breast cancer cases [2]. On the other side, a large proportion of breast cancer risk is attributed to polygenetic inheritance, involving numerous low-effect genetic variant that collectively influence disease onset [3]. Statistical models, like logistic regression and Cox proportional hazards models are used to estimate genetic risk [4]. However, these models often fail to capture complex interactions among genetic variants and environmental factors. This limitation points to the importance for more accurate and advanced computational methods. These challenges are addressed through the development of polygenic risk scores (PRS) and other genetic test [5]. They drastically improved genetic risk assessment. However, their predictive possibility remains inefficient, particularly in diverse populations [6]. At this point, the need for AI becomes evident. The integration of machine learning (ML) and deep learning (DL) techniques has demonstrated the capability to model single nucleotide polymorphism (SNP) interactions, gene–environment effects, and chromosomal-scale genomic variations [7].

Although the progress is being made, research specifically targeting genetic-based AI risk modelling is limited. Key challenges include lack of representation of under-studied populations, insufficient explainability of predictive models and limited integration of multi-omics data. In this paper, we present a review of intelligent approaches for genetic risk stratification and early detection of breast cancer. This study analyzes advancements in machine learning (ML) and deep learning (DL) methods aimed at enhancing polygenic risk score (PRS) computation, modeling single nucleotide polymorphism (SNP) interactions [8], leveraging chromosomal-scale genomic features [9], and integrating multi-omics data into predictive frameworks [10]. The review synthesizes findings from recent studies, demonstrating how AI-based techniques increasingly outperform traditional statistical models. While previous reviews have predominantly focused on imaging-based AI applications, our work specifically emphasizes genomics and genetics-oriented methodologies, providing a comprehensive understanding of current developments and emerging technical solutions. Key challenges include lack of representation of under-studied populations [11], insufficient explainability of predictive models [12], and limited integration of multi-omics data [13]. This paper contributes new insights by offering a structured framework that categorizes state-of-the-art AI methods for genetic-based breast cancer diagnosis. Future work should focus on developing explainable, federated, and multi-modal AI approaches that combine genetics, epigenetics and imaging data [13,24]. By providing this review, we aim to support ongoing research efforts and guide the development of AI solutions for precision oncology in breast cancer care.

2. Literature Review

The intersection of genetics, risk modeling, and artificial intelligence has revolutionized our understanding and management of breast cancer. This section provides a comprehensive overview of the genetic basis of the disease, examines the limitations of traditional risk models, and explores the integration of AI techniques for enhanced predictive analytics. By critically analyzing early machine learning applications, recent deep learning advancements, and novel approaches incorporating structural genomic features, we highlight both the progress and the persisting challenges in this rapidly evolving domain. In the following Section 2.1, we begin by examining the genetic foundations of breast cancer and the key variants implicated in hereditary risk. Section 2.2 outlines conventional statistical and polygenic models used for risk assessment. Section 2.3 explores how artificial intelligence is being employed to enhance genetic modeling, particularly through machine learning and deep learning approaches. Lastly, Section 2.4 discusses the current gaps, limitations, and ethical concerns associated with the application of AI in breast cancer genetics.

2.1 Genetic Basis of Breast Cancer

Genetic predisposition accounts for approximately 10–15% of all breast cancer cases. Yu et al. [2] demonstrated that high-penetrance mutations in BRCA1 and BRCA2 significantly increase lifetime breast cancer risk. Roberts et al. [3] identified additional pathogenic variants such as PALB2, TP53, and CHEK2. Amankwah et al. [18] reviewed the impact of polygenic susceptibility, highlighting that multiple low-effect variants collectively elevate disease risk. Cunningham et al. [19] discussed challenges of evaluating these variants. While significant progress has been made in AI-driven genetic risk modeling, other modalities such as histopathology imaging have also benefited from AI advancesIn our earlier work [32], we developed a parallel computing approach for efficient histopathological image classification using convolutional neural networks (CNNs). This method achieved 94.8% accuracy on the BreakHis dataset and demonstrated the potential of deep learning in automating breast cancer detection.

2.2 Traditional Risk Models

Polygenic risk scores (PRS), which aggregate numerous single nucleotide polymorphisms (SNPs), have been demonstrated by Badré et al. [20] and Lewis & Vassos [5] to improve predictive accuracy compared to single-gene testing. Yiangou et al. [6] emphasized that most PRS models are trained on European cohorts, leading to reduced accuracy in other ancestries.

2.3 AI in Genetics for Breast Cancer

While significant progress has been made in AI-driven genetic risk modeling, other modalities such as histopathology imaging have also benefited from AI advances. In our earlier work [38], we developed a parallel computing approach for efficient histopathological image classification using convolutional neural networks (CNNs). This method 94.8% accuracy on the BreakHis dataset and demonstrated the potential of deep learning in automating breast cancer detection.

2.3.1 Early Applications of Machine Learning

Applying random forest and gradient boosting algorithms, Behravan et al. [8] introduced the use of ML to detect SNP–SNP interactions that additive models miss. Similarly, Liu et al. [7] explored ML to identify higher-order gene–gene interactions, showing improvements over logistic regression.

2.3.2 Deep Learning for PRS Optimization

A deep neural network (DNN)–based approach by Badré et al. [20] improved PRS estimation and recall by capturing non-linear relationships among genetic variants, leading to more accurate risk stratification.

2.3.3 Chromosomal-Scale and Structural Variation Modelling

Beyond SNPs, chromosomal-scale genomic features provide complementary signal. Fatapour et al. [9] showed that length-variation–based features enhance risk prediction across populations, and Ko and Brody [17] proposed a chromosomal-scale risk score that generalized across cohorts, outperforming SNP-only models.

2.4 Current Gaps and Challenges

Although the results are promising, multiple challenges remain. Ding et al. [14] and Privé et al. [11] documented population bias and portability limitations in PRS, underscoring the need for inclusive model training and calibration. For variant interpretation, Ghasemi et al. [12] and Khandakji et al. [24] highlighted the limitations and opportunities of AI for classifying variants of unknown significance (VUS), emphasizing interpretable tools to support genetic counseling. The black-box nature of many deep learning models raises ethical and regulatory concerns; as noted by Ghasemi et al. [12] and Lekadir et al. [15], explainable AI (XAI) techniques and trustworthy-AI standards are essential for clinical adoption. Finally, regulatory validation remains crucial; recent FDA communications on the Clairity platform [16], [28] illustrate pathways to clearance while reinforcing the need for robust clinical evidence.

3. AI Methodologies for Genetic Risk Stratification

Artificial Intelligence has introduced transformative methodologies for breast cancer genetic risk prediction. These approaches go beyond traditional polygenic models by capturing complex interactions between genetic variants and integrating multi-dimensional data. This section reviews major AI techniques employed for risk stratification and early detection.

3.1 SNP Interaction Modeling

Machine learning models are frequently used to detect higher-order interactions between single nucleotide polymorphisms (SNPs) that are often missed by linear models. Behravan et al. [8] applied machine learning (including XGBoost) to detect interacting genetic variants significantly contributing to breast cancer risk. Liu et al. [7] further showed that ML-based gene–gene interaction networks outperform logistic regression for interaction discovery. These approaches highlight that modeling SNP–SNP interactions increases the predictive power of genetic risk assessments.

3.2 AI-Enhanced Polygenic Risk Scores (PRS)

Traditional PRS methods aggregate additive genetic effects but overlook complex variant relationships. Badré et al. [20] showed that deep neural networks (DNNs) significantly improve PRS estimation by capturing non-linear variant effects, resulting in better risk stratification. Roberts et al. [3] and Lewis & Vassos [5] also emphasized the clinical relevance of polygenic risk modeling, while Privé et al. [11] revealed that PRS portability remains a challenge across ancestry groups. Furthermore, privacy-preserving solutions such as federated learning proposed by Rieke et al. [21] enable distributed PRS model training without sharing raw genomic data.

3.3 Chromosomal-Scale Genomic Modeling

Recent advances extend beyond SNPs to chromosomal-scale genomic features. Fatapour et al. [9] demonstrated that chromosomal-scale length variation enhances breast cancer risk prediction. Similarly, Ko & Brody [17] proposed a chromosomal-scale genetic risk score that generalized across TCGA and UK Biobank cohorts, outperforming SNP-only approaches. These studies suggest that structural genomic variations provide complementary predictive information.

3.4 Multi-Omics Data Integration

Combining genomics with transcriptomics, proteomics, and epigenetics captures the biological complexity underlying breast cancer risk. Cai et al. [10] highlighted that machine learning-based integration of multi-omics datasets improves robustness and reveals pathway-level signals. Liu et al. [13] provided a systematic review of multi-omics integration strategies, emphasizing external validation. Furthermore, Hao et al. [23] proposed transformer-based survival models (PASSION) that outperformed conventional methods when applied to TCGA and METABRIC cohorts.

3.5 Explainable and Interpretable AI

Clinical adoption of AI-driven risk prediction requires explainable models. Ghasemi et al. [12] surveyed methods such as SHAP and attention-based explainability to visualize SNP contributions in risk models. Khandakji et al. [24] developed interpretable classifiers for BRCA variants of unknown significance, facilitating genetic counseling. Regulatory and ethical perspectives discussed by Lekadir et al. [15] reinforce the importance of integrating explainability frameworks into genetic AI models.

4. Applications in Early Detection and Screening

Artificial Intelligence has not only improved genetic risk stratification but also enabled practical applications in early breast cancer detection and clinical decision support. By integrating genetic insights into screening programs and leveraging advanced AI tools, healthcare systems can identify high-risk individuals earlier and personalize diagnostic pathways.

4.1 Risk-Based Screening Programs

AI-driven genetic models have been increasingly used to stratify women by their risk level and tailor screening schedules. Badré et al. [20] demonstrated that incorporating AI-enhanced polygenic risk scores (PRS) into clinical practice enables earlier identification of individuals with elevated lifetime risk. Ensemble models, such as those described by Chang et al. [28], outperform traditional clinical calculators, allowing more precise recommendations for mammography and MRI surveillance. Martin et al. [6] further highlighted that integrating PRS into population-based screening programs improves detection rates while reducing false positives. Importantly, studies by Ding et al. [14] and Zirpoli et al. [29] underscore that ancestry-specific or multi-ancestry PRS are critical to ensure equitable screening outcomes across diverse populations.

4.2 Predictive AI Tools for Future Cancer Development

AI applications have expanded to predicting future breast cancer development based on genetic profiles. Deep learning–based polygenic risk modeling reported by Badré et al. [20] and transformer survival models described by Hao et al. [23] illustrate the ability of AI to predict mid- to long-term cancer risk more accurately than conventional models. The recent approval of the Clairity Breast AI platform [16], [28] by the FDA marks a significant milestone for clinical adoption. Clairity applies machine learning to mammography images to predict individualized cancer risk, demonstrating how predictive AI tools can integrate into existing screening workflows for early intervention.

4.3 Integration with Imaging and Clinical Data

AI models that combine genetic risk factors with imaging modalities have shown promise for enhancing early detection. Huang et al. [30] demonstrated that transformer-based architectures integrating genomics and mammographic imaging outperform single-modality models. Similarly, Cai et al. [10] and Liu et al. [13] highlighted that adding multi-omics (proteomic and transcriptomic) data improves prediction robustness. Studies by Behravan et al. [8] and Shah et al. [13] emphasized that multi-modal integration can support radiologists in distinguishing benign from malignant lesions earlier in the disease course, potentially reducing unnecessary biopsies.

4.4 Personalized Prevention and Treatment Guidance

Beyond detection, AI-driven genetic risk models guide preventive measures and treatment decisions. Turner et al. [33] applied causal inference methods to model therapeutic responses based on genetic risk profiles, suggesting a pathway toward AI-driven treatment personalization. Khandakji et al. [24] developed interpretable AI classifiers for BRCA variants of unknown significance, enabling more informed clinical decision-making regarding prophylactic surgery or targeted therapies. Together, these applications illustrate AI's potential to shift breast cancer management from reactive treatment to proactive, personalized prevention and care.

5. Comparative Summary of Literature

To summarize the findings from existing studies, Table I summarizes AI-driven approaches for genetic risk stratification and early breast cancer detection. The table highlights the methodologies, data types, datasets, and major outcomes, providing a foundation for comparative analysis and discussion of the state-of-the-art.

Study / Authors	Year	AI Technique	Data Type	Dataset / Population	Key Findings
Liu et al. [7]	2022	Random Forest, ML	SNP interactions	Breast cancer cohorts (BMC Cancer study)	Identified gene–gene networks; improved over logistic regression.
Behravan et al. [8]	2024	XGBoost, ML	SNP interactions	Case-control genomic datasets	Detected interacting variants contributing to breast cancer risk.
Badré et al. [20]	2021	Deep Neural Networks	Polygenic Risk Scores	European women, PRS models	Improved PRS estimation, captured non-linear variant effects.
Privé et al. [11]	2022	PRS portability evaluation	Polygenic Risk Scores	Nine ancestry groups	Limited transferability of PRS across ancestries.
Yiangou et al. [6]	2024	PRS models	Polygenic Risk Scores	European populations	Showed variable PRS accuracy across European subgroups.
Ding et al. [14]	2023	Statistical + PRS comparison	Polygenic Risk Scores	Multi-ancestry cohorts	PRS accuracy declines in non-European groups.
Zirpoli et al. [29]	2024	PRS calibration	Polygenic Risk Scores	African American women (Black Women's Health Study)	Improved risk prediction with population-specific calibration.
Serrano-Gómez et al. [30]	2025	PRS calibration	Polygenic Risk Scores	Hispanic/Latina women	Demonstrated variability of PRS performance across Latina subgroups.
Fatapour et al. [9]	2025	ML classifiers	Chromosomal-scale variation	Multi-population cohorts	Structural variation improved prediction beyond SNP-only models.
Ko & Brody [17]	2023	Risk score modeling	Chromosomal-scale length variation	Population-scale genomic datasets	New risk score based on chromosomal features.
Cai et al. [10]	2022	ML frameworks	Multi-omics	Review (cancer datasets)	Highlighted consistent improvements via multi-omics integration.
Liu et al. [13]	2023	DL/ML integration	Multi-omics	Review (Trends in Genetics)	Recommended deep learning for heterogeneous omics.
Hao et al. [23]	2021	Transformer (PASSION)	Multi-omics survival	Cancer genomic cohorts	Improved survival prediction via transformer architectures.
Braytee et al. [31]	2024	Autoencoder + tensor integration	Multi-omics	Breast cancer cohorts	Combined transcriptomic, epigenomic, CNV data; improved predictions.
Khandakji et al. [24]	2022	ML classifier	BRCA1/2 variants	BRCA VUS dataset	Improved classification of variants of unknown significance.
Cunningham et al. [19]	2024	Statistical analysis	Low-effect variants	Breast cancer cohorts	Highlighted challenges in interpreting multiple low-effect variants.
Ghasemi et al. [12]	2024	Explainable AI (XAI)	Multi-omics & genetics	Review (scoping)	Summarized SHAP, attention-based models for clinical interpretability.
Lekadir et al. [15]	2023	AI standards & ethics	Cross-domain AI	Review	Defined requirements for trustworthy biomedical AI.
Rieke et al. [21]	2020	Federated Learning	Genomic + clinical data	Multi-institutional studies	Enabled privacy-preserving collaboration without centralizing data.
FDA [16], [28]	2025	Regulatory validation	AI platform (Clairity)	Mammography + genetics	First FDA-cleared predictive AI platform for breast cancer risk.
Turner et al. [33]	2025	Causal inference, XAI	Genetic models	AI-based genetic risk studies	Integrated causal inference to improve transparency and trust.

Table 1. Comparative Summary of AI-Driven Genetic Breast Cancer Risk Prediction Studies

The comparative analysis of recent studies highlights the evolution of AI techniques for genetic risk prediction in breast cancer. Early applications of machine learning, such as those by Liu et al. [7] and Behravan et al. [8], used random forest and gradient boosting approaches to detect SNP–SNP interactions that traditional additive models often missed. While these models demonstrated moderate improvements, their scalability and generalizability remained limited. Subsequent studies moved toward deep learning for polygenic risk score (PRS) optimization. Badré et al. [20] showed that deep neural networks could capture non-linear relationships among variants, providing more accurate stratification than linear PRS methods. However, portability remained a critical limitation, as documented by Privé et al. [11], Yiangou et al. [6], and Ding et al. [14], who each emphasized significant ancestry-related discrepancies in PRS performance. Population-specific evaluations by Zirpoli et al. [29] in African American women and Serrano-Gómez et al. [30] in Hispanic/Latina cohorts further reinforced the need for diverse calibration strategies. Beyond SNP-based models, chromosomal-scale and structural variation approaches have emerged as promising alternatives. Fatapour et al. [9] and Ko & Brody [17] demonstrated that large-scale genomic features improve prediction robustness, especially across populations underrepresented in standard GWAS data. The integration of multi-omics data has marked the most significant progress. Cai et al. [10] and Liu et al. [13] reviewed machine learning and deep learning frameworks that effectively combine genomic, transcriptomic, and proteomic information. Hao et al. [23] advanced this by introducing a transformer-based survival model, while Braytee et al. [31] used autoencoder-tensor integration to improve prognostic modeling. Together, these approaches show the power of multi-modal data fusion for robust risk assessment. Interpretability and explainability have also become central to clinical translation. Khandakji et al. [24] applied ML to BRCA variants of unknown significance (VUS), improving classification and supporting clinical counseling. Ghasemi et al. [12] reviewed explainable AI methods such as SHAP and attention-based visualizations, while Lekadir et al. [15] outlined standards for trustworthy biomedical AI. Turner et al. [33] further introduced causal inference frameworks that enhance interpretability, bridging a critical gap between predictive accuracy and clinician trust. Finally, regulatory progress underscores the movement toward implementation. The FDA's approval and clearance of the Clairity Breast AI platform [16], [28] represents the first clinical deployment of AI-driven genetic risk assessment in breast cancer. Complementing this, federated learning initiatives by Rieke et al. [21] show how data privacy and collaboration challenges can be addressed, supporting ethical and equitable deployment. Collectively, these studies illustrate a clear trajectory: from SNP-based machine learning models toward deep learning, multi-omics integration, explainable frameworks, and regulatory translation. Your own contribution [32] in parallel computing for histopathology imaging further complements this trend, demonstrating AI's versatility across genomic and imaging modalities.

6. Challenges and Limitations

Despite substantial progress in AI-driven genetic risk modeling and early detection of breast cancer, several challenges hinder widespread clinical implementation. These challenges relate to data quality and diversity, variant interpretation, model explainability, and ethical and regulatory barriers.

6.1 Data Diversity and Bias

Most genomic datasets used to train AI models are dominated by individuals of European ancestry. Yiangou et al. [6] and Ding et al. [14] showed that PRS trained in European cohorts often underperform in non-European populations, leading to reduced accuracy and generalizability. Privé et al. [11] further demonstrated limited portability of PRS across nine ancestry groups. To address these gaps, population-specific studies such as Zirpoli et al. [29] (African ancestry) and Serrano-Gómez et al. [30] (Hispanic/Latina women) emphasize the need for ancestry-tailored risk scores. Moreover, Fatapour et al. [9] and Ko and Brody [17] suggested incorporating structural genomic variation as complementary information, while Rieke et al. [21] highlighted that federated learning can help mitigate bias by enabling collaborative training across diverse cohorts without centralizing sensitive data.

6.2 Variants of Unknown Significance (VUS)

A persistent challenge in genetic testing is the classification of variants of unknown significance (VUS), particularly in BRCA1 and BRCA2. Khandakji et al. [24] developed machine learning classifiers to improve pathogenicity prediction of BRCA variants, but accuracy remains limited by incomplete functional data. Cunningham et al. [19] noted that the prevalence of VUS complicates counseling and clinical decision-making. Explainable AI frameworks, as reviewed by Ghasemi et al. [12], show promise for providing interpretable variant risk contributions, but large-scale clinical validation and standardization remain lacking.

6.3 Model Explainability and Clinical Trust

Deep learning models, while powerful, often operate as "black boxes," making their predictions difficult to interpret for clinicians. Ghasemi et al. [12] emphasized the importance of SHAP values and attention-based visualization for transparency in breast cancer AI models. Hao et al. [23] demonstrated that transformer-based survival models, although highly predictive, require additional causal inference approaches, such as those proposed by Turner et al. [33], to achieve clinician trust. Regulatory and clinical adoption are slowed by limited explainability, reinforcing the call for standards like those outlined by Lekadir et al. [15] in biomedical AI.

6.4 Ethical, Privacy, and Regulatory Concerns

Ethical concerns related to patient privacy, informed consent, and data sharing are central to AI-based genetic modeling. Rieke et al. [21] highlighted federated learning as a potential solution for privacy-preserving genomic analysis, though regulatory frameworks for such distributed approaches are still evolving. FDA reports [16], [28] on the Clairity platform underscore the importance of rigorous validation and regulatory oversight for AI systems in clinical use. Finally, Ghasemi et al. [12] and Lekadir et al. [15] stressed that international standards and ethical guidelines must accompany technical advances to ensure safe and equitable deployment.

7. Future Work

This study will serve as a foundation for future research aimed at integrating advanced Artificial Intelligence techniques with genetic data to enhance breast cancer risk prediction and early detection. Building on the findings of this review, our next steps will focus on developing privacy-preserving federated learning models to enable multi-institutional collaboration without compromising patient confidentiality [15], [21]. We plan to design multi-modal frameworks that combine genomics with transcriptomics, proteomics, and imaging data to capture the full biological complexity of breast cancer [10], [13], [30], [31]. Additionally, we will explore explainable and causal AI methods to improve model transparency, allowing clinicians to understand and trust AI-generated predictions [12], [33]. Ultimately, this future work aims to bridge the gap between AI research and clinical practice, contributing to equitable, personalized, and clinically validated breast cancer prevention strategies.

8. Conclusion

Breast cancer remains a leading cause of morbidity and mortality worldwide, with genetic factors playing a critical role in individual susceptibility and disease progression. Traditional statistical models and polygenic risk scores, while valuable, are limited in capturing the complex, non-linear interactions among genetic variants and environmental factors. This paper has provided a comprehensive review of recent Artificial Intelligence (AI) methodologies for genetic risk stratification and early detection of breast cancer, highlighting advances in SNP interaction modeling, AI-enhanced polygenic risk scores, chromosomal-scale genomic analysis, and multi-omics data integration. The review demonstrates that AI techniques, particularly machine learning, deep learning, and federated models, significantly outperform conventional approaches in predicting genetic risk while enabling more personalized screening strategies. Despite these advancements, challenges remain in addressing data diversity, interpreting variants of unknown significance, ensuring explainability, and achieving regulatory and clinical validation. Future research directions identified in this study emphasize the need for privacy-preserving federated learning, multi-modal data integration, and explainable AI to support trustworthy, equitable, and clinically actionable risk models. By bridging the gap between AI innovations and real-world clinical application, these next steps have the potential to revolutionize breast cancer prevention, diagnosis, and personalized treatment, ultimately improving outcomes for women worldwide.

9. References

[1] Sung, H., Ferlay, J., Siegel, R. L., Laversanne, M., Soerjomataram, I., Jemal, A., & Bray, F. (2021). Global cancer statistics 2020: GLOBOCAN estimates of incidence and mortality worldwide for 36 cancers in 185 countries. CA: A Cancer Journal for Clinicians, vol. 71, no. 3, pp. 209–249, May/June 2021. ISSN 0007-9235; eISSN 1542-4863. DOI: 10.3322/caac.21660

[2] Yu, S., et al. (2025). Breast cancer risk associated with BRCA1 and BRCA2 pathogenic variant carriers: penetrance estimates. The Breast, vol. 75, pp. 103128, March 2025. ISSN 0960-9776; eISSN 1532-3080. DOI: 10.1016/j.breast.2025.103128.

[3] Roberts, E., Howell, S., & Evans, G. (2023). Polygenic risk scores and breast cancer risk prediction. The Breast, vol. 67, pp. 60–67, March 2023. ISSN 0960-9776. DOI: 10.1016/j.breast.2023.01.003.

[4] Liu, L., et al. (2025). Development and performance of female breast cancer risk prediction models: a systematic review and meta-analysis. [In Press, 2025].

[5] Lewis, C. M., & Vassos, E. (2020). Polygenic risk scores: From research tools to clinical instruments. Genome Medicine, vol. 12, no. 44, pp. 1–11, June 2020. eISSN 1756-994X. DOI: 10.1186/s13073-020-00742-5.

[6] Yiangou, A., et al. (2024). Evaluating the performance of breast cancer polygenic risk scores in different European populations. Breast Cancer Research, vol. 26, no. 19, pp. 1–12, February 2024. eISSN 1465-5411. DOI: 10.1186/s13058-024-01947-x.

[7] Liu, Y., et al. (2022). Identification of gene–gene interaction networks for breast cancer using machine learning. BMC Cancer, vol. 22, no. 1043, pp. 1–14, October 2022. eISSN 1471-2407. DOI: 10.1186/s12885-022-10170-w.

[8] Behravan, B., et al. (2024). Machine learning identifies interacting genetic variants contributing to breast cancer risk. Scientific Reports, vol. 14, no. 1123, pp. 1–12, January 2024. eISSN 2045-2322. DOI: 10.1038/s41598-024-48291-y.

[9] Fatapour, Y., Brody, J. P., et al. (2025). Improved breast cancer risk prediction using chromosomal-scale length variation. Human Genomics, vol. 19, no. 42, pp. 1–13, April 2025. ISSN 1473-9542; eISSN 1479-7364. DOI: 10.1186/s40246-025-00776-z.

[10] Cai, Z., Poulos, R. C., Liu, J., & Zhong, Q. (2022). Machine learning for multi-omics data integration in cancer. iScience, vol. 25, no. 2, pp. 103798, February 2022. ISSN 2589-0042. DOI: 10.1016/j.isci.2022.103798.

[11 Privé, F., et al. (2022). Portability of 245 polygenic scores in 9 ancestry groups. Nature Communications, vol. 13, no. 554, pp. 1–12, January 2022. eISSN 2041-1723. DOI: 10.1038/s41467-022-28041-y.

[12] Ghasemi, A., Hashtarkhani, S., Schwartz, D. L., & Shaban-Nejad, A. (2024). Explainable AI in breast cancer detection and risk prediction: a scoping review. Cancer Innovation, vol. 3, no. 5, e136, May 2024. eISSN 2832-3517. DOI: 10.1002/cai2.136.

[13] Liu, Y., et al. (2023). Integrative multi-omics modeling for cancer risk prediction: a review. Trends in Genetics, vol. 39, no. 9, pp. 694–709, September 2023. ISSN 0168-9525; eISSN 1362-4555. DOI: 10.1016/j.tig.2023.05.007.

[14] Ding, Y., et al. (2023). Polygenic scoring accuracy varies across the genetic ancestry continuum. Nature, vol. 618, pp. 353–361, June 2023. ISSN 0028-0836; eISSN 1476-4687. DOI: 10.1038/s41586-023-06079-4.

[15] Lekadir, K., et al. (2023). Standards for trustworthy and explainable AI in biomedical applications. Nature Biomedical Engineering, vol. 7, pp. 1121–1132, December 2023. ISSN 2157-846X. DOI: 10.1038/s41551-023-01091-2..

[16] U.S. Food and Drug Administration (FDA). (2025). Clairity Breast AI platform for breast cancer risk prediction approved. FDA Press Release, January 2025. Available from: https://www.fda.gov/news-events. Accessed: 2025-07-25.

[17] Ko, C., & Brody, J. P. (2023). Genetic risk score from chromosomal-scale length variation. Human Genomics, vol. 17, no. 53, pp. 1–12, September 2023. ISSN 1473-9542. DOI: 10.1186/s40246-023-00482-8.

[18] Amankwah, L. T., et al. (2024). Polygenic susceptibility to breast cancer: A systematic review. Breast Cancer Research and Treatment, vol. 197, no. 3, pp. 521–536, February 2024. ISSN 0167-6806; eISSN 1573-7217. DOI: 10.1007/s10549-024-07281-3.

[19] Cunningham, A. P., et al. (2024). Challenges in interpreting multiple low-effect genetic variants for breast cancer risk assessment. Genetics in Medicine, vol. 26, no. 7, pp. 1402–1411, July 2024. ISSN 1098-3600; eISSN 1530-0366. DOI: 10.1016/j.gim.2024.01.009.

[20] Badré, A., et al. (2021). Deep neural networks improve polygenic risk score estimation for breast cancer. Journal of Human Genetics, vol. 66, pp. 359–369, March 2021. ISSN 1434-5161; eISSN 1435-232X. DOI: 10.1038/s10038-020-00832-7.

[21] Rieke, N., et al. (2020). The future of digital health with federated learning. npj Digital Medicine, vol. 3, no. 119, pp. 1–7, September 2020. eISSN 2398-6352. DOI: 10.1038/s41746-020-00323-1.

[22] Wang, H., et al. (2023). Application of convolutional neural networks to large-scale genomic variation data. Briefings in Bioinformatics, vol. 24, no. 1, bbad007, January 2023. ISSN 1467-5463. DOI: 10.1093/bib/bbad007.

[23] Hao, J., et al. (2021). PASSION: Path transformer for survival prediction using multi-omics data. Bioinformatics, vol. 37, no. 18, pp. 3437–3445, September 2021. ISSN 1367-4803. DOI: 10.1093/bioinformatics/btab339.

[24] Khandakji, M. N., et al. (2022). BRCA1-specific machine learning model predicts variant pathogenicity. Frontiers in Genetics, vol. 13, 982930, pp. 1–9, August 2022. ISSN 1664-8021. DOI: 10.3389/fgene.2022.982930.

[25] Louizos, C., et al. (2023). Causal effect inference with deep latent-variable models. In Advances in Neural Information Processing Systems (NeurIPS), pp. 6446–6456. arXiv:1705.08821.

[26] Huang, S., Chaudhary, K., & Garmire, L. X. (2020). More is better: recent advances in multi-omics data integration for cancer risk prediction. Annual Review of Biomedical Data Science, vol. 3, pp. 1–30, July 2020. ISSN 2574-3412. DOI: 10.1146/annurev-biodatasci-080917-013312.

[27] Mavaddat, N., et al. (2020). Polygenic risk scores for prediction of breast cancer and breast cancer subtypes. American Journal of Human Genetics, vol. 104, no. 1, pp. 21–34, January 2019. ISSN 0002-9297; eISSN 1537-6605. DOI: 10.1016/j.ajhg.2018.11.002.

[28] U.S. Food and Drug Administration (FDA). (2025). FDA clears Clairity Breast AI platform for predictive breast cancer risk. FDA News Release, February 2025. Available from: https://www.fda.gov/news-events. Accessed: 2025-07-25.

[29] Zirpoli, G. R., et al. (2024). Polygenic risk scores improve breast cancer risk prediction among African American women in the Black Women's Health Study. JNCI Cancer Spectrum, vol. 8, no. 5, pkad067, October 2024. eISSN 2515-5091. DOI: 10.1093/jncics/pkad067.

[30] Serrano-Gómez, S. J., et al. (2025). Performance of polygenic risk scores for breast cancer in Hispanic/Latina women: results from multi-cohort evaluation. Cancer Epidemiology, Biomarkers & Prevention, vol. 34, no. 3, pp. 515–526, March 2025. ISSN 1055-9965; eISSN 1538-7755. DOI: 10.1158/1055-9965.EPI-24-0701.

[31] Braytee, A., et al. (2024). Autoencoder-tensor integration of multi-omics data for breast cancer survival prediction. Scientific Reports, vol. 14, no. 11283, pp. 1–13, July 2024. eISSN 2045-2322. DOI: 10.1038/s41598-024-48291-y.

[32] Dupljak, E. (2025). Parallel computing for efficient histopathological image classification. Proc. Int. Conf. on Innovative Technologies in Biomedical and Translational Science (ICITTBT), in press, 2025.

[33] Turner, E., et al. (2025). Causal inference and explainability in AI-based genetic models. Frontiers in Artificial Intelligence, vol. 4, pp. 145–158, March 2025. eISSN 2624-8212. DOI: 10.3389/frai.2025.00145.

DOI: 10.2507/36th.daaam.proceedings.030

USING LUMINESCENCE SPECTROSCOPY TO MONITOR THE AGING OF LATENT FINGERPRINTS

Michal Indra, Michaela Mikuličová & Vojtěch Křesálek

This Publication has to be referred as: Indra, M[ichal]; Mikulicova, M[ichaela] & Kresalek, V[ojtech] (2025). Using Luminescence Spectroscopy to Monitor the Aging of Latent Fingerprints, Proceedings of the 36th DAAAM International Symposium, pp.0225-0230, B. Katalinic (Ed.), Published by DAAAM International, ISBN 978-3-902734-47-1, ISSN 1726-9679, Vienna, Austria
DOI: 10.2507/36th.daaam.proceedings.030

Abstract

The article deals with the estimation of the age of latent fingerprints using luminescence spectroscopy. For the purpose of measurement, 7 samples were analyzed (5 from men, 2 from women; age 24-37 years). Subsequently, the samples were measured for a period of twelve weeks using luminescence spectroscopy. Characteristic trends were selected from the measured values of luminescence intensity and emission spectra, with the help of which the luminescence behavior of latent fingerprints can be described over time. Two significant peaks were found in the regions of approximately 360 nm and 480 nm of the emission spectrum.

Keywords: Luminescence; Spectroscopy; Dactyloscopy; Latent Fingerprint; Fingerprint Aging

1. Introduction

Forensic dactyloscopy is one of the oldest and most important criminalistic disciplines. It examines the papillary lines found on the palms, soles and joints of the fingers. Thanks to three important properties, which are uniqueness, immutability and relative irremovability of papillary lines, they can be used for individual identification of a person. When examining and analyzing fingerprints, one usually works only with certain characteristic parts or patterns in which the papillary lines have specific shapes [1], [2]. According to the method of creation, dactyloscopic traces are divided into three-dimensional and planar. Three-dimensional traces are formed when pressure is applied to a soft material (e.g. wax), i.e. the print is pressed into the material. On solid materials, these are planar traces, which are of two types. Layered traces, which are composed of substances adhered to the skin (e.g. blood, sebum, dust particles, etc.), and delaminated traces, when, on the contrary, contact with an object causes part of the impurities from the object to adhere to the skin (e.g. a layer of dust, paint, varnish, etc.) [1], [3]. Dactyloscopic traces are also divided according to visibility, namely into visible traces, which can be clearly seen even by the human eye and are formed, for example, by blood or paint, and invisible or latent traces, which are the most common and are formed mainly by sweat and sebum [4], [5]. Latent fingerprints must be made visible before being photographed, secured (e.g. on fingerprint film) and subsequently examined. Various methods are used for this, the choice of which depends on the properties of the particular fingerprint. Physical methods use the adhesion or stickiness of a fingerprint powder (e.g. argentorate) applied with a brush to the fingerprint, to which it adheres.

Chemical methods use the reaction of chemical agents (e.g. silver nitrate) with various substances in the fingerprint. Physico-chemical methods include, for example, the application of cyanoacrylate vapors. The latter methods are special methods, working mainly with laser devices [4], [5], [6] It is not always possible to use a latent fingerprint for classical identification. The fingerprint is not suitable in situations where it is damaged or blurred in some way, and the papillary line cannot be distinguished accurately. Nevertheless, several pieces of information about the originator (e.g. gender, use of medications or drugs, age, etc.) can be determined from the fingerprint by analyzing its chemical composition. The imprint contains various substances found in human sweat (secreted by eccrine sweat glands) and sebum (secreted by sebaceous glands, which are not found on the palms or soles, but during normal activities when a person touches, for example, hair or face, the hands are also contaminated with sebum), such as fatty acids, amino acids or sugars. External contaminants (e.g. residues of food, cosmetics, inks, medications, etc.) may also be present in the fingerprint. The chemical composition of the imprint changes over time due to external conditions (e.g. temperature and humidity), which, among other things, cause the oxidation or evaporation of some substances [4], [7].

Currently, methods are being developed that can roughly estimate the age of latent fingerprints. In forensic investigations, this could be helpful, for example, in cases where it is necessary to determine whether a fingerprint found was created before a crime was committed. One option is to use dactyloscopic powders, which monitor their changing adhesion or the degree of visibility of the papillary terrain over time, as described, for example, in a study [8], where latent fingerprints were monitored over time when exposed to various conditions (e.g. light, humidity, etc.) and made visible using titanium dioxide-based powder. However, no specific conclusions were drawn. In [9], the prints were monitored for 90 days and after a certain period of time after the powder was applied, changes in color saturation were monitored. A more significant decrease in saturation occurred between days 3 and 7. The saturation did not change much over the last 30 days.

Another option is electrical methods that use the change in the electrical charge of the fingerprint over time. The study [10] dealt with contact electrification, where the electrical potential of the latent fingerprint was scanned by a scanning device with a contactless probe. A decreasing areal density of the electrical charge was recorded over time, while in very old fingerprints the charge was not detected at all.

For the analysis of the chemical composition of fingerprints, it is good to use destructive, but very accurate gas chromatography-mass spectrometry (GC-MS). This method was also used to determine the age of latent fingerprints in [11], [12]. In [11], 8 samples from different individuals were monitored for 28 days. The conclusions point to a faster decrease in some compounds over time (mainly squalene) in samples stored in the light. There was also an increase in some fatty acids over time. Other methods that can be used to roughly estimate the age of latent fingerprints include Raman spectroscopy, which was used, for example, in [13]. Among other things, it was found that the Raman bands of squalene degrade much faster in light, where they cease to be detectable after approximately 140 hours. If the samples were stored in the dark, the bands were still present in the spectrum after 700 hours. The FT-IR (Fourier Transform Infrared Spectroscopy) method was used, for example, in a study [14], which focused on fingerprints of children (2 to 11 years). The analysis showed that acid salts are suitable for studying children's fingerprints due to their stability. Some methods, in turn, deal with the physical degradation of the papillary terrain of the fingerprint over time (addressed, for example, in [15]), but the problem in this case is that it depends very much on the original quality of the fingerprint.

Luminescence spectroscopy has a wide range of applications in many areas, including forensic analysis, for example, in detecting counterfeit banknotes [16]. It can potentially be used to estimate the age of latent fingerprints. Since this method only works with the chemical composition, it is not necessary to address the original quality of the print, i.e. even a blurred or partial print can be analyzed. The method is based on the different luminescent properties of organic and inorganic substances found in the print. In [17], the change in luminescence intensity over time was monitored for compounds formed by the oxidation of unsaturated lipids. In the first three weeks, it was possible to estimate the age of the prints for approximately half of the male samples. In the study [18], two main peaks were found in the emission spectrum, approximately in the regions of 340 nm and 440 nm. In [19], faster degradation of latent prints in light or at higher air humidity was observed. It was also found that the intensity of the first peak at 340 nm decreases with time, and therefore it is more appropriate to focus on the region around 440 nm, in which the luminescence intensity increases with time. Luminescence spectroscopy is used to monitor the aging of latent fingerprints in this article as well. Due to the many advantages and simplicity of this method, it could potentially be used in forensic practice.

Seven fingerprints were measured over a twelve-week period using luminescence spectroscopy. Two significant peaks were found in the regions of approximately 360 nm and 480 nm of the emission spectrum. A tendency for the luminescence intensity to increase was observed for male samples, while for female samples the opposite was observed and the intensity decreased. The amount of material examined does not affect the measurement, but significant changes are caused by contamination of the samples, for example with cosmetics.

2. Instrumentation, Sample Preparation and Measurement

The emission spectra of latent fingerprints were measured with a Shimadzu RF-6000 spectrofluorimeter with a 150W high-pressure xenon lamp as the excitation source and a Hamamatsu R928 photomultiplier as the detector. LabSolutions RF software was used for setup and control. The most suitable excitation wavelength was experimentally determined to be $\lambda ex = 280$ nm. The emission spectra were measured in the range of 300–540 nm.

7 samples from 5 men and 2 women aged 24–37 years were analyzed. Sample collection and preparation took place in several steps. First, the individual washed his hands with soap and then dried them thoroughly. The individual ran a dry, arbitrarily selected finger over his face and forehead and made an impression on a clean glass slide. In order to maintain the same conditions, all samples were stored in the dark at room temperature. Fingerprint measurements were taken once a week for 12 weeks, with the first measurement taken immediately after sample collection.

3. Results and Discussion

For reasons of accuracy, all measured values are always reported after subtracting the luminescence intensity of the slide used as the base material. The emission spectra of the male sample M–1, which were measured over a period of 12 weeks, are shown in Fig. 1.

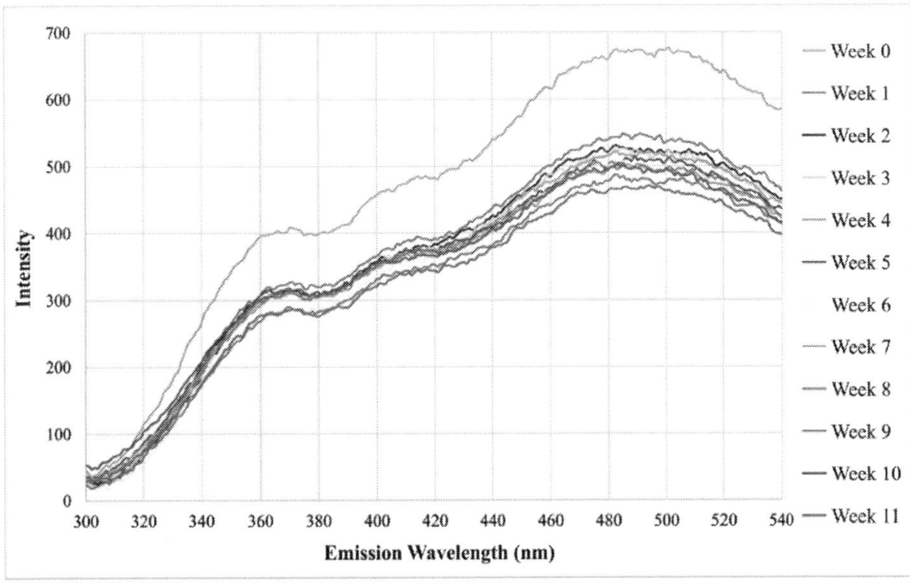

Fig. 1. Emission spectra of male sample M–1 over time.

Two main peaks can be observed in the emission spectrum. The first peak is located around 360 nm and is shown in more detail in Fig. 2a. The second peak is located around 480 nm and is shown in more detail in Fig. 2b. One smaller peak is also around 400 nm. The intensity values vary from sample to sample, but the location of the peaks and the shape of the spectrum are almost unchanged.

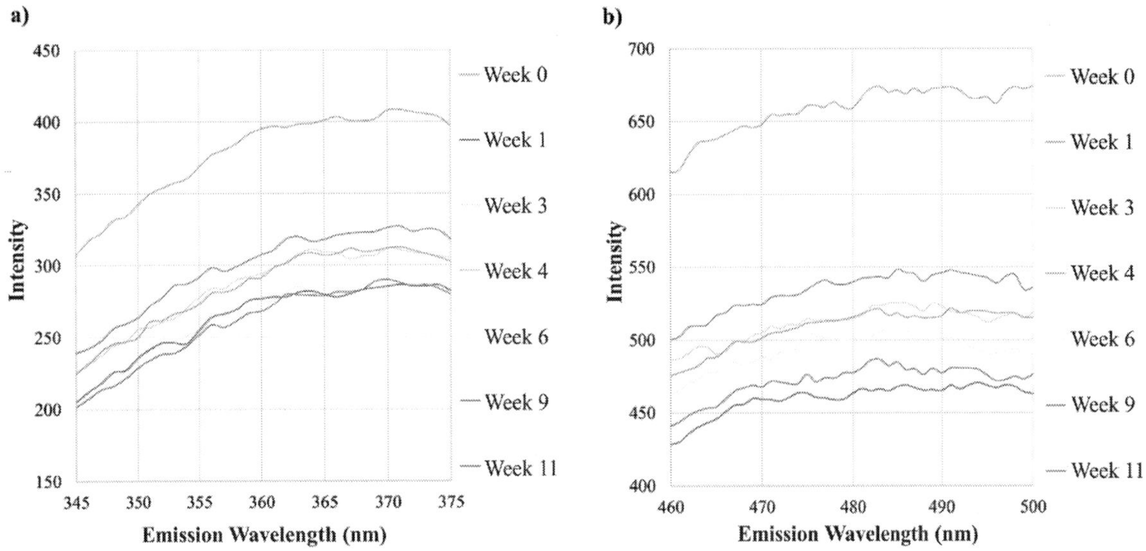

Fig. 2. Selected emission spectra of male sample M–1. a) region 345 nm to 375 nm, b) region 460 nm to 500 nm.

In 4 out of 5 male samples, the luminescence intensity decreased over time (most significantly during the first 2 weeks). In the following weeks, the intensity continued to decrease, but to a lesser extent and mainly at longer wavelengths (especially in the region around 480 nm of the emission wavelength). Fig. 3 shows the dependences of selected wavelengths of the emission spectrum of the male sample M–2 on time. The measurement uncertainty was determined by repeated measurements to be 3% and is shown by error bars. The graph shows that at 480 nm the greatest decrease in intensity occurs during the first week. In the following weeks, the intensity also decreases, but to a lesser extent. At wavelengths of 400 nm and 360 nm, the intensity also decreases during the first weeks, but the luminescence intensity did not change much in the remaining weeks.

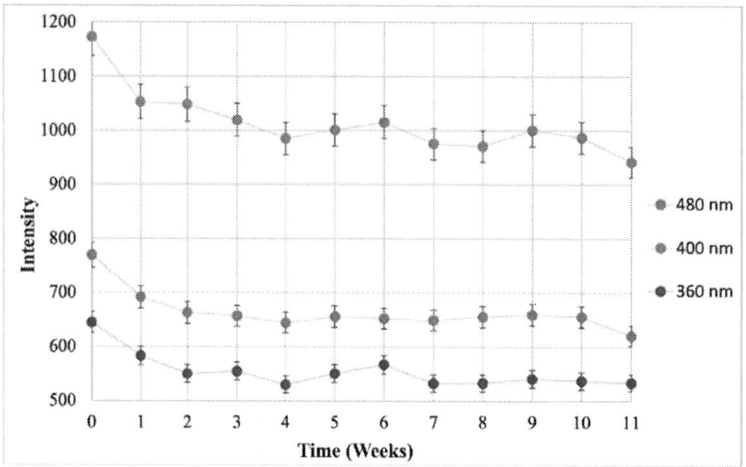

Fig. 3. Dependence of selected wavelengths of the emission spectrum of the male sample M–2 on time (12 weeks).

Unlike the male samples, the luminescence intensity of both female samples increased over time. The increase in intensity lasted for the first 9 weeks, with the most pronounced during the first 3 weeks. A decrease in intensity was recorded during the final 3 weeks. Figure 4 shows selected wavelengths of the emission spectrum for the female sample F–1. It is also possible to notice relatively high values of the luminescence intensity, which is caused by two identical and mutually overlapping prints on the glass slide. In practice, this situation occurs when, for example, a perpetrator leaves multiple prints on a specific object while manipulating it, which subsequently overlap. This overlap can significantly complicate classical dactyloscopic identification, but it is evident from the measurements that, apart from the higher luminescence intensity, the overlapping prints do not affect the shape of the spectrum.

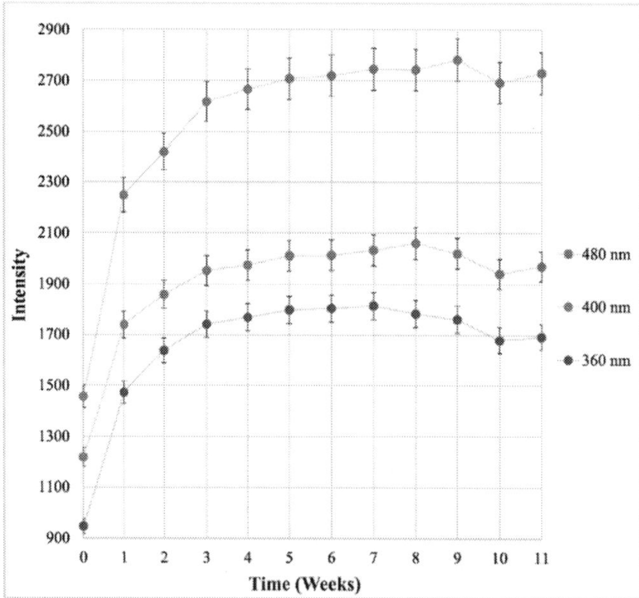

Fig. 4. Dependence of selected wavelengths of the emission spectrum of the female sample F–1 (two overlapping prints) on time.

Selected emission spectra of the female sample F–2 are shown in Fig. 5a. In the range from 350 nm to 430 nm (i.e. approximately in the region of the first peak), a significant change in the luminescence intensity is evident compared to other graphs, specifically its significant decrease. The first peak has practically disappeared and in the range between 380 nm and 410 nm, the emission spectra of all weeks almost overlap. However, the increasing tendency of the luminescence intensity remains, similarly to that of the female sample F–1. The attenuation of the intensity of this sample is caused by the admixture of external contaminants. The print contains trace amounts of cosmetics that got into it when the finger came into contact with the face during the sample collection phase. This is a simulation of a real situation that is relatively common in practice. A latent print is almost never composed only of human sweat and sebum. It also contains various external contaminants, such as the aforementioned cosmetics, varnishes, inks, food residues or gunpowder. The second peak in the region around 480 nm remains almost unchanged (see Fig. 5b), only the emission spectrum of week 0 is probably slightly affected by contaminants, when a steeper decrease in luminescence intensity is visible in the region from 475 nm to 500 nm. The remaining emission spectra do not contain any more visible changes. Similarly to the sample F–1, the luminescence intensity increases over time.

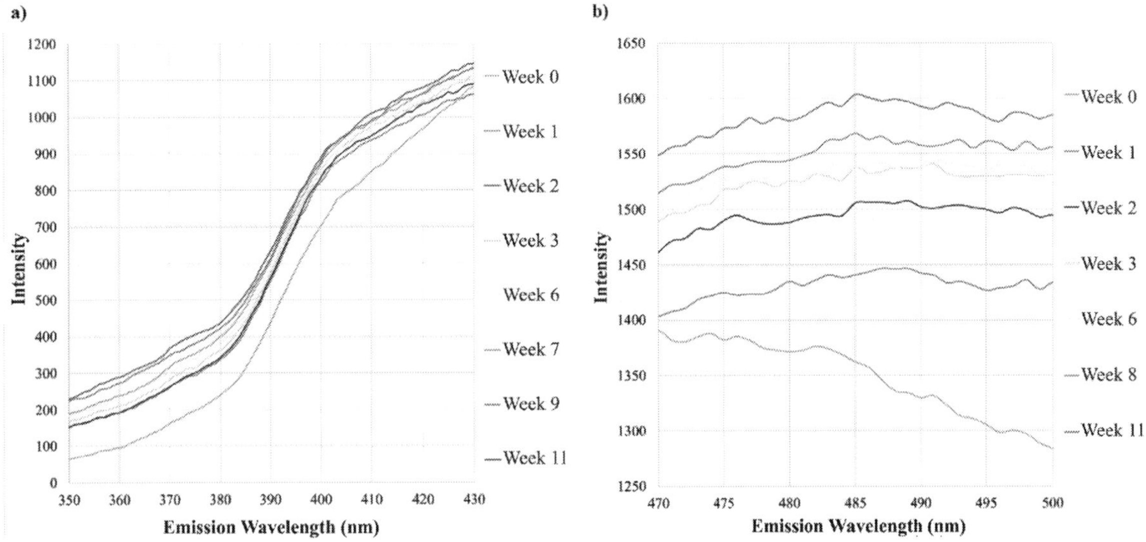

Fig. 5. a) Selected emission spectra of female sample F–2 (containing external contaminants – cosmetics) over time, b) selected emission spectra of female sample F–2 (containing external contaminants – cosmetics) over time (detail of the second peak in the region around 480 nm).

4. Conclusion

In this article, 7 latent fingerprint samples (5 from men, 2 from women) were analyzed for 12 weeks using luminescence spectroscopy to describe the luminescence behavior of the fingerprint over time. Two significant peaks were found in the emission spectrum (in the regions of approximately 360 nm and 480 nm), with the luminescence intensity mainly decreasing over time in male samples and increasing in female samples. It was also found that the amount of material in the fingerprint does not affect the shape of the spectrum itself, but only the value of the luminescence intensity. On the contrary, external contaminants (specifically cosmetics) significantly affect the shape of the spectrum, since absorption occurs at shorter wavelengths. According to the analyzed measurement results, luminescence spectroscopy could potentially be used to estimate the age of latent fingerprints, but the results need to be verified in further studies on a much larger set of samples.

5. Acknowledgments

This work was supported by the Internal Grant Agency of Tomas Bata University under the project No. IGA/CebiaTech/2024/003.

6. References

[1] Svoboda, I. (2016) Criminology, Key Publishing, ISBN 978-80-7418-259-4, Ostrava
[2] De Alcaras-Fossoul, J. (2021). Technologies for Fingermark Age Estimations: A Step Forward, Springer, ISBN 978-3-030-69336-7, Cham
[3] Porada, V. (2007). riminology: (introduction, technique, tactics), Aleš Čeněk Publishing, ISBN 978-80-7380-038-3, Plzeň

[4] Siegel, Jay A. & Saukko, P. & M. Houck, M. (2013) ed. Encyclopedia of forensic sciences. Second edition, Elsevier, ISBN 978-0-12-382165-2, Amsterdam

[5] Straus, J. & Porada, V. (2017). Theory, methods and methodologies of criminology. Aleš Čeněk Publishing, ISBN 978-80-7380-666-8, Plzeň

[6] Bumbrah, G. S., Sharma, R. M., & Jasuja, O. P. (2016). Emerging latent fingerprint technologies: a review. Research and Reports in Forensic Medical Science, pp. 39-50., ISSN 2230-2476, doi:10.2147/RRFMS.S94192

[7] Archer, N. E., Charles, Y., Elliott, J. A., & Jickells, S. (2005). Changes in the lipid composition of latent fingerprint residue with time after deposition on a surface. Forensic Science International, 154(2-3), 224-239, ISSN 03790738, doi:10.1016/j.forsciint.2004.09.120

[8] De Alcaras-Fossoul, J. & Mestres Patris, C. & Balaciart Muntaner, A. & Barrot Feixat, C. & Gené Badia, M. (2013) Determination of latent fingerprint degradation patterns—a real fieldwork study. International Journal of Legal Medicine, 127(4), pp. 857-870, ISSN 0937-9827, doi:10.1007/s00414-012-0797-0g

[9] Straus, J. & Hlavín, P. (2006). Contribution to the determination of the age of latent dactyloscopic traces. Criminology: a journal for criminological theory and practice. Praha: MV ČR, press department and public relations, 39(4), pp. 275–291, ISSN 1210-9150

[10] Watson, P., Prance, R. J., Beardsmore-Rust, S. T., & Prance, H. (2011). Imaging electrostatic fingerprints with implications for a forensic timeline. Forensic science international, 209(1-3), e41-e45, ISSN 0379-0738, doi:10.1016/j.forsciint.2011.02.024

[11] Frick, A. A., Chidlow, G., Goodpaster, J. V., Lewis, S. W., & Van Bronswijk, W. (2016). Monitoring compositional changes of the lipid fraction of fingermark residues deposited on paper during storage. Forensic Chemistry, 2, pp. 29-36, ISSN 24681709, doi:10.1016/j.forc.2016.09.001

[12] Girod, A., & Weyermann, C. (2014). Lipid composition of fingermark residue and donor classification using GC/MS. Forensic science international, 238, 68-82, ISSN 03790738, doi:10.1016/j.forsciint.2014.02.020

[13] Andersson, P. O., Lejon, C., Mikaelsson, T., & Landström, L. (2017). Towards Fingermark Dating: A Raman Spectroscopy Proof-of-Concept Study. ChemistryOpen, 6(6), pp. 706-709, ISSN 21911363, doi:10.1002/open.201700129

[14] Williams, D. K., Brown, C. J., & Bruker, J. (2011). Characterization of children's latent fingerprint residues by infrared microspectroscopy: Forensic implications. Forensic Science International, 206(1-3), pp. 161-165, ISSN 03790738, doi:10.1016/j.forsciint.2010.07.033

[15] Popa, G., Potorac, R., & Preda, N. (2010). Method for fingerprints age determination. Rom J Leg Med, 2, pp. 149-154, ISSN 12218618, doi:10.4323/rjlm.2010.149

[16] Mikulicova, M[ichaela]; Zimek, O[ndrej] & Kresalek, V[ojtech] (2019). Fluorescence of Selected Polymer Banknotes, Proceedings of the 30th DAAAM International Symposium, pp.0746-0752, B. Katalinic (Ed.), Published by DAAAM International, ISBN 978-3-902734-22-8, ISSN 1726-9679, Vienna, Austria

[17] DOI: 10.2507/30th.daaam.proceedings.102

[18] van Dam, A., Schwarz, J. C., de Vos, J., Siebes, M., Sijen, T., van Leeuwen, T. G., ... & Lambrechts, S. A. (2014). Oxidation monitoring by fluorescence spectroscopy reveals the age of fingermarks. Angewandte Chemie International Edition, 53(24), 6272-6275, ISSN 14337851, doi:10.1002/anie.201402740

[19] Saitoh, N., & Akiba, N. (2005). Ultraviolet fluorescence spectra of fingerprints. The Scientific World Journal, 5(1), 355-366, ISSN 1537-744X, doi:10.1100/tsw.2005.43

[20] Akiba, N., Kuroki, K., Kurosawa, K., & Tsuchiya, K. I. (2018). Visualization of aged fingerprints with an ultraviolet laser. Journal of forensic sciences, 63(2), 556-562, ISSN 0022-1198, doi:10.1111/1556-4029.13588

DOI: 10.2507/36th.daaam.proceedings.031

ENERGY-AWARE TASK SCHEDULING FOR SECURE AUTONOMOUS VEHICLES: AN ANALYSIS OF CURRENT SYSTEM DESIGNS

Suad Nesimi & Ervin Domazet

This Publication has to be referred as: Nesimi, S[uad] & Domazet, E[rvin] (2025). Energy-Aware Task Scheduling for Secure Autonomous Vehicles: An Analysis of Current System Designs, Proceedings of the 36th DAAAM International Symposium, pp.0231-0238, B. Katalinic (Ed.), Published by DAAAM International, ISBN 978-3-902734-47-1, ISSN 1726-9679, Vienna, Austria
DOI: 10.2507/36th.daaam.proceedings.031

Abstract

The adoption of heterogeneous architectures is driven by these demands, integrating CPUs, GPUs, and FPGAs to perform diverse tasks such as perception, planning, and control on different hardware. Such platforms promise enhanced performance and greater flexibility. This paper presents an overview of how contemporary industry and research systems act in task and energy scheduling on heterogeneous AV platforms. We analyse the solutions at hand - including NVIDIA DRIVE, Tesla's Full Self-Driving Computer, Mobileye EyeQ and AUTOSAR Adaptive to examine how they coordinate processing units and enforce security isolation. At the same time, we investigate how they balance energy consumption with real-time performance. The paper also highlights the trade-offs that these systems make and identifies key limitations like scheduling of mixed criticalities being lacklustre, for lack of a better phrase, integrated into the current schemes, as well as the absence of standardised thermal-aware mechanisms. Additionally, we must consider fragmented security models. By synthesising these existing findings and observations, as well as our research conclusions, we point out future research directions that aim to bring us closer to the system-level designed applications of energy-saving and secure autonomous driving platforms.

Keywords: autonomous vehicle; heterogeneous components; secure systems; task allocation; thermal management.

1. Introduction

Many computation-intensive tasks need to be conducted in modern AVs, including but not limited to multi-sensor fusion, perception, real-time control and decision making. These are high-performance, low-latency responsibilities that require task determinism to operate reliably and safely in dynamic environments. To satisfy these demanding conditions, AV platforms are more equipped with heterogeneous computation engines including general-purpose CPUs: massively parallel GPUs and hard- or software reconfigurable FPGAs [1], [3], [6]. The structure offers flexibility and processing capability to perform simultaneous parallel computation with scalability and real-time performance. But there is a catch to this flexibility.

In such environments, the project of task scheduling, energy efficiency and system security is very difficult. Energy-aware task scheduling has recently emerged as an important research thrust to reconcile the trade-off between performance requirements of safety-critical applications plans with limited energy budgets and heat dissipation constraints [2], [4], [5]. Such scheduling is charged with the obligations of consuming minimal energy while meeting time deadlines for critical perception and control tasks under unknown workloads and hardware. Recent advances in the industry, such as NVIDIA DRIVE AGX Orin [8], Tesla's Full Self-Driving (FSD) computer [9], and Mobileye EyeQ5 processors are also examples of the trend of heterogeneous architectures in AV, which integrates CPUs, GPUs and specialized accelerators for deep learning inference and sensor fusion [10]. These systems highlight the benefits of tightly integrated heterogeneous processing for enabling real-time decision-making, but they also open new challenges that come with sharing computing resources between workloads and ensuring isolation and security. Projects such as AUTOSAR Adaptive Platform further highlight the need for secure task coordination and communication in distributed automotive systems [11].

Although extensive academic research and industrial adoption have been achieved, to the best of our knowledge, there exists no comprehensive application which unifies the aspects (i.e., energy-efficiency; real-time scheduling; secure operation) during AV-heterogeneous SoC design. Previous work has investigated methods such as task-structure-aware scheduling [6] or GPU-based deep learning optimization [7], as well as dynamic system reconfiguration under energy constraints [2]. Nevertheless, the practical and theoretical challenges in incorporating such approaches into end-to-end autonomous driving stacks persist. The goal of this paper is to present a systematic literature review and critical study on the recent energy-aware task scheduling approaches for autonomous vehicular systems running on heterogeneous computing architectures. Rather than introduce new algorithms, we are concerned with observing how existing solutions, both from commercial trends and academic concepts, tend to organize task allocation and processor coordination AND secure execution under constraints of efficiency and real-time guarantee. The observations reported here are intended to recognize current challenges and suggest directions for designing future energy-aware and secure heterogeneous computing platforms in autonomous vehicle systems.

2. Background

As autonomous vehicles (AVs) have become smarter, the computational complexity of their software systems has grown with each new advance. These systems must process an enormous quantity of data from cameras, Light Detection and Ranging (LiDAR) sensors, radars and other sensors. For these systems to do this in a timely and safe manner, as well as (more difficult yet) meeting strict real-time criteria, is crucial, especially for brake control, obstacle avoidance and lane-keeping tasks. However, to meet these performance demands, today's AV platforms use a variety of processing elements. These range widely from more typical CPUs like ARM-architecture microprocessors to GPUs and often FPGAs (or ASICs). Each processor has its advantages:

1. High-performance sequential control tasks, logic-heavy processing and system orchestration are all handled well by CPUs.
2. With their massive parallelism, GPUs provide a great fit for computationally intensive tasks such as deep learning inference and point cloud processing.
3. FPGAs are commonly used in preprocessing sensor data or safety-critical modules. They offer low latency along with highly customised data path control facilities.

This wide range of processing units not only brings flexibility and power but also raises fundamental questions of how to match tasks effectively across them in ways which strive for a balance between categories such as performance, real-time constraints, and security.

2.1 Task Scheduling in Autonomous Systems

Task scheduling is the process of determining which tasks should run on which processor at any given time. In the context of AVs, computational tasks differ not only in complexity but also in their urgency and timing. Scheduling decisions that are made poorly will mean that a task misses its deadline for whatever reason; this usually results in unsafe behaviour. Worse still, if tasks are not isolated or lack security, then faults and cyber threats can spread from module to module unimpeded.

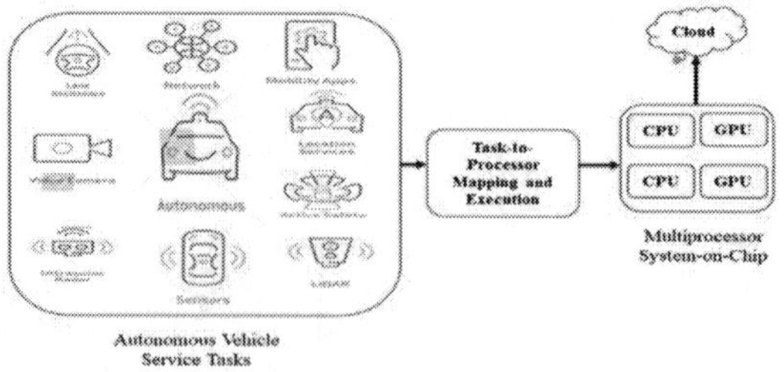

Fig. 1. Task scheduling in autonomous vehicles

The task-to-processor assignment used in this work is shown in Fig. 1, adapted from [14]. On the left, perception and actuation workloads generate input flows that enter the central Task-to-Processor Mapping and Execution stage, where they are scheduled onto heterogeneous units. The right-hand side shows the on-board multiprocessor SoC (CPUs/GPUs) executing the tasks, while a cloud uplink is reserved for non-safety-critical analytics and storage. Arrows indicate data flow and motivate why the placement fits the scheduling discussion.

2.2 Energy-Aware Scheduling

For AVs, especially those on electric vehicle (EV) platforms, energy efficiency is a crucial consideration. Each kind of processing unit is a glutton when it comes to power, and poor scheduling gives rise to overheating, diminished battery range or the requirement of under-clocking to respond to thermal pressure. Energy-aware scheduling involves organising tasks onto processors to minimise their energy consumption without degrading performance or safety. Techniques include dynamic workload balancing, power gating and keeping track of hot sections that are temperature sensitive.

2.3 Task Criticality and Mixed-Criticality Systems

The tasks in AV systems may be classified as critical or non-critical according to their influence on safety. Critical tasks include real-time control, braking and sensor fusion modules. System breakdowns or accidents are the direct result of malfunctions in these areas.

2.4 Non-critical Tasks

Non-critical tasks include user-interface processing, infotainment, and logging. If the system allows both types of tasks to share a hardware platform, it is a mixed-critical system. Careful scheduling and isolation are necessary to avoid situations where non-critical faults or delays interfere with the critical operations of safety.

2.5 Secure Execution and Isolation

Security is a major concern. Autonomous vehicles are subject to both physical and network-based threats. This necessitates secure execution through task isolation. Hardware features such as memory protection, privilege separation or trusted execution environments (e.g., ARM TrustZone, Intel SGX). Software containers and microkernels that enforce task separation and restrict access to shared resources. Effective task scheduling should take these requirements into account, ensuring that tasks from different security domains do not interfere with one another.

2.6 Purpose of this Paper

This paper is not going to introduce a new task scheduling algorithm or framework. Rather, it focuses on: Looking at current industry and research practices for energy-aware task scheduling on AVs. How these systems succeed or fail in managing criticality, coordinating processors to satisfy energy constraints and providing protection. Highlighting limitations, gaps and new opportunities for more integrated, intelligent scheduling strategies in the future.

3. Review of Existing Systems

In this part, we will try to review a selection of prominent industry platforms and academic systems that support AV functionality. Each is titled according to the target task scheduling, power efficiency, and secure execution on substrate platforms, which contain security processing units within central processing units implicitly or explicitly as part of their design.

3.1 NVIDIA DRIVE

NVIDIA's DRIVE [8] platform is the most widely used autonomous driving solution for commercial and research AVs. It is equipped with multi-core ARM CPUs, high-throughput GPUs and dedicated Deep Learning Accelerators (DLAs). Task scheduling is usually performed by applications using NVIDIA's SDKs (e.g., DriveWorks, TensorRT), and then developers are responsible for the detailed deployment of the workload CPU, GPU and DLA. While NVIDIA's design does support high-performance inference and real-time sensor processing, its ability to schedule for energy leads to less tightly organised scheduling behaviour. Developers may use Dynamic Voltage and Frequency Scaling (DVFS) or thermal-aware power domains, but these are hardware features and cannot be considered software-level schedulers in any meaningful sense of the term. Security is handled at the system level, with support for secure boot, root of trust, memory partitioning and so on. Little integration between the scheduling logic and isolation means little attention is paid to incorporating these principles into concrete mechanisms for practical enforcement of the security policy.

3.2 Tesla Full Self-Driving Computer

Tesla's FSD [9] computer has a custom-built SoC architecture that consists of two redundant chips. Inside are CPU clusters, GPU cores, and specialised Artificial Intelligence (AI) accelerators on each chip. The platform has been developed to be highly redundant for the parallel execution of perception and planning pipelines on different chips. Scheduling is an internal detail, but Tesla seems to do static partitioning and use the dedicated compute blocks for some workloads. Energy management is probably implemented at the level of silicon, with dedicated logic for reducing power consumption under idle or partial-load conditions. Security mechanisms such as fault detection and watchdogs are found, but due to the closed-end nature of the system, it is hard to evaluate how dynamic or adaptable the task scheduling model is, especially when different driving situations or processor loads are taken into consideration.

3.3 Mobileye EyeQ

Mobileye's EyeQ [10] series is dedicated to vision processing and deployed in many Levels 2–3 AVs. Its design is based not on GPGPUs but on highly optimised ASICs. This is very energy efficient, but less flexible. EyeQ systems are deterministically assigned hardware-pipelined (statically) such that performance can be guaranteed. Although it provides fine-grained control over when actions execute and how much energy they consume, it doesn't provide a way to either schedule actions dynamically during execution or to adapt to changes in the environment or the availability of resources. The platform is tailored for use in the context of ISO 26262 ASIL-D requirements with high safety integrity levels, which is less appropriate for full-stack autonomous systems making on-board decisions which require the reconfiguration of tasks

3.4 AUTOSAR Adaptive Platform

AUTOSAR Adaptive [11] is a software platform designed for new-generation ECUs, which are used in ADAS & AVs. It is capable of service-oriented communication, POSIX-compliant OS environments and deterministic application execution. AUTOSAR Adaptive does not support energy-aware scheduling. On the contrary, power management is commonly devoted to the ECU as OEM-dependent approaches. Also, task scheduling policies depend on the RTOS and the middleware selected by the designer. From the security point of view, AUTOSAR Adaptive provides some degree of support for role-based access control, secure communication and application separation. But these capabilities are not coordinated well with scheduling logic, and coordination between energy, timing, and security becomes suboptimal.

3.5 ROS2-Based Research Platforms

ROS2 [12] is commonly employed both in academic and open-source AV projects. It facilitates the distributed, modular design of AV functions, with nodes executing on various processing nodes. Real-time execution is also partially supported in ROS2 using the OpenSplice DDS middleware, and with Real-Time Operating Systems (RTOS) on systems such as RT Linux. There is no built-in support for energy-aware scheduling in ROS2. All energy optimisations must occur from the OS up or at a hardware level. Likewise, task criticality and fault isolation must be imposed manually, e.g. using containerization or external safety supervisors. Despite its flexibility, ROS2 is still constrained in system-level integration of scheduling, energy and security, deciding between the two rather than being a prototyping platform rather than a production-level AV solution.

4. Analysis and Observations

Based on these platforms, some trends and deficiencies can be identified, which indicate the status and immaturity of autonomous vehicle integrated energy-aware secure scheduling to a large extent.

4.1 Lack of Dynamic Scheduling Intelligence

Most of the platforms are based on a static or semi-static task assignment model. Real-time dynamic scheduling that considers the load on the network, available power, or the criticality of the task has not been well addressed. This may result in over-utilisation (and thermal strain) or under-utilisation (and wasted performance).

4.2 Disjoint Energy and Security Management

A lack of consideration is given to energy and security as a joint issue. Although energy principles can exist (DVFS, idle modes) and security features like secure boot can be supported, few systems integrate these during scheduling decisions. There is no known system that automatically schedules tasks to achieve both power efficiency and an isolation domain.

4.3 Minimal Support of Mixed-Criticality Systems

Even when these assumptions are not posed, most systems that are aware of mixed-criticality requirements (e.g., AUTOSAR) provide very limited actual runtime separation and priority management of such tasks. This is dangerous for system stability and functional safety in case of high calculating load, or partial failure.

4.4 Limited Interoperability and Openness

Closed-source providers such as Tesla and Mobileye provide minimal visibility into how they make scheduling decisions, so it is challenging to draw comparisons between solutions. At the same time, open platforms such as ROS2 are not yet mature and integrated enough for commercial deployment of AV (Table 1). To facilitate a qualitative comparison of the five AV platforms in this paper, Table 1 summarises them. The table compares each system according to 5 main characteristics: task scheduling mechanism, energy-awareness, security and isolation functions, support for mixed-criticality workloads and openness to the research/development community. The following table summarises the portability and integration maturity of these platforms. For example, to schedule in the power and task domain of developer-level control on NVIDIA DRIVE; however, it does not include something for criticality-based scheduling. Meanwhile, Mobileye's EyeQ supports extremely efficient computation, but is inflexible because of its fixed-function nature. This table aims to provide readers and reviewers with an overall view of the strengths, weaknesses and limitations for each system immediately, facilitated by detailed discussion presented in the previous section.

Platform	Task Scheduling Approach	Energy-Aware Support	Security & Isolation	Mixed-Criticality Support	Openness / Availability
NVIDIA DRIVE	Manual/ developer-driven	Hardware-level DVFS, power domains	Secure boot, memory partitioning	Partial (developer managed)	Partial (SDK access)
Tesla FSD	Likely static, internal only	Custom chip design, likely hardware-managed	Hardware watchdogs, dual redundancy	Via chip separation	Closed source
Mobileye EyeQ	Static/hardware pipeline	Extremely efficient ASIC-based design	Meets ASIL-D, strict isolation	Limited	Closed source
AUTOSAR Adaptive	OS/middleware dependent	Vendor-specific implementations	Role-based access, secure comms	Theoretical support	Open source
ROS2-based Systems	OS-level & node-based (flexible)	None natively	Optional via SROS2, containers	Developer-defined only	Fully open source

Table 1. Comparison of task scheduling, energy-awareness, security, and system openness across representative autonomous vehicle platforms.

5. Experimental Results

While the main topic of this work is a thorough overview of existing systems for energy-aware task scheduling in secure AVs, we present a simple simulated example below to demonstrate the challenges when allocating tasks to a target platform with a heterogeneous amount of processing capabilities. This is not intended as a performance benchmark. The purpose is to demonstrate the concept of how scheduling impacts energy and execution time in real AV systems. To emulate a realistic workload in an autonomous driving context, we define three representative tasks:

- T1: Object Detection (compute-intensive, latency-tolerant)
- T2: Emergency Braking (time-critical, safety-sensitive)
- T3: Lane Keeping Control (low latency, real-time loop)

These tasks are mapped to three processor types commonly used in AV platforms:

- CPU (deterministic control, lower power)
- GPU (parallel high-throughput, high energy)
- FPGA (low latency, energy-efficient)

Simulation results show that Scenario B (Energy-aware mapping) saves 45% of the total execution time and 41% of the energy consumption compared to Scenario A, which uses an energy-unaware allocation. Placing the safety-critical Emergency Braking on the CPU improves both energy efficiency and the task determinism at runtime, while transferring the Lane Keeping Control to the FPGA provides the low-latency processing.

While this is a step toward increasing the efficacy of both offline and online task scheduling policies based on complex real-world data, we posit this as an encouraging average case that should be used to vet other scheduling policies for intelligent frameworks in simulation, which, even in simulation, have the potential to provide large improvements in terms of both energy and response time in real time AV systems. Although actual results may vary according to hardware and task models, the direction demonstrated indicates the need for context-aware, dynamic scheduling schemes on heterogeneous AV platforms. This is a demonstrative simulation, not based on actual hardware experiments. However, it does help to confirm the usefulness of the design considerations presented in the earlier sections. We may also develop this simulation on a real-time middleware platform (e.g., ROS2 with DDS), and model it on machines (e.g., NVIDIA Jetson, Drive AGX, AUTOSAR Adaptive-based ECUs). We provide a light runtime-supervision layer, which subscribes to raw and pre-processed sensor topics, and performs inference with small neural models on the CPU side. The monitor emits raised events (anomalous, drifting and missing frames) that the scheduler treats as soft signals for admission control and mode switching without rejecting components responsible for controlling any safety-critical loops. To minimize overheads, inference is rate-limited and memory-pinned and put in a separate domain. This architecture is motivated by previous results indicating that NN-based monitoring of sensor signals can be performed with an acceptable latency and resource footprint for online applications [13].

Task	**Scenario A**	**Exec. Time (ms)**	**Energy (mJ)**	**Scenario B**	**Exec. Time (ms)**	**Energy (mJ)**
Object Detection	CPU	40	35	GPU	18	60
Emergency Braking	GPU	12	50	CPU	10	12
Lane Keeping Control	GPU	10	48	FPGA	6	7
Total		62	133		34	79

Table 2. Comparison of task execution time and energy consumption under naive vs. energy-aware task scheduling strategies.

Table 2 illustrates the impact of assigning perception, safety-critical tasks, and control to the appropriate processing units, such as CPU, GPU, and FPGA. The energy-aware scenario achieves significantly lower overall energy usage and improved response times through criticality-aware task allocation and hardware.

6. Challenges and Limitations

The overview of today's systems highlights the multiple challenges that the research community and industry face in paving the way towards more dependable, secure, and energy-efficient autonomous vehicle platforms.

6.1 Unified Scheduling Strategies

There is an urgent demand for schedulers which take timing constraints, energy budget, processor utilisation and task criticality into consideration. A unified approach will require running on multiple hardware domains (CPU, GPU, FPGA) on the fly.

6.2 Cross-Layer Security and Scheduling Integration

In the future, security should be part of the task scheduler rather than a bolt-on. e.g., the verification of the operation's origin, the assignment of rights to be executed, or the separation during execution of potentially unsafe or faulty code parts.

6.3 Middleware-Level Support

Native middleware-level APIs for measurement of power, task profiling and dynamic load balancing should be supported by AV stacks. It would have allowed application developers to write scheduling-friendly code without having to understand much about the hardware.

6.4 Evaluation Frameworks and Benchmarks

In such systems, since users' job experience makes their experiments more reproducible and results more reliable, we must procure open benchmarking environments to perform comprehensive performance evaluation of scheduling and energy management strategies under realistic workloads and fault conditions. They would facilitate academia-industry collaboration on best practices benchmarking.

6.5 Implications / Consequences

Enforcing energy and security-aware scheduling can further enhance AV range and diminish thermal throttling by smarter DVFS and load placement, better respect of deadlines for safety-critical loops through criticality-aware mapping, reduce lifetime costs by lowering the impact of thermal stresses on components, as well as lower attack surface area by enforcing isolation domains at scheduling time. Possible downsides: Complexity and potential overhead for the scheduler of added data that needs to be limited and justified (ISO 26262). Power, thermal, and isolation telemetry may demand more observability. Portability or certification issues involving various vendor APIs, Vendor lock-in if there are proprietary power/thermal/security interfaces.

6.6 Limitations

We have built a qualitative review, relying on what is publicly available documentation for such platforms as Tesla or Mobileye; this limits the ability to verify our claims through means such as telemetry or isolation mechanisms. Our table shows a simulated system. Although the numbers are easily tested directly on real hardware, they omit memory bandwidth, data-movement overheads, interference into interconnect contention latencies and sensing/actuation lag. Changes in environmental temperature, such as heating up and cooling down processes that respond to changes in how hard you keep both power transformer windings connected to one amplifier or another--these are not considered. Of course, with individual thermal simulations, no actual temperatures or powers reached the physical object being modelled. We also do not provide a single unifying cross-platform benchmark or support, so our work may only generalise to heterogeneous SoCs with similar accelerator mixes from different vendors and power/thermal APIs. Future work in this area must include hardware-in-the-loop experiments, standardised workloads, and physically measured power / temperature fluctuation traces.

6.7 Interpretation of Results

The toy experiment in Table 2 suggests that mapping, which is conscious of criticality and energy-wise lowers both. If you assign Emergency Braking to the CPU and Lane Keeping to the FPGA, the total time falls to (62 → 34 ms), and energy also dips down along the curve (133 → 79 mJ). However, if you move heavy perception into the GPU, it keeps the throughput unchanged. These developments align nicely with earlier findings that (i) heterogeneous offloading enhances speed through acceleration paths lined up well with data structures [6], [7]; and (ii) energy saving comes when safety-critical, tight-loop control goes on low-latency, low-power sources [2], [4]. One key point is that latency is minimised for all critical tasks while levels off are reached in terms of energy/thermal. But because in these benchmarks there is no bus contention or thermal effects penetrating standard workloads, readers might form their own judgment as to the values, as a clue at least. And because isolation-aware placement of real-time control tasks on (CPU, FPGA), and perception tasks on (GPU) is consistent with our thesis, this kind of tested hardware evaluations in Part II support our arguments.

7. Conclusion

The ongoing development of autonomous vehicles is conditioned largely on the progress made at the system level – beyond simply increasing numbers of sensors and increasing the complexity of the processing algorithms – and in the areas of task scheduling, energy efficiency, and secure execution. We had accordingly offered a systematic analysis of how state-of-the-art AVs deal with these factors in multi-processor settings. Although significant improvement of hardware performance and functionality has been achieved, our investigation discovers a huge space in integrated scheduling intelligence, energy-efficient task scheduling, and run-time security isolation. Existing systems generally address them in an isolated manner; thus, the adaptability and robustness under various practical constraints are restricted. Quantifying these gaps and surveying architectural trends, this paper establishes a foundation for future works that seek to narrow the gap between platform potential and software intelligence. Collaboration between industry and academia will be crucial in the pursuit of the subsequent generation of fast, smart autonomous systems that are also efficient, secure, and reliable. Current approaches often concentrate on improving one area while overlooking others. This can make vehicles inefficient or unsafe. The typical use of different architectures, such as CPUs, GPUs, and FPGAs, highlights the need for unified planning tools that can address power consumption, response time, and safety requirements. Dealing with these trade-offs through a combined, multi-layered design will be crucial for developing platforms that are both high-performing and reliable in real-world situations.

8. References

[1] Zou, A.; Liu, Q.; Xu, H. & Wang, Y. (2021). A survey of real-time scheduling on accelerator-based heterogeneous architectures for time-critical applications, Journal of Embedded Systems Research, 15(4), 233–250.

[2] Yi, S.; Kim, T. W.; Kim, J. C. & Dutt, N. (2020). Energy-efficient adaptive system reconfiguration for dynamic deadlines in autonomous driving, IEEE Transactions on Intelligent Vehicles, 5(2), 112–122.

[3] Miao, Z.; Chen, L. & Zhang, P. (2021). Review of task scheduling methods for heterogeneous chips, Journal of Computer Science & Technology, 36(7), 1258–1274.

[4] Li, R.; Feng, Y. & Hu, L. (2021). Energy-aware task scheduling on heterogeneous computing systems with time constraint, ACM Transactions on Embedded Computing Systems, 20(5), Article 45.

[5] Perin, G.; Rosa, L. & De Oliveira, A. (2023). EASE: Energy-aware job scheduling for vehicular edge networks with renewable energy resources, Frontiers in Computer Science, 7, 122–130.

[6] Zhang, X.; Li, C. & Wen, J. (2022). TSSA: Task-structure-aware scheduling of energy-constrained parallel applications in distributed embedded systems, Journal of Systems Architecture, 133, 102790.

[7] Kaur, R.; Singh, M. & Gupta, P. (2022). A survey of advancements in scheduling techniques for efficient deep learning computations on GPUs, Journal of Parallel and Distributed Computing, 165, 13–27.

[8] NVIDIA Corporation (2022). NVIDIA DRIVE AGX Orin Developer Kit: Product brief. Santa Clara, CA, USA.

[9] Greenhill, S. (2021). Inside Tesla's Full Self-Driving computer, IEEE Spectrum, 58(10), 34–41.

[10] Mobileye Ltd. (2021). EyeQ5 system-on-chip for ADAS and AV applications: Technical overview. Jerusalem, Israel.

[11] AUTOSAR Consortium (2021). AUTOSAR Adaptive Platform overview, Release R21-11. Munich, Germany.

[12] Maruyama, Y.; Kato, S. & Azumi, T. (2016). Exploring the performance of ROS 2, in Proceedings of the 13th International Conference on Embedded Software (EMSOFT '16), ACM, New York, 1–10.

[13] Topalova, I. & Radoyska, P. (2023). A model of an intelligent automation system for monitoring of sensor signals with a neural network implementation, Proceedings of the 34th DAAAM International Symposium, 0050–0055, B. Katalinic (Ed.), DAAAM International, Vienna, Austria. DOI: 10.2507/34th.daaam.proceedings.007. ISBN: 978-3-902734-41-9. ISSN: 1726-9679.

[14] Balasekaran, G.; Jayakumar, S.; Pérez de Prado, R. (2021). An Intelligent Task Scheduling Mechanism for Autonomous Vehicles via Deep Learning. Energies, 14(6), 1788. ISSN 1996-1073. DOI: 10.3390/en14061788.

DOI: 10.2507/36th.daaam.proceedings.032

CALIBRATION OF SEDIMENT TRANSPORT PARAMETERS IN THE XBEACH-G NUMERICAL MODEL ON ARTIFICIAL GRAVEL BEACH PLOČE

Hanna Milicevic, Dalibor Carevic & Damjan Bujak

This Publication has to be referred as: Milicevic, H[anna]; Carevic, D[alibor] & Bujak, D[amjan] (2025). Calibration of Sediment Transport Parameters in the XBeach-G Numerical Model on Artificial Gravel Beach Ploče, Proceedings of the 36th DAAAM International Symposium, pp.0239-0243, B. Katalinic (Ed.), Published by DAAAM International, ISBN 978-3-902734-47-1, ISSN 1726-9679, Vienna, Austria
DOI: 10.2507/36th.daaam.proceedings.032

Abstract

Artificial gravel beaches along the Croatian Adriatic coast are used for recreation and coastal protection but are subject to continuous morphological changes due to wave action and storm impacts. Numerical modelling of these processes requires precise calibration of sediment transport parameters. This study focuses on calibrating the inertia coefficient (*ci*), the angle of repose (Φ) , and the transport coefficient (γ) in the XBeach-Gravel numerical model to improve its accuracy under conditions characteristic of the Adriatic Sea. The calibration is based on field measurements using photogrammetry and unmanned aerial vehicles, enabling high-resolution monitoring of morphodynamic changes. By adjusting these parameters, the XBeach-G model, originally developed for high-energy waves, is adapted to better simulate berm formation and crest elevation processes characteristic of gravel beaches in the Adriatic. The research results enhance the model's reliability in predicting sediment transport dynamics and contribute to improving maintenance strategies for artificial gravel beaches in Croatia.

Keywords: XBeach-Gravel; berm formation; artificial beaches; gravel beaches

1. Introduction

Gravel beaches play a vital role in protecting the coastline from erosional processes and storm wave action. The Croatian coast is rich in gravel beaches, which can be of natural origin or artificially constructed to increase recreational capacity for tourism and to enhance coastal protection. Due to the intensity of hydrodynamic processes, artificial beaches often require continuous maintenance. Monitoring the morphodynamic changes of gravel beaches is essential for their long-term sustainability and management. Advanced methods such as drone-based photogrammetry [1] enable detailed monitoring of beach changes. In addition to field measurements, numerical models such as XBeach are increasingly used to simulate beach morphodynamics in the cross-shore direction, supporting predictions of coastal response under various hydrodynamic conditions [2].

Although the development of numerical models has significantly improved the understanding of sandy beach dynamics [3], [4], studies focusing on gravel beaches remain scarce. The XBeach-G model, a modified version of XBeach, was specifically developed to better simulate processes associated with gravel [2]. Its accuracy strongly depends on parameters such as the inertia coefficient (*ci*), sediment transport coefficient (γ), and angle of repose (Φ). Proper calibration of these parameters is crucial to obtaining reliable results. Previous studies have demonstrated that XBeach-G successfully reproduces cross-shore sediment transport processes, including erosion and accretion, during storm events with significant wave heights exceeding 5 m (Hs > 5 m) [5], [6] . However, conditions characteristic of the Adriatic Sea where significant wave heights rarely exceed 3 m and wave periods are shorter remain poorly investigated.

On the Croatian coast, berm formation and crest buildup are the most common morphological responses to storm events, driven by the relatively low wave heights and short periods. The aim of this study is to further improve the predictive capability of the XBeach-G model under Adriatic conditions. The analysis focuses on the influence of varying boundary conditions, such as significant wave height, wave period, and tidal level, while simultaneously optimizing the aforementioned model parameters (*ci*, γ, and Φ) to achieve a more realistic representation of berm formation and crest buildup.

2. Materials and Methods

2.1. XBeach-Gravel model Setup

Morphodynamic changes on the artificial gravel beach Ploče in Rijeka were monitored through 19 geodetic surveys conducted between January 17, 2020, and February 26, 2021. The methodology and results are described in detail in [1]. For the one-dimensional model simulations, four storm events with fixed wave parameters were selected: UAV3&4 (Hsmax = 1.5 m, Tp = 4.6 s, tide = 0.32 m), UAV4&5 (Hsmax = 1.6 m, Tp = 4.6 s, tide = 0.57 m), UAV5&6 (Hsmax = 2.1 m, Tp = 5.0 s, tide = 0.57 m), and UAV12&13 (Hsmax = 0.64 m, Tp = 3.52 s, tide = 0.49 m). Since the applied model is one-dimensional and specifically developed for simulating cross-shore sediment transport, profiles with minimal longshore transport influence were chosen (UAV3&4, UAV4&5, UAV5&6 – profile 10; and UAV12&13 – profile 20), as shown in Fig. 1.

For the calibration of the XBeach-Gravel numerical model, the following parameters were varied: sediment transport coefficient (γ) = 0.5, 1, and 3; inertia coefficient (*ci*) = 0.5, 1, and 2; and angle of repose (Φ) = 35°, 45°, and 55°. The calibration aimed to achieve a Brier Skill Score (*BSS*) as close as possible to 1, indicating a perfect agreement between simulated and field-measured beach profile changes. Additional parameters such as hydraulic conductivity (*kx*), median grain size (*D50*), and simulation duration were adopted according to the recommendations and calculation guidelines from previous studies [7].

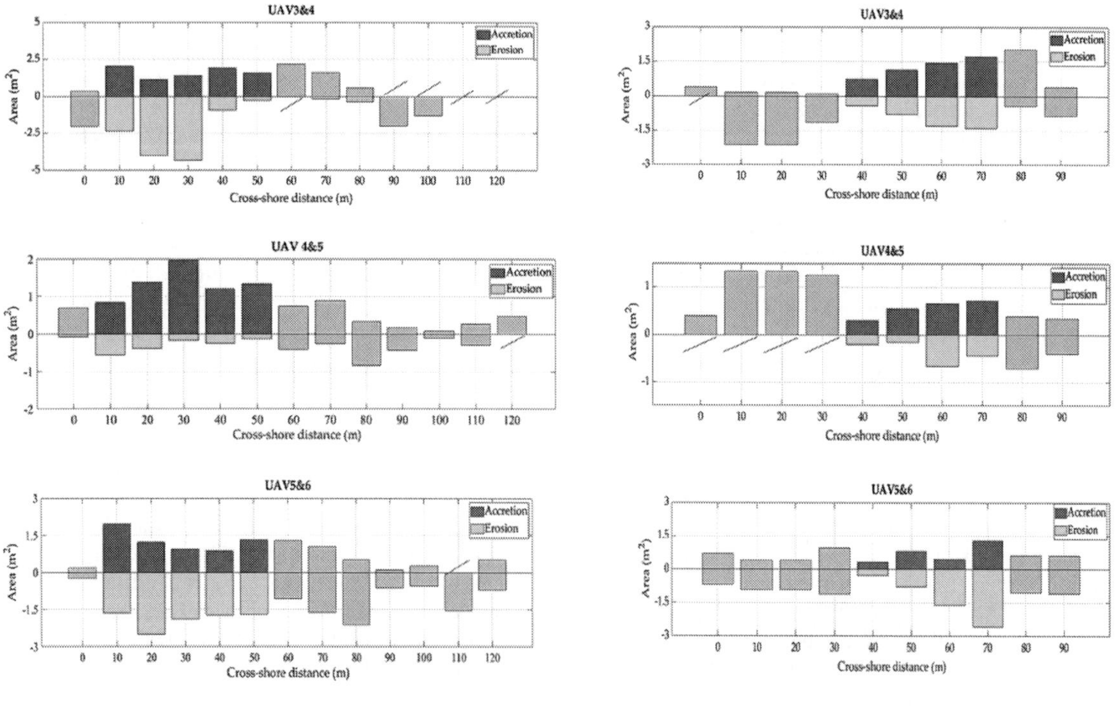

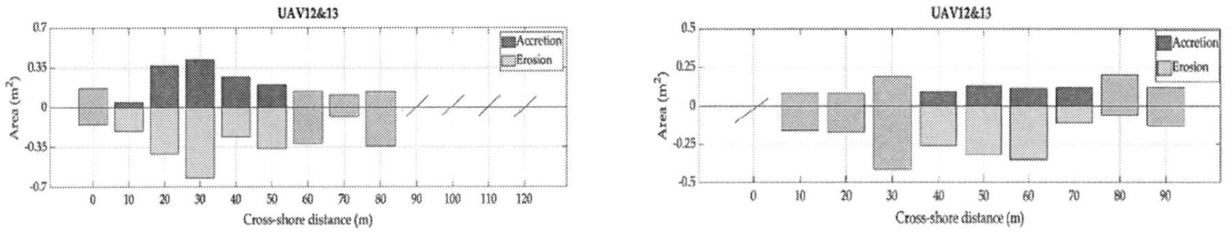

Fig. 1. Erosion and accretion diagrams on the western side of Ploče Beach for the selected storm events UAV3&4, UAV4&5, UAV5&6, and UAV12&13. Grey bars indicate profiles under the local influence of coastal groynes and the concrete promenade wall

3. Results and Discussion

A total of 108 simulations were carried out for four storm events on the western side of the artificial gravel beach Ploče in Rijeka. The BSS values for different combinations of inertia coefficient, sediment transport coefficient and angle of repose are shown in Tables 1, 2, and 3. The highest average BSS value of 0.89 was obtained with the parameter set (Φ = 35°, γ = 0.5, ci = 0.5) for which all analysed storm events were classified as "excellent" according to the Van Rijn criterion [8]. These results confirm the high sensitivity of the model to the selected transport parameters, especially for waves with low to moderate energy, which are characteristic of the eastern Adriatic. The strong decrease of the BSS values with increasing sediment transport coefficient (γ = 1 and γ = 3) to 0.87 and 0.38 indicates that too high transport rates lead to an overestimation of erosion or deposition processes. This tendency is consistent with the results of [2] and [6], which emphasised the importance of adjusting the transport coefficient to the local wave and sediment conditions. The combination (Φ = 45°, γ = 0.5, ci = 0.5) also delivered "excellent" results with a BSS value of 0.81. However, increasing the sediment transport coefficient (γ = 1 and γ = 3) led to a further decrease in model performance compared to Φ = 35°, with the average BSS values dropping to 0.67 and –0.46, respectively.

The largest decrease in performance was observed for Φ = 55° across all transport coefficient values (γ = 0.5, γ = 1, γ = 3), with average BSS values of 0.50, –0.05 and –2.23, respectively. These results suggest that higher angles of repose, although potentially valid for certain gravel sizes, may not accurately represent the dynamic equilibrium of coarse sediments under typical Adriatic storm conditions. Furthermore, increasing the inertia coefficient to ci = 1 and ci = 2 also led to a decrease in average BSS values. This suggests that lower inertia values better reflect sediment movement in short-period wave climates, where transport is driven by swash and uprush– backwash cycles rather than high momentum currents. This interpretation is consistent with [8], who observed similar behaviour on mixed and gravel beaches using process-based models. Overall, the parameter combination Φ = 35°, γ = 0.5 and ci = 0.5 was the most successful in replicating the observed morphological changes, confirming the ability of the model to capture the main sediment transport processes when calibrated appropriately. These results provide a clear answer to the research question: XBeach-G can be reliably calibrated for artificial gravel beaches under moderate wave energy conditions, provided that the model parameters are adjusted to the site-specific characteristics. The results not only improve the applicability of the model in the Adriatic context, but also contribute to broader efforts in developing sustainable beach maintenance strategies under changing climatic conditions. Despite the successful calibration of XBeach-G and its good performance under selected conditions in the Adriatic, several limitations of the current study should be recognised. First, the use of a one-dimensional model limits the analysis to cross-shore processes and does not account for potential longshore sediment transport, which may play a role in certain morphological responses. In addition, the calibration of the model was based on a limited number of storm events (four), which are representative but do not represent the full variability of wave and tidal conditions. The boundary conditions were selected based on available measurements and do not reflect a probabilistic or long-term climate spectrum. Furthermore, the performance of the model is highly site-specific, as the parameter values optimised for Ploče beach may not be directly applicable to other locations without further calibration.

BSS	Φ35°, γ 0.5	Φ35°, γ 1	Φ35°, γ 3	Φ45°, γ 0.5	Φ45°, γ 1	Φ45°, γ 3	Φ 55°, γ 0.5	Φ55°, γ 1	Φ55°, γ 3
UAV3&4	0.86	0.89	0.87	0.91	0.93	0.95	0.81	0.95	0.94
UAV4&5	0.84	0.83	0.51	0.66	0.42	-0.77	-0.06	-1.34	-4.68
UAV5&6	0.92	0.83	-0.83	0.80	0.58	-2.32	0.60	-0.02	-4.60
UAV12&13	0.93	0.96	0.95	0.85	0.73	0.32	0.63	0.20	-0.58
Average	0.89	0.87	0.38	0.81	0.67	-0.46	0.50	-0.05	-2.23

Table 1. Brier Skill Score (*BSS*) values for the calibration of seven storm events under different values of angle of repose (Φ), sediment transport coefficient (γ), and inertia coefficient ($ci = 0.5$)

BSS	Φ35°, γ 0.5	Φ35°, γ 1	Φ35°, γ 3	Φ45°, γ 0.5	Φ45°, γ 1	Φ45°, γ 3	Φ 55°, γ 0.5	Φ55°, γ 1	Φ55°, γ 3
UAV3&4	0.95	0.95	0.89	0.96	0.93	0.81	0.93	0.86	0.62
UAV4&5	0.60	0.51	0.16	-0.13	-0.45	-1.49	-1.26	-1.96	-4.75
UAV5&6	0.38	-0.08	-1.48	-0.20	-0.95	-2.93	-0.72	-2.03	-5.13
UAV12&13	0.91	0.85	0.82	0.74	0.48	0.28	0.36	-0.08	-0.58
Average	0.71	0.56	0.10	0.34	0.00	-0.83	-0.17	-0.80	-2.46

Table 2. Brier Skill Score (*BSS*) values for the calibration of seven storm events under different values of angle of repose (Φ), sediment transport coefficient (γ), and inertia coefficient ($ci = 1$)

BSS	Φ35°, γ 0.5	Φ35°, γ 1	Φ35°, γ 3	Φ45°, γ 0.5	Φ45°, γ 1	Φ45°, γ 3	Φ 55°, γ 0.5	Φ55°, γ 1	Φ55°, γ 3
UAV3&4	0.95	0.94	0.90	0.90	0.91	0.85	0.79	0.83	0.73
UAV4&5	-0.25	-0.28	-0.47	0.66	-1.11	-1.90	-2.05	-2.77	-4.14
UAV5&6	-0.52	-1.01	-1.56	-1.42	-2.02	-3.10	-2.05	-2.90	-4.56
UAV12&13	0.78	0.78	0.88	0.34	0.34	0.34	-0.08	-0.15	-0.35
Average	0.24	0.11	-0.06	0.12	-0.47	-0.95	-0.85	-1.25	-2.08

Table 3. Brier Skill Score (*BSS*) values for the calibration of seven storm events under different values of angle of repose (Φ), sediment transport coefficient (γ), and inertia coefficient ($ci = 2$)

4. Conclusion

This study confirms the reliability of the XBeach-Gravel numerical model in simulating the morphodynamic response of artificial gravel beaches exposed to moderate to low-energy waves, characteristic of the eastern Adriatic. Model calibration was conducted on the western side of Ploče Beach, analysing four storm events with significant wave heights (*Hs*) ranging from 0.64 m to 2.2 m and peak wave periods (*Tp*) between 3.5 s and 5.9 s. The best agreement between simulated and measured morphological changes was achieved with the parameter combination of inertia coefficient (*ci*) = 0.5, sediment transport coefficient (γ) = 0.5, and angle of repose (Φ) = 35°, yielding an average Brier Skill Score (*BSS*) of 0.89, classified as *"excellent"* according to Van Rijn's criterion. These results confirm the model's capability to realistically predict key sediment transport processes and beach profile transformations. Future research will aim to expand the analysed set of storm events and perform additional validation on the eastern side of the beach to further enhance the robustness of the findings.

5. References

[1] Tadić, A., Ružić, I., Krvavica, N. & Ilić, S. (2022). "Post-nourishment changes of an artificial gravel pocket beach using UAV imagery", J. Mar. Sci. Eng, Vol. 10, No. 3, pp. 358., ISSN: 2077-1312, https://doi.org/10.3390/jmse10030358

[2] Bogovac, T., Carević, D.,Bujak, D. & Miličević, H. (2023). "Application of the XBeach-Gravel Model for the Case of East Adriatic Sea-Wave Conditions", J Mar Sci Eng, Vol. 11, No. 3, pp. 680., ISSN: 2077-1312, https://doi.org/10.3390/jmse11030680

[3] De Beer, A.F., McCall, R.T., Long, J.W., Tissier, M.F.S. & Reniers, A.J.H.M. (2021). "Simulating wave runup on an intermediate–reflective beach using a wave-resolving and a wave-averaged version of XBeach", Coastal Engineering, Vol.163, pp. 103788., ISSN 0378-3839, https://doi.org/10.1016/j.coastaleng.2020.103788

[4] Marino, M., Nasca, S., Alkharoubi, A.I., Cavallaro, L., Foti, E. & Musumeci, R.E. (2025). "Efficacy of Nature-based Solutions for coastal protection under a changing climate: A modelling approach", Coastal Engineering, Vol. 198, pp. 104700., ISSN 0378-3839, https://doi.org/10.1016/j.coastaleng.2025.104700

[5] McCall, R. T.; Masselink, G.; Poate, T. G.; Roelvink, J. A. & Almeida, L. P. (2015). "Modelling the morphodynamics of gravel beaches during storms with XBeach-G", Coastal Engineering, Vol. 103, pp. 52–66., ISSN 0378-3839, https://doi.org/10.1016/j.coastaleng.2015.06.002

[6] McCall, R. T., Masselink, G., Poate, T. G., Roelvink, J. A., Almeida, L. P., Davidson, M. & Russell, P.E. (2014). "Modelling Storm Hydrodynamics on Gravel Beaches with XBeach-G", Coastal Engineering, Vol. 91, pp. 231–250, ISSN 0378-3839, https://doi:10.1016/j.coastaleng.2014.06.007

[7] Bujak, D., Ilic, S., Miličević, H. & Carević, D. (2023). "Wave Runup Prediction and Alongshore Variability on a Pocket Gravel Beach under Fetch-Limited Wave Conditions", J. Mar. Sci. Eng., Vol. 11, No.3, pp. 614., ISSN: 2077-1312, https://doi.org/10.3390/jmse11030614

[8] Van Rijn, L.C., Walstra, D.J.R., Grasmeijer, B., Sutherland, J., Pan, S. & Sierra, J.P. (2023). "The Predictability of Cross-Shore Bed Evolution of Sandy Beaches at the Time Scale of Storms and Seasons Using Process-Based Profile Models", Coastal Engineering, Vol.47, pp. 295–327., ISSN 0378-3839, https://doi.org/10.1016/S0378-3839(02)00120-5

DOI: 10.2507/36th.daaam.proceedings.033

DUAL-VIEW EXPLAINABLE AI FOR 12-LEAD ECG WITH OPENMP-ACCELERATED POSTPROCESSING AND EXTERNAL CALIBRATION

Bilgin Demir & Ervin Domazet

This Publication has to be referred as: Demir, B[ilgin] & Domazet, E[rvin] (2025). Dual-View Explainable AI for 12-Lead ECG With OpenMP-Accelerated Postprocessing and External Calibration, Proceedings of the 36th DAAAM International Symposium, pp.0244-0248, B. Katalinic (Ed.), Published by DAAAM International, ISBN 978-3-902734-47-1, ISSN 1726-9679, Vienna, Austria
DOI: 10.2507/36th.daaam.proceedings.033

Abstract

Accurate electrocardiogram (ECG) classifiers are increasingly deployed, but practical use requires interpretable outputs and CPU-class latency. This study presents a lightweight pipeline that couples a compact 1D-CNN with dual-view explanations-Grad-CAM for time-localised saliency and Integrated Gradients (IG) reduced to per-lead importance and accelerates explanation postprocessing via a C++/OpenMP module. Internal benchmarking adopts the official PTB-XL split. External assessment uses the St. Petersburg INCART database with an adapt → calibrate (isotonic) → test protocol. On PTB-XL, the model attains AUC = 0.808 at the window level and AUC = 0.833 at the record level on the official test fold. On INCART, uncalibrated scores thresholded using the calibration subset yield ROC-AUC $\approx$ 0.84 and PR-AUC $\approx$ 0.85 on a small test cohort; isotonic calibration shifts the operating point toward higher accuracy ($\approx$ 0.70) with ROC-AUC $\approx$ 0.70 and PR-AUC $\approx$ 0.68. The OpenMP postprocessing reproduces NumPy exactly (max $|\Delta| = 0$) while reducing CPU runtime. Results indicate that accuracy, interpretable rationales, and CPU-feasible explanation latency can be achieved simultaneously with a disciplined internal/external evaluation design.

Keywords: Explainable AI; Grad-CAM; 12-lead ECG; OpenMP; Integrated Gradients

1. Introduction

Electrocardiography (ECG) remains a first-line diagnostic for cardiac disorders, yet clinical deployment of learning systems still wrestles with three practical requirements: accurate discrimination on public benchmarks, interpretable rationales that clinicians can audit, and CPU-class latencies suitable for bedside devices. Public 12-lead corpora such as PTB-XL[1] have standardised internal evaluation through patient-stratified folds and SCP-ECG–based labels, enabling like-for-like comparisons across studies [2]. At the same time, post hoc explanation methods adapted from computer vision-Grad-CAM and Integrated Gradients (IG)-offer complementary views of model evidence; Grad-CAM localises salient time segments, while IG attributes importance in an axiomatic manner. However, external validity and calibrated operation remain unevenly addressed; deployment often requires probability calibration (e.g., isotonic regression) and threshold selection on a dedicated cohort before testing on a never-touched set.

Finally, most reports optimise classification throughput but rarely quantify the runtime cost of explanations themselves, despite many monitors running on CPUs. This study presents a compact 1D-CNN with dual-view explanations and an OpenMP postprocessing path, evaluated on PTB-XL (internal benchmark) and INCART [3] (external protocol).

2. Related Work

Recent work has converged on PTB-XL as the reference benchmark for 12-lead ECG modelling because its official patient-stratified folds (1-8/9/10) and SCP-ECG–based labels enable reproducible, like-for-like comparisons across architectures [2]. PTB-XL+ extends this ecosystem with harmonised feature sets for the same records, supporting hybrid modelling and ablation studies while reinforcing standardised evaluation practice [4]. In parallel, explainability has progressed from image-centric tools to time-series ECG. Studies demonstrate that Grad-CAM can highlight diagnostically relevant intervals on multi-lead signals for myocardial infarction detection, and ECG-tailored variants such as ECGradCAM generate attention maps adapted to 12-lead morphology [5], [6]. At the same time, reviews emphasise that visually persuasive saliency is not automatically reliable: many cardiac-AI papers lack quantitative checks of explanation stability or consistency with preprocessing, motivating paired use of complementary methods and simple, auditable aggregation steps [7]. Clinical-facing work increasingly stresses external validation and probability calibration; for example, the AIRE platform is evaluated across sites with attention to actionable, reliable risk estimation from a single ECG [8], while statistical guidance formalises isotonic regression and related post hoc calibration to correct probabilities and transport thresholds under class imbalance or prevalence shift, with explicit separation of calibration and test cohorts [9]. A complementary strand targets efficiency for edge/CPU deployment: compact or knowledge-distilled CNNs reduce latency and power draw and convincingly optimise classification throughput, but most reports do not measure the runtime cost of explanations themselves (e.g., saliency smoothing and per-lead aggregation) on CPUs, where many bedside systems actually run [10], [11]. High-performance ECG filtering demonstrates an emphasis on deployable, CPU-class signal-processing pipelines [12]. Saliency-based interpretability has also been explored in applied control settings, underscoring practical demands for transparent models [13]. Unlike prior work by the authors focusing on classification speed/accuracy [14], the present study integrates dual-view explainability, calibrated external testing, and CPU-feasible XAI postprocessing. Against this backdrop, this study adds value by integrating dual-view ECG-XAI (time-localised Grad-CAM with per-lead attributions), an explicit adapt → calibrate (isotonic) → test protocol on a different database, and a CPU-practical OpenMP implementation for explanation postprocessing, thereby closing the gap between interpretability, external validity, and deployable runtime that prior work typically treats in isolation.

3. Methods

Here we evaluate a lightweight, interpretable ECG pipeline under an internal–external design: internal benchmarking on the PTB-XL official split and external assessment on INCART[3]. Signals are standardised and windowed; a compact 1D-CNN produces window scores that are averaged to records, and explanations are dual-view Grad-CAM for time-localised saliency and Integrated Gradients aggregated into per-lead importance. To keep latency CPU-feasible, explanation postprocessing is implemented in C++/OpenMP with numerical parity to NumPy. External operation uses isotonic regression on a dedicated calibration split with a fixed threshold transported to the held-out test split.

3.1. Datasets and splits

Internal benchmarking used PTB-XL (12-lead ECG, 10 s, 500 Hz) with the official patient-stratified folds 1–8/9/10 for training/validation/test to ensure comparability with the literature [2]. External assessment used the St. Petersburg INCART database (12-lead ECG with beat annotations). Records were partitioned into three disjoint sets: adaptation (for distributional normalisation choices), calibration (for probability calibration and threshold selection), and test (held-out evaluation). The binary task on PTB-XL followed the dataset's diagnostic mapping: normal vs abnormal (any diagnostic class other than purely normal) [1].

3.2 Preprocessing and windowing

Signals were read via WFDB, reordered to a fixed 12-lead layout (I, II, III, aVR, aVL, aVF, V1–V6), band-pass filtered (0.5–40 Hz), and standardised by per-record z-scoring. Each record was segmented into overlapping windows of 512 samples with 256-sample stride. This corresponds to ≈1.02 s per window at 500 Hz (PTB-XL) and ≈1.99 s at 257 Hz (INCART). Inference operated at the window level; record-level scores were obtained by averaging window probabilities.

3.3. Model Architecture and Training

A compact 1D-CNN with residual connections was used: Conv1D-BatchNorm-ReLU blocks with skip connections, intermittent max-pooling, global average pooling, dropout (0.25), and a sigmoid output.

Optimisation used Adam (1×10^{-3}) with binary cross-entropy and class weighting computed on the training fold. Early stopping monitored validation ROC-AUC. Models were saved and reloaded using Keras' native serialisation for exact reproducibility.

3.4. Explainability

Two complementary post hoc methods were applied. Grad-CAM was computed from the final convolutional block to obtain time-localised saliency on 1-D signals. Integrated Gradients (IG) used a zero baseline with 64-128 integration steps to obtain attribution scores with axiomatic guarantees.

3.5. Parallel Postprocessing (OpenMP)

The module was compiled with -O3 -fopenmp (aggressive optimisation; OpenMP support) and run with OMP_PROC_BIND=true, OMP_PLACES=cores, and schedule(static) (core pinning and even work sharing to minimise overhead). The C++ outputs were verified to be numerically identical to the NumPy reference (maximum absolute difference = 0). Design choices for OpenMP parallelisation followed prior real-time ECG acceleration work by the authors and were adapted here to the XAI postprocessing stage.

3.6. Calibration and thresholding (external)

For INCART, raw scores were transformed using isotonic regression fit on the calibration split only. The decision threshold was selected on that same calibration split (F1 maximisation) and then applied unchanged to the held-out test split. Uncalibrated metrics at the calibration-chosen threshold were also reported for comparison.

3.7. Evaluation

Internal PTB-XL reporting included ROC-AUC and accuracy at the window level and after record-level averaging. External INCART reporting included ROC-AUC, PR-AUC, accuracy, precision, recall, and F1. To reflect the limited size of the external test cohort, bootstrap 95% confidence intervals were computed at the record level. All metrics were computed with scikit-learn under fixed seeds and batched inference settings. Isotonic regression (a monotonic mapping) was fit at the record level on the calibration split and applied to record-level scores on test.

4. Results

On the PTB-XL official split (training on folds 1-8, validation on fold 9, and testing on fold 10), the compact 1D-CNN achieved strong discrimination internally. At the window level, the model reached an AUC of 0.829 on validation and 0.808 on test, with corresponding accuracies of 0.744 and 0.732. Aggregating decisions to the record level by averaging window probabilities further improved performance: validation AUC rose to 0.854 and test AUC to 0.833, with accuracies of 0.764 and 0.759, respectively. Score polarity was consistently verified throughout the analysis, with no instances of inversion being necessary during the final evaluation. Each evaluation metric was computed on the officially established data folds, thereby maintaining full comparability with earlier studies conducted on the PTB-XL dataset.

For external evaluation, we assessed the frozen PTB-XL model on the INCART database using a three-way record split-adaptation, calibration, and test-to emulate deployment without re-training. Uncalibrated scores, thresholded using the calibration split, yielded a test ROC-AUC of approximately 0.840 and PR-AUC of 0.853, with accuracy around 0.50 on ten held-out records. Applying isotonic regression on the calibration split adjusted probability estimates and operating points: on the same test split the ROC-AUC was ~0.700 and PR-AUC ~0.683, with accuracy ~0.70 at the calibration-selected threshold. Bootstrap analysis at the record level (≈2000 resamples) reflected the small test set: the uncalibrated AUC had a mean of ~0.844 with a 95% CI [0.50, 1.00], and the isotonic-calibrated AUC had a mean of ~0.704 with a 95% CI [0.375, 1.00].

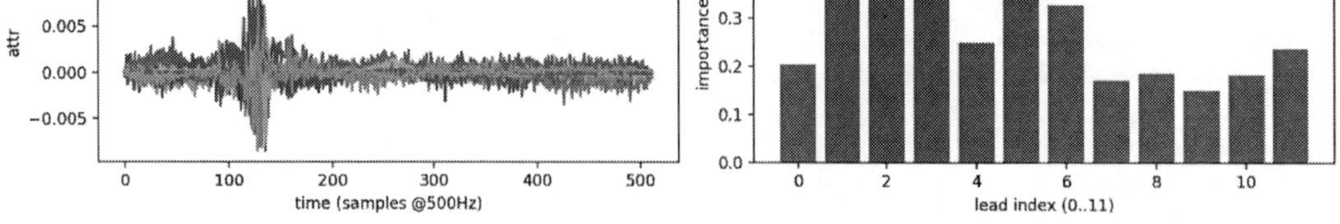

Fig. 1. Integrated Gradients (IG) on a representative record. Left: per-lead attributions over time after smoothing (kernel = 9); Right: per-lead contribution summary (IG summed across time). IG uses a zero baseline and 128 integration steps.

The explanation analysis combined Integrated Gradients (IG) and Grad-CAM to localise model evidence in time and across leads. IG bar plots of per-lead importance indicated lead-specific contributions that varied by case, while Grad-CAM heatmaps overlaid on the raw ECG highlighted short temporal segments around detected events; representative true-positive and false-positive examples illustrate these patterns. These views provide complementary insight-IG offers a global attribution per lead, and Grad-CAM pinpoints salient temporal intervals-without making clinical claims about causality. An example Integrated Gradients attribution is shown in Figure 1, with per-lead importance (right) and the time course of attributions (left). Complementary Grad-CAM temporal saliency is illustrated in Figure 2, highlighting short intervals that drove the decision.

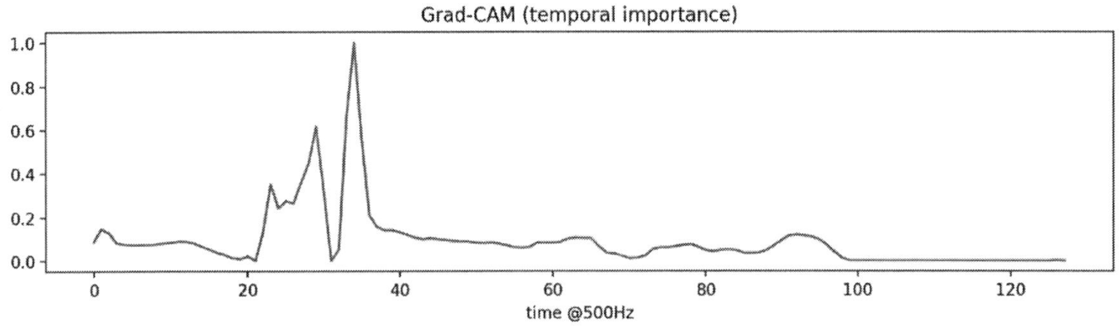

Fig. 2. Grad-CAM temporal importance for the same case as Figure 1. Saliency is computed from the final convolutional block and normalised to [0,1]; peaks align with short, decision-driving segments.

The OpenMP module reproduced NumPy exactly (max $|\Delta| = 0$) and met the sub-30 ms latency target on CPU. For the timing batch used here, single-thread execution was fastest; adding threads increased overhead and did not improve latency (Figure 3). Nevertheless, C++ delivered large gains over pure-Python loops (×49 at 1 thread; Figure 4). This indicates the explanation workload is small enough that parallel start-up and scheduling costs dominate; parallel benefits are expected only for larger batches.

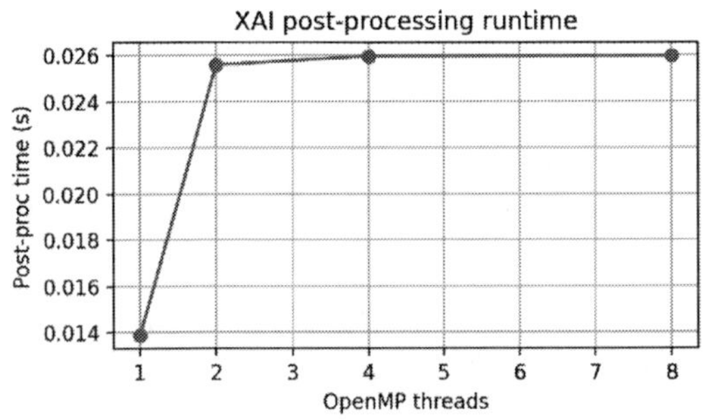

Fig. 3. Postprocessing latency vs OpenMP thread count (median over repeated runs; fixed affinity). Latency is minimised at 1 thread for this workload.

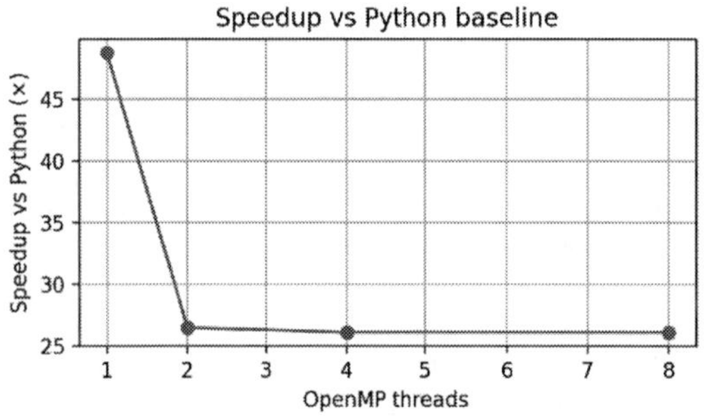

Fig. 4. Speedup relative to a pure-Python baseline. Single-threaded C++ already provides ×49; multithreading does not help at this batch size

5. Discussion

The internal PTB-XL evaluation shows that a compact 1D-CNN, trained with standard preprocessing and record-level aggregation, achieves strong discrimination on a widely used benchmark. On the external INCART cohort, calibration and strict cohort separation were essential. Uncalibrated scores paired with a threshold chosen on the calibration split yielded modest accuracy; isotonic calibration shifted the operating point toward a more balanced precision–recall trade-off. The dual-view explanation design provides complementary insight. Grad-CAM highlights short, decision-driving time segments, while Integrated Gradients-after simple smoothing-yields per-lead importance vectors, making both "when" and "which lead" auditable. From an engineering perspective, the OpenMP postprocessing reproduces the NumPy reference exactly (max $|\Delta| = 0$) and meets the latency target on a single CPU core. Limitations include a small external cohort, a binary PTB-XL task (normal vs abnormal) rather than multi-label diagnostics, and qualitative assessment of explanations; expanding external datasets, moving to multi-label targets, and conducting stability and reader studies are natural next steps.

6. Conclusion

This study consolidates three capabilities, typically approached in isolation for ECG machine learning: in situ transparent benchmarking, reasoned interpretation with explicit traceability, and deployment-ready execution within compact runtime constraints. Evaluated on the standardised PTB-XL division, our compact one-dimensional convolutional network exhibited robust class separation; decision sharpening through record-level averaging yielded marked stability against variances in window-based outputs. On INCART, strict cohort separation (adapt → calibrate → test) and isotonic calibration illustrated how probability mapping and threshold transport affect operating points under class imbalance-an essential step for deployment beyond the development dataset. Limitations include the small size of the external test cohort (wide uncertainty) and the binary task formulation on PTB-XL. Future work should expand to additional external databases, adopt multi-label diagnostic targets, quantify explanation stability (e.g., perturbation and test–retest analyses), and assess human factors through reader studies; exploring SIMD/AVX or GPU variants could further broaden deployment options. Overall, the results support a practical pathway wherein accuracy, calibrated external validity, and auditable explanations are jointly attainable on CPU-class hardware.

7. References

[1] PTB-XL Database, PhysioNet, (2020). Available from: https://physionet.org/content/ptb-xl/ Accessed: 2025-09-06

[2] Wagner, P.; Strodthoff, N.; Bousseljot, R.; et al. (2020). PTB-XL: A large publicly available electrocardiography dataset. Scientific Data, Vol. 7, No. 1, (May 2020) Article 154, ISSN 2052-4463, DOI:10.1038/s41597-020-0495-6

[3] INCART Database, St Petersburg Institute of Cardiological Technics 12-lead Arrhythmia Database, PhysioNet, (2008). Available from: https://physionet.org/content/incartdb/ Accessed: 2025-09-06

[4] Strodthoff, N.; Wagner, P.; Schaeffter, T.; et al. (2023). PTB-XL+: A comprehensive electrocardiographic feature dataset. Scientific Data, Vol. 10, No. 1, (May 2023) Article 279, ISSN 2052-4463, DOI:10.1038/s41597-023-02153-8

[5] Jahmunah, V.; Oh, S. L.; San, T. R.; et al. (2022). Explainable detection of myocardial infarction using deep learning models with Grad-CAM technique on ECG signals. Computers in Biology and Medicine, Vol. 146, (July 2022) Article 105550, ISSN 1879-0534, DOI:10.1016/j.compbiomed.2022.105550

[6] Hicks, S. A.; Isaksen, J. L.; Thambawita, V.; et al. (2021). Explaining deep neural networks for knowledge discovery in electrocardiogram analysis (ECGradCAM). Scientific Reports, Vol. 11, No. 1, (May 2021) Article 10949, ISSN 2045-2322, DOI:10.1038/s41598-021-90285-5

[7] Salih, A. M.; Lundervold, A.; Oehme, A.; et al. (2024). A review of evaluation approaches for explainable AI with applications in cardiology. Artificial Intelligence Review, Vol. 57, No. 9, (September 2024) Article 240, ISSN 1573-7462, DOI:10.1007/s10462-024-10852-w

[8] Sau, A.; Pastika, L.; Sieliwonczyk, E.; et al. (2024). Artificial intelligence-enabled electrocardiogram for mortality and cardiovascular risk estimation: A model development and validation study. The Lancet Digital Health, Vol. 6, No. 11, (November 2024) e791–e802, ISSN 2589-7500, DOI:10.1016/S2589-7500(24)00172-9

[9] Ojeda, F. M.; Kappert, P.; Schulte-Steinberg, S.; et al. (2023). Calibrating machine-learning approaches for probability estimation: A comprehensive comparison. Statistics in Medicine, Vol. 42, No. 29, (December 2023) pp. 5451–5478, ISSN 1097-0258, DOI:10.1002/sim.9921

[10] An, X.; Shi, S.; Wang, Q.; et al. (2024). Research on a lightweight arrhythmia classification model based on knowledge distillation for wearable single-lead ECG monitoring systems. Sensors, Vol. 24, No. 24, (December 2024) Article 7896, ISSN 1424-8220, DOI:10.3390/s24247896

[11] Lee, K.; Lee, J.; Shin, M. (2024). Lightweight beat score map method for electrocardiogram-based arrhythmia classification. Biocybernetics and Biomedical Engineering, Vol. 44, No. 4, (October 2024) pp. 844–857, ISSN 0208-5216, DOI:10.1016/j.bbe.2024.11.002

[12] Domazet, E.; Gusev, M. & Ristov, S. (2016). Optimizing High-Performance CUDA DSP Filter for ECG Signals, Proceedings of the 27th International Symposium, pp. 0623–0631. DOI: 10.2507/27th.daaam.proceedings.091

[13] Schwaiger, S.; Aburaia, M.; Aburaia, A. & Woeber, W. (2021). Explainable Artificial Intelligence for Robot Arm Control, Proceedings of the 32nd International Symposium. DOI: 10.2507/32nd.daaam.proceedings.090

[14] Demir, B. & Tafa, Z. (2025). Contribution to the development of intelligent systems for multi-lead ECG signal classification using convolutional neural networks. Proceedings of the 14th Mediterranean Conference on Embedded Computing (MECO), June 2025, Budva, Montenegro, ISBN 979-8-3315-1341-2, pp. 1–5, IEEE, DOI:10.1109/MECO66322.2025.11049271

DOI: 10.2507/36th.daaam.proceedings.034

RHEOLOGICAL MODEL OF ALUMINIUM ALLOYS AT HIGH TEMPERATURES

Domagoj Bendić, Neno Torić & Ivica Boko

This Publication has to be referred as: Bendic, D[omagoj]; Toric, N[eno] & Boko, I[vica] (2025). Rheological Model of Aluminium Alloys at High Temperatures, Proceedings of the 36th DAAAM International Symposium, pp.0249-0254, B. Katalinic (Ed.), Published by DAAAM International, ISBN 978-3-902734-47-1, ISSN 1726-9679, Vienna, Austria DOI: 10.2507/36th.daaam.proceedings.034

Abstract

The paper presents an overview of research on aluminium alloys and their applications as load-bearing structural components. It highlights the advantages and potential deficiencies of using elements made of aluminium alloys compared to steel, which remains the primary metal material in the construction industry, with the aim of encouraging more widespread use of aluminium. Particular attention is given to studies involving experimental testing of the mechanical properties of aluminium alloys at elevated temperatures, with a focus on the assessment of their reduced properties, especially time-dependent deformations - creep. Available rheological models are reviewed, including possibilities for their adequate adaptation and application to aluminium alloys. The intention is to develop a generalised creep model applicable to all aluminium alloys for structural purposes, also rheological model aimed at predicting mechanical properties following fire exposure.

Keywords: aluminium alloys; fire; constitutive behaviour; creep; rheological model.

1. Introduction

Although aluminium is the third most abundant element in the Earth's crust, it is a relatively modern material whose use in load-bearing structures was enabled by the development of suitable alloys. Today, beside steel, it is one of the most important metals in the construction industry. Its advantages include low density (about one-third that of steel), corrosion resistance without additional protective coatings, non-magnetic behaviour, and the retention of good ductility at low temperatures - unlike steel, which is prone to brittle fracture. It is highly recyclable, making it a sustainable material; estimates indicate that 70–98% of today's aluminium structures will be recycled [1]. Alloys from the 6xxx series suitable for structural applications include EN AW-6082, 6061, 6005A, 6106, 6063 and 6060 [2]. However, aluminium also has notable disadvantages. Its lower melting point and the more rapid degradation of mechanical properties at elevated temperatures make it less fire-resistant than steel. Degradation becomes noticeable above 100 °C and can become significant above 200 °C. Owing to its high thermal conductivity, heat spreads more rapidly through aluminium elements, which is a drawback for structural fire resistance. Furthermore, because of its crystalline lattice microstructure, whose vacancies enable atomic diffusion, the behaviour of aluminium alloys under sustained loading is characterised by time-dependent deformation (creep), particularly when exposed to elevated temperatures. Creep has been shown to significantly influence the reduction in collapse time of structural elements made out of aluminium.

Traditional rheological models (e.g., Burgers, standard solid models) represent only the primary and secondary creep stages. In addition to classical fire safety measures [3], and passive protections, mechanical properties and load-bearing capacity should be sufficient to allow all users to evacuate. Since the ability to predict the performance of aluminium structures exposed to fire is essential for ensuring sufficient load-bearing capacity, to address existing limitations, recent research aims to develop a model capable of incorporating all three creep phases and high-temperature behaviour of aluminium. If proven effective, it could be integrated into finite element-based codes.

2. Research Overview

2.1. Experimental Determination of Aluminium Alloy Properties

Numerous studies have focused on analysing the material characteristics and behaviour of aluminium alloys at ambient temperatures. However, their properties and reduced load-bearing capacity after exposure to elevated temperatures remain insufficiently investigated. Experimental investigations to determine the stress-strain relationships of aluminium alloys at elevated temperatures were conducted by Maljaars et al. [4], paying particular attention to the influence of creep. The shape of the stress–strain curve depends on the type of alloy; moreover, even for the same alloy, the curve at ambient temperature differs from that at elevated temperatures [5], [6]. Torić et al. [7] performed two distinct types of tests to quantify deformation caused by mechanical loading and that associated with creep - steady-state creep tests. They proposed an analytical model suitable for interpreting deformation test results, incorporating all three stages of creep. For alloys 6063-T5 and 6061-T6, Su et al. [8] carried out steady-state and transient tests at temperatures up to 600 °C. Liu et al. [9] conducted a detailed analysis using steady-state tensile tests on 6082-T6 specimens heated from 100 °C to 550 °C and subsequently either air-cooled or water prior to testing. Working on alloy 6063-T5 up to 550 °C, Sun et al. [10] confirmed the pronounced effect of elevated temperatures on mechanical properties, particularly above 200 °C, and proposed a series of models for estimating the post-fire properties of this alloy. Megalingam et al. [11] tested aluminium alloy 7075 at a constant temperature of 250 °C under high, sustained stress levels to analyse its creep behaviour under long-duration thermal loading.

2.2. Assessment of Aluminium Alloy Properties

Suzuki et al. [12] carried out fire loading tests on aluminium-alloy columns and beams to establish the link between stress and critical temperature. From these results, they formulated numerical expressions to predict temperature rise and to estimate the critical temperature of heated members. Disregarding the effects of initial imperfections and creep, Fogle et al. [13] proposed a simple analytical model for predicting failure under fire conditions. Maljaars et al. [14] highlighted limitations of Harmathy's constitutive model [15], particularly for 6xxx-series alloys, which exhibit an early onset of the tertiary creep phase. To verify constitutive behaviour of aluminium alloys in fire, they developed a FE model, and proposed extending the Dorn-Harmathy stress-strain model to transient state conditions for fire-exposed aluminium alloys [16]. Maljaars's extended model, with parameters adjusted for specific alloys, was also used by Soyal [17], who confirmed it via transient tests with rising temperature under constant or variable loading. By further augmenting Maljaars's model with additional parameters, Kandare et al. [18] enabled the assessment of failure in compression-loaded aluminium members under fire. Torić et al. [19] investigated creep deformation of aluminium at high temperatures. They identified dislocation climb to a neighbouring slip plane as the key high-temperature creep mechanism and developed a rheological model encompassing all three creep stages. Within the constraints set out in their study, the model is applied to metals susceptible to high-temperature creep, and was calibrated for S275 steel and for aluminium alloy 6082-T6. Zheng and Zhang [20] examined the behaviour of aluminium beams, specifically alloys 5083-H112 and 6060-T66, with and without protection, at ambient and elevated temperatures. From their results and developed models, they pointed to the conservatism of the Eurocode 9 approach to critical temperature and proposed alloy-dependent simplified equations for temperature rise and for determining critical temperature. Under diverse thermal and mechanical conditions, Li et al. [21] formulated a family of constitutive models for alloy 7050 by establishing the link between creep deformation and age hardening. These models enable reliable prediction of material properties regardless of initial states and loading history. Through tests on alloy 7075-T6, Sun et al. [22] determined reduction factors for material properties over 20–550 °C and proposed a set of models to estimate stiffness and strength of high-strength aluminium alloys during and after fire exposure. To capture the full range of elevated-temperature behaviour, the authors adapted the modified two-stage Ramberg–Osgood model previously proposed by Yun et al. [23], which is based on an extensive database of more than 700 ambient-temperature stress–strain curves. Spigarelli and Sandström [24] developed a model for estimating creep in pure aluminium, for which the chemical composition, stress and temperature must be determined experimentally.

3. Rheological Model of Aluminium

To model the mechanical behaviour of a material with sufficient accuracy, it is necessary to provide an analytical description of the relationship between stress and strain. The overall deformation process of metallic materials at elevated temperatures, consisting of three components of which one specifically represents creep deformation, was introduced in the 1980s in work of Anderberg [25].

$$\varepsilon_{tot} = \varepsilon_{th}(T) + \varepsilon_{\sigma}(\sigma, T) + \varepsilon_{c,r}(\sigma, T, t) \quad (1)$$

Where proposed total strain ε_{tot} formulation is described with thermal component $\varepsilon_{th}(T)$ dependent on temperature, a stress-related $\varepsilon_{\sigma}(\sigma, T)$ component governed by both applied stress and temperature, and a creep strain component $\varepsilon_{c,r}(\sigma, T, t)$. The latter is most complex, as it depends on stress, temperature, and time. The existing models like described ones in the previous section, models for predicting individual material characteristics, are necessary for defining a rheological model of aluminium. One of the first rheological model was developed in the 1970s: a Kelvin model composed of a spring and a dashpot with associated constants, devised to assess nonlinear deformations and load-bearing failure in concrete [26]. To represent viscoelastic responses in steel, Chindam et al. [27] used Kelvin-Voigt and Maxwell models - the parallel and the series connection of spring and dashpot (Figure 1). Analysing the influence of heat on the material during cyclic testing, they identified the Kelvin–Voigt model as the more suitable option for capturing the mechanical and thermo-mechanical behaviour of polycrystalline materials in the elastic regime, motivating its application to aluminium.

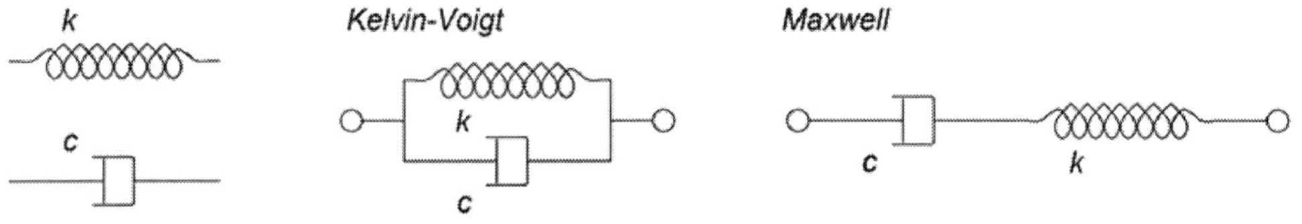

Fig. 1. Spring, dashpot, the Kelvin-Voigt and the Maxwell model

Models comprising multiple series-connected Kelvin–Voigt elements have been proposed, each describing a particular type of deformation response. Torić and Burgess developed a rheological model of two series-connected elements (Figure 2), in which the first represents the mechanical component of deformation and the second the viscous (creep) component, while the thermal deformation component, dependent solely on temperature, is omitted. By employing a serial connection of two Kelvin–Voigt elements, the development of creep deformation can be monitored under both stationary and transient heating conditions [28]. Torić and Burgess initiated their investigation of creep in S275 grade steel, providing a detailed description of the experimental procedures required to validate the analytical creep model in their study [29]. To develop an explicit creep model for S275 steel and an aluminium alloy 6082-T6, stationary-creep tests were conducted on cylindrical coupons and columns of the same materials. These tests highlighted the necessity of also accounting for the duration of exposure to elevated temperature.

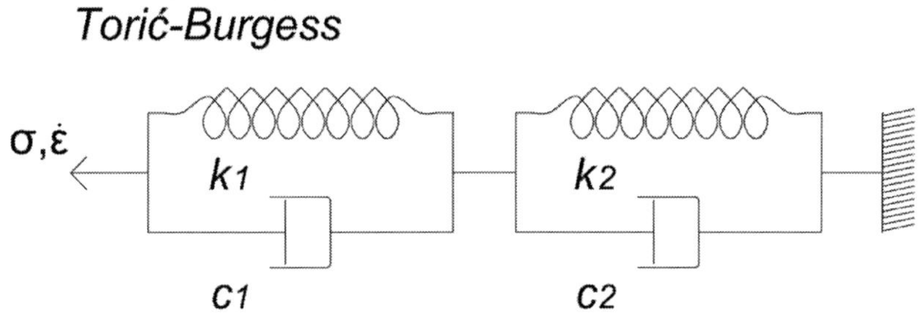

Fig. 2. Torić-Burgess model

The differential equation for each of the Kelvin-Voigt element can be defined as:

$$\frac{\sigma}{c_i} = \frac{k_i}{c_i}\varepsilon_i + \dot{\varepsilon}_i \quad (i = 1,2) \quad (2)$$

where total strain rate is constant and it can be defined:

$$\dot{\varepsilon} = \dot{\varepsilon}_1 + \dot{\varepsilon}_2 \quad (3)$$

Equation can be solved using Euler integration, where differential equations have to be integrated with respect to time by using small time increments. Type of solution (stress or strain-rate controlled) depends on type of loading.

Verification of each constitutive component, as well as of the overall rheological model, was carried out against experimental datasets from various studies [30] and against the authors' own tests on S275 steel over 400–600 °C and on aluminium alloy 6082-T6 over 200–300 °C [19], [30]. With appropriate calibration, the model is considered applicable to any grade of carbon steel (S235–S355) and may be regarded as a foundation for a universal rheological model for metals characterised by high-temperature creep. In order to explore the potential for application, it is necessary to address several new challenges that would provide a clearer understanding of the behaviour of individual alloys, thereby establishing a database of constitutive components that would reduce the number of experimental investigations required in the future. For example of adaptation to different steel type is the process of determining the exponents in the Ramberg–Osgood model [31], which represents the strain value as a function of stress, the modulus of elasticity, and the corresponding exponents that must be identified and calibrated for each alloy [23], [32], and it is essential for accurately assessing the material behaviour. The difficulty in predicting the modulus of elasticity lies in the remaining shortcomings concerning the required number of tests that would consistently capture the reductions associated with a given alloy and temperature. Eurocode [33] considers reduction coefficients for a single, unified material, whereas Figure 3 demonstrates that, even across a few examples of different alloys, varying ratios between reduced and initial values are obtained.

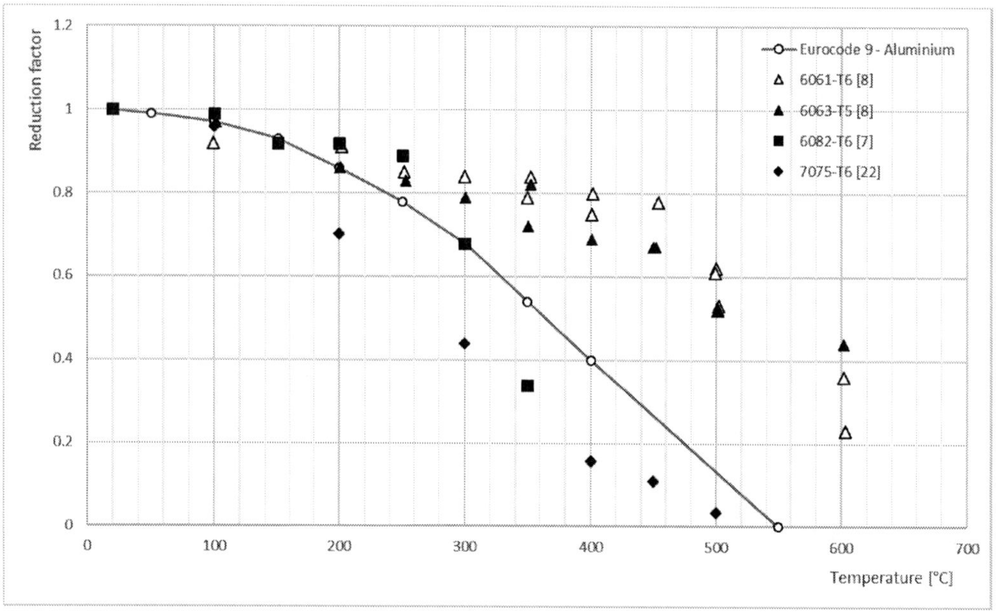

Fig. 3. Reduction factors of the in-fire elastic modulus test results and Eurocode 9 values

Various constitutive material models are already developed. Given the frequent use of the 7075 alloy in high-temperature applications, the behaviour of the stress exponent in relation to temperature and stress, as well as the creep deformation mechanisms expected to dominate under such conditions, have been examined [34], [35], [36]. Constitutive models at elevated temperatures have been developed [37], [38], along with a creep model for these high-performance aluminium alloys [39].

4. Conclusion

In recent decades, aluminium has been employed across engineering domains, primarily owing to its strength-to-weight ratio and excellent corrosion resistance. Given these benefits, the aim of this paper is to encourage its wider use in civil engineering. Aluminium comprises families of alloys with differing properties, with further distinctions between alloys within the same family. Owing to these numerous differences, in addition to studies on individual alloys, comparative analyses are required. To address the lack of critical information for load-bearing members, the creep behaviour of aluminium elements will be considered. In view of the identified challenges of fire resistance, particular attention will be devoted to these analyses. To reduce the need for costly and complex experimental programmes, parameters should be investigated through numerical analyses. The present review motivates the development of a rheological model for aluminium structures subjected to extreme conditions, encompassing all time-dependent deformation components. All critical thermomechanical variables (temperature and its rate of increase, stress, and strain rate) must be identified and accounted for. Following experimental characterisation of alloy behaviour at high temperatures, creep models tailored to those alloys will be developed, alongside a generalised model applicable to structural aluminium alloys used in civil engineering. Eventually, a rheological model will be defined to assess the extent of damage and the residual mechanical properties of aluminium elements after fire exposure.

5. References

[1] https://www.hazlemerecommercial.co.uk/blog/aluminium-the-green-metal/2012/04/, (2012). Aluminium - The Green Metal, Hazlemere, Accessed on: 2025-09-23

[2] EN 1999-1-1 (2007) (English): Eurocode 9: Design of aluminium structures - Part 1-1: General structural rules [Authority: The European Union Per Regulation 305/2011, Directive 98/34/EC, Directive 2004/18/EC]

[3] Malatinsky, A. & Hromada, M. (2022). Preventive and Repressive Measures for Fire Safety, Proceedings of the 33rd DAAAM International Symposium, pp.0397-0403, B. Katalinic (Ed.), Published by DAAAM International, ISBN 978-3-902734-36-5, ISSN 1726-9679, Vienna, Austria DOI: 10.2507/33rd.daaam.proceedings.056

[4] Maljaars, J.; Soetens, F. & Snijder, H. H. (2009). Local buckling of aluminium structures exposed to fire. Part 1: Tests, Thin-Walled Structures, 47-11, pp. 1404-1417., DOI 10.1016/j.tws.2009.02.008

[5] Maljaars, J.; Soetens, F. & Katgerman, L. (2008). Constitutive model for aluminum alloys exposed to fire conditions, Metallurgical and Materials Transactions A: Physical Metallurgy and Materials Science, 39A-4, pp. 778-789., DOI 10.1007/s11661-008-9470-0

[6] Sun, Y.; Fu, Z.; Song, Y. & Xia, J. (2023). Cross-sectional behavior of aluminum alloy channel section stub columns after exposure to fire, Journal of Structural Engineering, pp. 149-7, DOI: 10.1061/jsendh.steng-12383

[7] Torić, N.; Brnić, J.; Boko, I.; Brčić, M.; Burgess, I. W. & Uzelac, I. (2017). Experimental analysis of the behaviour of aluminium alloy EN 6082AW T6 at high temperature, Metals, 7, DOI 10.3390/met7040126

[8] Su, M. N. & Young, B. (2019). Material properties of normal and high strength aluminium alloys at elevated temperatures, Thin-Walled Structures, 137, pp. 463–471., DOI 10.1016/j.tws.2019.01.012

[9] Liu, Y.; Liu, H. & Chen, Z. (2019). Post-fire mechanical properties of aluminum alloy 6082-T6, Construction and Building Materials, 196, pp. 256–266., DOI 10.1016/j.conbuildmat.2018.10.237

[10] Sun, Y.; Zhang, K. & Gong, G. (2023). Material properties of structural aluminium alloys after exposure to fire, Structures, 55, pp. 2105–2111., DOI 10.1016/j.istruc.2023.07.027

[11] Megalingam, A.; Ahmad, A. H.; Alang, N. A.; Alias, J. & Naher, S. (2024). Creep behaviour of aluminium 7075 feedstock billet globular microstructure at high processing temperature, Journal of Failure Analysis and Prevention, 24-3, pp. 1324–1332., DOI 10.1007/s11668-024-01917-7

[12] Suzuki, J.; Ohmiya, Y.; Kohno, M.; Yusa, S.; Harada, K. & Wakamatsu, T. (2002). Evaluation of fire resistance for aluminum alloy members, AIJ Journal of Technology and Design, 16, pp. 151-156., DOI 10.3130/aijt.8.151_2

[13] Fogle, E. J.; Lattimer, B. Y.; Feih, S.; Kandare, E.; Mouritz, A. P.; Case, S. W. & Lattimer, B. Y. (2011). Compression load failure of aluminum plates due to fire, Engineering Structures Journal, 34, pp. 155-162., DOI 10.1016/j.engstruct.2011.09.014

[14] Maljaars, J.; Soetens, F. & Snijder, H. H. (2009). Local buckling of aluminium structures exposed to fire. Part 2: Finite element models, Thin-Walled Structures, 47-11, pp. 1418–1428., DOI 10.1016/j.tws.2008.06.003

[15] Harmathy, T.Z. (1967). A comprehensive creep model, Journal of Basic Engineering, 89-3, pp. 496-502, DOI 10.4224/40001531

[16] Maljaars, L.; Twilt, L.; Fellinger, J. H. H.; Snijder H. H. & Soetens, F. (2010). Aluminium structures exposed to fire conditions-an overview, Heron, 55-2, pp. 85-122.

[17] Soyal, Z. (2016). Creep of aluminum alloys exposed to fire conditions, Master thesis, University of technology Eindhoven, Eindhoven, Netherlands

[18] Kandare, E.; Feih, S.; Kootsookos, A.; Mathys, Z.; Lattimer, B. Y. & Mouritz, A. P. (2010). Creep-based life prediction modelling of aluminium in fire, Materials Science and Engineering: A, 527-4-5, pp. 1185–1193., DOI 10.1016/j.msea.2009.10.010

[19] Torić, N.; Glavinić, I. U. & Burgess, I. W. (2018). Development of a rheological model for creep strain evolution in steel and aluminium at high temperature, Fire and Materials, 42-8, pp. 879–888., DOI 10.1002/fam.2643

[20] Zheng, Y. Q. & Zhang, Z. (2016). The fire performance and fire-resistance design of aluminium alloy I-beams, Fire and Materials, 40-1, pp. 141–157., DOI 10.1002/fam.2275

[21] Li, Y.; Hou, T.; Huang, X.; Gao, Y.; Zeng, Y. & Li, D. (2023). Constitutive modelling of coupled creep deformation and age hardening behavior of aluminum alloys under various thermal and mechanical loadings, Journal of Materials Research and Technology, 25, pp. 333–353., DOI 10.1016/j.jmrt.2023.05.183

[22] Sun, Y.; Cheng, W. & Chen, K. (2024). Experimental study of in-fire and post-fire material response of high-strength aluminium alloys, Journal of Building Engineering, 91, DOI 10.1016/j.jobe.2024.109581

[23] Yun, X.; Wang, Z. & Gardner, L. (2021). Full-range stress–strain curves for aluminum alloys, Journal of Structural Engineering, 147-6, DOI 10.1061/(asce)st.1943-541x.0002999

[24] Spigarelli, S. & Sandström, R. (2018). Basic creep modelling of aluminium, Materials Science and Engineering: A 711, pp. 343–349., DOI 10.1016/j.msea.2017.11.053

[25] Anderberg, Y. (1988). Modelling steel behavior. Fire Safety Journal, 13-1, pp. 17-26., DOI 10.1016/0379-7112(88)90029-X

[26] Helman, H. & Creus, G. J. (1975). A non-linear rheological model describing time-dependent deformations and failure, International Journal of Non-Linear Mechanics, 10, pp. 167-172, DOI: 10.1016/0020-7462(75)90034-7

[27] Chindam, C.; Venkata, K. C.; Balasubramaniam, K. & Prakash, R. V. (2013). Thermomechanical response of metals: Maxwell vs. Kelvin-Voigt models, Materials Science and Engineering: A 560, pp. 54–61, DOI 10.1016/j.msea.2012.08.152

[28] Torić, N. & Burgess, I. W. (2016). A unified rheological model for modelling steel behaviour in fire conditions, Journal of Constructional Steel Research, 127, pp. 221–230, DOI 10.1016/j.jcsr.2016.07.031

[29] Torić, N.; Brnić, J.; Boko, I.; Brčić, M.; Burgess, I.W. & Uzelac Glavinić, I. (2017). Development of a high temperature material model for grade s275jr steel, Journal of Constructional Steel Research, 137, pp. 161-168., DOI 10.1016/j.jcsr.2017.06.020

[30] Torić, N. & Burgess, I. (2021). Rheological modelling of high-temperature stationary creep tests of Grade S275JR steel, ce/papers, 4, pp. 546-550., ISSN 2509-7075, DOI 10.1002/cepa.1328

[31] Ramberg, W. & Osgood, W.R. (1943) Description of stress-strain curves by three parameters, Advisory Committee for Aeronautics (NACA), Washington, Technical Note No. 902.

[32] Mišović, M.; Tadić, N.; Lučić, D. (2016). Deformacijske karakteristike aluminijskih legura, Građevinar, 68-3, pp. 179-189., DOI 10.14256/JCE.1457.2015

[33] EN 1999-1-2 (2007) (English): Eurocode 9: Design of aluminium structures - Part 1-2: Structural fire design [Authority: The European Union Per Regulation 305/2011, Directive 98/34/EC, Directive 2004/18/EC]

[34] Spigarelli, S. (1999). Creep of aluminum and aluminium alloys, TALAT Lecture 1253, European Aluminium Association, Italy

[35] Mahathaninwong, N.; Zhou, Y.; Babcock, S.E.; Plookphol, T.; Wannasin, J. & Wisutmethangoon, S. (2012). Creep rupture behavior of semi-solid cast 7075-T6 Al alloy, Materials Science and Engineering: A, 556, pp. 107-113., DOI 10.1016/j.msea.2012.06.064

[36] Thongkam, S.; Wisutmethangoon, S.; Wannasin, J.; Chantaramanee, S. & Plookphol, T. (2013). Creep of rheocast 7075 aluminum alloy at 300 °C. Applied Mechanics and Materials, 372, pp. 288–291, DOI 10.4028/www.scientific.net/AMM.372.288

[37] Wu, J.F.; He, D.G.; Lin, Y.C.; Chen, Z.J.; Ding, H.B. & Qiu, Y.L. (2024). A unified constitutive model and microstructure evolution of a solution treated 7075 aluminum alloys under warm forming condition, Journal of Materials Research and Technology, 30, pp. 4159-4168., DOI 10.1016/j.jmrt.2024.04.101

[38] Omer, K.; Butcher, C. & Worswick, M. (2020). Characterization and Application of a Constitutive Model for two 7000-Series Aluminum Alloys Subjected to Hot Forming, International Journal of Mechanical Sciences, 165, DOI 10.1016/j.ijmecsci.2019.105218

[39] Lin, Y.C.; Jiang, Y.Q.; Zhou, H.M. & Liu G. (2014). A new creep constitutive model for 7075 aluminum alloy under elevated temperatures, Journal of Materials Engineering and Performance, 23 (12), pp. 4350-4357., DOI 10.1007/s11665-014-1191-5

DOI: 10.2507/36th.daaam.proceedings.035

ANALYSIS OF A TYPICAL 1960S LARGE PANEL BUILDING SEISMIC RESISTANCE

Romano Jevtić Rundek & Mario Uroš

This Publication has to be referred as: Jevtic Rundek, R[omano] & Uros, M[ario] (2025). Analysis of a Typical 1960s Large Panel Building Seismic Resistance, Proceedings of the 36th DAAAM International Symposium, pp.0255-0259, B. Katalinic (Ed.), Published by DAAAM International, ISBN 978-3-902734-47-1, ISSN 1726-9679, Vienna, Austria
DOI: 10.2507/36th.daaam.proceedings.035

Abstract

Construction of standardised precast large panel buildings intensified during the 1960s in the city of Zagreb. Many were built before the first significant seismic code was introduced (1964). Modern understanding shows that non typical and potentially dangerous failure modes are possible in these large panel buildings. In this article, some methods for analysis are presented, using software ETABS and Abaqus. High fidelity models of structural details are created in Abaqus and used to approximate their capacity curves. These curves are then used to calibrate a simpler lumped plasticity model in ETABS. The calibration process is validated by comparative analysis between wall assemblies in both software's to confirm appropriate interaction between elements of the lumped plasticity model. As a result, a simplified numerical model is provided capable of estimating complex failure mechanisms associated with large panel buildings to a reasonable degree.

Keywords: seismic analysis; Large panel buildings; Failure mechanism; Numerical model calibration

1. Introduction

The construction of standardized large-panel buildings intensified in the 1960s in Zagreb and across the Balkans. The first significant seismic code in the region was introduced in 1964, after many such buildings had already been constructed. Large-panel systems built in Zagreb before 1964 are characterized by a low amount of reinforcement, smooth reinforcement bars, a possible lack of longitudinal walls, and details prone to non-standard failure mechanisms. Resistance of a building portofolio can have a significant impact to seismic risk as shown in (1). Research into the seismic resistance of Large panel buildings gained momentum in the 1980s, notably with a report from the National Technical Information Service – NTIS (2), which consolidated experimental data, engineering recommendations, computational methods, and guidelines for numerical models. Among the research conducted in the Balkans, the master's thesis of Professor Zamolo (3) stands out. Large-panel buildings were predominantly constructed as semi-prefabricated reinforced concrete structures, composed of precast reinforced concrete panels. Typical 1960s large-panel buildings exhibit weak and brittle connections between elements, low reinforcement ratios, and the use of smooth rebar. Relevant failure mechanisms vary across individual buildings but generally include shear failure along horizontal or vertical panel joints, uplift of horizontal joints, and panel crushing at horizontal joints.

When analysing such structures, it is necessary to account for the nonlinear behaviour of the connections to identify the governing failure mechanism of a large panel building. In the studies conducted at the time, due to limited computational capacity, analyses were restricted to smaller models and 2D simulations. Today, more advanced computational software such as Abaqus (4) and ETABS (5) is available. Simulations of kinematic and plastic behaviour of load-bearing elements and element interactions in assemblies can be seen in the literature, conducted using Abaqus (6)(7). Furthermore, examples of nonlinear building models applying concentrated plasticity can be found, which are used in modern quantification of seismic performance in the form of Probabilistic Seismic Hazard Analysis (PSHA) (8). From this, it can be concluded that a well-designed and calibrated model with concentrated plasticity would enable the application of PSHA to large-panel buildings. Additionally, it can be observed that there are significant variations in the shape and dimensions of large-panel elements, as well as in the overall geometry of the buildings themselves. Consequently, there are substantial differences in the resistance of details across different types of large-panel buildings. This paper describes the calibration procedure of a numerical model of one large panel building. Detailed local models are applied, along with a simplified global model calibrated against the detailed model. This approach allows sufficiently accurate modelling of the entire structure on a desktop computer, due to the significantly lower numerical complexity of the simplified model. This approach is likely to enable a more accurate PSHA analysis of buildings of this type, as well as global failure mechanism analysis and development of subsequent retrofit solutions.

2. Model definition

Two modelling approaches are employed: a detailed and computationally demanding 3D volumetric finite element model developed in Abaqus (4), and a simplified model based on concentrated plasticity elements developed in ETABS (5). The Abaqus model (Figure 1, left) is constructed using volumetric finite elements to simulate concrete and beam elements to simulate reinforcement, inserted as "embedded elements." For the constitutive material model, the Concrete Damaged Plasticity (CDP) model (9) is selected. Its calibration, based on a chosen set of experiments, corresponds to a concrete with compressive strength of 20 MPa and is not described in this paper. The reinforcement is modelled using a bilinear constitutive law, equivalent to steel grade S235. The interaction between precast elements is defined either by friction alone or by friction and cohesion, depending on the model. The friction coefficient is taken as 0.6, with a sensitivity analysis performed.

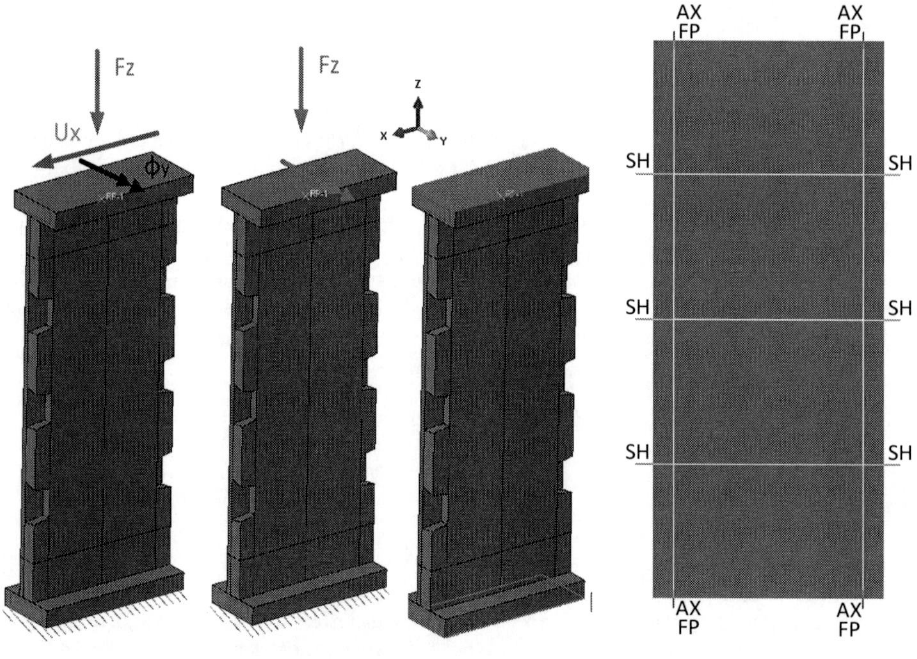

Fig. 1. One panel modelled in Abaqus (left) and ETABS (right)

The ETABS model (Figure 1, right) consists of linear shell and frame elements, and nonlinear link elements. Shell and frame elements form individual panels that behave elastically, while the links are arranged as shown in Figure 1 and simulate the behaviour of connections between adjacent elements. Links labelled SH simulate shear-keys, links labelled AX simulate panel uplift and crushing, and links labelled FP simulate panel sliding along horizontal joints. Links AX and SH are defined as bilinear (multilinear plastic) in the relevant degrees of freedom, while the FP link is defined as a connection with a friction coefficient that differs under tension and compression states (T/C friction pendulum).

3. Analysis definition

Two levels of analysis are performed. The first level involves multiple analyses of individual elements, and the second level involves multiple analyses of assembled walls. The first level is used for the initial calibration of nonlinear connection behaviour, while the second level is used to verify the interaction of calibrated connections within an assembled wall. Specifically, the first level includes pushover analyses of a single RC panel under different axial force levels, as well as analysis of a shear keys, allowing the simulation of all relevant failure mechanisms mostly independently. This level of analysis is conducted in both Abaqus and ETABS. Once the ETABS model is calibrated so that the behaviour of a single panel adequately matches the Abaqus results, the second level of analysis is performed to check the interaction of panels in an assembled wall. Pushover analyses are performed on several wall configurations with varying slenderness, thereby inducing different failure mechanisms (shear failure, flexural failure). The shape of pushover curves, failure mechanisms, and force distributions within elements is examined, assuming the more complex Abaqus model is accurate, while the simpler ETABS model is calibrated to achieve equivalent results.

4. Analysis results

Results are presented for a wall three panels wide and three panels high, as this configuration exhibits two failure mechanisms, enabling the demonstration of both shear and flexural failure. Figure 2 shows the Abaqus model (left) and ETABS model (right) after pushover analysis. Figure 3 presents the pushover curves of the entire walls in both models. Furthermore, Figure 4 shows the diagrams of horizontal forces in individual panels during pushover analysis for the Abaqus model, and Figure 5 for the ETABS model.

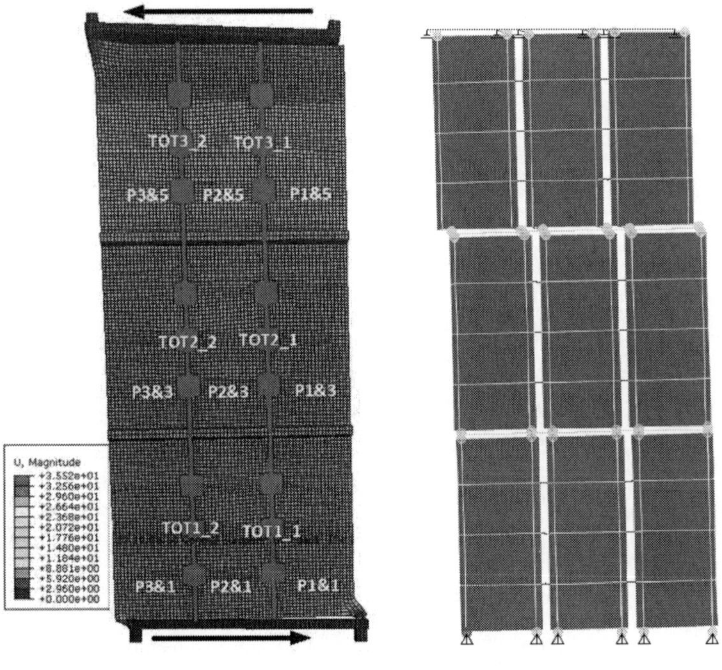

Fig. 2. Failure mechanism in Abaqus (left) and ETABS (right)

The pushover curve of the entire wall indicates good agreement in stiffness and strength of the assembled wall, as seen in Fig 3. Therefore, at minimum, good predictions of story drifts can be expected in the global model. The following discussion focuses on the distribution of forces and damage in individual elements of the assembled wall. In Figure 2, panel uplift and tie-beam sliding on the top can be observed in both models. Observed only in the ETABS model are concrete crushing (bottom left of panel P3&1) and sliding of the top panels, while in the Abaqus model sliding of the bottom of the wall is observed. While there is good overlap in the failure mechanisms, differences in the observed failure mechanism do exist and are attributed to the simplification of the ETABS model. There are two significant intrinsic limitation of the ETABS model as shown here. One is the inability to accurately simulate the location of the resultant axial force at the end of the panel, which will in reality vary depending on the axial loading. In the lumped model this is fixed. The second is the axial resistance of the axial link being one fixed value. Axial resistance of the link depends on the stress distribution, which depends on the axial force. This is modelled conservatively in the ETABS model. This approach is the reason for compressive axial failure in the ETABS model, while no failure occurred in the Abaqus model.

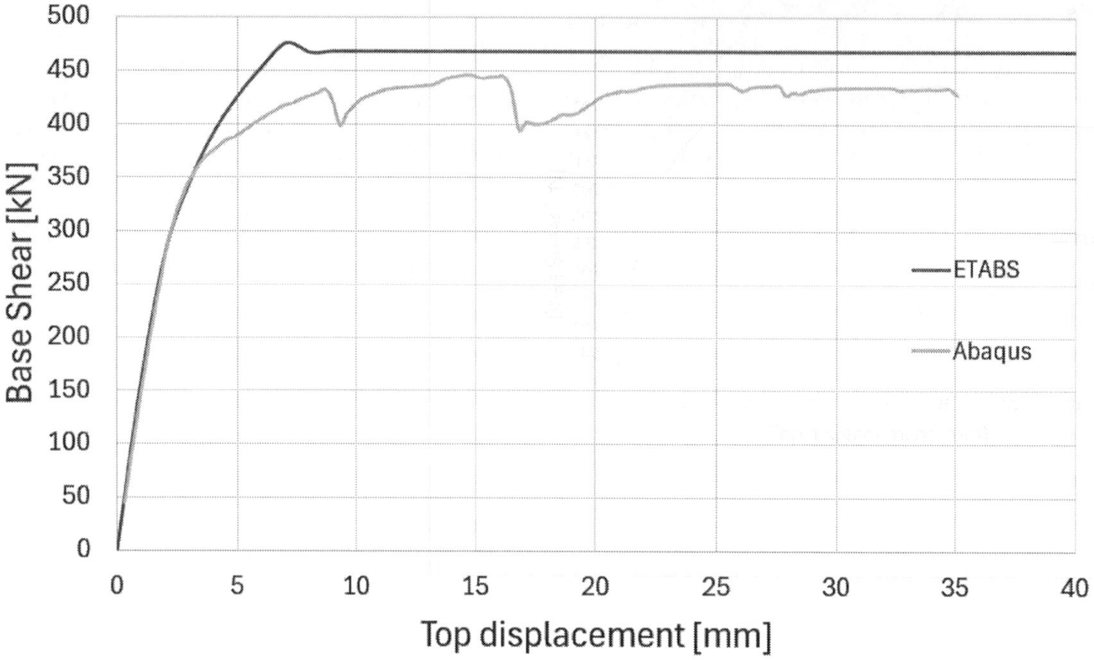

Fig. 3. Pushover curves of the assembled walls

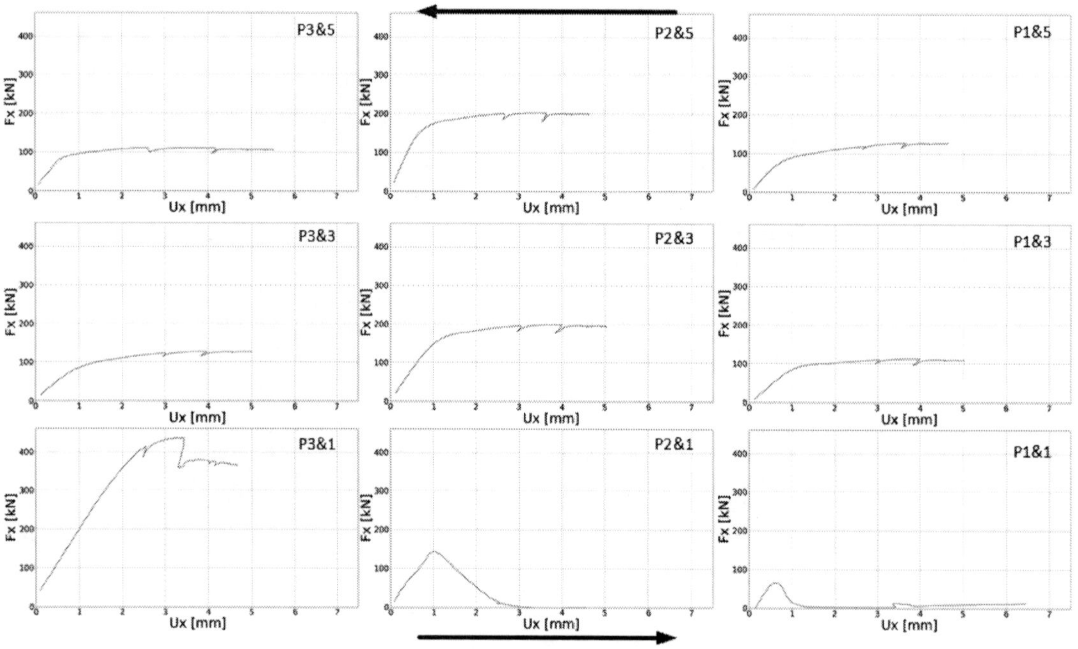

Fig. 4. Horizontal components of forces in individual panels (Abaqus)

Comparison of force distribution between Abaqus (Fig 4) and ETABS (Fig 5) shows almost identical distributions in the bottom row of panels, while differences appear in the upper layers of the wall. The Abaqus model shows the typical distribution of shear stress, with peak in the middle of the wall, indicating the wall is acting like a single deformable body, while the ETABS model shows a different distribution, more in line with shear transfer correlated to the axial loading of individual elements. The main cause of this discrepancy lies in the simulation of cohesion in the Abaqus model, which ETABS does not replicate, relying only on a friction coefficient. This should result with conservative results, and it seems that effect on failure mechanisms is small enough. In this example, differences with regards to internal forces are most pronounced in the top of the wall, with visible sliding of the top wall panel row in the ETABS model and no sliding in the ABAQUS model, which is also visible in the force distribution.

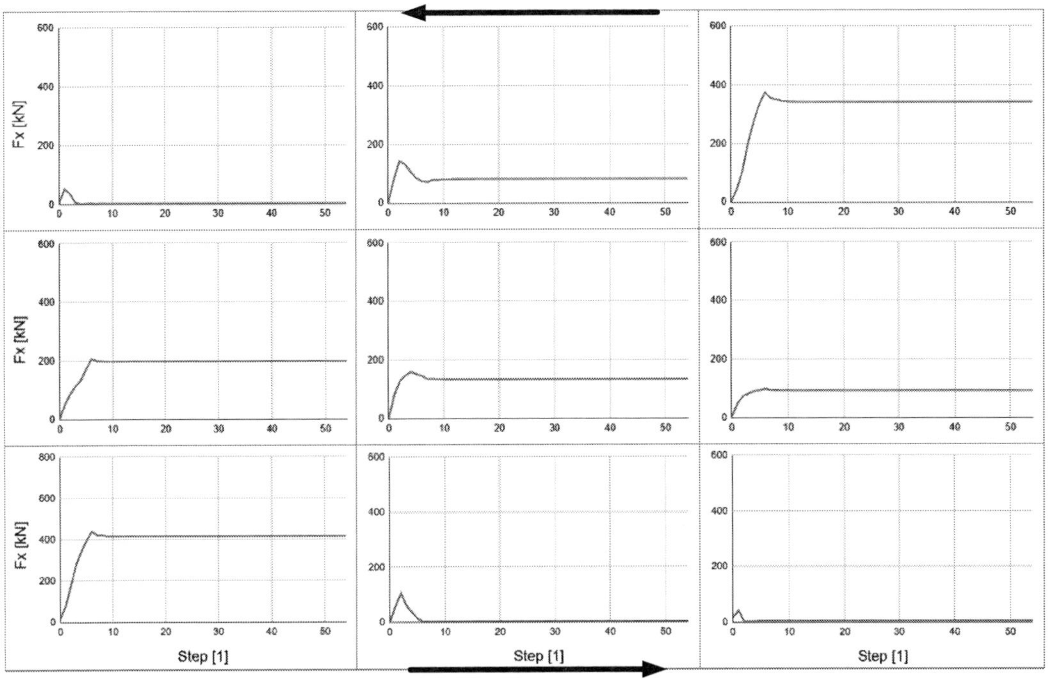

Fig. 5. Horizontal components of forces in individual panels (ETABS)

5. Conclusion

A set of pushover analyses has been conducted for various wall configurations, with results shown for a wall configuration which exhibits both relevant failure mechanisms and maximises differences between the models. Certain differences in force distribution and damage between the two models are observed, but these differences are acceptable given the simplified nature of the ETABS model and are generally conservative. Despite relying solely on a friction coefficient and bilinear behaviour, the ETABS model is able to reproduce the complex behaviour of large panel building walls reasonably well, enabling conducting a PSHA or similar analyses on a global 3D model of the entire Large Panel building. Limitations of this general approach are dependent on the choices in modelling used by the researcher and available data for calibration. Limitations of the model shown in this article mainly stem from assumptions made during simplification of the wall panel model, and its interaction of resistance and axial force as described in previous chapters. Furthermore, while the simplified model exists to make a global analysis of such buildings feasible, the computational load can still be very significant. This will be further developed, and a comprehensive PSHA analysis will be conducted in future work. Further research on simplification of complex numerical models is warranted.

6. References

[1] Zelinschi, G. A. (2011). The fiscal impact of the direct losses caused by a future earthquake to the public buildings in Romania. Annals of DAAAM & Proceedings, 1129-1131.

[2] Schricker, V.; Powell, GH. (1980). Inelastic Seismic Analysis of Large Panel Buildings.

[3] Mihaela, Zamolo. (1988). Ponašanje spojeva krupnopanelnih konstrukcija u seizmičkim područjima [Magistarski rad]. [Zagreb]: Građevinski fakultet sveučilišta u Zagrebu;

[4] Abaqus 2016 Documentation, link (http://130.149.89.49:2080/v2016/index.html), pristupljeno 2025 May 12

[5] Users Guide ETABS 2016, link (https://ottegroup.com/wp-content/uploads/2021/02/ETABS2016-Users-Guide.pdf), accessed 2025 May 12

[6] She, J.; Zou, Y.; Liu, Y.; Li, ZH.; Li, KW. (2014). Nonlinear Numerical Analysis of Precast Concrete Shear Wall, Applied Mechanics and Materials, 1192–6, 651–653, DOI: 10.4028/www.scientific.net/AMM.651-653.1192

[7] Henry, RS.; Sritharan, S.; Ingham, JM. (2016). Finite element analysis of the PreWEC self-centering concrete wall system, Eng Struct.; 115:28–41, https://doi.org/10.1016/j.engstruct.2016.02.029

[8] Paudel, S.; Ilham, Maulana, TIiM.; Prayuda, H. (2024). Seismic Vulnerability Assessment of Regular and Vertically Irregular Residential Buildings in Nepal, Journal of the Civil Engineering Forum, 20;199–208, DOI: 10.22146/jcef.10316

[9] Michał, S.; Andrzej, W. (2015). Calibration of the CDP model parameters in Abaqus, Advances in Structural Engineering and Mechanics, Incheon, Korea, 25-29.8.

DOI: 10.2507/36th.daaam.proceedings.036

MACHINABILITY COMPARISON OF CONVENTIONAL INCONEL AND 3D PRINTED INCONEL 718

Michal Straka

This Publication has to be referred as: Straka, M[ichal] (2025). Machinability Comparison of Conventional Inconel and 3D Printed Inconel 718, Proceedings of the 36th DAAAM International Symposium, pp.0260-0265, B. Katalinic (Ed.), Published by DAAAM International, ISBN 978-3-902734-47-1, ISSN 1726-9679, Vienna, Austria
DOI: 10.2507/36th.daaam.proceedings.036

Abstract

Inconel 718 is among the most widely used nickel-based superalloys due to its high tensile strength, corrosion resistance, and creep resistance at elevated temperatures. However, these same properties also make its machining both technically challenging and costly. With the growing adoption of additive manufacturing for functional components, it has become increasingly important to understand potential differences in machinability between additively manufactured and conventionally produced Inconel 718 to optimize production processes. This article presents a comparative study of the machinability of conventionally manufactured Inconel 718 and its counterpart produced by Direct Metal Laser Sintering (DMLS). The experimental results indicate that DMLS Inconel 718 generally exhibits lower cutting forces and slower tool wear progression compared to wrought material, suggesting improved machinability. Surface roughness values were comparable between the two materials. These findings can support a more efficient selection of cutting parameters when machining additively manufactured components, helping to reduce costs and increase production efficiency.

Keywords: Inconel 718; Additive Manufacturing; Machining; Machinability; DMLS

1. Introduction

Inconel 718 is a precipitation-hardened, nickel-based superalloy extensively employed in aerospace applications. Engineered to retain high strength at elevated temperatures while exhibiting excellent corrosion and creep resistance. It performs reliably in service environments subject to severe thermal and mechanical loading. Consequently, its use extends beyond aerospace to nuclear and petrochemical systems, where structural components must maintain integrity under extreme operating conditions [1]. Strengthening is achieved via fine precipitates γ' $Ni_3(Al,Ti)$ and γ'' Ni_3Nb distributed in an austenitic γ-matrix, along with solid-solution strengthening by elements such as Mo and Nb. Inconel 718 also contains dispersed carbides, which contribute to its hardness [2], [3]. Transformation of γ'' to orthorhombic δ Ni_3Nb can occur during prolonged exposure at elevated service temperatures or under unsuitable heat-treatment schedules. Although a limited, discontinuous amount of δ at grain boundaries may help grain stabilization, its presence diminishes alloy strength because it depletes the Nb available for γ'' precipitation hardening. Another undesirable feature of the microstructure is the formation of Laves phases, typically $(Ni,Fe,Cr)_2(Nb,Ti,Mo)$, which develop in interdendritic regions during solidification. These brittle intermetallic compounds reduce toughness, promote crack initiation, and consume Nb needed for the precipitation of the strengthening γ'' phase [4], [5].

However, the very attributes that give Inconel 718 outstanding in-service performance also render it notoriously difficult to machine. Its high strength, tendency to work-harden during cutting and low thermal conductivity lead to extreme conditions at the tool–chip interface. Machining Inconel 718 generates high cutting forces and heat in the cutting zone. These aggressive conditions promote rapid tool wear [9]. The main tool wear mechanisms observed during the machining of Inconel 718 include adhesive wear leading to built-up edge, abrasive wear driving flank wear and notch wear due to hard particles, high-temperature diffusion and oxidation wear, frequently accompanied by crater wear on the rake face. As a result, cutting tools suffer from short tool life when machining 718. To control heat and tool wear, industry practice for machining Inconel 718 is to employ low cutting speeds and feeds, sometimes with high-pressure coolant. Machined Inconel 718 surfaces can develop tensile residual stresses and a thin work-hardened layer, which, if excessive, may degrade fatigue performance of critical components [8]. In response, numerous studies have explored methods to extend tool life through optimization of tool geometry (edge radius preparation), tool texturing, and development of advanced tools. Despite these efforts, Inconel 718 remains challenging to machine, and manufacturers often face short tool life [6], [7]. Additive manufacturing (AM) has emerged as a promising way to alleviate these challenges by producing near-net-shape Inconel 718 components that require less machining. In particular, laser powder bed fusion (LPBF) techniques such as Direct Metal Laser Sintering (DMLS) allow fabrication of complex Inconel 718 parts with minimal material waste. This approach can significantly reduce the amount of material that must be removed by machining, which is attractive given the difficulty of cutting Inconel 718. However, the geometrical near-net capability of additive manufacturing does not remove the need for subtractive finishing. As-built surfaces are typically rough, and dimensional accuracy generally falls short of functional tolerances. Consequently, finish machining remains essential to achieve the required surface integrity and precision [10]. Additively manufactured Inconel 718 differs significantly from conventionally produced material. AM builds often show anisotropic grain orientations, internal residual stresses, and in some cases higher porosity. Microstructural features such as columnar grains aligned with the build direction and finely dispersed Laves and δ phases can influence machinability, while embedded defects and pronounced anisotropy may lead to unpredictable tool wear or non-uniform surface finish [3]. Both build orientation and process parameters directly influence the final part's properties and machinability. Process parameters such as laser power and scan speed, hatch spacing, layer thickness and scan strategy are relevant to formation of porosity, lack-of-fusion, grain morphology, and phase distribution [12]. In combination, anisotropy and the process-induced defects can manifest during machining as variations in cutting forces, differences in chip formation, and changes in surface integrity [13]. Standard post-build treatments such as solution annealing, ageing, and in many cases hot isostatic pressing (HIP), can relieve residual stress, close pores, and drive precipitation toward the desired γ' and γ'' distribution, but differences in grain size, texture, and inclusion content commonly persist relative to wrought material [11]. The aim of this paper is to provide a comparative investigation of the machinability of conventional and additively manufactured Inconel 718 under controlled turning conditions. In particular, the study focuses on dry turning. The evaluation focuses on three indicators those being cutting-tool wear progression, cutting forces, and machined-surface roughness, measured under identical cutting conditions to isolate the effect of the production route. The objective is to determine whether established machining practices for wrought Inconel 718 translate to AM Inconel 718, or whether adjustments in tool material, edge preparation, geometry, or cutting parameters are required to achieve comparable or improved performance with the AM material.

2. Materials and Methods

Two material variants of Inconel 718 were investigated. A conventionally produced wrought bar and an additively manufactured specimen produced by Direct Metal Laser Sintering (DMLS) on an EOS M 290 system. The printed specimen was built in the vertical orientation and subsequently subjected to standard stress-relief annealing and precipitation hardening.

Fig. 1. Machining setup

Turning experiments were carried out on an EMCO MAXXTURN 25 CNC lathe using an SCLCL 1212F-09 toolholder equipped with CCMT 09T304-SM IC807 carbide inserts coated with TiAlN+TiN. Cutting forces were recorded using a KISTLER 9119AA2 multi-component piezoelectric dynamometer, connected to a KISTLER 5080 amplifier and 5691A data acquisition card, with data logged and processed in DynoWare software. Tool wear was monitored after each cutting pass using a KEYENCE VHX-6000 digital microscope. Flank wear (VB) was measured using image analysis software. Surface roughness (Ra) was measured in three positions on each machined surface using a Surtronic Duo II portable profilometer. Figure 1 shows the workspace prior to machining. The test specimen was secured in the chuck jaws and additionally supported by a live center in order to provide adequate rigidity of the setup. The machining tests were carried out under dry conditions without the use of cutting fluid. Each cut lasted approximately two minutes, and the machining process continued until the flank wear of the tool reached the predefined wear limit of VB = 0.25 mm. This procedure was repeated for three different cutting speeds, while the feed rate and depth of cut remained constant. The same experimental procedure was applied to both the conventionally produced and the additively manufactured Inconel 718 samples. The applied cutting conditions are summarized in Table 1.

Measurement	V_c [m/min]	f [mm/rev]	ap [mm]
1	40	0.15	0.5
2	55	0.15	0.5
3	75	0.15	0.5

Table 1. Cutting conditions

3. Results and Discussion

Figure 2 presents a comparison of flank wear progression at all investigated cutting speeds for both conventionally manufactured and DMLS produced Inconel 718. At the lowest cutting speed Vc = 40 m/min, the tool life when machining the DMLS material was approximately four times longer than in the case of the conventional material. This indicates a significantly improved machinability of the additively manufactured alloy under these conditions. With increasing speed, the time to reach the wear criterion decreased for both materials. At Vc = 55 m/min, the difference between DMLS and conventional Inconel 718 was still evident, though less pronounced compared to the lowest speed. At the highest cutting speed Vc = 75 m/min, the machining of both materials resulted in very rapid tool wear, indicating the severe impact of elevated cutting speed on tool life. Interestingly, the tool life when machining DMLS material at 75 m/min was only slightly longer than when machining the conventional alloy at 55 m/min. During all tests, the formation of built-up edge and abrasive flank wear of the cutting inserts were observed. As expected, higher cutting speeds led to accelerated tool wear for both materials. Nevertheless, additively manufactured Inconel 718 consistently provided longer tool life compared to its conventionally produced counterpart.

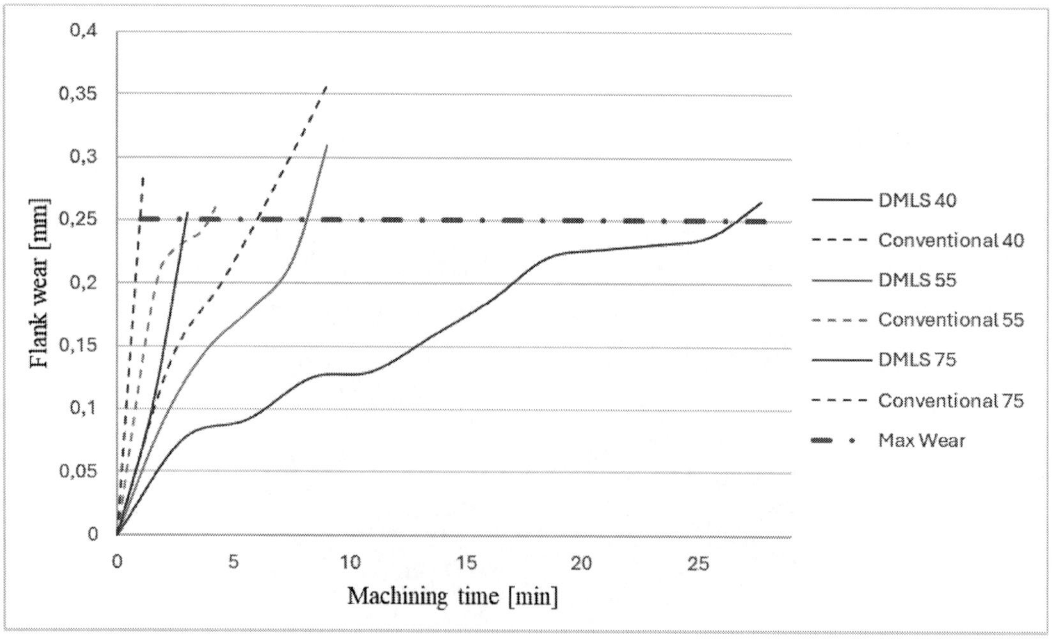

Fig. 2. Flank wear progression

The cutting forces during the experiments were measured using a KISTLER dynamometer connected to the DynoWare software for data acquisition and analysis. To ensure reliable results, the measured data were processed by excluding entry and exit regions of the cut, where vibrations of the system could occur. Average values for each force component were extracted from the remaining data. For evaluation, only the main cutting force (Fc) was considered. Figure 3 shows the forces measured at a cutting speed of Vc = 40 m/min. For the DMLS sample, the cutting force remained relatively stable over the entire machining time, with only minor deviations, even as tool wear gradually progressed. This indicates that the cut was stable throughout the test, with occasional small fluctuations likely caused by built-up edge formation. In contrast, machining of the conventional material resulted in a rapid increase in cutting force, closely correlated with faster tool wear.

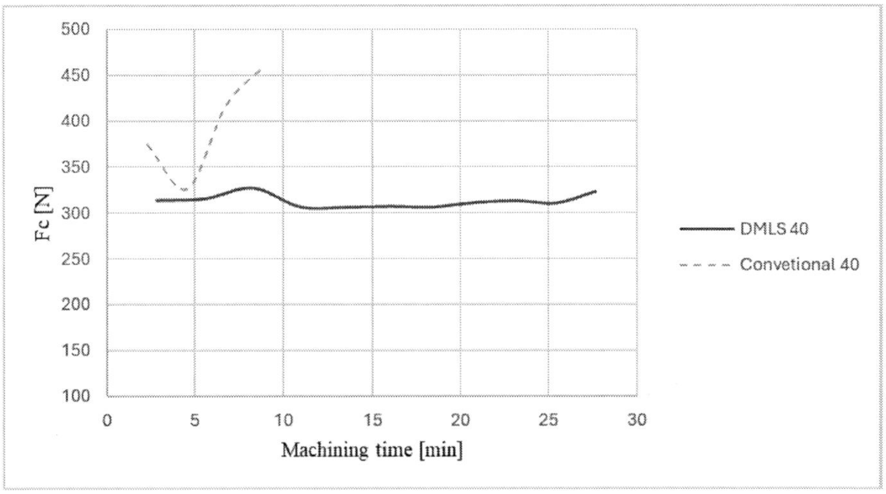

Fig. 3. Cutting forces for Vc = 40 m/min

As shown in Figure 4 the behavior of cutting forces at the intermediate speed of Vc = 55 m/min followed the same general trend as at 40 m/min. For the DMLS material, the cutting force remained stable, within 300–320 N, and comparable to the lower-speed test. By contrast, the conventional material generated higher forces from the start, which continued to rise during machining, ultimately reaching a value about 1.75 times greater than for the DMLS material and also higher than those recorded at the lower cutting speed.

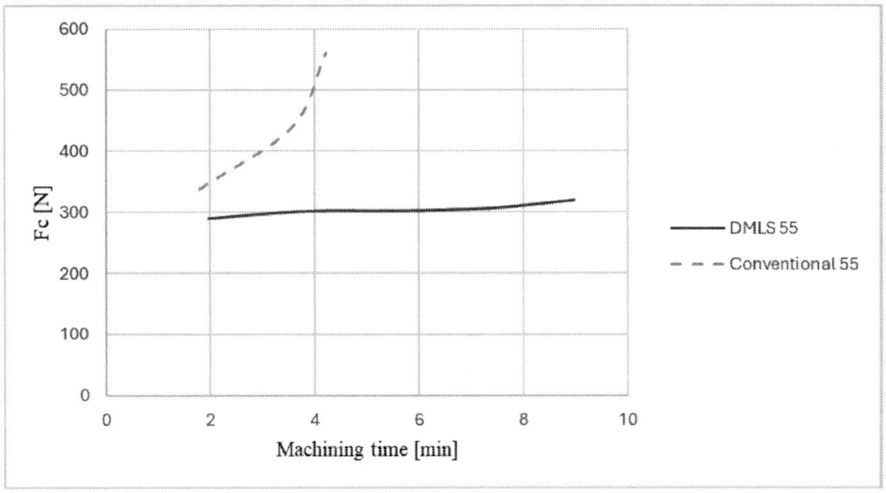

Fig. 4. Cutting forces for Vc = 55 m/min

At the highest cutting speed of Vc = 75 m/min, the number of valid data points was limited due to the rapid progression of tool wear. Nevertheless, a short evolution of cutting force could be recorded before tool failure. Figure 5 shows that, unlike the previous tests, the initial cutting forces for both DMLS and conventional Inconel 718 were nearly identical. As machining progressed, however, the forces began to diverge, with the conventional material again producing consistently higher cutting forces than the DMLS material, in line with the general trend observed at lower speeds.

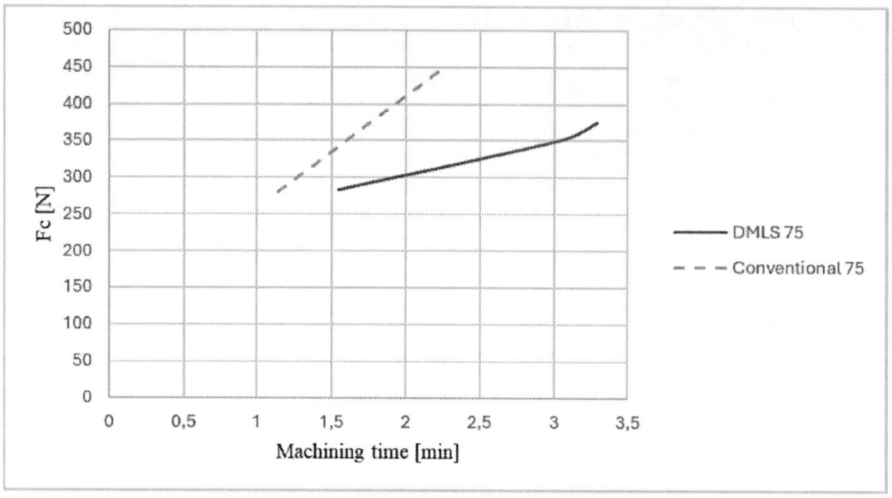

Fig. 5. Cutting forces for Vc = 75 m/min

The surface roughness was evaluated during machining for both materials. The results obtained are not fully conclusive, as the measured values were very close to each other. For both materials, roughness values remained in the range Ra 1.5 – Ra 2.5.

4. Conclusion

This study compared the machinability of conventionally wrought and DMLS Inconel 718 under identical dry-turning conditions. Across all cutting speeds, the additively manufactured material consistently exhibited lower main cutting forces and a slower progression of flank wear than the wrought alloy, while surface roughness values were comparable. The difference was most pronounced at the lowest speed Vc = 40 m/min. At higher speeds, tool wear accelerated for both materials, but the DMLS Inconel 718 maintained a measurable benefit. These results indicate that finish machining strategies established for wrought Inconel 718 can be applied to DMLS 718, potentially enabling longer tool life or higher productivity at comparable surface quality. Given that the experiments were conducted under dry-cutting conditions with a single tool type and a specific build orientation, future work should examine coolant application, alternative tool geometry, and different AM build orientations and parameters to generalize the findings and further optimize cutting parameters for AM components.

5. Acknowledgments

This article has been prepared under the project SGS-2025-025: Research and development for innovation in Machining, additive technology and quality assurance.

6. References

[1] Reed, R. C. (2006). The superalloys: Fundamentals and applications. Cambridge University Press, Cambridge, UK, ISBN 978-0-521-85904-2.

[2] Deng, D. (2018). Additively manufactured Inconel 718: Microstructures and mechanical properties. Ph.D. Dissertation, Department of Management and Engineering, Linköping University, Linköping, Sweden, ISBN 9789176853832.

[3] Hosseini, E. & Popovich, V. A. (2019). A review of mechanical properties of additively manufactured Inconel 718. Additive Manufacturing, Vol. 30, Article 100877, ISSN 2214-8604, https://doi.org/10.1016/j.addma.2019.100877

[4] Sundararaman, M.; Nalawade, S.; Singh, J. B.; Verma, A.; Paul, B.; Ramaswamy, K. (2012). Evolution of δ phase microstructure in Alloy 718. In: Superalloy 718 and Derivatives, pp. 737–750, Wiley, https://doi.org/10.1002/9781118495223.ch57

[5] Wilson, A. S. (2017). Formation and effect of topologically close-packed phases in nickel-base superalloys. Materials Science and Technology, Vol. 33, No. 9, pp. 1108–1118, ISSN 0267-0836, http://dx.doi.org/10.1080/02670836.2016.1187335

[6] Vopat, T.; Straka, R.; Kuruc, M. & Peterka, J. (2022). The effect of cutting edge radius sizes on tool life in machining nickel alloy Inconel 718. Proceedings of the 33rd DAAAM International Symposium, pp. 0180–0187, B. Katalinić (Ed.), DAAAM International, Vienna, Austria, ISBN 978-3-902734-36-5, ISSN 1726-9679, https://doi.org/10.2507/33rd.daaam.proceedings.025

[7] Mezulianik, J.; Povolny, M. & Kroft, L. (2023). Development of a high-performance tool for superalloy machining (HRSA). Proceedings of the 34th DAAAM International Symposium, pp. 0537–0543, B. Katalinić (Ed.), DAAAM International, Vienna, Austria, ISBN 978-3-902734-41-9, ISSN 1726-9679, https://doi.org/10.2507/34th.daaam.proceedings.071

[8] Al-Maskari, F. et al. (2022). Tool wear in nickel-based superalloy machining: An overview. Processes, Vol. 10, No. 11, 2380, ISSN 2227-9717, https://doi.org/10.3390/pr10112380

[9] Liang, X.; Liu, Z. & Wang, B. (2019). State-of-the-art of surface integrity induced by tool wear effects in machining process of titanium and nickel alloys: A review. Measurement, Vol. 132, pp. 150–181, ISSN 0263-2241, https://doi.org/10.1016/j.measurement.2018.09.045

[10] Ben Said, L.; Shevchenko, N.; Tiwari, A.; Sundaram, M. M. & Silwal, B. (2025). Recent advances in additive manufacturing: A review of current developments and future directions. Machines, Vol. 13, No. 9, 813, ISSN 2075-1702, https://doi.org/10.3390/machines13090813

[11] Mostafa, A.; Picazo Rubio, I.; Brailovski, V.; Jahazi, M. & Medraj, M. (2017). Structure, texture and phases in 3D printed IN718 alloy subjected to homogenization and HIP treatments. Metals, Vol. 7, No. 6, 196, ISSN 2075-4701, https://doi.org/10.3390/met7060196

[12] Bean, G. E.; McLouth, T. D.; Witkin, D. B.; Sitzman, S. D.; Adams, P. M. & Zaldivar, R. J. (2019). Build orientation effects on texture and mechanical properties of selective laser melting Inconel 718. Journal of Materials Engineering and Performance, Vol. 28, pp. 1942–1949, https://doi.org/10.1007/s11665-019-03980-w

[13] Schneider, S. A. M.; Hermsen, S.; Kirchmann, S.; Ganser, P.; Bergs, T. & Schleifenbaum, J. H. (2022). Machinability analysis for milling of additively manufactured Inconel 718 with specifically induced porosity. SSRN preprint, https://doi.org/10.2139/ssrn.4259357

DOI: 10.2507/36th.daaam.proceedings.037

URBAN PLANNING CRITERIA FOR RIVERFRONT CATEGORISATION

Dora Ivančan & Sanja Gašparović

This Publication has to be referred as: Ivancan, D[ora] & Gasparovic, S[anja] (2025). Urban Planning Criteria for Riverfront Categorisation, Proceedings of the 36th DAAAM International Symposium, pp.0266-0270, B. Katalinic (Ed.), Published by DAAAM International, ISBN 978-3-902734-47-1, ISSN 1726-9679, Vienna, Austria
DOI: 10.2507/36th.daaam.proceedings.037

Abstract

Riverfronts, as a transitional zone, have been recognised as a catalyst of urban regeneration. Their revitalisation has become a widespread global urban phenomenon and a frequently implemented strategy in urban planning. Despite numerous successful examples, riverfront transformation in some cities failed due to unsuitable interventions and a tendency to replicate globally popular models without contextual adaptation. To identify cities with the greatest potential for successful transformation and revitalisation, the research aims to determine the systematic framework for assessing the transformation potential of riverfronts. Derived from a literature review, the paper establishes a set of spatial criteria for the categorisation of riverfronts. These criteria are systematised and categorised according to four observation scales: (1) general spatial characteristics, (2), city, (3) riverfront, and (4) waterline. The result emphasises the importance of criteria defined at the city and riverfront level, which prove to be the most influential in determining transformation potential from the urban planning point of view. The established framework will guide further research on Croatian continental cities to evaluate their riverfront transformation potential and identify context-appropriate urban planning models.

Keywords: river-city relationship; blue urban space; riverfront transformation; urban regeneration

1. Introduction

Riverfronts, defined as the border area/transition zone between urban fabric and water [1], have been recognised as a "spark of urban regeneration" [2]. Their transformation can affect the economic, environmental, aesthetic, and social conditions of the entire city [1], [3]. Also, their revitalisation is one of the most widespread urban phenomena [4] and a major tool in urban planning since the 1970s, with the potential to create a new image of the city, attract investment, and prevent the deterioration of space [5]. The phenomenon of urban revitalisation, which began with the transformation of American post-industrial cities, has spread to all continents, affecting cities of various sizes and importance, cities by the sea and those inland [6]. In many cities, the transformation of riverbanks is a global phenomenon that has had a positive impact on the development of the entire city. The popularity of such projects and the desire to replicate their success have, in many cases, encouraged urban transformations that are not appropriate in terms of location, budget, programme, time, and type of transformation, resulting in unsuccessful developments [7]. This paper establishes the criteria for the categorisation of riverfronts based on a review of previous research.

The literature was selected by reviewing the Web of Science and Scopus scientific databases based on keywords ("waterfront" OR "riverfront" OR "urban blue space") AND ("classification" OR "typology" OR "categorisation") in the fields of urban science (WOS) and technical sciences, social sciences, and arts and humanities (Scopus). The review of databases was limited to works in English with open access, resulting in a total of 26 publications. From the mentioned 26 papers, ten papers that state the categorisation of riverfront areas were singled out, namely eight scientific papers, one article from the proceedings of a scientific conference and one book. The criteria defined in the mentioned literature are analysed, systematised, and categorised in order to highlight the urbanistically relevant ones for riverfront transformations. The established criteria serve as a foundation for further research - establishing a method of evaluating the urban transformation potential of riverfronts in Croatian cities.

2. Review of the research criteria for the categorisation of riverfronts

Previous research has dealt with the criteria for categorising riverfronts from different perspectives and at different scales. Petrtýlová and Jaššo, in their categorisation of riverfronts, highlight the specific characteristics of the city and rivers [8] (Table 1). They point to relevant aspects such as the size of cities (the status and area of the city, as well as the number of inhabitants) and the size of the river (the ratio of city area to river width, the hydrological significance of the watercourse, and the way the river is used) as an important indicator of the degree of urbanisation and the character of the space.

Criterion	Source	Categorisation
City status	[8]	Capital city; Federal state/municipality/regional capital; City; Town
City area	[8]	-
Number of inhabitants	[8]	-
The width of the river in the area closest to the city center	[8]	-
Ratio of river width to city size	[8]	<1; 1; >1
Size of the river in relation to the city	[8]	The river is not significant in size for the city; The relationship between the city and the river is in balance; The size of the river exceeds the importance of the city
Hydrological significance of water	[8]	Hydrologically significant watercourses, Small watercourses
River use	[8]	Electricity production and operation; Farming of freshwater fish and other aquatic organisms; Navigation, rafting, canoeing and other watercraft; Installation of floating or floating objects

Table 1. Criteria for categorisation of riverfronts according to general spatial characteristics of the city and the river

A number of criteria have been established to categorise the relationship between a city and a river (Table 2). Cities are categorised according to waterfront settlement type (city divided by a river, city on the edge of the river, city divided by several rivers and tributaries, swamp city, coastal city divided by a river, coastal city adjacent to a river, mountain town divided by a river, city on a lake divided by a river) [8], [9] and the significance of the city by the river (waterfront settlements, small to medium-sized waterfront towns, waterfront cities as lifestyle destinations, commuter waterfront settlements, waterfront regional cities) [9]. The location of the river within the city is categorised according to the location of the city in relation to the river, the location of the urbanised area in relation to the river, and the location of the river in relation to the historic centre [8].

Criterion	Source	Categorisation
City-river layout	[8], [9], [19]	City divided by a river; City on the edge of the river; City divided by several rivers and tributaries; Swamp city; Coastal city divided by a river; Coastal city adjacent to a river; Mountain town divided by a river; City on a lake divided by a river
The importance of the city by the river	[9]	Waterfront settlements; Small to medium-sized waterfront towns; Waterfront cities as lifestyle destinations; Commuter waterfront settlements; Waterfront regional cities
The position of the city in relation to the river	[8]	Right bank; Left bank; Both banks
Location of the riverfront in relation to the urbanized area	[8]	Riverfront outside of the urbanised area; Riverfront within the urbanised area
The position of the historic center in relation to the river	[8]	Historical city centre divided by several rivers and tributaries; Historical city centre divided by river; Historical city centre on the edge of the river; Historical city centre is not located near the river

Table 2. Criteria for categorisation of riverfronts on a city scale

At the riverfront scale (Table 3), research has focused on the categorisation of riverfronts by planned land use [8], [9], [10], [11], [12], character and morphology [13], the categorisation of public open spaces – parks, promenades and recreation [1], [14] and brownfield areas [15]. Breen and Rigby categorised riverfronts by land uses (commercial, cultural, educational, environmental, entertainment, industrial, business, residential and historic ports) [16], while Vallega adds categories that more closely describe their existing uses, such as enjoyment of the ecosystem, fishing, congresses, media, transport and navigation, research and education and training [10]. In the categorisation, Breś and Krośnicka place more emphasis on the river area, dividing the land use into the natural environment, urban environment for housing (traffic, infrastructure, housing, culture, education, commercial and recreational space) and urban environment for industry and infrastructure (defence, port, industry, transport, infrastructure, energy production and waste disposal) [17]. In addition to land use, the coast is characterised by its morphological shape, which Aouissi et al. divide into convex, concave and linear [13], and the character of the coastal area defined by human influence (natural, regulated, and built) [1], [8]. Lyu et al. categorise urban coastal parks according to landscape character - meadows with a type of forest area, squares, recreational areas, and promenades along the coast [14]. Duran Vian, Pons Izquierdo and Serrano Martinez establish criteria for the categorisation of public open spaces along the coast based on accessibility and position in relation to the river. According to accessibility, they distinguish between physical and visual accessibility, which may be total, partial, occasional, or completely absent. They establish that the location and accessibility of public spaces are greatly influenced by the morphology of the river valley and coast (vertical bluff and floodplain, large and broad floodplain, terraced floodplain, broad and braided river, or gently-sloping V-shaped valley) and position of public open spaces relative to the river (near the river, in the river or on the river) [1]. Brownfield areas, often located near rivers, in areas that are both attractive and valuable within the context of the entire city, represent an important element in the categorisation of the riverfronts. Rey, Laprise, and Lufkin cite brownfield categorisations according to location, use, ownership, size, level of contamination, and primary use [15].

Criterion	Source	Categorisation
Planned land use of the riverfront	[8]	Commercial; Cultural; Educational; Environmental; Recreational; Residential; Working; Transportation
Existing use of the riverfront	[10]	Ecosystem enjoyment; Fishing; Tourism; Recreation; Entertainment; Congress; Media; Transport and navigation; Trade and finance; Research areas; Education and training; Cultural heritage
Morphological shape of the riverfront	[13]	Convex; Concave; Linear
Morphological shape of the river valley	[1]	Vertical bluff and floodplain; Large, broad floodplain; Terraced floodplain; Broad, braided river; Gently sloping V-shaped valley
Position of public spaces in relation to the river	[1]	Near the river (at the edge, In a setback position, Multileveled open space); In the river; On the river
Physical and visual accessibility	[1]	(Almost) Total; Partial, Occasional; No
Brownfield type by location	[15]	Urban; Suburban; Rural
Brownfield type according to use	[15]	Under-utilised; Vacant; Derelict; Dangerous
Brownfield type by ownership	[15]	Private ownership; public ownership; mixed ownership; unsettled ownership rights
Size of the brownfield area	[15]	Small (<1ha); Middle-sized (1-10ha); Large (10-100ha); Extensive (>100ha)
Contamination level	[15]	Without ecological burden; with ecological burden; unknown degree of ecological burden; Without contamination; Contaminated, Unknown degree of contamination
Primary use of brownfield areas	[15]	Primary sector (agricultural, mining, and extractive); Secondary sector (industrial and construction, transport, and infrastructure); Tertiary sector (logistic, commercial, and administrative facilities, former public facilities, residential and housing facilities, others); Others (religious buildings, military, former chateaus)

Table 3. Criteria for categorisation of riverfronts in the scale of the riverfront

At the waterline scale (Table 4), Breś and Krośnicka propose criteria for the categorisation of urban blue space - an area that includes water and at least one edge between water and land. As an important element of urban blue space, they emphasise its edge, i.e. the waterline, which they categorise according to the quantity (one, two, more, complex edge) and type (fixed - vertical, over water; flexible – sloping, slanted, terraced, floating). They further categorise the relationship between land and waterline using the criteria of connectivity (not connected to water, connected with waterline, connected with more than one waterline) [17] and permeability (near the water, on the water, in the water) [18].

The edge element is also observed through the criteria of flood protection, where the categories of hard and soft edges are recognised [1].

Criterion	Source	Categorisation
Number of waterlines	[17]	Single edge; Double edge; Multiple edge; Complex edge
Type of waterlines	[17]	Fixed (Vertical, Over water); Flexible (Sloping, Slanted, Terraced, Floating)
The relationship between land and waterline	[17]	Not connected with waterline, Connected with waterline (longitudinal, perpendicular, independent from waterline), Connected with more than one waterline
Waterline in terms of permeability	[1],[18]	Near the water, In the water, On the water
Waterline in terms of flood protection	[1]	Hard edge (Flood wall, Hardened bank, Filled bank, Embankment wall); Soft edge (Natural, Floodplain, Fluvial park, Vertical bluff, Terraced plains, Levee)

Table 4. Criteria for categorisation of riverfronts at the scale of the waterline

From the conducted analysis and systematisation of previous research, it can be concluded that the criteria for the urban riverfront categorisation can be divided into four basic groups according to observation scale:

1. General spatial characteristics of the city and the river – quantitative characteristics that may have an impact on the categorisation of riverfronts - globally observed
2. The relationship between the position of the city and the river observed on a city scale
3. Accessibility and land use characteristics observed at the scale of the riverfront
4. Characteristics of the immediate city/water contact edge – observed at the waterline scale.

The first category is focused on descriptive and analytical data related to cities and their rivers. While this information is important to consider, it is not directly relevant when researching spatial data from an urban planning point of view. Moreover, the criteria used for categorisation at the waterline scale are entering the domain of architecture and urban design. On the other hand, the second and third groups of criteria are expected to have a greater influence on the urban perspective, as they most directly capture the relationship between the riverfront area and the surrounding urban fabric. Therefore, the next phase of the research will primarily focus on the city and riverfront scales.

3. Starting points for exploring the riverfronts of cities in Croatia

In all Croatian continental cities, it is possible to find watercourses of different types, significance, and character. Examples of successful urban transformation based on the riverfront revitalisation suggest that the same approach could be used in the context of Croatian cities. In order to select cities with the greatest potential for transformation and to use transformation methods that best suit each city, it is necessary to conduct a systematic study of the interrelationship between the city and the river from an urban planning perspective. For such research, a criteria systematised on the basis of previous research at the scales of the city, and the riverfront area will be used. By applying these criteria, it is necessary to examine whether the established categorisation encompasses all Croatian continental cities and, possibly, supplement the categorisation based on the observed shortcomings. In addition, it is necessary to implement criteria related to the level of preservation of natural and cultural heritage [18], navigability of rivers and port activity, which are not considered in the reviewed literature but can have a great impact on the potential for urban transformation.

4. Conclusion

With the primary aim to determine the systematic framework for assessing the transformation potential of riverfronts, the conducted research provided the first step - a synthesis of existing categorisation of riverfronts and identifying key criteria that influence their differentiation. The criteria were analysed and categorised into four basic groups according to observation scale: (1) general spatial characteristics of the city and the river, (2) the spatial relationship between the city and the river, (3) riverfront scale, and (4) the characteristics of the immediate city–water interface at the waterline scale. The scale of the city and the scale of the riverfront are recognised by this research as the most relevant, from the urban planning perspective, for determining the potential of the transformation of the city based on the transformation of its riverfront. In further research, a comparative analysis of riverfront transformation in European cities will be conducted. Criteria established in this research will be supplemented by additional factors that can influence the transformation potential of the riverfront but were not addressed in the reviewed literature, such as natural and cultural heritage, as well as port activities. Through this approach, a holistic methodology for evaluating transformation potential and identifying site-specific urban planning models will be developed, which can also be applied in Croatian cities.

5. References

[1] Durán-Vian, F.; Pons-Izquierdo, J. J. & Serrano-Martínez, M. (2021). River-city recreational interaction: a classification of urban riverfront parks and walks, Urban Forestry & Urban Greening, Vol. 59, 2021, pp. 1–12, ISSN 1618-8667, https://doi.org/10.1016/j.ufug.2021.127042

[2] Carta, M. (2016). The fluid city paradigm: a deeper innovation, In: The fluid city paradigm: waterfront regeneration as an urban renewal strategy, pp. 1–10, Springer, ISBN 978-3-319-28004-2, https://doi.org/10.1007/978-3-319-28004-2_1

[3] Breen, A. & Rigby, D. (1994). Urban values: a new appreciation, In: Waterfronts: cities reclaim their edge, pp. 1–8, McGraw-Hill, ISBN 0-07-068458-8, New York

[4] Giovinazzi, O. & Moretti, M. (2010). Port cities and urban waterfront: transformations and opportunities. TeMA Journal of Land Use, Mobility and Environment, Vol. 3, 2009, pp. 57–64, ISSN 1970-9870, https://doi.org/10.6092/1970-9870/123

[5] Dowey, K. (2005). Flows, In: Fluid city, pp. 9–15, Routledge, ISBN 9780415359238, London

[6] Sieber, T. (1991). Waterfront revitalization in postindustrial port cities of North America, City & Society, Vol. 5, No. 2, 1991, pp. 120–136, ISSN 0893-0465, https://doi.org/10.1525/city.1991.5.2.120

[7] Pinto, P. J. & Kondolf, G. M. (2020). The fit of urban waterfront interventions: matters of size, money and function, Sustainability, Vol. 12, No. 4079, 2020, pp. 1–17, ISSN 2071-1050, https://doi.org/10.3390/su12104079

[8] Petrtýlová, R. & Jaššo, M. (2024). Towards a new typology for classifying urban riverfront character sections: insights from European cities, Europa XXI, Vol. 47, 2024, pp. 31–53, ISSN 1429-7132, https://doi.org/10.7163/Eu21.2024.47.14

[9] Aulia, D.; Marpaung, B. & Zahrah, W. (2019). Typology of livable waterfront settlement and how to manage the community, IOP Conference Series: Materials Science and Engineering, Vol. 505, 2018, Medan City, Indonesia, pp. 1–7, IOP Publishing, ISSN 1757-8981, https://doi.org/10.1088/1757-899X/505/1/012136

[10] Vallega, A. (2001). Urban waterfront facing integrated coastal management, Ocean & Coastal Management, Vol. 44, 2001, pp. 379–410, ISSN 0964-5691, https://doi.org/10.1016/S0964-5691(01)00056-4

[11] Keyvanfar, A.; Shafaghat, A.; Mohamad, S.; Abdullahi, M. M.; Ahmad, H.; Mohd Derus, N. H. & Khorami, M. (2018). A sustainable historic waterfront revitalization decision support tool for attracting tourists, Sustainability, Vol. 10, 2018, pp. 1–23, ISSN 2071-1050, https://doi.org/10.3390/su10020215

[12] Tolnai, G. (2018). Budapest's fragmented riverfront renewal: Western trends interspersed with post-socialist characteristics, Belgeo, Vol. 4, 2018, pp. 1–16, ISSN 1377-2368, https://doi.org/10.4000/belgeo.21210

[13] Aouissi, K. B.; Madani, S.; Hein, C. & Benacer, H. (2023). Morphological approach for the typological classification of waterfront revitalization, Journal of the Geographical Institute Jovan Cvijic SASA, Vol. 73, 2023, pp. 109–122, ISSN 0350-7599, https://doi.org/10.2298/IJGI2301109A

[14] Lyu, M.; Wang, S.; Shi, J.; Sun, D.; Cong, K. & Tian, Y. (2025). Visual satisfaction of urban park waterfront environment and its landscape element characteristics, Water, Vol. 17, 2025, pp. 3–18, ISSN 2073-4441, https://doi.org/10.3390/w17060772

[15] Rey, E.; Laprise, M. & Lufkin, S. (2022). Neighborhoods in transition: brownfield regeneration in European metropolitan areas, Springer, ISBN 978-3-030-82207-1, Cham

[16] Breś, J. & Krośnicka, K. A. (2021). Evolution of edges and porosity of urban blue spaces: a case study of Gdańsk, Urban Planning, Vol. 6, No. 3, 2021, pp. 90–104, ISSN 2183-7635, https://doi.org/10.17645/up.v6i3.4108

[17] Dubinina, A.; Wawrzyńska, A. & Krośnicka, K. A. (2022). Permeability of waterfronts: contemporary approach in designing urban blue spaces, Sustainability, Vol. 14, 2022, pp. 1–16, ISSN 2071-1050, https://doi.org/10.3390/su14159357

[18] Cenar, I. (2011). Heritage Assets in the Accounting of Public Institutions, in Katalinic, B. (ed.), DAAAM Proceedings, DAAAM International, Vienna, 2011, pp. 993–994, ISSN –, https://doi.org/10.2507/22nd.daaam.proceedings.485.

DOI: 10.2507/36th.daaam.proceedings.038

SELF-CURING CONCRETE: A MODERN APPROACH TO OPTIMISING HYDRATION AND DURABILITY IN EUROPEAN CONDITIONS

Adam Hopocký & Marek Ďubek

This Publication has to be referred as: Hopocky, A[dam] & Dubek, M[arek] (2025). Self-Curing Concrete: A Modern Approach to Optimising Hydration and Durability in European Conditions, Proceedings of the 36th DAAAM International Symposium, pp.0271-0275, B. Katalinic (Ed.), Published by DAAAM International, ISBN 978-3-902734-47-1, ISSN 1726-9679, Vienna, Austria
DOI: 10.2507/36th.daaam.proceedings.038

Abstract

Self-curing concrete (SCC) represents a significant innovation in the field of building materials, optimizing the cement hydration process without the need for traditional curing methods. This technology is particularly relevant for concrete structures exposed to extreme climatic conditions or for elements where access to standard curing procedures is limited. The implementation of SCC can bring significant benefits in terms of durability, reduced labour costs, and time efficiency of construction processes. This article focuses on introducing the principles of self-curing concrete, evaluating its technical and economic benefits, and discusses potential areas for improvement. Particular emphasis is placed on identifying suitable materials and additives applicable to European climatic and standard conditions, thereby promoting sustainable development and more efficient use of resources in construction.

Keywords: Self-curing concrete; internal concrete treatment; cement hydration; treatment medium; Lightweight aggregates

1. Introduction

Concrete is the most widely used building material, which is why research and testing are constantly underway to improve and optimize different types of concrete that are suitable for all kinds of application [1]. Traditional concrete curing is one of the key steps that determine the quality, durability, and strength of concrete structures. Traditional concrete curing methods (external surface curing) are based on minimizing water loss or compensating for this loss by wet curing. The problem, however, is that traditional methods are not very effective in delivering water to deeper parts of the structure. However, in conditions of accelerated construction, climatic extremes, and rising labour costs, however, its reliable implementation is becoming increasingly difficult. Self-curing concrete (SCC, also known as internally cured concrete) is an innovative approach that allows the necessary moisture to be supplied directly from within the cement matrix, thereby eliminating or significantly reducing the need for traditional surface treatment [2]. Self-curing concrete is a technological solution that ensures that cement hydration takes place thanks to the availability of added water inside the concrete, which is separate from the mixing water.

The key to implementing SCC was to incorporate curing water into a carrier that can bind it physically or chemically. These water carriers (media) act as water reservoirs that are evenly distributed throughout the concrete volume. This article summarizes the principles of SCC technology, presents a suitable material for European conditions, and discusses the technical, economic, and environmental benefits and limitations of this solution.

2. Basic principles of Self-curing

First, it is necessary to define and clarify the basic terms. Self-curing is an innovative technology that can bring new solutions to established curing processes and improve them – the primary task is to prevent problems that arise during the cement hydration process in concrete paste. This type of self-curing concrete differs significantly from self-healing concrete, which is designed to repair cracks that have formed within the structure [3]. It also differs from self-compacting concrete, which is designed for optimal compaction so that external sources of vibration, impact, and other compaction methods are not necessary. Self-curing concrete is a technology based on internal curing (IC). The primary idea is to ensure the supply of curing water directly from within the concrete, where it was incorporated during mixing [4]. SCC ensures the supply of water directly from inside the concrete via a carrier (medium) to compensate for moisture loss, whether through autogenous drying (water consumption for hydration) or external evaporation. However, the curing water must not be free. It must be bound in the medium (carrier) to prevent an increase in the water-cement ratio (w/c) of fresh concrete and thus a negative impact on its mechanical properties. The aim is to distribute the curing water (in the curing media) evenly throughout the entire cross-section of the structure and to release it at the appropriate time when it is necessary for hydration to continue.

Sufficient and effective curing of concrete during the initial setting phase is a key factor in ensuring its required mechanical and durability properties. A lack of water during hydration leads to a slowdown or complete cessation of cementation reactions, which negatively affects not only the development of strength, but also the homogeneity and integrity of the resulting microstructure. Reduced hydration causes the formation of pores and microdefects, which contribute to higher permeability and reduced resistance of concrete to long-term degradation effects. Self-curing concrete solves this problem by releasing water reserves from internal sources, thereby stabilizing moisture conditions and promoting continued hydration even in later stages of setting. In addition, excessive and rapid evaporation of water from the surface of fresh concrete causes volume changes in the form of shrinkage, which leads to internal stresses and the subsequent formation of microcracks and macrocracks. These defects significantly accelerate the penetration of aggressive substances into the concrete matrix, thereby shortening the service life of the structure and increasing the need for subsequent interventions during operation. A properly designed and applied curing system is therefore not only a technological step, but also a strategic measure aimed at ensuring the long-term durability of the structure. Minimizing the formation of defects and limiting the occurrence of cracks reduces the frequency of maintenance and the need for remediation, which ultimately brings significant economic savings and environmental benefits. The longer service life of structures directly contributes to reducing the consumption of building materials, energy, and emissions associated with repairs or renovations, thereby supporting the concept of sustainable construction. [5]

At the microstructural level, internal water reserves reduce capillary porosity and improve Interfacial Transition Zone (ITZ) between the aggregate and cement paste, limiting microcrack growth and improving overall durability [6]. The effectiveness of SCC can be evaluated using calorimetry to monitor hydration, internal relative humidity sensors, dilatometers to measure shrinkage, or electron microscopy (SEM) and micro-CT analysis, which provide an insight of the development of porosity and microcracks.

3. Requirements for successful application

The successful implementation of SCC technology requires strict compliance with several material and technological requirements that determine the system's ability to retain and release water during cement hydration. The basic prerequisite is the selection of a suitable curing medium. The material used as a water carrier must not only ensure the required physical and chemical properties of the system, but also reflect the locally available resources, economic conditions, and environmental aspects of the specific region. For this reason, a number of studies are currently being conducted worldwide to identify and optimize materials that could effectively replace or supplement traditionally used commercial products such as lightweight aggregates [7] or superabsorbent polymers (SAP) [8]. Individual research teams are therefore investigating various types of porous materials – from natural ones (e.g., pumice, diatomite, expanded clays, naturally absorbent fibres) to industrial by-products such as fly ash, slag, or recycled ceramic aggregates [9].

This approach not only allows internal curing to be adapted to the specific climatic and technological conditions of a given area, but also reduces the ecological footprint of concrete production through the use of local raw materials or waste materials. Regional optimization of materials for SCC is therefore a key step towards increasing the efficiency, sustainability, and economic competitiveness of this technology on a global scale. In Central Europe, where climatic influences are characterized by significant temperature differences between seasons and the periodic alternation of wet and dry cycles, the use of lightweight absorbent aggregate appears to be one of the most suitable alternatives for internal curing. This type of aggregate can effectively act as an internal treatment medium, but its functionality is directly dependent on several physical parameters, in particular adequate porosity, sufficient absorbency capacity, and suitable grain size distribution.

The size and morphology of the pores in the water carrier significantly affect its functionality. In order for water to be released from the pores efficiently, the pores in the lightweight absorbent aggregates (SLWA) must be larger than the pores in the cement mortar. An overly fine porous structure causes strong capillary binding of water, while excessively open pores can lead to premature water release and reduced process efficiency [10]. Figure 1 schematically shows water absorption into the structure of lightweight aggregate – Liapor, the most accessible aggregate of this type on the Slovak and Czech markets.

Fig. 1. Illustration of water binding in the structure of lightweight aggregate [Author]

The water release process proceeds as follows:

- Gradient formation: Water bound in the curing medium does not participate in the chemical reaction of hydration until a moisture gradient is formed during the initial hydration phase.
- Start of release: During the setting and hardening of concrete, especially in concretes with a low water coefficient (w/c), the cement paste consumes the available mixing water for hydration (autogenous drying), which causes a decrease in relative humidity (RH) inside the cement paste.
- Capillary tension: This decrease in RH creates tensile stresses (capillary tension) in the pore solution.
- Desorption: As a result of these capillary stresses, water begins to gradually empty from the pore structure of the curing medium. The principle of conservation of minimum internal energy of the system dictates that the largest pores of the curing medium are emptied first.
- Compensation: Water is released from the reservoirs into the drier parts of the cement paste (into the capillary pores). This compensates for the lost moisture, prolongs the hydration of the cement, and maintains an almost constant RH in the concrete. [10]

In the context of local material availability, it is also advantageous that suitable types of lightweight aggregates – such as expanded clay, pumice, or recycled ceramic fractions – are relatively readily available within the region. Their use can thus contribute not only to improving the technological properties of concrete, but also to reducing the environmental impact and overall economic costs associated with the production and maintenance of concrete structures. Equally important is the precise design of the self-curing agent dosage. The amount of water released must be sufficient to compensate for autogenous drying and moisture loss from the surface, while excessive water content can lead to segregation or a reduction in strength. When using dry porous aggregate, there is a risk that water absorption from the mixture will cause it to dry out and bleed; therefore, pre-saturating the aggregate is necessary. These characteristics allow for uniform dispersion of the aggregate in the cement matrix and ensure a stable and controlled supply of water during the hydration reactions.

4. Limits of technology

Despite the proven effectiveness of internal treatment in reducing the risk of autogenous shrinkage and improving the microstructure of cement paste, this technology, like most technologies, has its limitations, which must be taken into account in the process of experimental research, formulation design, and practical implementation in construction. One of the main limitations of SCC is its time-limited effectiveness. The effect of internal curing depends on the amount of water bound in the self-curing medium (e.g., SLWA) and on the conditions of concrete setting and hardening. Once this water supply is exhausted, SCC ceases to be active, with the time span of this effect being relatively short and concentrated mainly in the early stages of hydration. This means that internal curing can effectively eliminate defects associated with early drying, but is not sufficient as a standalone long-term protection of the concrete surface.

Another limitation is the decreasing diffusion range of the curing medium over time. With increasing density and reduction of capillary pores during hydration, the ability of water to diffuse from SLWA pores to more distant areas of the cement matrix is significantly reduced. Experiments have shown that while in the first 24 hours, lightweight aggregate is able to supply water to the cement paste at a distance of approximately 15–20 mm, after one week this can decrease to less than 1 mm. This phenomenon highlights the need for proper design of the fraction distribution and uniform distribution of the curing medium throughout the bulk mixture. From a technological point of view, internal treatment can also affect the mechanical properties of concrete. When superabsorbent polymers (SAP) are used, the release of water causes micro-pores to form in the cement matrix, which can locally reduce compressive strength or modulus of elasticity. However, in most cases, this negative effect is compensated by improved hydration of cement grains and greater homogeneity of the resulting microstructure. In the case of lightweight aggregate (SLWA), some types of materials can have a similar effect to air-entraining admixtures, which must be taken into account when optimizing the mixture design.

Particular attention must be paid to the precise dosing of the treatment medium, especially in the case of SAP. An excessive amount of polymer can lead to excessive porosity, a decrease in strength and, in extreme cases, the formation of cracks when drying in a dry environment. Properly designed dosing must therefore respect the balance between the water needed for hydration and the water supplied by the medium to ensure that the process is stable and controlled. From the point of view of application in Central European conditions, a significant obstacle to the implementation of SCC technology is the fact that its use is not currently common practice in commercial concrete plants. The main reasons for this are the lack of demand from the construction sector, the lack of practical experience with the technological processing of self-curing mixtures, and the absence of standardized methodologies for industrial production. Nevertheless, research activities at universities and professional institutions around the world suggest and, in some cases, prove that the technology has considerable potential, especially in the production of high-strength concretes and concretes with low and very low water coefficients. ($w/c < 0.40$). The technology is most effective in concretes with a very low water coefficient, as there is a natural shortage of water for complete cement hydration. In mixtures with a higher water coefficient, the effectiveness of internal curing decreases because the excess water in the system already partially covers the hydration requirements. Ultimately, it can be concluded that internal treatment is a very promising but technologically sensitive method, the success of which depends on proper design, controlled implementation, and combination with other forms of treatment depending on the required durability of the structure.

5. Conclusion

This article addressed the problem of insufficient effectiveness of traditional concrete curing methods, especially in conditions of accelerated construction and extreme climatic influences, where external curing often fails to provide sufficient internal moisture necessary for continuous cement hydration. A review of the literature and previous research has demonstrated the potential of SCC to overcome these limitations through internal water reservoirs. This article provides a critical analysis of existing approaches, focusing on the identification of materials that can ensure both technical efficiency and environmental compatibility under European climatic conditions. The research focuses on evaluating lightweight aggregates as effective self-curing media, taking into account their pore structure, absorption capacity, and desorption behaviour during hydration. The proposed concept confirms that, with the right composition, pre-wetting, and precise dosing, SLWA can effectively ensure internal water supply, stabilize the hydration process, and limit the formation of shrinkage cracks. The results point to several key findings. First, the internal treatment method significantly improves the microstructural integrity of concrete by reducing capillary porosity and improving the interface between aggregate and cement paste (ITZ). Second, regional materials may be available, cost-effective, and still suitable for use in SCC. Third, SCC is suitable for combination with available external curing methods, which can extend the effect of internal curing, and thus contribute to environmentally friendly and economical construction with long durability. Despite these benefits, however, the technology has certain limitations, including time-limited water release efficiency, non-homogeneous distribution of the treatment medium in the mixture, and the need for precise control of the formulation design. These limitations could be the basis for further experimental testing and further development of this technology. The theoretical findings will serve as the basis for planned experimental tests on Liapor lightweight aggregate, and the test results should provide important parameters that must be taken into account when designing concrete with self-curing technology in the construction sector.

6. Acknowledgments

This article was funded by the Program for the Support of Young Researchers (Program na podporu mladých výskumníkov)

7. References

[1] Kozar, I. & Sulovsky, T. (2023). Numerical model for identification of fiber-reinforced concrete parameters from the crack-mouth opening displacement, Proceedings of 34th DAAAM International Symposium on Intelligent Manufacturing and Automation, DAAAM Proceedings, Vienna, Austria, Katalinic, B. (Ed.), pp. 0544–0550, DAAAM International Vienna, Vienna, DOI: 10.2507/34th.daaam.proceedings.072

[2] Bentz, D. P. & Weiss, W. J. (2011). Internal curing: A 2010 state-of-the-art review, National Institute of Standards and Technology, NIST IR 7765, pp. NIST IR 7765.

[3] Meraz, M. M.; et al. (2023). Self-healing concrete: Fabrication, advancement, and effectiveness for long-term integrity of concrete infrastructures. Alexandria Engineering Journal, Vol. 73, No., (July 2023) 665–694, DOI: 10.1016/j.aej.2023.05.008

[4] American Concrete Institute (2016). Standard specification for curing concrete, American Concrete Institute, ACI 308

[5] Cusson, D.; Lounis, Z. & Daigle, L. (2010). Benefits of internal curing on service life and life-cycle cost of highperformance concrete bridge decks – a case study. Cement and Concrete Composites, Vol. 32, No. 5, (2010) 339-350

[6] Ma, X.; et al. (2019). A review on the use of LWA as an internal curing agent of high performance cement-based materials. Construction and Building Materials, Vol. 218, No., (September 2019) 385–393, DOI: 10.1016/j.conbuildmat.2019.05.126

[7] Murugan, K.; et al. (2023). Experimental studies on light weight concrete using LECA material. Materials Today: Proceedings, Vol. 74, No., (2023) 1035–1041, DOI: 10.1016/j.matpr.2022.11.467

[8] Mechtcherine, V.; Wyrzykowski, M.; Schröfl, C.; Snoeck, D.; Lura, P.; De Belie, N. & Igarashi, S. I. (2021). Application of super absorbent polymers (SAP) in concrete construction—update of RILEM state-of-the-art report. Mater. Struct., Vol. 54, No., (2021) 1–20,

[9] Hamzah, N.; et al. (2022). A review on the use of self-curing agents and its mechanism in high-performance cementitious materials. Buildings, Vol. 12, No. 2, (February 2022) 152, DOI: 10.3390/buildings12020152

[10] Abd Elrahman, M.; et al. (2019). Effect of different expanded aggregates on the properties of lightweight concrete. Magazine of Concrete Research, Vol. 71, No. 2, (January 2019) 95–107, DOI: 10.1680/jmacr.17.00465

DOI: 10.2507/36th.daaam.proceedings.039

THE IMPACT OF THE UKRANIAN CRISIS ON ENERGY PRICES AND CAPITAL MARKET PROSPECTS

Dušan Vještica, Ilija Ćosić, Nikola Gradojević & Vladimir Đaković

This Publication has to be referred as: Vjestica, D[usan]; Cosic, I[lija]; Gradojevic, N[ikola] & Dakovic, V[ladimir] (2025). The Impact of the Ukranian Crisis on Energy Prices and Capital Market Prospects, Proceedings of the 36th DAAAM International Symposium, pp.0276-0284, B. Katalinic (Ed.), Published by DAAAM International, ISBN 978-3-902734-47-1, ISSN 1726-9679, Vienna, Austria
DOI: 10.2507/36th.daaam.proceedings.039

Abstract

The research in this paper provides a significant contribution to the professional public in the subject field, especially in exogenous shocks, including wars, natural disasters and their impact on capital markets. The survey includes more than 1,600 companies in the energy sector. The paper presents evidence that in the short period after the onset of the Ukraine crisis on February 24, 2022, companies in the energy sector experienced positive cumulative average abnormal returns (CAAR) on average. The survey shows that companies in the energy sector exceeded the expected growth in the stock market at that time. The effect is different for different regions and is greater for North American companies than for European and Asian companies. Energy companies performed above expectations in export markets that compete with Russian suppliers of renewables sources, fossil fuels and uranium. Significance of results (CAAR) is verified by statistical methods and the results of the research confirm that none of the model specifications generated results that contradict the main results of the research. The robustness of the research results is checked by estimating abnormal yields with a different number of risk factors, with a focus on companies in the leading oil and gas exporting countries.

Keywords: energy prices; energy sector; CAAR; renewable energy sources; capital markets.

1. Introduction

This paper analyses how the Russia-Ukraine crisis has had implications for supply chains and the availability of fossil fuels, and how this event affects the profitability of companies in the energy sector. The analysis was carried out by conducting a case study and an analysis of the return on equity (Cumulative Average Abnormal Returns (CAAR)). The business model of energy companies focuses, among other things, on companies that buy and sell energy products, as well as components used during the electricity generation process. The aim is to clarify the impact of the Russian-Ukrainian military conflict on these firms by conducting a case study in which the start of the military operation on February 24, 2022, acts as the date of the event. The Russia-Ukraine crisis has had significant consequences for energy markets, including disruptions in supply chains while generating geopolitical tensions [1], [2], [3], [4]. Political instability in key regions has disrupted energy supply, driven up commodity prices, and reduced consumer purchasing power [4].

Overall, we find clear evidence that the shares of companies in the energy sector have risen significantly on the stock market around the event. This finding is valid for all types of energy segments, i.e. for; renewable energy, fossil fuels and uranium. However, there are differences related to the location of the company. North American energy companies show significantly higher performance than European and Asian firms. We consider the Russia-Ukraine crisis as an event in our empirical analysis (similar to [5],[6],[7],[8],[9], [10],[11]). This event allows us to observe how an unexpected change in the market environment (exogenous factors) affects the value of shares of companies in the energy sector. One example of an unexpected change in the market environment was a significant change in commodity prices, Figure 1. As a key exporter of commodities such as crude oil, natural gas and coal, Russia's actions have caused reverberations in energy markets around the world. While the price of natural gas and oil has risen to some extent, the price of coal has seen an increase of almost 150% immediately after February 24, 2022. The political and economic sanctions against the Russian Federation were responded to by Russia's announcement of a possible shortage or even halt of gas supplies [12]. If capital market participants expect significant corrections in energy production plans, the stock exchanges will react accordingly. Tests for different regions (Asia, Europe, North America) reveal that these positive CAAR in the short event window of three days around the event mostly come from the North American sample, while European firms show negative CAAR. Companies in a subgroup of the renewable energy industry experienced a short-lived upward movement a few days after the crisis, and then fell short of fossil and nuclear power. In particular, companies in the uranium industry subgroup show high CAAR. These findings may also contribute to the discussion [13], about whether the crisis in Ukraine will disrupt the green energy transition. With fossil fuels and uranium-based energy technology companies watching outperform the stock market, investors seem to still believe in these business models. These results are robust for a set of additional tests, such as estimating abnormal returns with a different number of risk factors, a different approach to matching risk factors, a focus on large stocks, a focus on firms located in the leading oil and gas exporting countries, as well as changes in the length of the appraisal period.

2. Data and methodology

The analysis uses a global sample of stocks that includes all firms in the Energy (50) economic sector according to the Refinitiv Business Classification (TRBC). The survey sample consists of over 1,600 firms from 75 countries with a sufficient number of observations on returns related to the event and the timeframe of the assessment. All data points are from Refinitiv Datastream and Worldscope. The analysis uses the Industrial Subgroup Classification (TRBC) to divide our sample into five groups according to their main energy production stream: coal (501,010), oil and gas (501,020), oil and gas-related equipment and services (501,020), renewable energy (502,010), and uranium (503,010). The distribution by industry group and country is shown in Table 1. We apply a standard case study with the following parameters: The main event is defined as February 24, 2022, i.e. the beginning of the Russian-Ukrainian crisis. Our 250 daily yield valuation period ends 20 trading days of the event submitted. The event frame ranges from –20 to +20 days around the day of the event (day 0). For each firm, we estimate the daily abnormal yield actions during the event period using [14]. A three-factor model based on global factors obtained from the site [15] focuses on time series regression. We follow a common case study approach and define average abnormal returns (ADR).

The average abnormal returns of a stock portfolio on a given day is the average of the abnormal returns on that day. Cumulative Average Abnormal Return (CAAR) is the sum (AAR) over a period of time and is a metric. This metric is used in finance, especially in case studies, to estimate the impact of a particular event on the returns of a group of securities. It represents the average cumulative abnormal yield (CAR) for multiple securities over a defined period of time, called the event window. In simpler terms, it shows how, on average, a set of stocks has outperformed or underperformed in the market due to a particular event, such as a merger or earnings announcement. Event Studies (CAAR) are used to test hypotheses about how certain events affect stock prices and to compare the impacts of different events. The statistical significance of the results is shown using the t-test. Results that have not been transmitted show that our findings are robust for testing statistics such as the Patell test [16], the Custom Patell test [17], and the Vilcoxon signed ranks test [18]. The setting of the event study is adequate to cause whether the outcome variable (CAAR) of the sample firm reacts to an event (crisis) in a short period of time. The effect can be attributed to the event because we consider the same firms in two environments (before and after the event). Since the (CAAR) periods are short (3 days, 7 days, 12 days, 22 days), it is likely that the Russia-Ukraine crisis is the only significant change in the sample of firms. Changes in the characteristics of the firm that could affect the (CAAR) are unlikely to be observed in these short periods. Moreover, we consider the current share price of a company as the value that investors attribute to the future profitability of the company. Normal returns are calculated using data from the assessment period, i.e. period of time prior to the event. Among all models, the OLS market model provides superior results for event analysis [19] and is therefore used here to estimate yields.

The expected yield $E(R_{i,t})$ is calculated using equation (1):

$$E(R_{i,t}) = \alpha_i + \beta_i R_{mt} + s_i SMB_t + h_i HML_t + \varepsilon_i \qquad (1)$$

Where "α_i"is the risk-free interest income, "R_{mt}" is the return of the market on day "t". "SMB_t" is the yield on a diversified portfolio of the difference between small and large stocks, "HML_t" is the difference between the yield on diversified portfolios of high and low B/M stocks. The coefficients "β", "s", and "h" are the coefficients of sensitivity. "$R_{i,t}$"is the return of the stock index "i" on day "t". The robustness of the research results is checked using equation (2):

$$E(R_{i,t}) = \alpha_i + \beta_i R_{mt} + s_i SMB_t + h_i HML_t + r_i RMW_t + c_i CMA_t + \varepsilon_i \tag{2}$$

Where "RMW_t" is the difference in returns on diversified portfolios of stocks with robust and weak profitability, and "CMA_t " is the difference in returns on diversified portfolios of stocks of low and high investment, "conservative and aggressive" firms.

Equation (3) is used to calculate the real daily return of a sample of stock indexes:

$$R_{i,t} = \frac{P_{i,t} - P_{i,t-1}}{P_{i,t-1}} \tag{3}$$

Where "$P_{i,t}$" is the price of the selected stock index "i" on day "t", and "$P_{i,t-1}$", is the price of the index "i" one day before "t". The abnormal return for each index is equal to the actual return calculated from Equation (3) minus the expected return obtained from Equation (1), which gives Equation (4):

$$AR_{i,t} = R_{i,t} - E(R_{i,t}) \tag{4}$$

Where "$AR_{i,t}$"is the abnormal return of the index "i" on day "t", "$R_{i,t}$" is the actual return of the index "i" on day "t", and E(Ri,t) is the expected return on day "t". Through the event window, we can calculate the Cumulative Abnormal Returns (CAR) as follows:

$$CAR_i(t_1, t_2) = \sum_{t=t_1}^{t_2} AR_{it} \tag{5}$$

Where CARi shows the CAR index "i" from day "t_1" to day "t_2", and "AR_{it}" is the abnormal yield estimated in Equation (4). We can estimate the effect of the Russia-Ukraine conflict by calculating the (CAR) of each index. To analyze the impact of conflicts in the energy markets in the sample, we calculated the daily abnormal returns of all indices and determined the response of each market to the event, we use the average abnormal return (AAR), which is calculated using Equation (6):

$$AAR_t = \frac{1}{N} \sum_{i=1}^{N} AR_{i,t} \tag{6}$$

Where AARt is the daily average abnormal return on day "t" relative to the sample, and "N" is the total number of indices in the energy market. In addition, we can use (AAR) to calculate the Cumulative Average Abnormal Returns (CAAR) using Equation (7):

$$CAAR(t_1, t_2) = \sum_{t=t_1}^{t_2} AAR_t \tag{7}$$

We follow a common case study approach and define average abnormal returns (AAR). The Average Abnormal Returns (AAR) of a given stock portfolio date is the average of the abnormal returns of the return on that day. Cumulative Average Abnormal Return (CAAR) is the sum (AAR) over a period of time. We show the statistical significance of the results using the t-test. Results that have not been transmitted show that our findings are robust for testing statistics such as the Patell test (Patell, 1976), the Adapted Patell test (Kolari and Pynnonen, 2010), and the Vilcoxon test signed ranks (Wilcoxon, 1945). Since the (CAAR) periods are short (3 days, 7 days, 12 days, 22 days), it is likely that the Russia-Ukraine crisis is the only significant change in the sample of firms. A positive (CAAR) indicates that a company's share price has risen more than a comparable company's share price. According to the price-building mechanisms of capital markets, a higher stock price indicates that more investors are showing demand for the firm, possibly due to the belief that the firm's profitability has increased [20].

3. Empirical results and discussion

3.1. Main results

The main results of the analysis show that energy firms have experienced positive (CAAR) around the Russia-Ukraine crisis. Table 2 contains (CAAR) all energy firms in the sample, as well as five industry subgroups of the energy sector for the periods [–1; 1], [–5; 5], [–1; 2], [–1; 5], [–1; 10] and [–1; 20] around the date of the event.

With the exception of two (CAAR) (CAAR groups of coals in [−1; 1] and [−1; 2] periods), all (CAAR) are significantly higher than zero. As a result, the shares of companies in the energy sector outperformed the market during the period around the event in general. Moreover, the question arises whether the Russian-Ukrainian crisis has had a different impact on the share prices of energy companies with different energy production technologies.

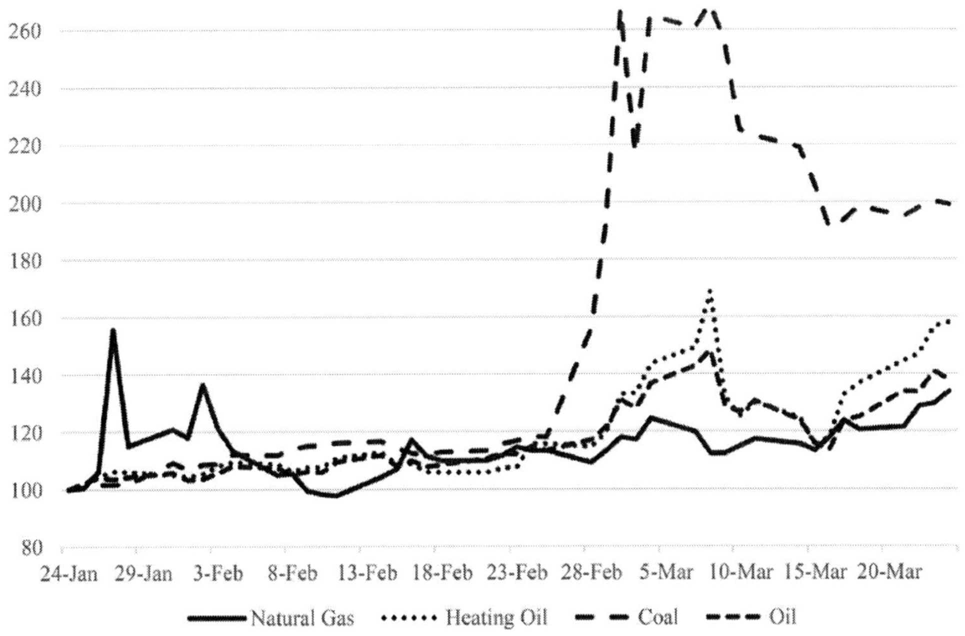

Fig. 1. Energy prices around the event.

This figure shows the energy commodity indices for natural gas, heating oil, coal, and oil in the period from January 24, 2022 to March 24, 2022 around the date of the event [20].

Industry Group/Continent	Numerički iznos	Percentage %
Coal	140	8,67
Oil and gas	750	46,47
Oil and Gas Equipment & Services	408	25,28
Renewable energy sources	252	15,61
Uranium	64	3,97
Total	**1614**	**100**
Africa	21	1,30
Asia	580	35,94
Australia	104	6,44
Europe	290	17,97
North America	591	36,62
South America	27	1,73
Total	**1614**	**100**

Table 1. Displays the industry group and continent of the global stock sample, shown as a number and percentage

Table 2., shows the positive CAARs of all energy firms in the sample, as well as the five subsystems of the energy sector industry for the periods [−1; 1], [−5; 5], [−1; 2], [−1; 5], [−1; 10] and [−1; 20] about the date of the event. Except for the two CAAR (CAAR groups for coal in periods [−1; 1] and [−1; 2] which are negative), all CAAR are significantly greater than zero. As a result, shares of companies in the energy sector generally outperformed the market during the period around the event. While the short-term reaction in the capital market for renewable energy companies was positive with a CAAR of 0.021, (0.040) to [−1; 1] ([−1; 2]) period of the event, market participants favored shares of uranium companies with a CAAR of 0.051, (0.102). Moreover, in the event periods spanning the 5, 10, and 20 trading days following the event, renewable energy companies achieved lower CAAR compared to companies in the coal, oil and gas, oil and gas equipment and services industries, and uranium companies. Uranium companies have outperformed the market with the highest CAARs compared to other technologies.

	CAAR [-1;1]				CAAR [-5;5]			
Industry Group	FF3	CAPM	Carhart	N	FF3	CAPM	Carhart	N
All Firms	0.010***	0.005*	0.012***	1606	0.059***	0.059***	0.055***	1562
Coal	-0.018**	-0.022***	-0.015***	140	0.078***	0.079***	0.074***	121
Oil and gas	0.009**	0.000	0.011***	750	0.056***	0.053***	0.054***	729
Oil and gas Equipment	0.014***	0.005	0.015***	403	0.058***	0.053***	0.056***	400
Renewable energy sources	0.021***	0.022***	0.021***	249	0.028**	0.046***	0.020*	248
Uranium	0.051***	0.042***	0.049***	64	0.152***	0.165***	0.153***	64
	CAAR [-1;2]				CAAR [-1;5]			
Industry Group	FF3	CAPM	Carhart	N	FF3	CAPM	Carhart	N
All Firms	0.029***	0.027***	0.031***	1545	0.051***	0.049***	0.053***	1590
Coal	-0.005	-0.007	-0.003	123	0.075***	0.073***	0.076***	126
Oill and gas	0.026***	0.018***	0.027***	726	0.044***	0.038***	0.047***	747
Oil and gas Equipment	0.029***	0.019***	0.029***	392	0.052***	0.041***	0.053***	401
Renewable energy sources	0.040***	0.055***	0.042***	240	0.028***	0.048***	0.032***	252
Uranium	0.102***	0.107***	0.101***	64	0.158***	0.166***	0.159***	64
	CAAR[-1;10]				CAAR[-1;20]			
Industry Group	FF3	CAPM	Carhart	N	FF3	CAPM	Carhart	N
All Firms	0.059***	0.012***	0.062***	1614	0.071***	0.070***	0.073***	1612
Coal	0.044***	-0.017***	0.048***	140	0.075***	0.076***	0.077***	140
Oil and gas	0.049***	0.010***	0.052***	750	0.062***	0.055***	0.066***	750
Oil and gas Equipment	0.059***	0.015***	0.062***	408	0.070***	0.057***	0.073***	406
Renewable energy sources	0.054***	0.022***	0.060***	252	0.064***	0.097***	0.069***	252
Uranium	0.208***	0.049***	0.210***	64	0.194***	0.211***	0.195***	64

Table 2. Cumulative average abnormal returns in different groups of the energy industry

Table 2., shows the cumulative average abnormal returns (CAAR) across different energy industry groups and event frameworks around the date of the event. Daily abnormal returns are estimated using the Fama and French models with three factors, (CAPM) and (Carhart model) with four factors. A framework of events ranging from the shortest interval of the day [–1; 1] to the longest [–1; 20] are calculated for all firms in the sample and for the industry subgroups: coal, oil and gas, oil and gas equipment and services, renewable energy and uranium. Significance (CAAR) is determined by the t-test of equations (8) and equations (9). ***, ** and * denote significance at the levels of 0.1%, 1% and 5%, respectively.

$$t = \frac{x_1 - x_2}{\sqrt{\frac{s_1^2}{n_1} + \frac{s_2^2}{n_2}}} \qquad (8)$$

$$d_f = n_1 + n_2 - 2 \qquad (9)$$

3.2. Returns differences between different regions

Since the Russian Federation is the world's largest oil producer and the second largest producer of natural gas [21], it is of great importance for the supply of oil, gas and coal to European [22] and Asian [23] countries. While oil and gas are one of the main sources in European countries, energy production in Asian countries relies heavily on coal. North America is largely independent of Russian goods.

These different dependencies of different regions on certain commodities (and the volume of their supply by Russia) raise the question of whether firms of different regions exhibit different (CAAR) patterns accordingly. We observe (CAAR) at intervals of [–1; 1] and [–1; 10] and the division of "All firms" and five industrial subgroups into regions of Asia, Europe and North America, following the classification of developed market factor countries Fama and French. These three regions make up a total of 90% of the stocks in this sample and we ignore other regions due to their small sample size. Table 3., contains the (CAAR) of each region/group, as well as the result of the t-test with two samples. For example, in the [–1; 1] period, the "All firms" group in Europe (CAAR=-0.023) performed significantly worse than the North American group "All firms" (CAAR=0.040). The difference (-0.063) is statistically less than zero at the level of 0.1%. Table 3 reveals that the performance of the North American sample is significantly lower than that of the Asian and European samples. Therefore, the share prices of the North American energy sector did not see a significant correction after the onset of the Russia-Ukraine crisis. While European stocks showed a negative reaction in the three trading days around the event, in the long term, (CAARs) of European energy firms were also positive. The CAAR [–1; 1] sample of renewables in Europe (-0.004) documents the "highest" effect among European industrial subgroups in the short term. However, the (CAAR) sample of renewables in Europe (0.048) is the third largest compared to other industrial subgroups in the [–1; 10] window. In particular, uranium showed high positive CAAR values (0.185) in the two weeks following the event. In North America (Asia), uranium (coal) energy firms recorded the highest CAAR values [–1; 10] Compared to the sub-group of the industry. In all three samples, the sample of renewable energy firms was among the top 3 subgroups of the industry that recorded the highest CAAR values [–1; 10].

Panel A. Europe vs. North America	CAAR[-1;1]	CAAR[-1;1]		CAAR[-1;10]	CAAR[-1;10]	
Industry Group	Europe	N. America	Difference	Europe	N.America	Difference
All Firms	- 0.023	0.040	-0.063***	0.035	0.128	-0.092***
Coal	- 0.111	0.029	-0.141***	-0.060	0.167	-0.226*
Oil and gas	- 0.022	0.036	-0.058***	0.022	0.119	-0.096***
Oil and gas Equipment	- 0.012	0.031	-0.042***	0.056	0.111	-0.054
Renewable energy sources	- 0.004	0.074	-0.078***	0.048	0.171	-0.122*
Uranium	- 0.039	0.054	-0.093	0.185	0.205	-0.020
Panel B. Europe vs. Asia	CAAR[-1;1]	CAAR[-1;1]		CAAR[-1;10]	CAAR[-1;10]	
Industry Group	Europe	N. America	Difference	Europe	N.America	Difference
All Firms	-0.021	-0.003	-0.017**	0.037	-0.006	0.042***
Coal	–0.112	–0.016	-0.096***	-0.060	0.036	-0.096
Oil and gas	-0.022	-0.005	-0.017	0.023	-0.016	0.040*
Oil and gas Equipment	-0.011	0.007	-0.018	0.056	-0.022	0.078***
Renewable energy sources	-0.004	0.005	-0.009	0.046	-0.002	0.048*
Uranium	-0.039	0.009	-0.047	0.185	-0.152	0.337
Panel C. North America vs. Asia	CAAR[-1;1]	CAAR[-1;1]		CAAR[-1;10]	CAAR[-1;10]	
Industry Group	Europe	N. America	Diffference	Europe	N.America	Diference
All Firms	0.039	-0.004	0.043***	0.128	-0.006	0.134***
Coal	0.030	–0.016	0.044*	0.167	0.038	0.127***
Oil and gas	0.037	-0.005	0.043***	0.119	-0.016	0.135***
Oil and gas Equipment	0.030	0.005	0.025*	0.111	-0.022	0.132***
Renewable energy sources	0.076	0.006	0.070***	0.170	-0.001	0.171***
Uranium	0.054	0.008	0.045	0.205	-0.153	0.358

Table 3. Regional differences in cumulative average abnormal yields

Table 3., shows the differences in cumulative average abnormal returns (CAARs) across different regions, energy industry groups, and event windows around the date of the event. Daily abnormal yields are estimated using the Fama and French models with three factors. Event windows ranging from the shortest window of the day [–1; 1] to the longest [–1; 10] are calculated for all sampled firms and for the industry subgroups: coal, oil and gas, oil and gas equipment and services, renewable energy and uranium in the regions of Europe, North America and Asia.

The significance of the difference in CAAR is determined by means of a t-test. ***, ** and * indicate significance at the levels of 0.1%, 1% and 5%, respectively. In summary, the results show evidence supporting the assumption that the pattern of dependence on the supply of goods coming from the Russian Federation affects CAAR. North America, with low exposure to goods coming from the Russian Federation, has the highest CAAR (which are even significantly higher than those in Europe and Asia) during the event period. A possible explanation for the slightly higher CAAR of Asian coal companies compared to European ones (significantly in the short term) is the high share of coal-based power generation in Asia [24]. The expected shortage of coal supplies from the Russian Federation is causing coal prices to rise, and local Asian coal-fired energy companies may make higher profits due to higher selling prices. The same explanation applies to the significantly higher CAAR of the European sample of oil and gas companies (compared to Asia). Nevertheless, future research may consider this and possible other reasons for the presented pattern of CAAR in more detail. As exogenous factors weigh heavily on supply chains, energy prices, and capital markets, companies in the energy sector need to take business continuity plans (BCM) and disaster recovery plans (DRP) seriously for business stability [25].

3.3. Robustness of research results

We ran various specifications of our tests to determine the robustness of the results. The robustness tests (summarized in Table 4) show that our main results are not driven by a specific empirical setup. In detail, we also estimated abnormal yields based on the FF model of five factors (see Table 4, Panel A.1).

Variation of factors	Panel A.1 Fama French Five–Factor		Panel A.2 Factor Matching	
	CAAR [–1;1]	CAAR [–1;10]	CAAR [–1;1]	CAAR [–1;10]
All Firms	0.010***	0.058***	0.015***	0.068***
Coal	–0.017**	0.045***	-0.005	0.065***
Oil and gas	0.008**	0.048***	0.012***	0.056***
Oil end gas Equipment	0.014***	0.060***	0.017***	0.067***
Renewable energy sources	0.018***	0.052***	0.024***	0.069***
Uranium	0.047***	0.209***	0.044***	0.216***
Window Variation	Panel B.1 Pre–Event Period		Panel B.2 Placebo Test	
	CAAR [–20;2]	CAAR [–10; –2]	CAAR [–366:–364]	CAAR [–366:–365]
All Firms	0.014**	0.007*	-0.035	-0.015*
Coal	0.074***	0.010	-0.005	-0.036**
Oil and gas	0.020**	0.013**	0.004	-0.020**
Oil and gas Equipment	0.010	0.015**	0.008	-0.019*
Renewable energy sources	-0.011	-0.014	-0.019*	-0.122***
Uranium	-0.051	-0.037	-0.029	-0.073*
	Panel B.3 Double Est. Window		Panel B.4 Half Est. Window	
	CAAR [–1;1]	CAAR [–1;10]	CAAR [–1;1]	CAAR [–1;10]
All Firms	0.011***	0.056***	0.011***	0.053***
Coal	-0.019**	0.048***	-0.015*	0.048***
Oil and gas	0.007*	0.045***	0.007**	0.042***
Oil and gas Equipment	0.015***	0.060***	0.013***	0.053***
Renewable energy sources	0.017**	0.045***	0.018***	0.053***
Uranium	0.047***	0.198***	0.039**	0.168***
Firm Variation	Panel C.1 Large Firms		Panel C.2 Oil & Gas Exporters	
	CAAR [–1;1]	CAAR [–1;10]	CAAR [–1;1]	CAAR [–1;10]
All Firms	0.014***	0.065***	0.031***	0.111***
Coal	-0.028***	0.055***	-0.003	0.071
Oil and gas	0.009***	0.047***	0.032***	0.100***
Oil and gas Equipment	0.017***	0.072***	0.022***	0.099***
Renewable energy sources	0.038***	0.075***	0.062***	0.160***
Uranium	0.087***	0.290***	0.046***	0.189***

Table 4. Robustness tests of research results

We have also replaced the global risk factor yield with the return of the risk factors of the respective developed market (Asia-Pacific, Europe, North America, USA, Japan and developed for all other countries; see Table 4, Panel A.2). In the second step, we moved the event window before the event date (Pre-Event, Panel B.1) and one year in the past (Placebo Event, Panel B.2). Since we do not find similar, very significant CAARs in these periods, it is likely that the observed CAARs in the "real" window of events are a consequence of the Russian invasion of Ukraine. In addition, we used an estimation window with two times (half) the number of trading days and presented the results in panel B.3 (B.4) of Table 4. Finally, we analyzed the CAARs of different subsamples: a subsample of large (mid-sized) companies (Panel C.1) and a subsample of companies located in the top 10 exporting countries [26] of oil and gas (Panel C.2). Table 4., shows the cumulative average abnormal returns (CAARs) across different energy industry groups and event windows around the date of the event. Event windows ranging from the shortest window of the day [–1; 1] to the longest [–1; 10] are calculated for all companies in the sample and for the industry subgroups: coal, oil and gas, oil and gas equipment and services, renewable energy, and uranium [27]. In Panel A.1, we replaced the FF three-factor model with the FF five-factor model. In Panel A.2, we compare factor returns based on equity regions with regional factors instead of using global FF factors. In Panel B.1 (B.2), we use an evaluation window with twice (half) the number of days. In Panel C.1, we repeat our main analysis using only large (above the median) firms in our sample, and in Panel C.2, we use only firms located in large oil and gas exporting countries. The significance of the difference in CAAR is determined by means of a t-test. ***, ** and * indicate significance at the levels of 0.1%, 1% and 5%, respectively.

4. Conclusion

The results of the research systematize the existing knowledge in this area and are cited in recent literature to provide a more precise answer to the research question, whether the Russia-Ukraine crisis, starting on February 24, 2022, caused an abnormal reaction of share prices in energy companies. The results of the survey also show that energy firms experienced positive cumulative average abnormal returns (CAARs) around the date of the event. The results of the industry subgroup indicate that stock market participants expect conventional energy segments, such as uranium technology, to be profitable in the future. Regarding the necessary and expected transition to green energy with a focus on renewables (and therefore the expectation that investors believe that renewable energy companies benefit the most from the event under consideration), the reasons for our results may be challenges such as long planning horizons for such an energy transition. The Glasgow Climate Change Conference (COP26) will take place at the end of 2021. The U.S. has highlighted the need for a rapid transition to a zero-emission economy in order to meet the goals of the Paris Agreement. An essential part of this transition is the energy transition from fossil fuels to renewable energy sources, as electricity and heat generation account for a quarter of global greenhouse gas emissions (EPA, 2022). The claim at COP26 was that "the transition from coal to clean energy must be about five times faster" (COP26, 2022). In this context, the results of the research indicate that capital markets react very quickly to changes in supply chains caused by unique events such as wars, and that these events result in competitive markets that react to new market opportunities in order to take advantage of disruptions in supply chains. Statistical significance of results (CAAR-s) is demonstrated by means of a t-test at a predefined significance level and it confirms that none of the model specifications generated results that contradict the main results of the research in the paper. This confirms the significance of the research results as well as their validity. This could be an open question for future research to examine how long-term pressure on traditional supply chains supports the transition from carbon-intensive to zero-emission energy.

5. Reference

[1] Ferriani, F., Gazzani, A., 2023. The Impact of the War in Ukraine on Energy Prices: Consequences for Firms' Financial Performance. International Economics.

[2] De Villa, M.A., 2023. Assessing geopolitical risk: a multi-level approach for top managers of multinationals. AIB Insights 23 (1), 1–7.

[3] Sauadaoui, F., Jabeur, S.B., 2023. Analyzing the influence of geopolitical risks on European power prices using a multiresolution causal neural network. Energy Econ. 106793.

[4] Goodell, J.W., Gurdgiev, C., Paltrinieri, A., Piser`a, S., 2023. Global energy supply risk: evidence from the reactions of European natural gas futures to Nord stream announcements. Energy Econ. 106838.

[5] Basnet, A., Blomkvist, M., Galariotis, E., 2022. The decision to remain or leave Russia amid the Ukrainian invasion: a study on the stock market reaction and the role of ESG. Work. Pap. https://doi.org/10.2139/ssrn.4078143 .

[6] Boubaker, S., Goodell, J.W., Pandey, D.K., Kumari, V., 2022. Heterogeneous impacts of wars on global equity markets: evidence from the invasion of Ukraine. Finance Res. Lett., 102934 https://doi.org/10.1016/j.frl.2022.102934 .

[7] Boungou, W., Yati´e, A., 2022. The impact of the Ukraine-Russia war on world stock market returns. Econ. Lett. 215, 110516 https://doi.org/10.1016/j. econlet.2022.110516. COP26 (2022). Accelerating the transition from coal to clean power. Proceedings of the Website of the UN Climate Change Conference UK 2021, last access on May 6, 2022. https://ukcop26.org/energy/ .

[8] Deng, M., Leippold, M., Wagner, A.F., Wang, Q, 2022. Stock prices and the Russia-Ukraine war: sanctions, energy and ESG. Work. Pap. https://doi.org/10.2139/ssrn.4080181EPA (2022). Global greenhouse gas emissions data. Website of the United States Environmental Protection Agency, last access on May 6, 2022. https://www.epa.gov/ghgemissions/global-greenhouse-gas-emissions-data.

[9] Huang, L., Lu, F., 2022. The cost of Russian sanctions on the global equity markets. Work. Pap. https://doi.org/10.2139/ssrn.4060927

[10] Mohamad, A., 2022. Safe flight to which haven when Russia invades Ukraine? A 48-hour story. Econ. Lett. 216, 110558 https://doi.org/10.1016/j.econlet.2022.110558

[11] Tosun, O.K., Eshraghi, A., 2022. Corporate decisions in times of war: evidence from the Russia-Ukraine conflict. Finance Res. Lett. 48, 102920. https://doi.org/10.1016/j.frl.2022.102920

[12] International Energy Agency, 2021. Net Zero by 2050, IEA, Paris (last accessed 15 June 2022). https://www.iea.org/reports/net-zero-by-2050

[13] Hook, L. and Hume, N. (2022). Will the Ukraine war derail the green energy transition? Financial Times, last access on March 8, 2022. https://www.ft.com/content/93eb06ec-ba6c-4ad2-8fae-5b66235632b2

[14] Fama, E.F., French, K.R., 1993. Common risk factors in the returns on stocks and bonds. J. Financ. Econ. 33 (1), 3–56. https://doi.org/10.1016/0304-405X(93)90023-5

[15] website: https://mba.tuck.dartmouth.edu/pages/faculty/ken.french/data_library.html

[16] Patell, J.M., 1976. Corporate forecasts of earnings per share and stock price behavior: empirical test. J. Account. Res. 246–276. https://doi.org/10.2307/2490543

[17] Kolari, J.W., Pynnonen, ¨ S., 2010. Event study testing with cross-sectional correlation of abnormal returns. Rev. Financ. Stud. 23 (11), 3996–4025. https://doi.org/10.1093/rfs/hhq072

[18] Wilcoxon, F., 1945. Some uses of statistics in plant pathology. Biom. Bull. 1 (4), 41–45. https://doi.org/10.2307/3002011

[19] Dyckman, T., Philbrick, D., & Stephan, J. (1984) A comparison of event study methodologies using daily stock returns: A simulation approach. Journal of Accounting Research, 22, 1–30.

[20] Martin Nerlinger and Sebastian Utz, Finance Research Letters 50 (2022) 103243. Contents lists available at ScienceDirect. Finance Research Letters journal homepage: www.elsevier.com/locate/frl https://doi.org/10.1016/j.frl.2022.103243

[21] IEA, 2022. Energy Fact Sheet: Why Does Russian Oil and Gas Matter? IEA. March 21, 2022. https://www.iea.org/articles/energy-fact-sheet-why-does-russian-oil-andgas-matter

[22] Singh, K.K. (2022) Why are Russia's oil and gas important to the world? Daily Pioneer, April 4, 2022. https://www.dailypioneer.com/2022/columnists/why-arerussia—s-oil-and-gas-important-to-the-world-.html

[23] Reuters (2022). Factbox: Asian Buyers of Russian oil, gas and coal, Reuters, April 6, 2022. https://www.reuters.com/business/energy/asian-buyers-russian-oil-gascoal-2022-02-22/

[24] IndexMundi (2022). Electricity production from coal sources (% of total), IndexMundi, last access on July 25, 2022. https://www.indexmundi.com/facts/indicators/EG.ELC.COAL.ZS

[25] Dakic, V., Mikulic P., Katja & Petrunic, R. (2024). Business Continuity Planning (BCP) and Disaster Recovery Planning (DRP), Proceedings of the 35th DAAAM International Symposium, Published by DAAAM International, ISBN 978-3-902734-44-0, ISSN 1726-9679, Vienna, Austria doi.org/10.2507/35th.daaam.proceedings.007.

[26] Statista (2022). Leading global crude oil and oil products exporters worldwide in 2020. https://www.statista.com/statistics/280972/global-oil-exporters-by-region /. And leading gas exporting countries in 2021. https://www.statista.com/statistics/217856/leading-gas-exporters-worldwide

[27] Tollefson, J., 2022. What the war in Ukraine means for energy, climate and food. Nature 604. https://doi.org/10.1038/d41586-022-00969-9

DOI: 10.2507/36th.daaam.proceedings.040

THE IMPACT OF CHANGING THE SENSING SYSTEM ON MEASUREMENT ACCURACY

Vaclav Blabol & Jan Kutlwaser

This Publication has to be referred as: Blabol, V[aclav] & Kutlwaser, J[an] (2025). The Impact of Changing the Sensing System on Measurement Accuracy, Proceedings of the 36th DAAAM International Symposium, pp.0285-0292, B. Katalinic (Ed.), Published by DAAAM International, ISBN 978-3-902734-47-1, ISSN 1726-9679, Vienna, Austria
DOI: 10.2507/36th.daaam.proceedings.040

Abstract

The work includes an experiment aimed at determining whether repeated replacement of a contact probe with a non-contact probe affects the measurement accuracy of form or geometric characteristics. Each tool qualification typically takes several hours. For this reason, an experiment was conducted to establish whether it is necessary to qualify the probes after every probe change, or how frequently this should be done. The experiment was carried out on a gauge block. It was demonstrated that probe qualification is not required after every probe change, resulting in time and cost savings. When operating the machine, it is essential to consider the required measurement accuracy. If only the maximum and minimum permissible measurement error of the gauge are considered, contact measurement requires more frequent calibration than non-contact measurement, as shown by the experiment. "The test results will be used to make the measurement plans for measurements in the Metrology Laboratory of the University of West Bohemia in Pilsen more effective [1]."

Keywords: CMM; Coordinate Measuring Machine; repeated probe head replacement; contact and non-contact probe; qualification

1. Introduction

Metrology is a scientific and technical discipline that deals with the analysis and measurement of shapes and geometric characteristics of objects. In industry and industrial production, metrology is a key element for manufacturing and quality control. Measurement on Coordinate Measuring Machines (CMM) was developed primarily for use in the automotive and later the aerospace industries. This was driven by the increasing need for fast and precise measurement of more complex parts. The development of CMM measurement took place in the second half of the 20th century, with the first CMM being developed in 1950 [2]. CMM belongs to the most productive and highly accurate methods for measuring geometric features. Today, manufacturing—especially in the industrial sector—faces continuously increasing demands for precision in production and subsequent inspection. At the same time, due to high competition and ongoing efforts to maximize profit, there is a need to reduce production and inspection times. Therefore, fast and accurate machines are required, which CMMs fulfill [3]. CMMs allow for the measurement and scanning of 3D objects. Scanning is conducted either by a tactile or non-contact probe. The scanned results are processed by software, providing very precise and repeatable measurements. Because they can record a 3D model, they are also used for reverse engineering.

Reverse engineering enables scanning a product and creating a 3D model from the scanned data, usable in CAD software for further use or processing. This is applied for parts without existing documentation or for rare historical artefacts. Another significant advantage of CMMs is their ability to be integrated into automated operations. Due to these properties, they are an integral part of many industries, such as the automotive industry, aerospace, and healthcare [4]. Measurement is always subject to error. For this reason, each repeated measurement varies to some extent. Therefore, the deviation of the measuring instrument, machine, or tolerance on the measured part must always be specified and adhered to. Achieving an exact dimension is almost impossible, and the goal is to get as close as possible [5]. Tolerances and deviations are calculated relative to the theoretically exact dimension or ideal model. This deviation can be caused by various factors, such as the non-transferability of the measuring system, failure to maintain consistent conditions like ambient temperature, machine temperature and varying humidity, as well as the influence of the machine operator and the program [a reference to an article discussing measurement errors would be useful here]. Measurement uncertainty needs to be minimized or, at best, quantified precisely because it influences decision-making regarding whether the measured part meets specifications or not [6].

1.1. Aim of the Study

The aim of this study is to determine whether repeated replacement of tactile and non-contact probes affects the measurement accuracy of shape or geometric characteristics. This article addresses the deviation caused by this replacement on the linear dimension. Each sensor qualification is very costly in terms of time and finances, typically lasting several hours; in the case of more complex scanning systems with multiple positions, it can take up to six hours unless an accelerated qualification method is used [7]. For this reason, an experiment was conducted to investigate whether repeated probe replacement influences measurement accuracy. If sensor qualification on the calibration sphere would not be necessary after each probe replacement, it would save machine time, operator time, and ultimately reduce costs associated with non-productive machine operation. The results of this study could be useful for metrology laboratories using the Carl Zeiss Prismo 7 Navigator machine together with the Zeiss probes and sensors employed.

1.2. Used Machine:

CMM Carl Zeiss Prismo 7 Navigator The machine used for the experiment is from Carl Zeiss. It is the Prismo 7 Navigator model, shown in Figure 1. Its advantage lies in high-speed scanning while maintaining high accuracy. It is suitable for measuring precise and complex components. The machine can be equipped with both tactile probes for contact measurement and probes for non-contact measurement. This machine represents a key piece of equipment in the workshop metrology laboratory of the Regional Technology Institute, which is part of the Faculty of Mechanical Engineering at the University of West Bohemia in Pilsen.

Fig. 1. Carl Zeiss Prismo 7 Navigator [11]

Maximum Permissible Error	0,9 + L / 350 μm
Temperature	19–22°C

Table 1. Prismo 7 Navigator properties [8]

1.3. Sensor Used in the Experiment

The sensor used during the experiment is shown in Figure 4. It is a direct sensor equipped with a ruby sphere with a diameter of 3 mm, the shaft material is carbon fibre, and the overall sensor length is 50 mm.

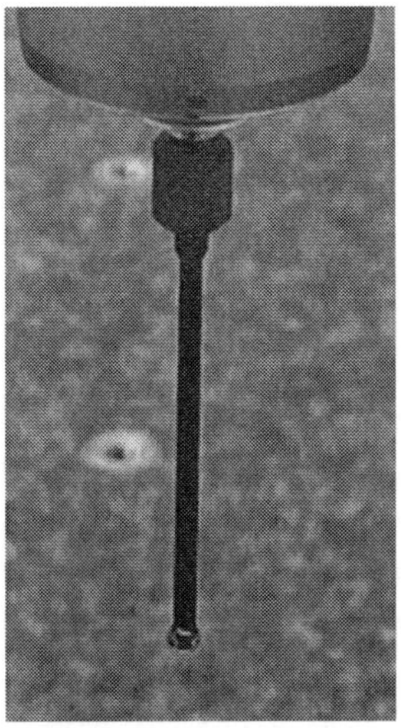

Fig. 2. Sensor Used in the Experiment [11]

1.4. The Zeiss LineScan sensor head

The Zeiss LineScan sensor head was used during the experiment and can be seen in Figure 3.

Fig. 3. ZEISS LineScan [11]

Scanning speed	250000 [points per sec]
Scanning accuracy	± 0,025 mm

Table 2. ZEISS LineScan properties [9]

2. Experiment Description

The experiment was carried out to evaluate the influence of changing the sensor head between individual measurements.

Experiment Procedure:

1. Creation of a 3D model of the gauge block and preparation of the measurement plan in the Calypso software.
2. Cleaning of the gauge block and its placement into the measuring area of the machine.
3. Measurement of the gauge block using discrete points with the previous probe qualification from February 9, 2024. (Repeated 3 times)
4. Replacement of the contact sensor head with a non-contact sensor head, followed by scanning of the gauge block using the qualification from October 16, 2023. (Repeated 5 times)
5. Replacement of the non-contact sensor head with the contact sensor head and subsequent probe qualification.
6. Measurement of the gauge block using discrete points with the new qualification from March 1, 2024.
7. (Repeated 3 times)
8. Replacement of the contact sensor head with the non-contact one and subsequent qualification of the LineScan sensor on the corresponding qualification sphere.
9. Scanning of the gauge block using the new qualification from March 1, 2024. (Repeated 5 times)
10. Replacement of the non-contact sensor head with the contact one again, followed by contact measurement of the gauge block (repeated 3 times) after the replacement, with the March 1, 2024 qualification.
11. Finally, the contact sensor head was replaced by the non-contact sensor head once more, and the gauge block was scanned (repeated 3 times) after the replacement with the March 1, 2024 qualification.
12. Data processing and subsequent evaluation.

2.1. Experimental Procedure

A 3D model of the gauge block (100 mm gauge) was created in a CAD program. In the Calypso software, which is used to control the coordinate measuring machine (CMM), planes were defined first, followed by the coordinate system, ensuring alignment with the machine's coordinate system. Contact measurement of the working surfaces of the gauge block was performed using a five-point discrete measurement strategy, with points arranged in a pattern resembling the five-dot side of a dice (see Figure 4). As part of the experiment, two dimensional characteristics were evaluated:

- Distance with caliper-like evaluation, determined using two points located along a straight line,
- Simple distance characteristic along the gauge block's main axis.

For non-contact measurement, the **wbScan** software was used for scanning and storing the measured points acquired from the functional surfaces of the gauge block. The scanned data was then processed in the specialized software **Geomagic Control X**. The measured point cloud was polygonised and then it was aligned to the pre-created CAD model using the **best-fit** alignment function, and deviations from the CAD model were subsequently evaluated. Figure 4 shows the CAD model of the gauge block with numbered points marked on it. These points are used for determining dimensions at specific locations. The functional surfaces are also indicated.

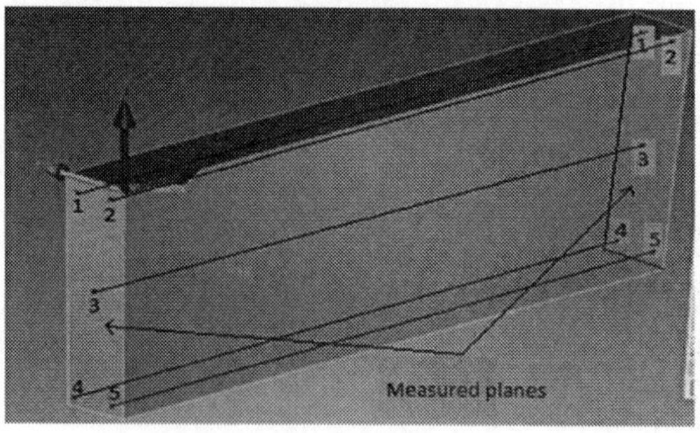

Fig. 4. End gauge model with predefined points [11]

2.2. Laser Sensor Qualification Procedure

The qualification is performed on a qualification sphere specifically designed for laser sensor calibration. This qualification sphere has a matte, grey surface to ensure the qualification process can be completed successfully. A standard ceramic qualification sphere is unsuitable for laser sensors due to its colour and glossy surface, which could lead to undesirable reflections of the laser beam. Before starting the qualification, the qualification sphere is fixed to a plate within the machine's working area using screws in pre-drilled threaded holes. Figure 5 illustrates the threaded hole used. The sphere is then aligned with a reference sensor, and the laser sensor qualification can proceed.

Fig. 5. Laser sensor qualification [11]

2.3. Procedure for Qualification of the Contact Sensor

The qualification is performed on a sphere fixed to the machine table. The qualification sphere is secured with a screw in pre-drilled threads, as shown in Figure 6. The qualification sphere for contact qualification is made of ceramic material. The qualification process follows a predefined procedure [10], in which the sphere is scanned by contact. The scanning occurs twice during the qualification, with different applied forces, in order to determine the compliance/stiffness, especially in the case of long sensor shafts.

Fig. 6. Contact sensor qualification [11]

3. Results discussion

In the experiment, several data sets were obtained, including measurements with the contact sensor using both the previous qualification (performed some time ago) and the new qualification. The same two sets of data were also collected for the non-contact laser sensor. The data were processed into graphs and tables to facilitate evaluation and improve clarity in the measured values. The measured values in the table are color-coded for better clarity. In the graphs, the MPE (Maximum Permissible Error) is indicated, which represents the maximum allowable error of the machine and is shown with an error bar. MPE applies to specific measured values. The diagrams display averaged values from the respective measurements. The value of 100 mm in the graphs represents the size of the measured part, which is the gauge block. In this evaluation, the mean values of the plane distances in graphs 7 and 9 are compared. In graphs 8 and 10, the distances between points "2" and "2", which are marked on the end gauge model in Figure 4, are compared. The planes for the evaluation are calculated from the measured points using the Gauss method. In the case of contact measurement, the calculation is based on 270 measured points (measured by contact scanning method). For non-contact measurement using the scanner, the plane is calculated from 55,000 points.

For contact measurement, all average measured values with the new/fresh qualification from March 1, 2024, have about a 50% smaller deviation compared to the average measured values with the old/original qualification from February 9, 2024, as shown in Figure 7 and 8. The mean measured values can distort the results of the measured distances between individual points. However, as seen in graphs 7 and 8, the mean average values and the distance between points "2" and "2" behave similarly. It has also been shown that it is not necessary to re-qualify after every exchange. The values after the first exchange were still within the tolerance of ±0.0016 mm, with no significant changes, or only minimal changes, as seen in graphs 7 and 8. For this reason, further experiments would be advisable to determine the exact time/number of exchanges after which requalification is necessary.

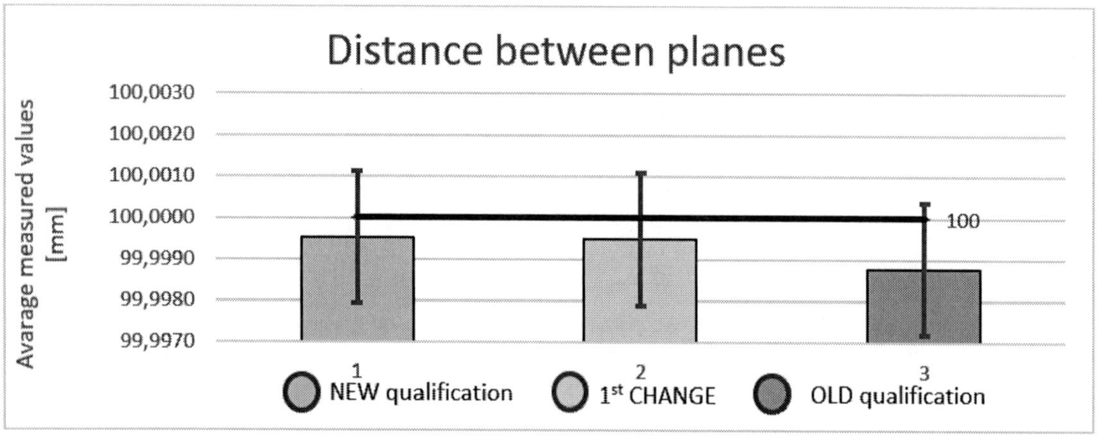

Fig. 7. Contact measurement – distance between planes

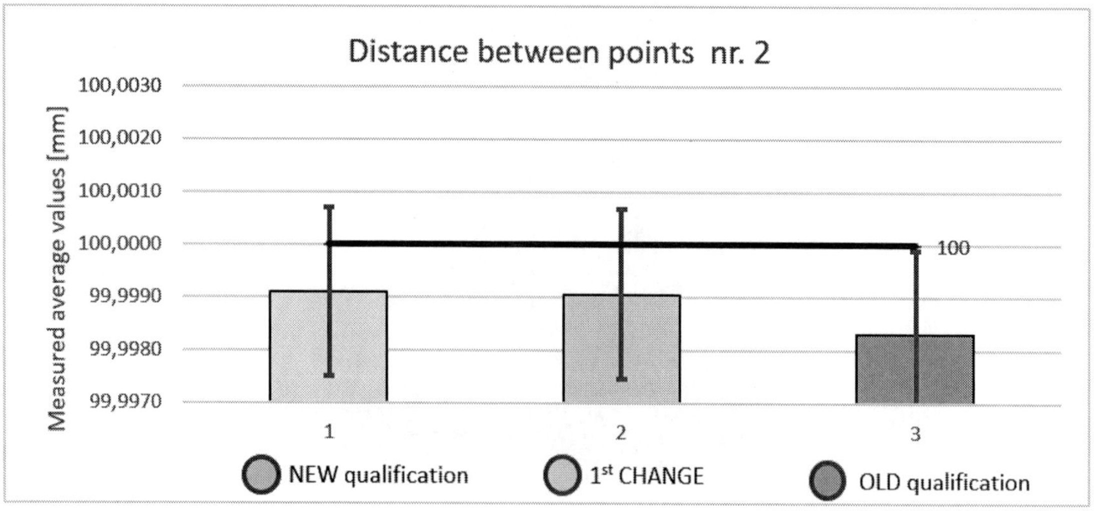

Fig. 8. Contact measurement – distance between points

In the case of non-contact measurement, the progression of the average measured value and the distance between points "2" and "2" appears differently, as shown in graphs 9 and 10. The average measured values may slightly distort the results due to extremes or measurement errors. The measured values remain within tolerance, as seen in graphs 9 and 10. The distances between points behave similarly, and it is noticeable that the measured values increased after the first exchange compared to the values immediately after calibration. It has also been shown that the measured values with the original/old qualification increased compared to the new calibration and even compared to the values after the first exchange, as seen in graph 10. For this reason, further experiments would be advisable to determine the exact time/number of exchanges after which requalification is necessary.

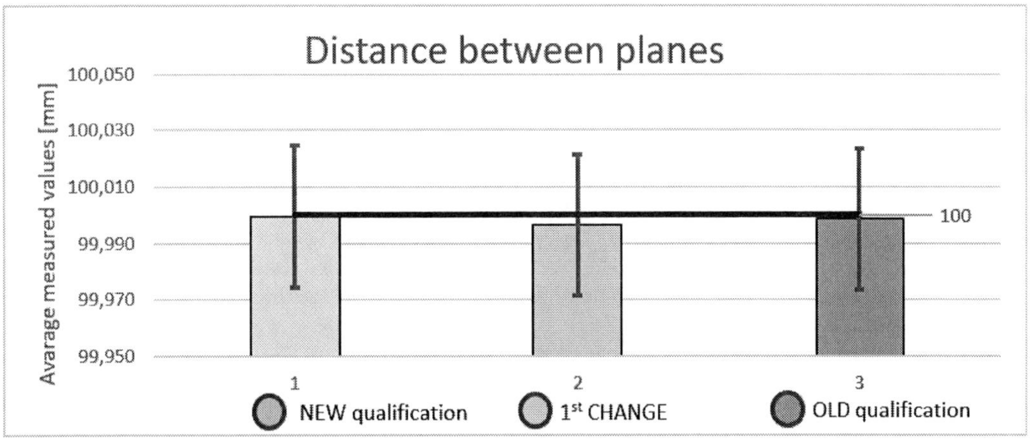

Fig. 9. Non–Contact measurement - distance between planes

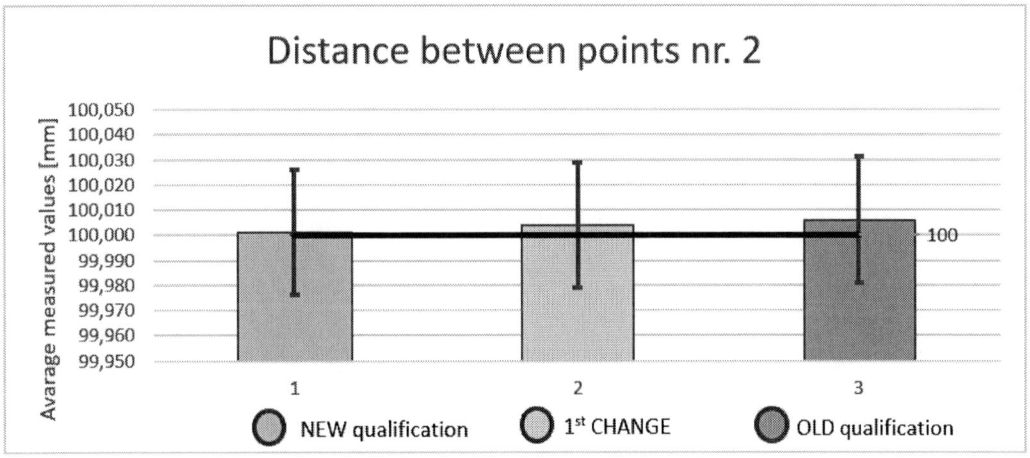

Fig. 10. Non-Contact measurement – distance between points

4. Conclusion

Each tool qualification costs 75 Eur per hour and in case of complex styluses with lot of qualified positions can take up to 6 hours. For this reason, an experiment was conducted to determine whether requalification is necessary after each measurement or how often it should be performed. The results demonstrated that requalification is not required after every probe exchange. This leads to significant time and cost savings. In machine operation, it is crucial to consider the required accuracy. When considering only the maximum and minimum permissible errors of the measuring instrument, contact measurements require more frequent qualification than non-contact measurements, as shown by the experiment. For future work, it would be beneficial to determine the exact time frame or number of exchanges after which requalification becomes necessary. This would allow for more efficient machine time usage and further cost savings. The ideal scenario would be if no additional qualifications were needed; however, this experiment has already disproven that. Therefore, it would be advantageous to create a table that links the number of exchanges and the time since qualification to measurement accuracy. Such a table could be used as a guide in future machine operations.

5. Acknowledgments

This article has been prepared under the project SGS-2025-025: Research and development for innovation in Machining, additive technology and quality assurance.

6. References

[1] Dana Kubátová, Martin Melichar. Influence of the Number of Points on the Evaluated. DAAAM International Symposium. 2019, str. 0476.

[2] Stojkic, Z[eljko] a Culjak, E[va] & Saravanja, L[uka] (. 3D Measurement - Comparison of CMM and 3D Scanner. Proceedings of the 31st DAAAM International Symposium. 2020, stránky 780-787.

[3] Šárka, TICHÁ. Strojírenská metrologie. (Mechanical engineering metrology.) Ostrava : autor neznámý, 2006. ISBN 80-248-1209-6.

[4] J., Pernikář. Assessment of the competence of control means. [Online] http://gps.fme.vutbr.cz/STAH_INFO/31_Pernikar_VUTBR.pdf.

[5] Pavel, Němeček. Nejistoty měření. (Measurement uncertainties) Praha : Česká společnost pro jakost, 2008. ISBN 978-80-02-02089-9.

[6] Runje, B[iserka] a Horvatic Novak, A[malija] & Keran, Z[denka]. Impact of the Quality of Measurement Results on Conformity Assessmen. [editor] B. Katalinic. Proceedings of the 29th DAAAM International Symposium. 2018.

[7] Long, Chen et al. Discussion on Fast Calibration Method of Coordinate Measuring Machine. Journal of Physics: Conference Series. 2022, Sv. 2206, 1.

[8] ZEISS. [Online] https://www.zeiss.cz/metrologie/systemy/souradnicove-merici-stroje/portalove-merici-stroje/rada-prismo/zeiss-prismo.html.

[9] ZEISS. Operating Instructions PRISMO navigator. 2014. 61211-1120802.

[10] Kiška, Roman. Determining the Measurement Accuracy of a Coordinate Measuring Machine Zeiss UPMC Carat. Master's thesis. Brno: Brno university of technology, Faculty of mechanical engineering, Institute of production machines, systems and robotics, 2021.

[11] Blábol, Václav. The Impact of Changing the Sensing System on Measurement Accuracy. Bachelor's thesis. Pilsen, 2025.

DOI: 10.2507/35th.daaam.proceedings.041

DESIGN AND IMPLEMENTATION OF AN ADDITIVELY MANUFACTURED, ALUMINIUM-REINFORCED 4-AXIS PICK-AND-PLACE ROBOT

Markus Sicha, Kemajl Stuja, Mohamed Aburaia & Ali Aburaia

This Publication has to be referred as: Sicha, M[arkus]; Stuja, K[emajl]; Aburaia, M[ohammed] & Aburaia, A[li] (2025). Design and Implementation of an Additively Manufactured, Aluminium-Reinforced 4-Axis Pick-and-Place Robot, Proceedings of the 36th DAAAM International Symposium, pp.0293-0297, B. Katalinic (Ed.), Published by DAAAM International, ISBN 978-3-902734-47-1, ISSN 1726-9679, Vienna, Austria
DOI: 10.2507/36th.daaam.proceedings.041

Abstract

This paper presents the design and implementation of a four-axis pick-and-place robot, manufactured using a combination of additive manufacturing techniques and aluminium reinforcements. The robot is intended for educational purposes and serves as a foundation for future research and development projects. Furthermore, it demonstrates the potential applicability of lightweight robotic manipulators in small and medium-sized enterprises (SMEs). The design process followed the VDI 2221 methodology and included conceptual design, simulation, and physical implementation. Structural simulations conducted in SolidWorks enabled the identification and resolution of mechanical weaknesses through iterative design optimization. The final prototype validated the simulation results and provided valuable insights for further refinement. The outcome is a cost-effective, lightweight robot with high precision and load-bearing capacity, suitable for a range of industrial and educational applications

Keywords: Additive Manufacturing; SCARA-Robot; SolidWorks-Simulation; Lightweight Design; Topology Optimization.

1. Introduction and Problem Statements

In recent years, industry has increasingly focused on sustainability and resource efficiency. This shift demands the reduction of emissions, minimization of waste, and greater reliance on renewable energy sources [1]. Additive manufacturing offers significant advantages in this context, enabling the rapid and cost-effective production of complex geometries [2]. As a result, development cycles are shortened, and time-to-market is reduced [3], Expanding on earlier research and its findings [4], a four-degree-of-freedom robot [5] was created for pick-and-place tasks, including use cases like shelf handling. The robot was produced through additive manufacturing, supplemented with aluminium elements in areas requiring high dimensional accuracy. To maintain a methodical and consistent development process, the project adhered to the VDI 2221 design guidelines [6].

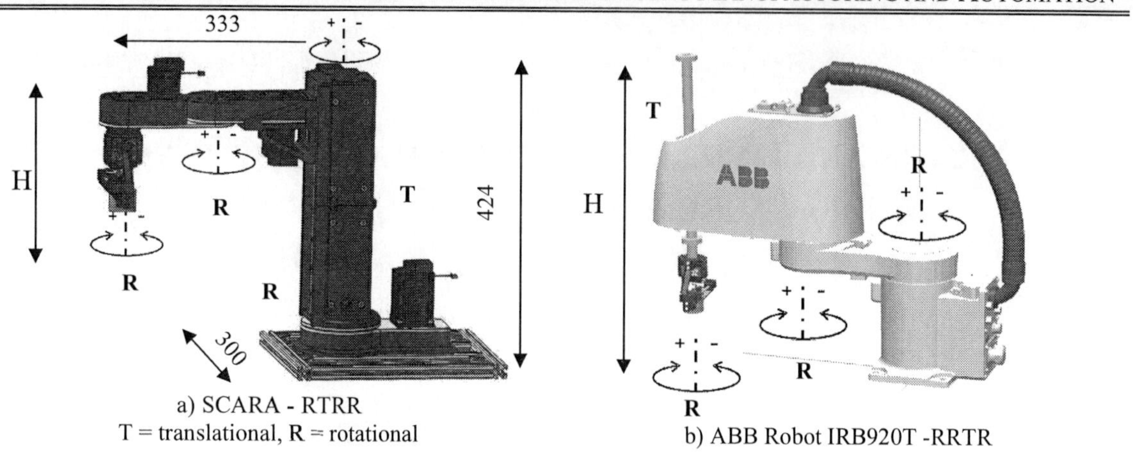

Fig. 1. SCARA Robot for Pick and Place applications with different structure a) RTRR and b) ABB Robot-RRTR.

Pick-and-place operations involving workpieces are among the most common applications in industrial automation [7]. These tasks are typically characterized by high-speed execution and high positional repeatability. Among the most established kinematic architectures for such applications are the SCARA (Selective Compliance Assembly Robot Arm) and Delta robots. Both are specifically optimized to meet the demands of precision and energy efficiency in high-speed environments. However, certain application scenarios require alternative kinematic solutions.

One such case is shelf handling, where the spatial dimensions of the robot arm — particularly its vertical reach — are more critical than maximum operating speed. As illustrated in Figure 1, the height (H) and overall form factor of the robot arm differ significantly between robot (a) — a customized unit developed at UAS Technikum Wien (SICHA robot) and robot (b) — a conventional SCARA robot (ABB IRB 920T) [8]. To address the specific spatial constraints of the shelf-handling task, the translational joint typically located at the third axis in the traditional SCARA configuration was repositioned to the second joint in the customized design. Consequently, the conventional RRTR axis sequence (R = rotational, T = translational) was modified to RTRR. While the modified robot architecture offers improved adaptability for the specific application scenario, it exhibits limitations in terms of dynamic performance. As a result, it is inherently less suitable for high-speed pick-and-place operations. Another relevant constraint concerns the arrangement of objects: if this arrangement is predefined by the customer, a 4-axis robot can be more energy-efficient than a conventional 6-axis robot, due to its simpler kinematics and reduced degrees of freedom.

This observation leads to the central research question addressed in this work:nWhat are the implications of this kinematic configuration on the robot's design process? How can the robot poses be efficiently computed within this framework and control strategy?

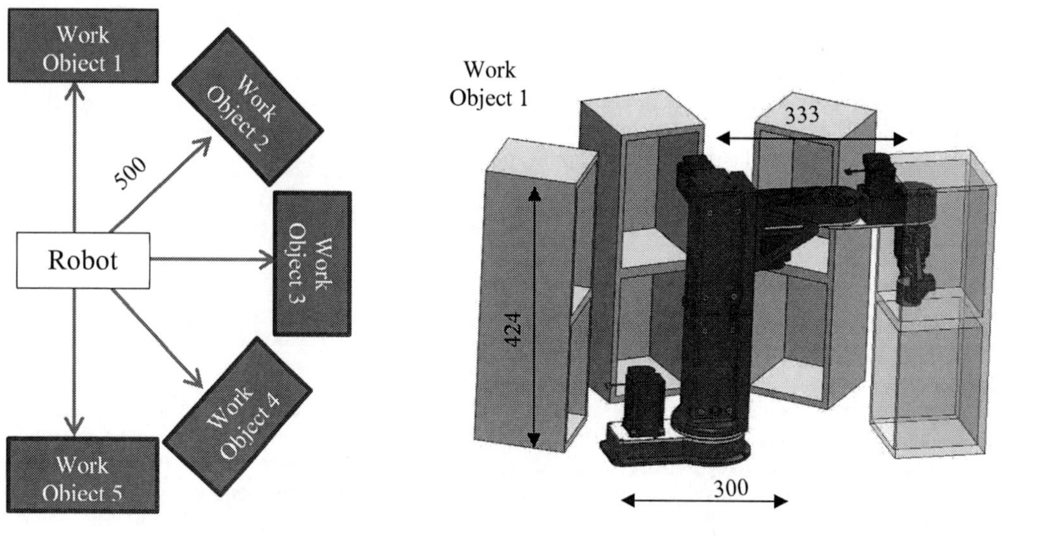

Fig. 2. Customer applications using SICHA-Robot.

2. Materials and Methods

The methodological framework of this study is based on the VDI 2221 guideline [6], which outlines a systematic approach to the development of technical products. The objective is to design a functional, stable, and lightweight 4-axis robotic system that can be manufactured using a combination of additive and subtractive production methods. The development process is divided into four sequential phases: requirements analysis, concept development, design and detailing, and validation.

2.1 Requirements Analysis

The first phase involves identifying and clearly defining the customer's requirements as well as the technical constraints. The main criteria include:

- the use of a 4-axis kinematic configuration,
- high structural stability and stiffness,
- low production costs,
- compatibility with FDM 3D printing and CNC machining,

2.2 Concept Development

Based on the defined requirements, three alternative design concepts for the 4-axis robot were developed. Each concept adopts a hybrid manufacturing approach that combines Fused Deposition Modelling (FDM) for the primary structural components with aluminium reinforcements to increase mechanical rigidity and durability. To identify the most promising solution, a structured evaluation was conducted using a weighted scoring method. The assessment considered key factors such as overall weight, structural stability, manufacturability, cost, and energy efficiency. This systematic approach ensures an objective comparison and supports a rational decision-making process. The evaluation results indicate that the concept featuring an RTRR kinematic structure best meets the customer requirements and the defined criteria. Additionally, it allows a high degree of modularity and simplifies future design modifications or scaling. Therefore, this concept was selected for further detailed development and subsequent simulation studies.

2.3 Design and Detailing

The chosen concept was further developed in the CAD software SolidWorks, following the iterative design process defined in the VDI 2221 guideline. Throughout this phase, particular attention was given to mechanical integrity, manufacturability, and functionality. The design was optimized for additive manufacturing by minimizing material usage and printing time while ensuring sufficient structural strength. In addition, tolerances were explicitly defined within the CAD model to guarantee precise fitting and reliable functionality of the components after production. In Figure 3, the iterative development process of the last robot chain according to VDI 2221 is illustrated using simulation and finite element analysis (FEA). The left side of the figure represents the initial design simulation, while the right side shows the final design simulation. In both cases, the assembly was fixed at the screw connections to the preceding arm, while the contact surfaces of the bearing seats were constrained using a *Sliding Fixture*. The loads were applied at the locations where the respective components are supported—namely at the interfaces of the stepper motor, the end effector, the workpiece, and the remaining attachments. As shown in Figure 3a, the maximum displacement in the initial design amounts to 2.211×10^{-1} mm, whereas in the final design (highlighted in red) it was successfully reduced to 4.113×10^{-2} mm. This corresponds to an improvement by a factor of five compared to the original design and confirms that the resulting displacement is negligible for the intended application.

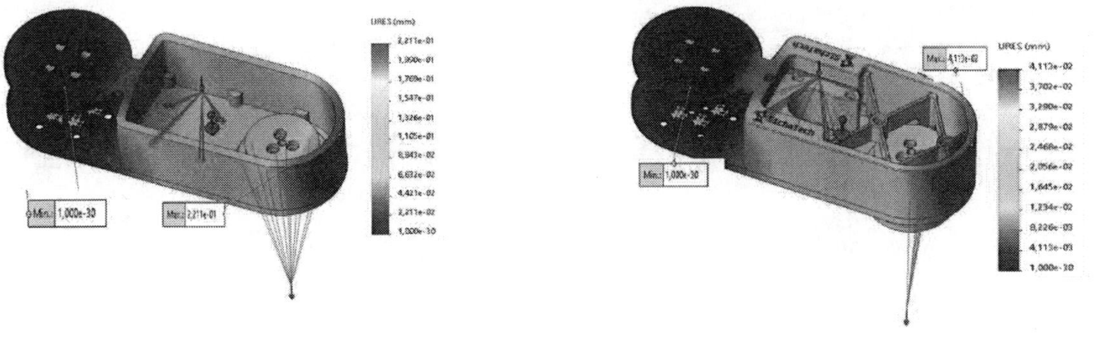

Fig. 3. Iterative design of the SICHA-Robot using the FMEA tool in SolidWorks.

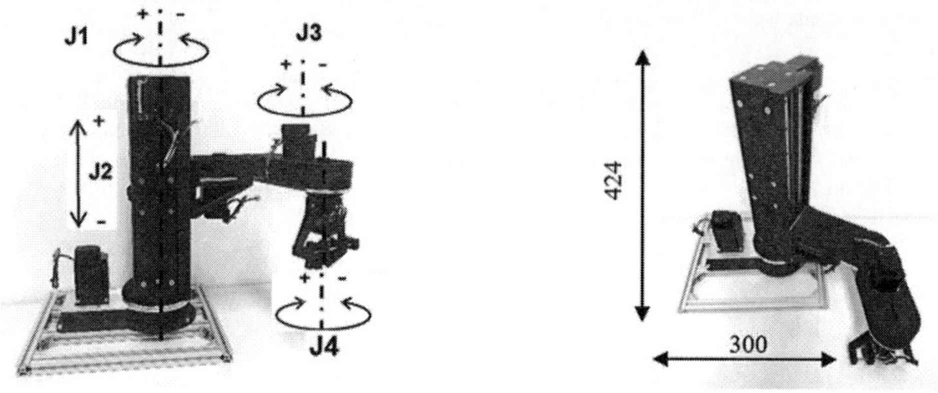

Fig. 4. Manufacturing and assembly of SCARA robot.

2.4 Manufacturing and Assembly

The robotic components are fabricated using a combination of FDM 3D printing and CNC machining. Prior to production, the 3D printer is calibrated, and tolerance tests are performed to ensure dimensional accuracy. PA6-CF - Polyamide 6 (CF = Carbon Fiber) filament is used for the printed parts, while aluminium elements are manufactured using CNC technology. Following fabrication, all components are assembled into a functional prototype. Special attention is paid to joint precision, alignment, and overall structural stability during assembly. To verify mechanical behaviour under load, the manipulator is statically loaded with a payload of 0.5 kg for a duration of 30 seconds. The resulting deflection is measured using a dial gauge. These measurements are then compared with simulation results to identify and analyse deviations between the virtual model and real-world behaviour.

2.5 Simulation and Validation

Prior to any physical experimentation, the unconventional 4-axis robot with an RTRR kinematic arrangement is subjected to a comprehensive analysis in a virtual simulation environment. This digital evaluation serves to examine the kinematic workspace, positioning precision, and the system's ability to sustain operational loads. The numerical results obtained from these simulations are subsequently compared with experimental measurements carried out on the prototype. Differences between the predicted and the observed behaviour are carefully documented and interpreted in order to detect possible inaccuracies in design, modelling assumptions, or control strategy. A key step in this process involves topology optimization, which helps to adjust the mass distribution within the structure according to the expected load paths. Material is selectively reduced in regions where the mechanical stresses are minimal, thereby lowering the overall mass without compromising structural strength. The resulting geometry leads to a more efficient structure with reduced inertia and enhanced dynamic properties. For the load-bearing components, the high-performance composite PA6-CF is employed. This material offers an advantageous balance of stiffness and weight, enabling the design of a slim and lightweight robotic structure. The reduced mass contributes directly to improved responsiveness, lower energy consumption, and increased dynamic performance during operation. This combination of virtual validation, material selection, and structural optimization allows critical aspects of the robot's performance to be fine-tuned before physical testing begins. As a result, development time can be reduced, while the likelihood of major design modifications at later stages is significantly reduced.

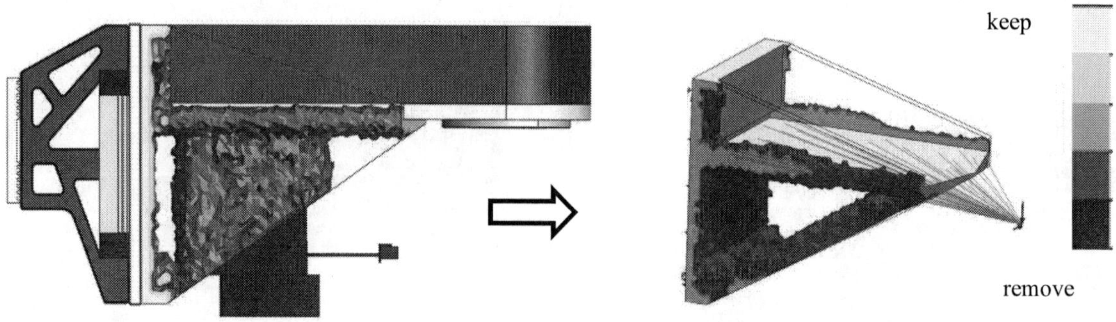

a) Assembly Simulation

b) Part Simulation

Fig. 5. Topology optimization of robot components.

3. Results and Discussion

The simulation results demonstrate that the robot meets the required load-bearing capacity and precision through the combination of additive manufacturing and aluminium reinforcements. Under a maximum acceleration of 2 m/s^2 and the specified payload of 0.5 kg, the simulation yielded a maximum displacement of the TCP (Tool Centre Point) of 0.16 mm. These results highlight the critical role of aluminium reinforcements in ensuring the mechanical stability of the structure. During the static load tests with a 0.5 kg payload, a displacement of 0.43 mm was measured at the TCP, with the robot's tower (vertical axis module) mechanically locked. The tower was fixed to obtain more accurate results, as the bearing between the base and the tower was identified as a minor but noticeable structural weak point. This measure was taken to isolate the influence of this specific weakness and analyse the robotic arm independently. The resulting deviation between simulation and real-world measurements was 0.27 mm. After removing the mechanical lock, the displacement increased to 0.53 mm, underscoring the relevance of the tower bearing as a critical factor in the overall structural stability. Furthermore, the experimental measurements indicate that external influences—such as manufacturing tolerances, material inconsistencies, and the variable properties of the 3D printed components—play a significant role. These factors were not fully captured in the simulation model, which explains the observed discrepancy between simulated and actual behaviour. For the calculation of Robot Target and Inverse kinematics Law of Cosines can be used.

4. Summary and outlook

This study demonstrates the feasibility of developing a lightweight, energy-efficient 4-axis robotic system through a systematic design approach based on the VDI 2221 guideline. By combining additive manufacturing with aluminium reinforcements, a structurally stable prototype was realized that meets the defined functional and mechanical requirements. Simulation results showed that the robot maintained high precision under dynamic loading conditions, with a maximum TCP displacement of 0.16 mm. The subsequent experimental validation confirmed the general accuracy of the simulation, though minor deviations were observed. In particular, the static tests revealed a displacement of 0.43 mm with the tower locked, and 0.53 mm when unlocked. These findings emphasize the relevance of the tower bearing as a critical structural component and highlight the impact of real-world factors such as material inconsistencies, manufacturing tolerances, and 3D printing variability—none of which were fully accounted for in the simulation model. The methodology applied—combining conceptual design, CAD modelling, hybrid manufacturing, and physical testing—proved effective in bridging the gap between digital simulation and physical implementation. It also revealed the limitations of simplified simulation models in predicting mechanical behaviour with full accuracy. Overall, the results confirm that the proposed robotic concept is viable for applications requiring moderate payloads and precision. Further optimization—particularly in the bearing design and selection of printing materials—could enhance system stability and performance. Future work may include dynamic testing under varying load conditions, long-term durability assessments, and integration of advanced materials or sensor-based feedback systems.

5. References

[1] Klein, B. & Gänsicke, T. (2019), Leichtbau-Konstruktion, Light construction design, Springer, ISBN: 978-3-658-26845-9, Wiesbaden.

[2] Richard, H. A.; Schramm, B. & Zipsner ,T. (2019), Additive Fertigung von Bauteilen und Strukturen- Additive Manufacturing of Components and Structures, Springer, ISBN: 978-3-658-17779-9, Wiesbaden.

[3] Grunwald, A. & Hillerbrand, R. (2021), Handbuch Technikethik - Handbook of Ethics of Technology, Metzler Verlag, ISBN: 978-3-476-04900-1, Stuttgart.

[4] Stuja K.; Bruqi, M.; Markl E. & Aburaia, M. (2016). Lightweight 4 -Axis Scara Robot for Education and Research, Proceedings of the 27th DAAAM International Symposium, pp. 0102-0108, B. Katalinic (Ed.), Published by DAAAM International, ISBN 978-3-902734-08-2, ISSN 1726-9679, Vienna, Austria.

[5] https://www.igus.at/product/21278?artNr=RL-A9.0216, (2025), SCARA-Robot 3, IGUS® polymer Innovation Accessed on: 2025-10-10.

[6] Sattinger, V.; Papa, M.; Stuja, K. & Kubinger, W. (2019). Methodik zur Entwicklung sicherer kollaborativer Produktionssysteme im Rahmen von Industrie 4.0.- Methodology for the Development of Safe Collaborative Production Systems in the Context of Industry 4.0 Elektrotech. Inftech. 136, pp.318–325, https://doi.org/10.1007/s00502-019-00744-1.

[7] Stuja, K.; Katalinic, B.; Aburaia, M. & Pajaziti, A. (2021). Design of a Robot Application with Regard to Energy Efficiency. Vol. 1. first: DAAAM International Vienna. http://www.daaam.info/Downloads/Pdfs/proceedings/proceedings_2021/005.pdf, Accessed on: 2025-10-10.

[8] ABB AG. (2023). RobotStudio Manual. Robotics Products SE-721 68 Västerås Sweden. https://new.abb.com/products/robotics/robotstudio. Accessed on: 2021-08-21.

DOI: 10.2507/36th.daaam.proceedings.042

CHALLENGES OF INTEGRATING VIRTUAL REALITY INTO POLYGRAPH TESTING

Lucie Sousedikova & Milan Adamek

This Publication has to be referred as: Sousedikova, L[ucie] & Adamek, M[ilan] (2025). Challenges of Integrating Virtual Reality into Polygraph Testing, Proceedings of the 36th DAAAM International Symposium, pp.0298-0304, B. Katalinic (Ed.), Published by DAAAM International, ISBN 978-3-902734-47-1, ISSN 1726-9679, Vienna, Austria
DOI: 10.2507/36th.daaam.proceedings.042

Abstract

The use of virtual reality (VR) is on the rise due to technological advances. One way to utilize VR is by incorporating it into polygraph testing. The main focus of this research paper is to investigate the possibility of integrating these two technologies to improve the polygraph testing process, whether for lie detection or individual profiling. Important factors of polygraph testing in VR are analysed in this work, with an emphasis on movement, as traditional polygraph examinations are conducted in a neutral, quiet environment and do not involve any movement. For effective implementation of VR into polygraph testing, it is therefore necessary to analyse the impact of movement on the accuracy of evaluation. The research also presents and describes VR technology and the entire polygraph examination process, including the specialized LX software.

Keywords: polygraph testing; virtual reality; physiodetection; movement; analysis.

1. Introduction

The pursuit of a method for verifying truthfulness has a long history. But only since the 19th century, there have been efforts to find scientific ways to expose lies, including polygraph testing. With the advancement of technology in recent years and the growing interest in virtual reality (VR), the integration of polygraph devices with VR has become a topic of discussion, but it has been rarely researched. The use of polygraph technology in VR environments holds promise as a powerful tool for criminal investigation and assessment of credibility, and has the potential to significantly advance the field of polygraph examination. Current research papers focus on the effects of 2D and 3D environments on individuals and on the potential for VR environments to elicit a higher degree of emotional arousal. Although these studies are important for the integration of VR and polygraph, there are still many aspects to be explored in this promising application for the successful implementation of these two technologies together. One such aspect is the impact of navigating in virtual environment on polygraph test results. The traditional polygraph examinations are conducted in a quiet, distraction-free setting, with the examinee seated and answering questions posed by the polygraph examiner. However, in the environment of virtual reality, it needs to be researched whether it is possible to accurately detect physiological responses during basic movements related to VR control. As movement can cause physiological changes, it is necessary to investigate whether movements during navigation in VR have a significant impact on polygraph results. Therefore, research devoted to the determination whether important data can be distinguished from distractions caused by movement is relevant.

2. Virtual Reality Technology

Virtual reality (VR) is an immersive simulation of an artificial environment, that is modelled by a computer [1]. The user of VR usually sees created scenes with a headset and hears environmental sounds through headphones placed on this headset. However, it is possible to enhance immersion exploring the other human senses [2]. For interaction with the environment are used VR controllers that tracks user movement in 3D space and for more interaction users can use gestures and press buttons on wearables. Typical example of VR headset with controllers can be seen in Fig. 1.

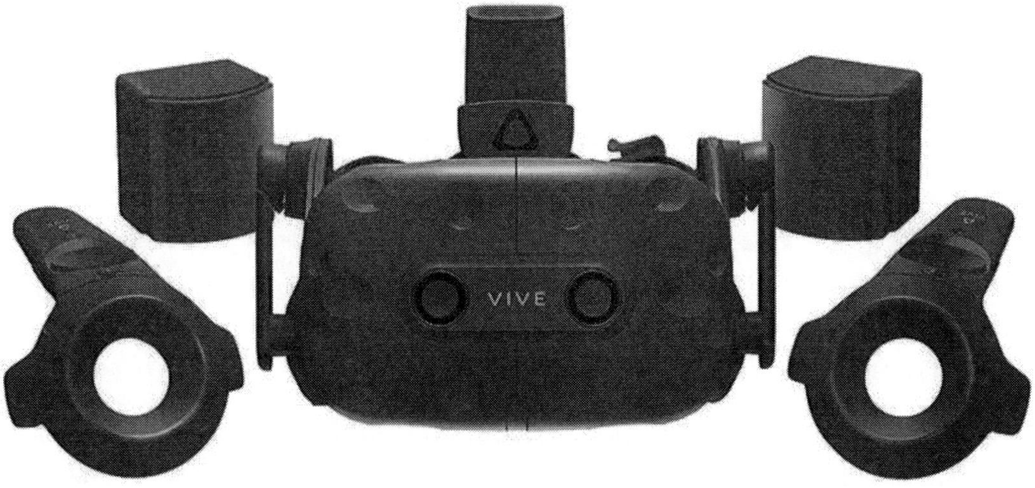

Fig. 1. VR HTC Vive Pro Full kit [18]

Tracking of movement can be done using tracking base stations or inside-out tracking systems. Tracking base stations scan the whole room with synchronization beams many times per second. Controllers' relative position to base stations is then found using simple triangulation which is based on the time when controllers were hit by beam signal. Inside-out tracking systems use cameras located on the headset and detect infrared light from LEDs on wearables. With this system, the position is triangulated relatively to the headset position. For best simulation results, VR needs high precision of feedback with low latency, high resolution of the headset, and high sound quality. The main driver for improving VR quality was mainly the military, space, and entertainment industries [3], [4], [5].

3. Polygraph System

The polygraph measures and records physiological changes in a person's body as they answer questions from a polygraph examiner. The whole polygraph system is composed of a a Data Acquisition System (DAS), software, and several sensors:

- Blood pressure cuff that records blood pressure and its changes.
- Pneumograph tubes placed over the chest and abdomen to record breathing.
- Electrodermal (EDA) electrodes attached to fingers for measuring the skin conductance.
- Photoelectric plethysmograph that records rapid changes in pulse blood volume.
- Activity sensor pads used to detect minor movements of the arms, feet, or bottom.

3.1. Polygraph Screening Process

The polygraph examination procedure consists of four stages: preparation and consultation, polygraph examination, evaluation of results, and drawing conclusions. First, it is necessary to assess and decide when to conduct the polygraph examination. It has been found that the most effective time for the use of physiodetection is at the beginning of a criminal investigation in the initial stages of clarifying the crime, as the individual being investigated does not have access to other sources of information at this time. The next step in the preparation phase is the creation of a series of questions, which are tailored to each specific case. At the start of the examination, the entire process is explained to the examinee and he/she is asked to confirm his/her written consent for voluntary participation in the examination. Furthermore, the examined person is asked about his current situation and health problems. Any information that could negatively impact the objectivity of the examination must be taken into account. Subsequently, the individual is asked to provide information about the case. At the end of the examination, the person is asked to sign the graphic record which confirms the date and duration of the examination and the person´s voluntary consent. After the physiodetection examination phase, the obtained results are evaluated and the conclusions are drawn [6], [7], [8].

3.2. LX Software

During the examination, physiological data are controlled, displayed and recorded by LXSoftware that is a specialized software platform designed for analysis of polygraph examinations. Fig. 2 presents an example of a polygraph measurement taken using LXSoftware.

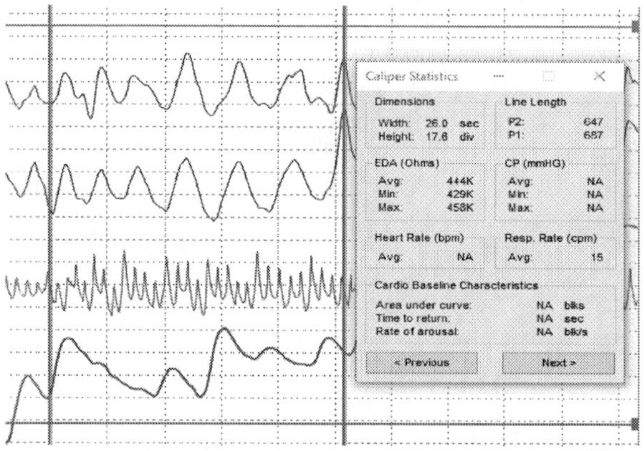

Fig. 2. Sample of polygraph measurement from LXSoftware

The bottom line represents time, with markers indicating each task that the examinee performed. The blue line depicts the signal from the pneumograph sensor, the red line illustrates the signal from the photoelectric plethysmograph, and the green line denotes the signal from the EDA sensors. This software seamlessly integrates with the polygraph instrument, facilitating real-time monitoring and capturing of various physiological parameters. The software ensures precise and synchronized data acquisition, enabling examiners analysing physiological reactions in relation to specific stimuli using on-screen calipers, as depicted in Fig. 3. These calipers provide the measurement of minimum, maximum, and average EDA, heart rate, respiration rate, cuff pressure, and cardio baseline characteristics. Respiration line lengths are also automatically measured.

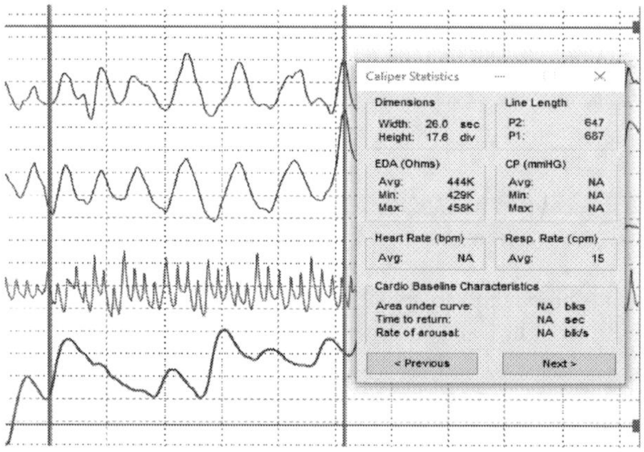

Fig. 3. The Caliper Statistics window displays the dimensions of the chart area selected by the calipers. This feature enables the measurement of distinct patterns in the obtained data.

4. Challenge of Integrating Virtual Reality into Polygraph Testing

The polygraph, or lie detector, is a widely used technology for assessing credibility. It is frequently employed by law enforcement and federal agencies for detecting deception and profiling individuals. It is used in criminal investigations, security clearance processes, and managing deviant sexuality. The recent study [9] has firmly established the validity and effectiveness of lie detection methods, demonstrating an impressive accuracy rate that surpasses 90%. However, a significant concern persists regarding the relatively high occurrence of false positive responses, estimated to be around 25%. It is suggested that the changes in physiological responses immediately after the act of lying are not adequately distinguishable. Virtual reality is a rapidly growing field of research with expanding applications across multiple disciplines. VR is utilized not only in the entertainment and gaming industries but also in sectors such as information technology, biomedical engineering, and training aids.

VR visualization offers an enhanced simulation environment with the potential to improve and accelerate human learning, treat various mental health conditions and behavioural disorders, and explore human immersion in artificial environments. Therefore, new simulators and therapy solutions have been developed, with ongoing research into their effectiveness and long-term impact on individuals.

There are a few potential issues that could arise when integrating virtual reality into polygraph testing:

- technical difficulties: There may be challenges with setting up and running the virtual reality equipment, as well as ensuring that it is functioning properly during the test;
- participant discomfort: Some individuals may experience discomfort or dizziness when using virtual reality, which could affect their ability to complete the test;
- ethical concerns: There may be ethical concerns to consider when using virtual reality in polygraph testing, such as participant privacy and consent;
- validity of test results: It is important to ensure that the virtual reality environment does not interfere with the accuracy of the polygraph test.

All of the aforementioned points must be considered for future use. While the first three points may not be difficult to address, the validity of the test must be thoroughly examined. The following section addresses individual research studies and analyses the aforementioned issues.

5. Analysis of Polygraph Screening in VR

5.1. VR as an Amplifier of Presence Levels

In the field of aerospace training, research [10], [11] has been introduced that presents new simulators developed by combining the use of VR and realistic computational models. These studies comparing VR simulator training and real flight training suggest that the strong feeling of immersion generates high presence levels. It showed higher heart rate and lower heart rate variability in the real flight condition compared to the VR one. However, similar patterns of subjective rating and cardiac activity were found across different segments of the scenarios for both conditions. Despite lower physiological values in VR due to the lack of real risk, the same physiological changes occurred as in a real environment. However, the research only evaluated cardiovascular activity and none of these studies take into account the influence of movement on physiological responses. It is necessary to assess not only deviations in measured values but also the influence of movement in VR on these values. This approach was used in [12] where the body's response to movement was measured, however the study was more focused on gradually increasing physical activity until reaching very high intensity. The researchers used an electrodermal activity (EDA) sensor, but their measured values were very low when measuring during rest. The low measured EDA values that the researchers achieved in the resting phase are most likely due to the size of the electrodes or it may also be due to incorrect settings during some of the tests.

Human immersion in artificial environments has also been used in psychological treatment [13], [14] to treat various mental health conditions and behavioural disorders. Interactive and immersive realistic 3D graphic scenes allow patients to experience different emotions or become accustomed to fearful feelings. Due to the immersion and strong sense of presence, VR amplifies emotional responses to stimuli, showing positive effects such as overcoming fear. From this study, it is clear that using VR in polygraph examination can enlarge physiological reactions to emotional stimuli. This knowledge could help to streamline profiling or criminal investigations by assuming that if there is an artificial realistic visualization, the physiological reaction is then stronger, and it is more difficult to deny a physiologically false statement. These studies primarily focus on the impact of VR on psychological health. But they do not provide any information how to distinguish between emotional and movement physiological responses which is important for polygraph testing.

5.2. Enhancing Polygraph Accuracy with VR Technology

Owing to its potential, the integration of virtual reality and polygraph technology could prove effective in criminal investigations by enhancing the exposure of suspects and increasing the reliability of polygraph testing. However, there is only one isolated study [15] investigating the use of VR technology in conjunction with polygraph testing. This study showed that the use of VR technology in forensic investigations can increase recognition of crime-related stimuli by over 25 %, enhancing accuracy in detecting concealed information. However, in this study, the VR environment did not include independent object motion and only sensors measuring electrodermal activity and heart rate were used in just one artificially created crime – theft. Therefore, despite the significance of these results, all factors of polygraph testing in VR have not been included and further research is needed to fully understand this implementation.

5.3. Physiological changes caused by movement¨

The evaluation of the impact of movement on measurement accuracy was achieved by thoroughly testing volunteers who had to perform predetermined movements while attached to polygraph sensors. The types of movements that examinees had to perform were chosen to simulate movements that might be necessary in virtual reality.

The aim of the study [16] was to investigate whether it is possible to correctly detect physiological responses during basic movements related to VR control and to determine whether important data, such as emotional arousal, can be distinguished from the changes caused by this movement. Due to movement constraints, only some sensors were used in this study, including pneumograph tubes, EDA electrodes to record electrodermal activity, and a photoelectric plethysmograph to measure heart rate.

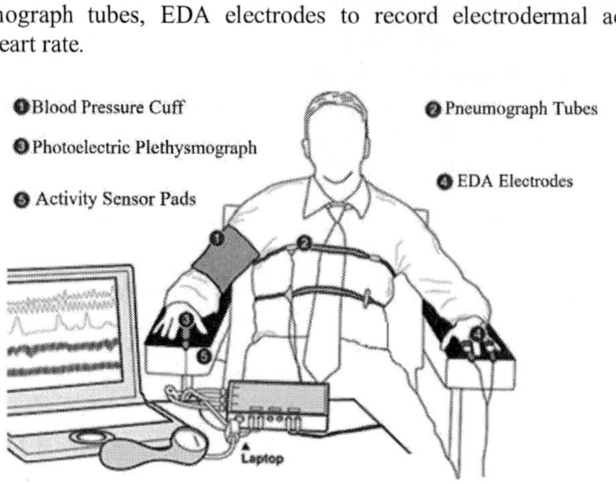

Fig.4. Attached polygraph sensors [17]

During the measurement process, five situations were chosen to encompass the primary movements observed during VR navigation. Each situation lasted for approximately 20 seconds, representing the time between initiating a movement and commencing the subsequent movement. This duration was predetermined within LXSoftware and corresponds to the minimum time between question onsets required for valid scoring results in traditional polygraph testing. All tested individuals without any health problems or other issues completed the following tasks:

1. Sitting in a relaxed position on a chair.
2. Head movement in a 3D scene.
3. Standing up and remaining still without any other movement.
4. Slow walking in VR – simulation of movement in VR.
5. Hand movement with controllers.
6. Falling in VR - inducing fear of a fall.

The fall mentioned in the sixth point was performed in the VR environment. The examinee was instructed to hold onto a desk, and the timing of the drop was unknown to the examinee to induce a sense of increasing tension. The obtained data indicated that physical activity has a significant effect on polygraph measurement. This is likely due to changes in physiological responses that occur during different levels of physical activity. Surprisingly, the findings showed that the physiological differences between sitting in a relaxed position and head movement are similar, in contrast to the comparison between any other movement and a relaxed position. The standard deviations also indicated a significant amount of variation in the measured values, which could be attributed to the individual physical differences among the examinees. For some individuals, the tasks were simple, and responses were not as high as for others. The results suggest that emotions can elicit more pronounced physiological reactions compared to movement. However, changes caused by movement were also significant and, in some cases, could be mistaken for emotional arousal.

6. Discussion of Results of the Analysis Focused on the Use of the Polygraph in VR

From the study [16], it is obvious that each movement has a certain physiological impact on people, which usually increases together with amount of movements. The movement induce increased values of physiological reactions mostly during the movement itself or shortly after it. This research confirms the results achieved in [12], which, however, emphasized higher physical load and did not use a pneumograph and photoelectric plethysmograph. This study also established the precise behaviour of EDA values during movement. Typically, EDA values range from 0.5 μS to 20 μS, depending on factors such as skin dryness, individual physiological differences, and electrode site. This is in contrast to the aforementioned study, where these values were even negative. The results of this study demonstrate that during simple movement, values cannot be negative and fall within the standard range. In addition, the measured values show an increase in values even at low load caused by simple movements. Apart from head movement all values increased on all measured sensors. As can be seen from [16], [18] physiological reactions caused by emotions were very pronounced in observed persons. It shows that emotional arousal in reaction to virtual reality can be really pronounced and it also means that the emotions can sometimes evoke more pronounced physiological reactions than the movement itself. The study [16] excluded movements that require increased physiological activity and are not essential for controlling VR. Other research, as cited in [12], has shown that such movements can lead to elevated physiological responses.

The main disadvantage of VR-polygraph testing could be the confusion of physiological reactions caused by movement with those caused by emotional reactions. This could reduce the main advantage of VR-polygraph testing, which is the enhancement of physiological reactions to emotional stimuli. This disadvantage can be eliminated by limiting the movement of the test subject to necessary minimum, for example, to only head movements. As highlighted in [16] the physiological values associated with sitting in a rest position and head movement demonstrate similarity. Consequently, head movement can be considered as an effective movement to be employed in the context of polygraph screening. To avoid any other movements than those by head, the examinee would be placed within a 3D scene without any possibility of self-navigation and polygraph examiner would maintain control over the VR movement within the designated scene. If movements in VR were used, it could lead to the emergence of false positives. However, with limited movement, this will not occur. The amplification of physiological responses to emotional stimuli also facilitates the differentiation between false positives and detected lies, as larger differences make it easier to distinguish, potentially improving accuracy.

It is important to note that the movement testing [16] was performed on a small sample of volunteers and usage of VR-Polygraph screening method proposed by this research would require analysis of precise scenarios tailored for specific polygraph usage, such as criminal case investigations. Additionally, the VR-Polygraph is a high-priced technique that requires the cooperation of multiple experts. These include a polygraph specialist with the appropriate qualifications to conduct polygraph examinations and experts in virtual reality for 3D scene reconstruction. These two factors are the main limitations of this study. When using virtual reality, it is necessary to consider the ethical implications of its use. Virtual reality can enhance emotional experiences through visual stimuli, and it is therefore important to prevent negative impacts on the mental well-being of tested individuals. Polygraph testing itself can be stressful, and insensitive VR-Polygraph testing, such as in a recreated environment depicting a crime scene, can evoke strong emotional responses and cause psychological harm. Therefore, when creating VR scenes for polygraph testing, these considerations must be taken into account. The results of this research should serve for the future application of virtual reality in the polygraph testing process. However, as previously mentioned, future research must conduct a broader study for specific cases of using this new method. Once ethical and legal obstacles are overcome, this method could potentially be used to improve the efficiency of criminal investigations or individual profiling.

7. Conclusion

It should be mentioned that a more pronounced movement observed in the examinee during a conventional polygraph examination may indicate their effort to conceal a lie. In this way, the examinee may attempt to undermine the results by employing deliberate movement during control or non-relevant questions, but the mere fact that they are making an effort through movement can be valuable information. And as demonstrated, the detection of even minor movements can be achieved, even in the absence of activity sensor pads. Based on the analysis, for the effective use of the polygraph in VR environment, the movement of the examined person in this environment should be avoided, as it distorts the results, and it is difficult to distinguish the reactions caused by movements from changes caused by emotions. Therefore, VR should be used in fixed scenes where examinee movement is restrained, and he can only explore the space in head-rotation. The transitions between scenes should be then done by examiner.

8. Acknowledgment

This publication was supported by the Internal Grant Agency of Tomas Bata University in Zlin, the Department of Security Engineering, Faculty of Applied Informatics, under project No. IGA/FAI/2023/003.

9. References

[1] Hu, Z., Liu, L. (2023). Research on the application of virtual reality technology in 3D animation creation. Optik, 272, doi: 10.1016/j.ijleo.2022.170274. Available at: https://www.researchgate.net/ publication/ 365600084_ Research_on_the_application_of_virtual_reality_technology_in_3D_animation_creation

[2] Haiyan, Ch. (2020). Research on 3D Animation Design Based on VR Technology. Journal of Physics, Conference Series, 1607, doi: 10.1088/1742-6596/1607/1/012016. Available at: https://www.researchgate.net/publication/ 343716975_Research_on_3D_Animation_Design_Based_on_VR_Technology

[3] Mittal, H. (2020). Virtual Reality: An Overview. Journal CSI Communications, 44. Available at: https://www.researchgate.net/publication/343095838_Virtual_Reality_An_Overview

[4] Mathivanan, K., Swathi, T., Ashapriya, B., Sruthi, R. (2017). A Study of Virtual Reality. International Journal of Trend in Research and Development, 4(3), 374-377, ISSN 2394-9333. Available at: https://www.researchgate.net/publication/320517587_A_Study_of_Virtual_Reality

[5] Greengard, S. (2019). Virtual Reality. The MIT Press, USA, Massachusetts, ISBN 978-0262537520.

[6] Kohout, J. (2022). Lie detector examination in the process of clarifying criminal activity. Criminalistics Journal, Praha, MV CR, ISSN 1210-9150.

[7] Warner, W. J. (2018). Going Knee to Knee with FBI Polygraph. The Peppertree Press, USA, ISBN 978-1614935766.

[8] Rastogi, P., Kanchan, T., Das, S., Daya, P. (2008). The Truth behind Lie Detector. Journal of Indian Academy of Forensic Medicine, 30(3), 161-163, ISSN 0971-0973.

[9] Sousedikova, L., Adamek, M. (2022). Security Clearance and Polygraph Examination, Proceedings of 33rd DAAAM International Symposium, Published by DAAAM International, ISSN 1726-9679, Vienna, Austria. Available at: https://www.daaam.info/Downloads/Pdfs/proceedings/proceedings_2022/working_papers/ dpn76042_b_1_Sousedikova.pdf

[10] Labedan, P., Darodes-de-Tailly, N., Dehais, F., Peysakhovich, V. (2021). Virtual Reality for Pilot Training: Study of Cardiac Activity, 81-88, doi:10.5220/0010296700810088. Available at: https://www.researchgate.net/publication/349389668_Virtual_Reality_for_Pilot_Training_Study_of_Cardiac_Activity

[11] Bosch Bruguera, M., Ilk, V., Ruber, S., Ewald, R. (2019). Use of Virtual Reality for astronaut training in future space missions – Spacecraft piloting for the Lunar Orbital Platform – Gateway (LOP-G). Proceedings of 70th International Astronautical Congress, USA, Washington D.C. Available at: https://www.researchgate.net/publication/336750035_Use_of_Virtual_Reality_for_astronaut_training_in_future_space_missions_-_Spacecraft_piloting_for_the_Lunar_Orbital_Platform_-Gateway_LOP-G

[12] Posada – Quintero, H. F., Reljin, N., Mills, C., Mills, I., Florian, J. P., VanHeest, J. L., Chon, Ki H. (2018). Time – varying analysis of electrodermal activity during exercise. PLoS One, 13(6), doi: 10.1371/journal.pone.0198328. Available at: https://www.ncbi.nlm.nih.gov/pmc/articles/PMC5983430/

[13] Guazzaroni, G. (2019). Virtual and Augmented Reality in Mental Health Treatment. IGI Global, USA, ISBN 9781522571698.

[14] Abdullah, M., Shaikh, Z. A. (2018). An Effective Virtual Reality based Remedy for Acrophobia. International Journal of Advanced Computer Science and Applications, 9, doi:10.14569/IJACSA.2018.090623. Available at: https://thesai.org/Publications/ViewPaper?Volume=9&Issue=6&Code=IJACSA&SerialNo=23

[15] Norman D., Wade K., Williams M., Watson D. (2020). Caught Virtually Lying – Crime Scenes in Virtual Reality Help to Expose Suspects' Concealed Recognition. Journal of Applied Research in Memory and Cognition, 9, doi: 10.1016/j.jarmac.2019.12.008. Available at: https://www.sciencedirect.com/science/article/pii/S2211368119301895

[16] Svarc, V. (2021). Polygraph measurement. Available at: https://digilib.k.utb.cz/bitstream/handle/10563/46090/%c5%a1varc_2021_dp.pdf?sequence=1&isAllowed=y. [cit. 2024-12-08].

[17] Cincinnati Polygraph. (2018). Do Polygraphs Detect Lies. Available at: https://lafayettepolygraph.com/images/lx6-515.png.

[18] Haiyan, Ch. (2020). Research on 3D Animation Design Based on VR Technology. Journal of Physics, Conference Series, 1607, doi: 10.1088/1742-6596/1607/1/012016. Available at: https://www.researchgate.net/publication/343716975_Research_on_3D_Animation_Design_Based_on_VR_Technology

DOI: 10.2507/36th.daaam.proceedings.043

EVALUATING DEXTERITY LIMITATIONS OF BIONIC HANDS VIA REINFORCEMENT LEARNING

Yussef Abdel-Wahab, Mohamed Aburaia & Ali Aburaia

This Publication has to be referred as: Abdel-Wahab, Y[ussef]; Aburaia, M[ohammed] & Aburaia, A[li] (2025). Evaluating Dexterity Limitations of Bionic Hands via Reinforcement Learning, Proceedings of the 36th DAAAM International Symposium, pp.0305-0316, B. Katalinic (Ed.), Published by DAAAM International, ISBN 978-3-902734-47-1, ISSN 1726-9679, Vienna, Austria
DOI: 10.2507/36th.daaam.proceedings.043

Abstract

In-hand manipulation is a key benchmark for dexterous robotic control, especially in bionic hands where mechanical simplicity competes with functional complexity. This work uses Isaac Sim and reinforcement learning (RL) to evaluate the dexterity of a custom bionic hand trained to manipulate a cube in 3D space. The goal is to assess whether the design provides sufficient degrees of freedom (DoF) to complete the task under domain randomization, enabling potential zero-shot sim-to-real transfer. Despite high-fidelity simulation and robust policy optimization, the bionic hand failed to perform the task, suggesting that limitations stem from its mechanical design rather than learning or modeling errors. To verify this, the same setup was tested on the Shadow Hand, which succeeded. Three bionic hand variants were then evaluated: the baseline, one with extended supination, and one with an added wrist joint. Their performance was compared to the 24-DoF Shadow Hand. The modified designs showed clear improvement, isolating the effects of joint range and placement. Results indicate that moderate increases in motion range or adding even a single DoF can substantially enhance manipulation ability, emphasizing the importance of joint-level design in achieving dexterous behavior in bionic and robotic hands.

Keywords: Dexterous Robotics; Reinforcement Learning; In-Hand Manipulation; Sim-to-Real

1. Introduction

Modern robotic systems are typically designed with a specific use case in mind and deployed in controlled, structured environments. While this design has enabled reliable performance in industrial automation and service robotics, it shows clear limitations in human-centric settings, which are characterized by variability, uncertainty, and unstructured interactions [1] [2]. To address this challenge, research has increasingly turned toward humanoid and bionic robotic systems that can operate in environments originally built for humans. Within this area of research, dexterous robotic hands and arms represent a particularly promising but difficult path. A bionic arm with multiple degrees of freedom can, in principle, replicate human-like manipulation and enable a wide range of tasks. However, the very complexity that makes such systems versatile also creates challenges. High-dimensional control spaces, nonlinear joint couplings, and intricate contact dynamics make classical control methods difficult to apply effectively [2]. As a result, controllability and robustness remain major bottlenecks in the deployment of dexterous robotic systems [1].

Previous attempts to solve the complex nature of high dimensional action spaces required the use of model predictive training or contact-invariant optimization [3] [4]. The inherent issue with in-hand manipulation stems most likely not from the action space alone but also the observation space, including contact points [5]. In recent years, reinforcement learning (RL) has emerged as a compelling alternative for solving these challenges [2], [6]. Unlike traditional controllers, RL agents can learn policies directly from interaction data, making them well-suited for tasks with complex dynamics such as in-hand object manipulation. Landmark studies have demonstrated that RL can control high-degree-of-freedom robotic hands to perform nontrivial manipulation tasks in simulation and, with the help of techniques like domain randomization, even transfer these skills to real-world hardware [7]. These advances suggest that simulation-based RL provides a viable pathway for training dexterous bionic arms to achieve robust, human-like manipulation in diverse environments.

2. State of the Art

Dexterous in-hand manipulation requires coordinated control across many coupled degrees of freedom. In practice, however, many robotic and prosthetic hands provide only a small number of actuated joints. This mismatch forces controllers to achieve outcomes that normally require complex hands with only simplified mechanisms. The following overview highlights how research has approached this problem through contact-based planning, learning-based control, the exploitation of external forces, and design strategies that rely on either compliance or rigid linkages. Early approaches to dexterous manipulation treated it as a problem of motion planning with explicit contact reasoning. Contact invariant optimization planned trajectories in combination with contact events, which produced finger gaits and regrasp motions in simulation [4], [5]. Predictive control schemes similarly synthesized rapid behaviors for dynamic manipulation tasks [3]. These methods established as an important foundation that, with accurate models of friction and contact, coordinated motions could be planned even when actuation was limited. However, in real scenarios such approaches were often not sufficient, since minor errors in modeling or friction parameters could destabilize the resulting plans [1], [2]. For hands with few actuators, the dependence on subtle contact exploitation made the gap between simulation and reality especially significant. This behavior that emerged from these limitations is referred to as extrinsic dexterity. Instead of relying only on the internal joints of the hand, a controller can make use of gravity, inertia, and environmental contact surfaces to reorient an object [8]. Some studies demonstrated how external forces can increase the set of reachable object configurations with minimal finger motion [8]. More recently, learning-based methods rediscovered these principles, with policies that pressed or rolled objects against planes or the palm to create rotations that could not be commanded directly [9], [10]. This work makes clear how a simple hand can still perform meaningful reorientation by creating contact situations that substitute for missing degrees of freedom. Parameterized manipulation primitives now provide a structured way to represent such strategies [11]. The rise of deep RL introduced a powerful alternative to analytic control. With appropriate reward design and exploration, RL agents discovered non-intuitive solutions for in-hand reorientation tasks [2]. A landmark example is the OpenAI Shadow Hand project, in which a policy optimized with Proximal Policy Optimization (PPO) [6] successfully reoriented a cube and later solved the Rubik's Cube [7]. Success depended on extensive domain randomization, which randomized masses, frictions, and joint gains to improve robustness [12]. Later work extended this pipeline to more agile behaviors on the same platform, showing that realistic simulation combined with massive randomization can enable direct transfer to hardware [13]. Theoretical analyses now provide guarantees that domain randomization can narrow the gap between simulated and real environments under broad conditions, and that memory-dependent policies can further stabilize performance when dynamics vary [14].

Although many of these advances were demonstrated on hands with many actuated joints, the same principles can be applied to simpler devices. Tendon-driven designs with rolling contacts, for example, provide natural compliance that makes it easier for policies to exploit contact geometry [15]. At the opposite end, dynamic two-handed throw-and-catch experiments highlight how momentum and timing can replace internal joints with well-orchestrated external effects [16]. Recent studies also demonstrate end-to-end visual policies for humanoid manipulation tasks, showing that learning directly from camera observations can reduce the mismatch between simulation and sensor inputs [17]. Other work combines RL with demonstration-led curricula, which improves stability during training without altering the mechanical complexity of the platform [18]. The physical mechanism itself ultimately determines what strategies are possible. Tendon-driven fingers allow a form of compliant following that can be approximated by proportional–derivative joint controllers, where dependent joints naturally lag behind and adjust smoothly. By contrast, rod-coupled or gear-coupled mechanisms behave as stiff kinematic loops. These must be modeled with extremely high stiffness values and minimal damping, and backlash needs to be represented explicitly to avoid unrealistic rigid couplings. Studies that successfully transferred from simulation to real Shadow Hands emphasize the importance of aligning simulator contact models, joint limits, and controller gains with the actual mechanism [13]. In summary, low-actuator hands can still achieve tasks that normally require complex devices, provided that they exploit extrinsic contacts, are trained under diverse simulation conditions, and are modeled with accurate representations of their mechanical couplings. High-actuator hands remain the most capable, but targeted design choices such as compliant tendons or fast global wrist joints can raise the capabilities of simpler hands. This understanding motivates the following study, in which a rod-coupled bionic hand with limited joints is evaluated against a high-degree Shadow Hand baseline. The simulations explicitly account for stiff linkage behavior and backlash, and training is performed with PPO in Isaac Sim with extensive domain randomization following established practice.

3. Materials and Methodes

In this work, the Bionic Arm developed in [19] was evaluated in the in-hand cube manipulation task and compared against the Shadow Hand [20]. An altered version of the Bionic Arm was also tested, incorporating an additional axis in the wrist. To improve computational efficiency during simulation, the CAD models of both versions of the Bionic Arm were simplified. Fasteners such as screws, nuts, and bolts were intentionally excluded. The outer geometry, particularly the surfaces that may come into contact with the cube, was preserved to avoid altering interaction behaviour. Regions previously occupied by washers were filled to prevent edge gaps that could catch the cube and lead to non-transferable behaviors. Internal cavities not involved in contacts were also filled where appropriate. One simplification that may influence sim-to-real transfer is the exclusion of the metallic linkage rods that couple the distal phalanx as depicted in Fig. 1. This rod can collide with the cube, since it is mounted on the inside of each finger.

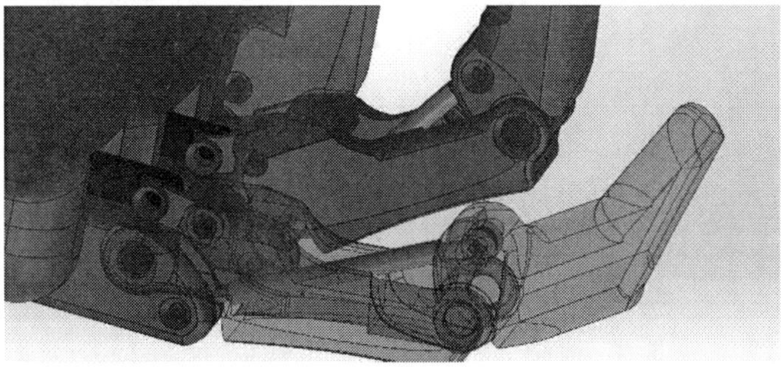

Fig. 1. teel linkages (yellow) connecting the distant phalanx with the driven middle phalanx.

It was left out because it forms a closed-loop mechanism that the Unified Robot Description Format (URDF) does not natively support. URDF is an XML-based description language widely used in robotics to specify a robot's kinematic and dynamic properties, including joints, links, and sensor configurations. However, URDF supports only tree-structured kinematic chains and does not represent closed-loop mechanisms such as the four-bar linkage in the Bionic Hand. As a workaround, URDF provides the mimic joint feature. In this approach, one joint is defined as dependent on another, its position expressed as a linear function of the parent joint's position, as shown in Equation 1.

$$q_{distal} = \alpha q_{middle} + \beta \tag{1}$$

Where α is a scaling factor and β an offset. Since four-bar linkages in robotic fingers constrain the distal joint to a single degree of freedom (DoF) [21], and the motion of the distal phalanx in the Bionic Arm is a linear function of the middle phalanx, the mimic joint feature can be used as an approximation. In this implementation, the distal phalanx joints were set to mimic the middle phalanx joints with the appropriate scaling factor, effectively reproducing the closed-loop motion in simulation. The assembly can then be exported from SolidWorks using the ROS add-on **sw_urdf_exporter**. The resulting URDF was then imported as a Universal Scene Description (USD) into NVIDIA Isaac Sim for simulation.

Inside Isaac Sim, further properties of the joints and linkages can be configured. Since the connection to the fingertips in the Bionic Hand prototype is established with a stiff metal rod instead of compliant tendons (Fig. 1), the mimic-follow behavior requires a more detailed description. Behind the mimic tag, Isaac Sim implements a standard Proportional–Derivative (PD) controller, where the stiffness parameter corresponds to the proportional gain (P-gain) and the damping parameter corresponds to the derivative gain (D-gain) [22]. This formulation is appropriate for tendon-driven linkages, as it allows the dependent joint to give in and lag slightly behind during movement. Since most dexterous robotic hands, such as the Shadow Hand, employ tendon-driven actuation, the PD-based mimic joint is typically sufficient [7]. In contrast, the Bionic Hand prototype uses a rigid rod, linear motor mechanism, where each finger is directly actuated along a single axis. For such stiff kinematics, a conventional PD controller is inadequate, since the mechanism requires an instantaneous response of the driven linkage. To approximate rigid coupling in simulation, the stiffness must be set to a very high (near-infinite) value, while damping is set to zero. This ensures that mimic joints behave as stiff connections, causing the fingertips to follow the driving joint without delay.

However, to enable zero-shot transfer from simulation to the real world, it is also necessary to simulate the backlash introduced by the drive mechanism as accurately as possible. In this case, a Proportional–Threshold (PT) controller would be more appropriate than a PD controller [23], [24]. A PT controller is a type of non-linear control system. Unlike a standard proportional controller, that applies a corrective force directly proportional to the position error, a PT controller incorporates a dead zone (threshold). Within this dead zone, the controller output remains zero, meaning no corrective action is applied. Only when the error exceeds this predefined threshold does the controller activate, applying a force proportional to the remaining error [25]. To overcome this issue, empirical values for the P- and D-gains were set to approximate the backlash observed on each finger.

3.1 Reinforcement Learning Configuration

Isaac Sim was used with the PhysX backend to simulate the physics in the environment, while Isaac Lab was employed to manage the RL tasks. For training, the RL library **skrl** [26] was used together with a PPO agent [6]. The only difference in training between the Shadow Hand and the Bionic Hand was the size and depth of the neural network policy. The Shadow Hand was trained with a four-layer fully connected network of sizes [512, 512, 256, 128], whereas the Bionic Hand employed a deeper five-layer network of sizes [1024, 1024, 512, 256, 128]. The larger architecture was chosen to better handle the increased complexity of the Bionic Hand, which included nonlinearities and backlash effects in its joints, making the policy optimization problem more challenging [27], [28]. The observation space consisted of 120 dimensions, including joint positions and velocities, the cube's position, orientation, and velocity, the goal position and orientation, fingertip positions and velocities, as well as the previous action values. The action space, in contrast, was limited to seven dimensions, corresponding to the commanded positions of each actuated DoF. A summary of the observation and action space is provided in Table 1.

Space	Description
Observations (120D)	Joint positions and velocities; Cube position, orientation, velocity; Goal position and orientation; Fingertip positions and velocities; Previous actions
Actions (7D)	Target positions for each DoF

Table 1. Observation and Action Space Configuration

Each training run consisted of 80,000 simulation steps with 8192 parallel environments. To promote robustness and enable zero-shot transfer from simulation to the real world, domain randomization was applied [12], [29]–[31]. For each environment instance, random values were sampled for the following parameters:

- static and dynamic friction between cube and hand,
- joint stiffness and damping values,
- upper and lower joint limits,
- mass distribution of the cube,
- gravity vector applied to the scene.

Parameter	Value
Distance reward scale	−10.0
Rotation reward scale	1.0
Action penalty scale	−0.001
Goal bonus	250
Success tolerance (rad)	0.05
Fall penalty	0
Fall distance	0.24
Velocity obs. scale	0.2
Force/torque obs. scale	10.0
Averaging factor (α)	0.1
Action moving average	0.3

Table 2. Reward Function Parameters

The reward function was designed to encourage successful object manipulation. The most important components were:

- Rotation reward: encouraging alignment between cube and goal orientation.
- Action penalty: discouraging unnecessarily large or unstable joint actions.
- Goal bonus: rewarding successful alignment within tolerance.

Additional shaping terms (e.g., velocity scaling, force/torque scaling, and averaging factors) were tuned empirically to balance stability and exploration. The key reward scaling parameters are summarized in Table 2.

4. Results

To evaluate the performance of each agent, several parameters were tracked across episodes to compare the different configurations. In particular, the number of successful trials and dropped episodes was recorded for each run. Each episode consisted of 600 steps and was terminated either when the object was dropped or when the maximum step count was reached. Episodes that did not result in a success or a drop were classified as other.

A distinction was made when classifying cases in which both the success and drop conditions were triggered. For the analysis of episode outcomes, any episode in which the cube was successfully reoriented at least once was counted as a success. In contrast, when evaluating reward values or episode length, such cases were labeled as drops. In addition, the rotational error was computed and monitored throughout each episode, along with the corresponding return values. To ensure stability of the simulation and avoid introducing artifacts, the training of the Bionic Hand in its standard configuration, as presented in [19], was conducted in stages. The process began with the kinematic model and joint limits, while actuator dynamics were left unconstrained. In the next step, stiffness and damping values were explicitly defined for the joints. Subsequently, joint velocity limits were enforced to better approximate the physical system. Finally, simulations were repeated with backlash enabled on the joints and domain randomization applied to capture additional mechanical characteristics of the real hardware. With this staged approach, the standard configuration was able to sequentially solve the cube rotation task. However, once actuator speed limits were introduced, the same level of performance was no longer achievable. It was observed that the agent initially relied on rapid, momentum-based reorientation motions, a strategy that became infeasible once actuator velocity limits were enforced. Training the 7-DoF Bionic Hand with all limitations and domain randomization enabled proved unsuccessful without further modifications. The system appeared to lack the ability to maneuver the cube into more favorable positions within the workspace of each finger, which prevented complex reorientations. As a result, the agent became trapped in a reward plateau. Instead of attempting reorientation, which often led to dropping the cube, it learned to balance the cube on the palm to maximize reward. To address this, the wrist rotation was extended to include an additional 90° of supination beyond the typical human anatomical range, resulting in a full 270° range of motion. This modification was motivated by observations from training the 8-DoF version of the Bionic Hand that also included dorsiflexion and palmar flexion of the wrist. In that case, although the added DoFs enabled more varied strategies, the limited supination range still caused difficulties: depending on the randomized initial pose and velocity, the cube often rolled toward the hypothenar region, from which recovery was impossible. Consequently, the agent adopted an alternative strategy of balancing the cube on the back of the thumb, where the available 90° of supination could be exploited to stabilize, recover, and even reorient the cube.

4.1 Bionic Hand (7-DoF)

The Bionic Hand was also evaluated under full physical constraints, including domain randomization, joint backlash, and velocity limits. Under these conditions, the task could not be solved. In rare cases, when the randomized initial state was favorable, the hand was able to balance the cube on its palm. However, complex reorientation maneuvers were not attempted, as the limited actuation strategies provided few options for moving the cube into configurations that enabled further manipulation. As illustrated in Figure 2, the system converged to a steady state after approximately 80,000 training timesteps, with no further manipulations attempted. Based on these results, it can be concluded that the initialization of the cube led to frequent early drops before the hand could react. This effect was reinforced by the velocity limits and by initializing the robotic hand in a random configuration. Without the additional supination range, the palm's orientation was often misaligned with the cube's initial dropping velocity vector, causing deflections on impact and leading to premature episode termination. Figure 3 shows that in 80 out of 200 episodes the cube remained balanced on the palm. Of the 200 episodes, eight were registered as successful completions. However, seven of these eight cases were immediately followed by a drop, indicating that they were likely favorable initializations rather than genuine solves. This interpretation is supported by the time-to-success distributions in Figure 4, where most successes occurred unrealistically early. Only one episode reached completion at around 100 steps, which aligns more closely with the mean time-to-success observed in the extended supination configuration. Overall, these results demonstrate that without the additional supination range, the Bionic Hand was unable to reliably achieve stable reorientation.

Fig. 2. Steady state of the 7-DoF Bionic Hand after 80,000 timesteps of training, where the cube remains balanced on the palm without further manipulation.

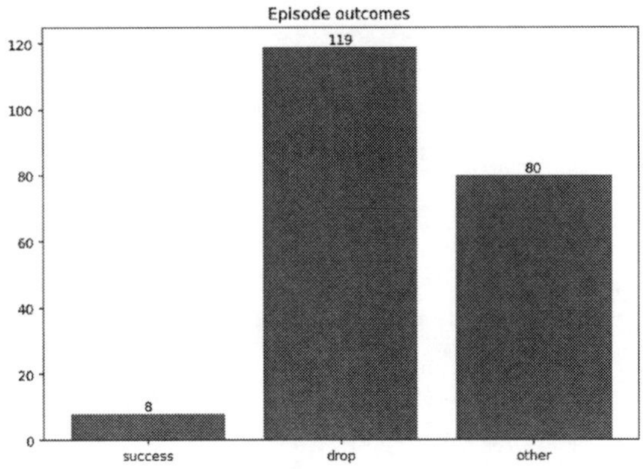

Fig. 3. Outcome distribution for 200 evaluation episodes of the 7-DoF Bionic Hand under full constraints.

4.2 Bionic Hand (7-DoF with Extended Supination)

The reorientation strategy of the 7-DoF Bionic Hand with Extended Supination relied strongly on extrinsic dexterity [8]–[10]. Whenever possible, the cube was allowed to rest on the palm while the distal phalanx of the thumb was positioned against the proximal phalanx of either the middle or index finger, creating a confined space that prevented the cube from slipping (Figure 9a). Within this configuration, wrist flicks combined with coordinated finger positioning enabled clockwise or counterclockwise rotations of the cube. For flipping maneuvers, the thumb in conjunction with the index and middle fingers was used to push the cube against the metacarpal region of the thumb, after which a simultaneous wrist flick and thumb swivel produced the rotation. Once a coarse orientation was achieved, the hand attempted fine alignment by forming a firmer grasp [32]. The configuration illustrated in Figure 9a represents the firmest grasp achievable by the Bionic Hand, where the cube is lifted from the palm and stabilized in the target orientation.

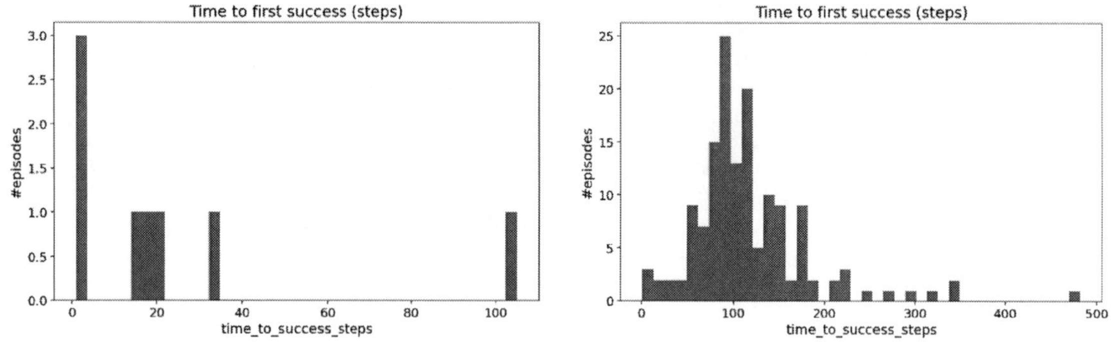

Fig. 4. Comparison of time-to-success distributions. Left: 7-DoF Bionic Hand under full constraints. Right: 7-DoF Bionic Hand with extended supination.

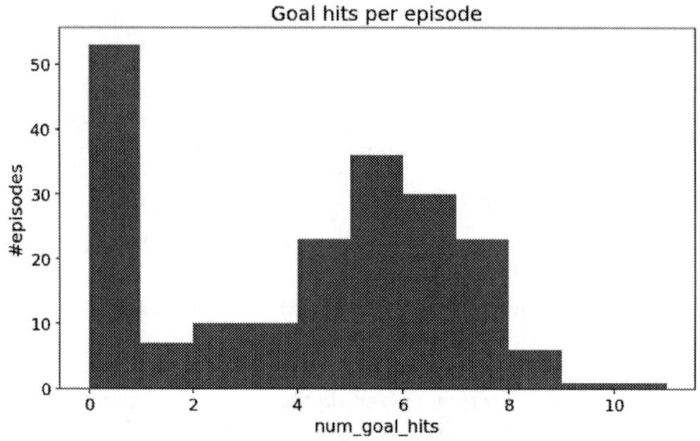

Fig. 5. Distribution of goal hits per episode for the 7-DoF Bionic Hand with extended supination.

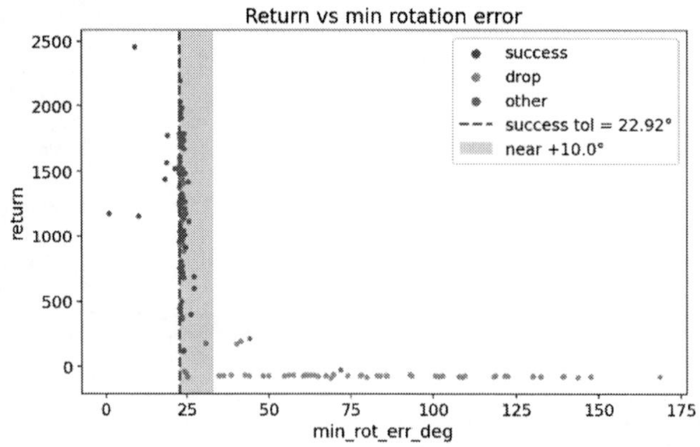

Fig. 6. Return as a function of minimum rotation error per episode

In contrast, the Shadow Hand relied less on palm support and more on direct finger coordination. Its higher number of DoFs enabled stable in-hand reorientation without the need to exploit extrinsic forces to the same extent, allowing for more consistent manipulation strategies [7] [1]. Figure 5 shows the distribution of goal hits per episode during evaluation. While many episodes achieved zero goal hits, most successful episodes clustered around 4 to 8 hits, with a peak at 5 to 6. This indicates that the agent, when successful, was able to repeatedly reorient the cube within the same episode. The relationship between return and minimum rotation error per episode is illustrated in Figure 6. Successful episodes (blue) dominate the region close to the defined success tolerance (22.9°). Dropped episodes (orange) are associated with high rotation errors and consistently low returns, meaning that fewer drops occur after the task has been solved once. This confirms that the return function strongly correlates with orientation accuracy, and that the additional supination increased the likelihood of stabilizing the cube within the tolerance region.

Finally, Figure 7 shows the distribution of minimum rotation errors across episodes. Most successful runs achieved errors very close to the success threshold, with a dense concentration between 20° and 25°. A smaller fraction of episodes resulted in much higher errors (> 40°), typically corresponding to dropped cubes. The shaded region in the plot highlights the tolerance and near-miss zones, confirming that the agent was able to reliably achieve cube orientations near the success threshold, though occasional instability remained. Overall, the introduction of an additional 90° of supination improved the agent's ability to stabilize the cube and avoid manipulation of dead-ends. The results suggest that increased wrist mobility is a critical factor for achieving reliable in-hand manipulation, as it reduces the dependence on momentum-driven reorientation strategies.

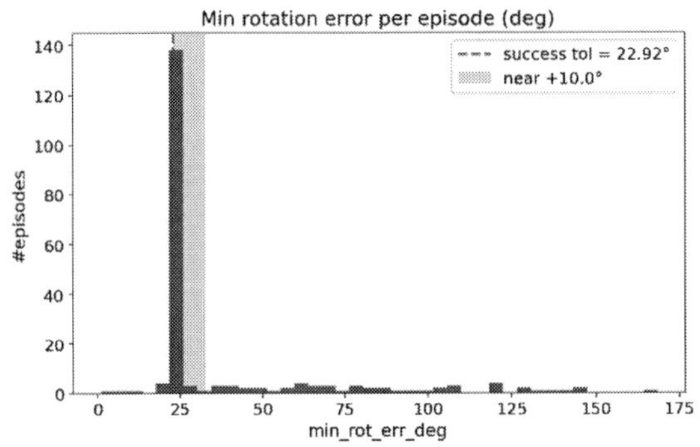

Fig. 7. Distribution of minimum rotation error per episode.

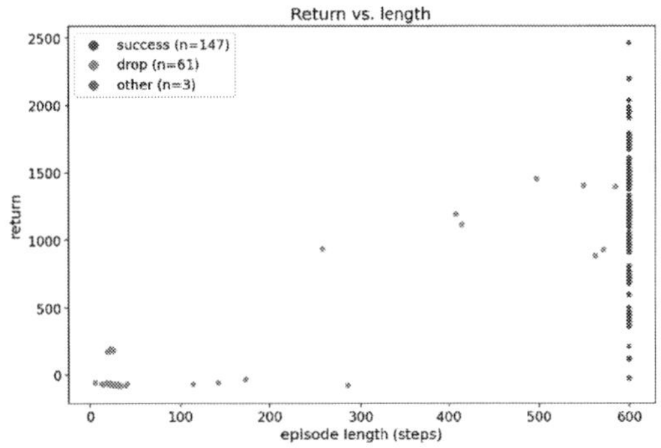

Fig. 8. Return as a function of episode length. Instances where the cube converged close to the target orientation but failed to complete the reorientation task are highlighted in green.

As illustrated in Figure 8, most cube drops occurred early in the episodes, suggesting that the initialization of cube states could be further optimized for the Bionic Hand. Additionally, several episodes showed instances where the agent achieved initial success but subsequently dropped the cube before completion. Across all evaluated episodes, 147 were classified as successful, 61 as drops, and 3 as near-misses.

a) 7-DoF with extended supination Bionic Hand failing to reach the goal

b) 7-DoF with extended supination Bionic Hand failing to reach the goal due to domain randomisation.

Fig. 9. Performance of the 7-DoF extended supination Bionic Hand in two scenarios.

As shown in Figure 9, the agent occasionally came very close to fully solving the cube reorientation task. However, in the example on the right (Figure 9b), the full 90° supination range had already been utilized, constraining the agent in a configuration from which further reorientation was not possible. Careful inspection reveals that the achieved range does not perfectly align with the intended 90° supination. This discrepancy arises from domain randomization: during training, joint limits were randomized across environments, and the agent learned behaviors under slightly different constraints. When evaluated in this instance, the learned strategy, effective in an environment with looser limited joint parameters, led the agent to attempt a rotation that exceeded the randomized limit of the current environment. As a result, the policy was trapped at the boundary of the joint's feasible range, unable to complete the reorientation.

4.3 Bionic Hand (8-DoF with Wrist Flexion/Extension)

Even in this configuration, without enabling the extension of the supination, the results were still lacking. Since this version was designed as an exaggerated configuration of the Bionic Hand, the extra 90° of supination was also unlocked. With this additional freedom, the agent focused on solving the task in the most efficient way possible. The added axis allowing for flexion/extension of the wrist, together with the extended supination, provided relatively fast actuation speeds (180°/s for flexion/extension and 230°/s for supination, compared to only 80°/s for the fingers). As a result, the policy relied mainly on wrist motions to orient the cube, while the fingers were mainly used for stabilization and small corrective adjustments.

Fig. 10. 8-DoF Bionic Hand with wrist flexion/extension using the fast axes to orient the cube and keeping it in place with finger joints

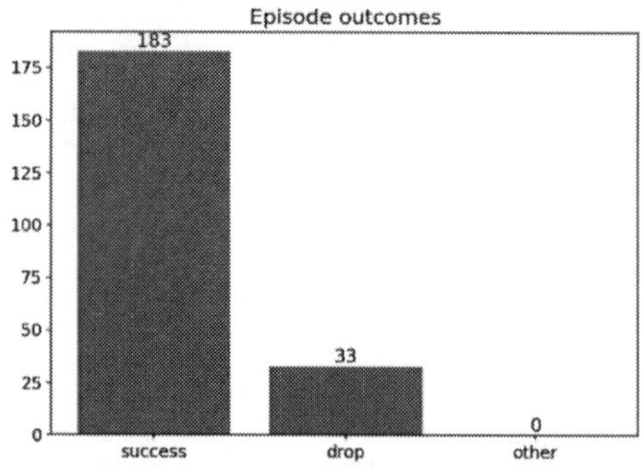

Fig. 11. Outcome distribution for the 8-DoF Bionic Hand with wrist flexion/extension

The evaluation outcomes are summarized in Figure 11. Out of all episodes, 183 were classified as successful, while 33 resulted in drops. The distribution of minimum rotation errors per episode is shown in Figure 12. Most successful trials achieved errors tightly clustered around the 22.9° tolerance, with a sharp peak between 20° and 25°. A smaller fraction of episodes produced much higher errors (> 40°), which typically corresponded to drops. Figure 13 shows the time-to first-success distribution. Most successful episodes converged within 50–100 steps, indicating that the additional wrist DoFs enabled rapid alignment strategies. Only a small number of cases required more than 200 steps, typically when the cube was initialized in an unfavorable pose.

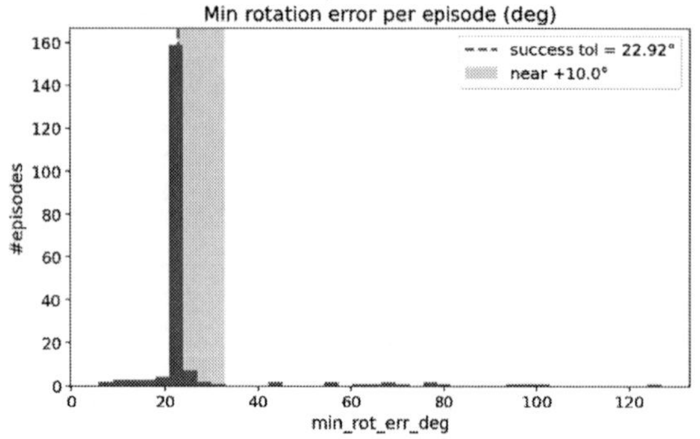

Fig. 12. Distribution of minimum rotation errors for the 8-DoF Bionic Hand.

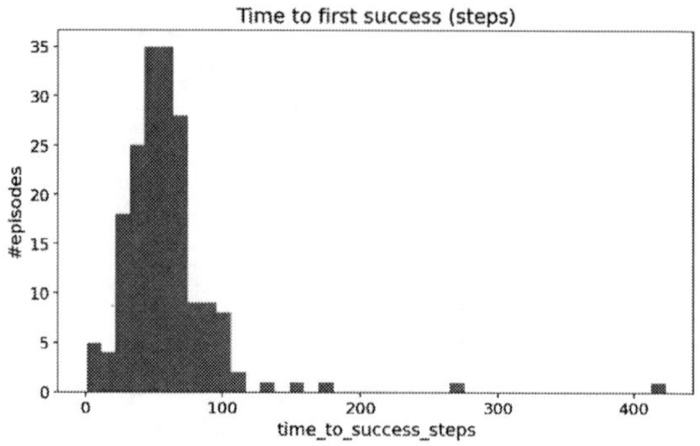

Fig. 13. Time to first success in the 8-DoF Bionic Hand.

Finally, Figure 14 shows return as a function of episode length. Most successful episodes cluster near the maximum length, reflecting policies that achieved repeated reorientations. In contrast, drops typically occurred earlier and yielded lower returns, whereas later drops were associated with higher returns, indicating that the agent often lost the cube only after several successful maneuvers.

Overall, the 8-DoF configuration demonstrated a clear improvement in stability and efficiency compared to the 7-DoF versions. The fast wrist axes dominated the manipulation strategy, enabling rapid coarse reorientations, while the fingers provided fine control.

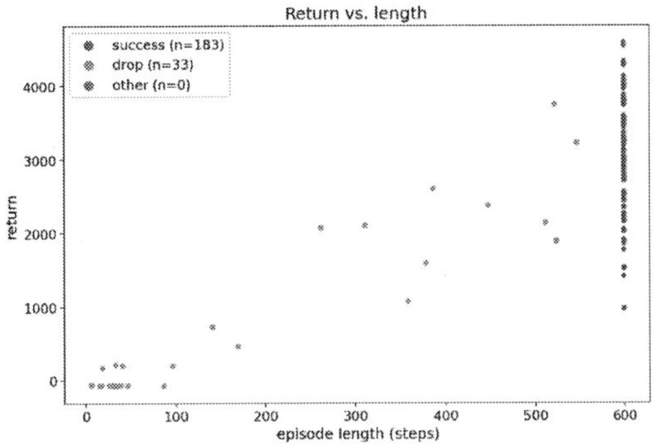

Fig. 14. Return versus episode length for the 8-DoF Bionic Hand

4.4 Shadow Hand (24-DoF)

Figure 15 presents the distribution of goal hits per episode during evaluation. In contrast to the Bionic Hand, the Shadow Hand achieved a broader spread of goal hits, with a significant portion of episodes exceeding ten successful reorientations. This suggests that the higher number of DoFs allow for more frequent and consistent in-hand reorientation cycles within a single episode.

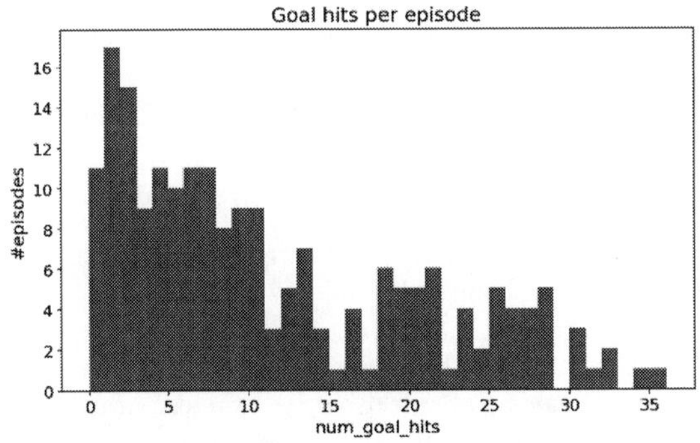

Fig. 15. Distribution of goal hits per episode for the Shadow Hand.

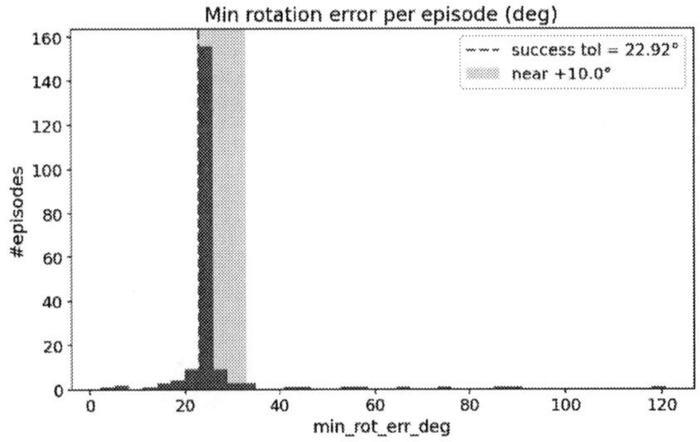

Fig. 16. Distribution of minimum rotation error per episode for the Shadow Hand.

The distribution of minimum rotation errors (Figure 16) further supports this observation. Most successful episodes achieved errors tightly clustered near the success threshold, with relatively few cases exceeding 40°. This concentration highlights the Shadow Hand's ability to consistently bring the cube into near-goal orientations without relying on boundary strategies.

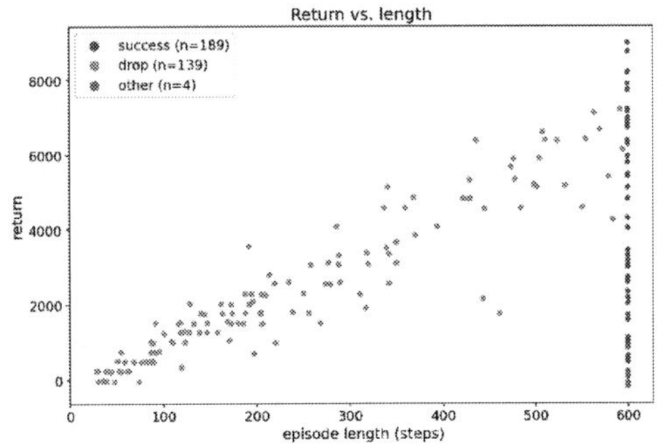

Fig. 17. Return as a function of episode length for the Shadow Hand. Episodes that converged to kinematic deadlock states are highlighted in green.

Finally, Figure 17 illustrates the relationship between return and episode length. Most episodes terminated with either a clear success or a failure, suggesting that the Shadow Hand policy rarely exhibited unstable or oscillatory behavior. In comparison to the Bionic Hand, the distribution of dropped cases is more dispersed, with only a small fraction associated with low returns. Drops occurring later in the episode correspond to higher returns, indicating that the agent often completed several successful reorientations before ultimately losing the cube.

This behavior was observed only when using the modified inhand_manipulation_env.py in combination with the adapted play.py script that enabled explicit tracking of successes. In contrast, when evaluated with the original environment configuration used for training and the un-edited play.py, drops were virtually absent but could not be systematically recorded. Overall, the Shadow Hand achieved more frequent and stable reorientation cycles than the Bionic Hand, with tighter rotation error distributions and fewer near-miss outcomes. Some dropped episodes emerged, however these typically followed several successful reorientations. The results highlight that higher degrees of freedom increase robustness in cube stabilization but also expand the policy search space, making training outcomes more sensitive to reward design and exploration.

5. Discussion

The results indicate that the primary bottleneck is mechanical rather than algorithmic. The 7-DoF baseline plateaued under realistic actuator limits, favoring balance-and-hold over true reorientation. The findings suggest that wrist mobility plays a central role in enabling stable in-hand manipulation while minimizing reliance on high-velocity, impulse-based motions. As expected for lower intrinsic dexterity, the bionic variants leaned on extrinsic contacts (palm and thumb–finger corridor), while the Shadow Hand required less environmental support.

Simulation fidelity limits remain: linear mimic with very high P and near-zero D suppresses backlash, some initializations left little corrective margin and broad domain randomization ranges can teach strategies that overfit to joint-limit “slack.” Although a deeper network was used for the bionic hand to account for nonlinearities, training both hands with the same architecture would rule out network size as a factor in the performance gap. Practically, modest and well-placed DoF/range changes can deliver large gains. Fast orthogonal wrist axes plus finger stabilization proved especially effective. Further steps include modelling the four-bar as parallel kinematics with a dead-zone (backlash) controller in USD, switch from simulator pose to vision estimates for sim-to-real, diversify objects/contacts, and report stability-oriented metrics (time-in-tolerance, multi-success streaks, actuation costs), culminating in zero-shot hardware validation.

6. Summary and Outlook

In this work, a custom bionic hand was evaluated to determine whether it can be trained to perform dexterous in-hand cube manipulation using RL. A reproducible pipeline was established in Isaac Lab: CAD models were exported to URDF (with linear mimic couplings), converted to USD, and used to train PPO agents via skrl. Since URDF cannot represent closed kinematic loops, the distal–middle phalanx coupling was approximated with linear mimic joints. Within Isaac Sim, this induces a PD controller with stiffness and damping mapped to P /D gains. For stiff rod-driven fingers, near-infinite stiffness and minimal damping were required to emulate rigid linkages; however, this configuration removes backlash. As a result, a Proportional–Threshold (PT) controller with a dead-zone was identified as a promising alternative for future work to better capture transmission free-play. From a learning perspective, the 7-DoF baseline could balance the cube but failed to consistently reorient it once actuator speed limits were imposed. In the absence of actuator constraints, the policies exploited high-velocity, momentum-transfer strategies, but performance plateaued once these were restricted. Extending wrist supination by 90° (to 270° total) enabled recovery postures and repeatable reorientation cycles, improving success rates and reducing kinematic deadlock situations. Adding wrist flexion/extension (resulting in an 8-DoF hand) further enhanced stability and efficiency.

For comparison, the Shadow Hand (24-DoF) achieved the tightest rotation-error distributions and frequent multi-success episodes, though it occasionally exhibited late drops under the evaluation protocol that explicitly logged successes. Future work can improve both the modeling and evaluation of the bionic hand. On the modeling side, the four-bar linkage with a stiff rod could be represented as parallel kinematics directly in USD, therefore capturing contact interactions more accurately. A custom backlash-controller may also be integrated to more accurately reproduce the true mechanical free-play. On the learning side, training could be extended with convolutional neural networks that estimate the cube pose from virtual camera input, rather than relying on ground-truth data from Isaac Sim, to facilitate sim-to-real transfer. Beyond single-object tasks, future experiments could incorporate multiple objects, varied masses, and different friction properties. Reporting standardized metrics, such as success rate, median rotation error, time-to-success, multi-success streaks, and actuation costs (energy/torque), would further strengthen comparability across studies. Finally, validation on real hardware with zero-shot transfer would provide the ultimate benchmark for the proposed approach. Overall, the results show that extending supination significantly improved success and recovery, while adding flexion/extension increased efficiency. The 7-DoF hand relied heavily on external forces such as gravity to achieve reorientation, whereas the Shadow Hand required less environmental support.

7. References

[1] G. Li, R. Wang, P. Xu, Q. Ye, and J. Ch<2en, "The developments and challenges towards dexterous and embodied robotic manipulation: A survey," 2025. [Online]. Available: https://arxiv.org/abs/2507.11840

[2] C. Yu and P. Wang, "Dexterous manipulation for multi-fingered robotic hands with reinforcement learning: A review," Frontiers in Neurorobotics, vol. Volume 16 - 2022, 2022. [Online]. Available: https://www.frontiersin.org/journals/neurorobotics/articles/10.3389/fnbot.2022.861825

[3] V. Kumar, Y. Tassa, T. Erez, and E. Todorov, "Real-time behaviour synthesis for dynamic hand-manipulation," in 2014 IEEE International Conference on Robotics and Automation (ICRA), 2014, pp. 6808–6815.

[4] I. Mordatch, Z. Popovi´c, and E. Todorov, "Contact-invariant optimization for hand manipulation," in ACM SIGGRAPH/Eurographics Symposium on Computer Animation. Eurographics Association, 2012, pp. 137–144.

[5] I. Mordatch, E. Todorov, and Z. Popovi´c, "Discovery of complex behaviors through contact-invariant optimization," ACM Trans. Graph.,vol. 31, no. 4, Jul. 2012. [Online]. Available: https://doi.org/10.1145/2185520.2185539

[6] J. Schulman, F. Wolski, P. Dhariwal, A. Radford, and O. Klimov, "Proximal policy optimization algorithms," CoRR, vol. abs/1707.06347, 2017. [Online]. Available: http://arxiv.org/abs/1707.06347

[7] OpenAI, M. Andrychowicz, B. Baker, M. Chociej, R. Jozefowicz, B. McGrew, J. Pachocki, A. Petron, M. Plappert, G. Powell, A. Ray, J. Schneider, S. Sidor, J. Tobin, P. Welinder, L. Weng, and W. Zaremba, "Learning dexterous in-hand manipulation," 2019. [Online]. Available: https://arxiv.org/abs/1808.00177

[8] N. C. Dafle, A. Rodriguez, R. Paolini, B. Tang, S. S. Srinivasa, M. Erdmann, M. T. Mason, I. Lundberg, H. Staab, and T. Fuhlbrigge, "Extrinsic dexterity: In-hand manipulation with external forces," in 2014 IEEE International Conference on Robotics and Automation (ICRA), 2014, pp. 1578–1585.

[9] W. Zhou and D. Held, "Learning to grasp the ungraspable with emergent extrinsic dexterity," 2022. [Online]. Available: https://arxiv.org/abs/2211.01500

[10] C. Ma, H. Yang, H. Zhang, Z. Liu, C. Zhao, J. Tang, X. Lan, and N. Zheng, "Dexdiff: Towards extrinsic dexterity manipulation of ungraspable objects in unrestricted environments," 2024. [Online]. Available: https://arxiv.org/abs/2409.05493

[11] S.-M. Yang, M. Magnusson, J. A. Stork, and T. Stoyanov, "Learning extrinsic dexterity with parameterized manipulation primitives," 2024.

[12] J. Tobin, R. Fong, A. Ray, J. Schneider, W. Zaremba, and P. Abbeel, "Domain randomization for transferring deep neural networks from simulation to the real world," 2017. [Online]. Available: https://arxiv.org/abs/1703.06907

[13] A. Handa, A. Allshire, V. Makoviychuk, A. Petrenko, R. Singh, J. Liu, D. Makoviichuk, K. Van Wyk, A. Zhurkevich, B. Sundaralingam, Y. Narang, J.-F. Lafleche, D. Fox, and G. State, "Dextreme: Transfer of agile in-hand manipulation from simulation to reality," 2024.

[14] X. Chen, J. Hu, C. Jin, L. Li, and L. Wang, "Understanding domain randomization for sim-to-real transfer," 2022, publisher Copyright: © 2022 ICLR 2022 - 10th International Conference on Learning Represen-tationss. All rights reserved.; 10th International Conference on Learning Representations, ICLR 2022 ; Conference date: 25-04-2022 Through 29-04-2022.

[15] Y. Toshimitsu, B. Forrai, B. G. Cangan, U. Steger, M. Knecht, S. Weirich, and R. K. Katzschmann, "Getting the ball rolling: Learning a dexterous policy for a biomimetic tendon-driven hand with rolling contact joints," in Humanoids, 2023.

[16] B. Huang, Y. Chen, T. Wang, Y. Qin, Y. Yang, N. Atanasov, and X. Wang, "Dynamic handover: Throw and catch with bimanual hands," 2023.

[17] T. Lin, K. Sachdev, L. an, J. Malik, and Y. Zhu, "Sim-to-real reinforcement learning for vision-based dexterous manipulation on humanoids," 2025, project: https://toruowo.github.io/recipe.

[18] M. Bauza, J. E. Chen, V. Dalibard, N. Gileadi, R. Hafner, M. F. Martins, J. Moore, R. Pevceviciute, A. Laurens, D. Rao, M. Zambelli, M. Riedmiller, J. Scholz, K. Bousmalis, F. Nori, and N. Heess, "Demostart: Demonstration-led auto-curriculum applied to sim-to-real with multifingered robots," 2024.

[19] F. Dannereder and P. H. Pachschwoll, "Development of a 3d-printed bionic hand with muscle- and force control," in Austrian Robotics Workshop 2018, 2018, pp. 59–66.

[20] Shadow Robot Company, "Shadow Robot Dexterous Hand," https://www.shadowrobot.com/products/dexterous-hand/, 2005, accessed: [18.09.2025].

[21] M. West, "Four-bar linkages - dynamics," https://dynref.engr.illinois.edu/aml.html, 2015, accessed on 26.06.2025. NVIDIA. (2025) Tuning gains. NVIDIA Corporation. [Online]. Available: https://docs.isaacsim.omniverse.nvidia.com/latest/robot_setup/ext_isaacsim_robot_setup_gain_tuner.html

[22] R. Bruns, J. Diepstraten, X. Schuurbiers, and J. Wouters, Motion control of systems with backlash, ser. DCT rapporten. Technische Universiteit Eindhoven, 2006, dCT 2006.075.

[23] F. Cursi, W. Bai, E. Yeatman, and P. Kormushev, "Model learning with backlash compensation for a tendon-driven surgical robot," IEEE Robotics and Automation Letters, vol. 7, pp. 1–8, 07 2022.

[24] J. V¨or¨os, "Modeling and identification of systems with backlash," Automatica, vol. 46, no. 2, pp. 369–374, 2010. A. Serrano-Mu ˜A±oz, D. Chrysostomou, S. B ˜A,gh, and N. Arana-Arexolaleiba, "skrl: Modular and flexible library for reinforcement learning," Journal of Machine Learning Research, vol. 24, no. 254, pp. 1–9, 2023. [Online]. Available: http://jmlr.org/papers/v24/23-0112.html

[25] S. Levine, C. Finn, T. Darrell, and P. Abbeel, "End-to-end training of deep visuomotor policies," 2016. [Online]. Available: https://arxiv.org/abs/1504.00702

[26] V. Mnih, K. Kavukcuoglu, D. Silver, A. A. Rusu, J. Veness, M. G. Bellemare, A. Graves, M. Riedmiller, A. K. Fidjeland, G. Ostrovski, S. Petersen, C. Beattie, A. Sadik, I. Antonoglou, H. King, D. Kumaran, D. Wierstra, S. Legg, and D. Hassabis, "Human-level control through deep reinforcement learning," Nature, vol. 518, no. 7540, pp. 529–533, 2015. [Online]. Available: https://doi.org/10.1038/nature14236

[27] T. Dai, K. Arulkumaran, T. Gerbert, S. Tukra, F. Behbahani, and A. A. Bharath, "Analysing deep reinforcement learning agents trained with domain randomisation," Neurocomputing, vol. 493, pp. 143–165, 2022. [Online]. Available: https://www.sciencedirect.com/science/article/pii/S0925231222003708

[28] M. Ma, H. Li, G. Hu, S. Liu, and D. Zhao, "Comparison of different domain randomization methods for policy transfer in reinforcement learning," in 2023 IEEE 12th Data Driven Control and Learning Systems Conference (DDCLS), 2023, pp. 1818–1823.

[29] A. Shakerimov, T. Alizadeh, and H. A. Varol, "Efficient sim-to-real transfer in reinforcement learning through domain randomization and domain adaptation," IEEE Access, vol. 11, pp. 136 809–136 824, 2023.

[30] C. Pehoski, A. Henderson, and L. Tickle-Degnen, "In-hand manipulation in young children: rotation of an object in the fingers," American Journal of Occupational Therapy, vol. 51, no. 7, pp. 544–552, 1997.

DOI: 10.2507/36th.daaam.proceedings.044

TECHNOLOGIES FOR BUILDING COMPLEX SENSOR SYSTEMS FOR COLLABORATIVE SERVICE ROBOTS

Valentin Pryanichnikov & Viatcheslav Iastrebov

This Publication has to be referred as: Pryanichnickov, V[alentin] & Iastrebov, V[iatcheslav] (2025). Technologies for Building Complex Sensor Systems for Collaborative Service Robots, Proceedings of the 36th DAAAM International Symposium, pp.0317-0320, B. Katalinic (Ed.), Published by DAAAM International, ISBN 978-3-902734-47-1, ISSN 1726-9679, Vienna, Austria
DOI: 10.2507/36th.daaam.proceedings.044

Abstract

In this paper, we study the issues of developing an interface, that ensures optimized operator interaction with robotic systems. The interface should have a high degree of intuitiveness and provide the user with access to up-to-date information about the current state of the system through feedback sensors. These sensors record parameters such as torques, forces, and positions, which allows monitoring the relative position of the robot's and operator's work elements. Special attention is paid to the visualization of algorithms designed to identify potential drive overloads and changes in technological parameters. Software and hardware have been developed to provide a dialogue with collaborative robots using laser graphics, taking into account the individual needs of the operator and realizing the recognition of sign language and voice input. This is aimed at freeing the operator's hands from the need to enter instructions using the keyboard or remotes. The results of experimental studies aimed at indicating mutual intentions between the operator and robotic systems are presented.

Keywords: Mobile service collaborating robots; common video-streaming web-cam field; laser and immersive technologies; indication of intentions; synesthesia

1. Introduction

Modern robotic systems, that operate in cooperation with humans (known as collaborative robots or cobots), enable the combination of machine efficiency in handling repetitive tasks and human creativity in resolving ambiguous situations. The traditional safety approach for such complexes involves completely excluding the operator from the robot's workspace during its operation, which often hinders the full automation of processes requiring human intervention. A key objective is building effective collaboration where humans and machines share the same workspace. In this context, controlling parameters such as movement speed and applied force becomes critically important. Safety systems are built using machine vision, force-torque sensors, and specific control algorithms, that reduce the robot's speed and force upon detecting a human. Such solutions pave the way for automating new types of tasks, including assembly, quality control and maintenance. Significant attention is paid to designing user-friendly interaction interfaces.

They must be intuitive and provide the operator with data from sensors (torque, force, position) to assess mutual positioning with the robot. Visualizing the operation of algorithms, that identify actuator overloads and adjust system parameters is also necessary. The adaptive control systems being developed must flexibly adapt to changing conditions and individual user needs. Besides standard programming methods, interfaces based on gestures and voice commands are gaining popularity; their main goal is to free the user from using keyboards or buttons for input. This report describes our experiments on demonstrating the mutual intentions of the robot and the operator using laser projections.

2. Technologies for Collaborative Interaction Under Development

Our research focuses on integrating heterogeneous sensor systems, creating universal software [1–10], and expanding the service zone of robotic complexes through the use of dual manipulators [11–14]. We have developed a method for identifying intentions in the human-robot system based on a compact laser prototype, that projects pointer arrows, text prompts, or pictograms onto surfaces (Fig. 1, 2). This led to the concept of a unified information space based on webcam video streams for cobot-human interaction. If the operator uses wearable sensors (electromyography and/or inertial IMU sensors), it will enable the creation of algorithms, that dynamically coordinate mutual movements. Compared to known methods like Kinect, this approach offers the potential to account for the dynamic characteristics of human motion. A promising direction is cooperation in the "robot-operator" system through the indication of emotions, facial expressions and schematic drawings to demonstrate motivation, based on the principles of synesthesia. This research area is just beginning to be explored in collaborative robotics.

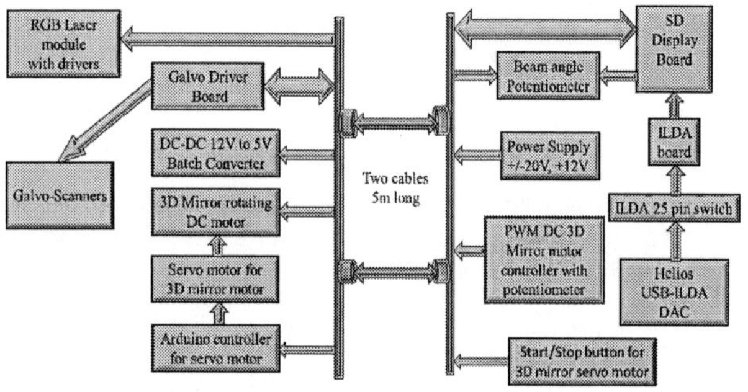

Fig. 1. Functional diagram. On the left is the scanning and emitting module, on the right is the interface and control.

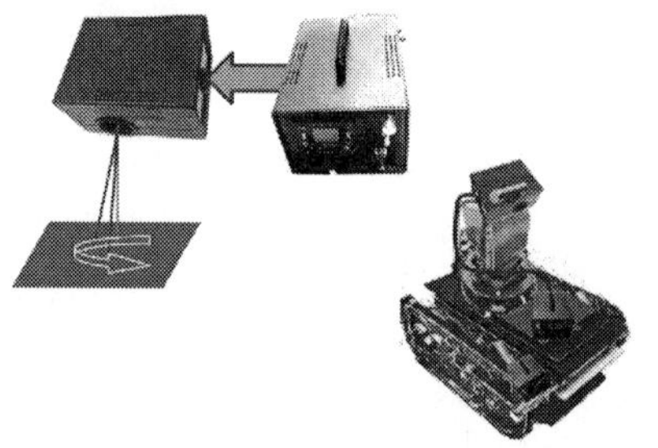

Fig. 2. Mock-up of a laser scanning communication system installed on the Cobot.

It should be noted, that using immersive technologies (AR/VR) often leads to increased operator fatigue, making "glasses-free" approaches more preferable. However, the issue of high cost for such devices and other challenges remain. The figure 3 shows a schematic example of human-cobot interaction through the ManyCam program using a webcam, stereo camera, Kinect 3D scanner, and virtual reality glasses.

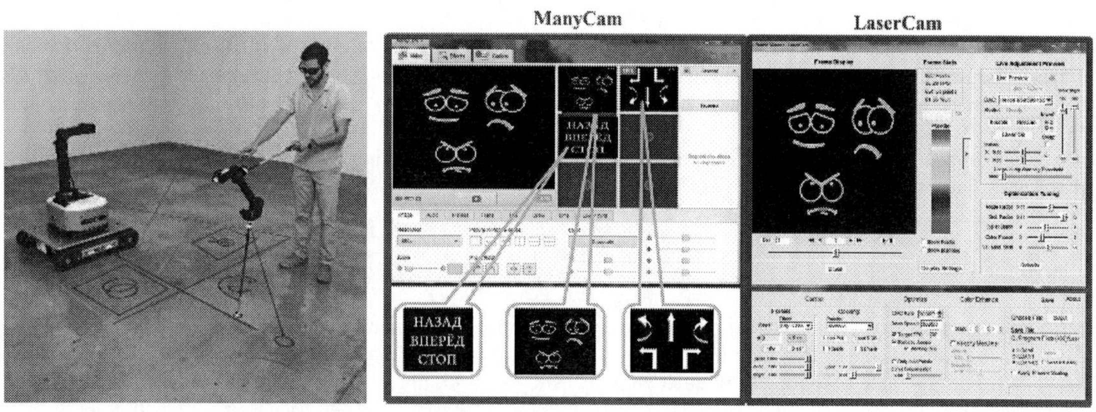

Fig. 3. An example of human-cobot interaction through the ManyCam program.

Simultaneously, the LaserCam application intercepts the video stream from the ManyCam application and controls the laser projector, displaying emotions. This is what we call the common a “video-streaming common webcam field” (Figure 4).

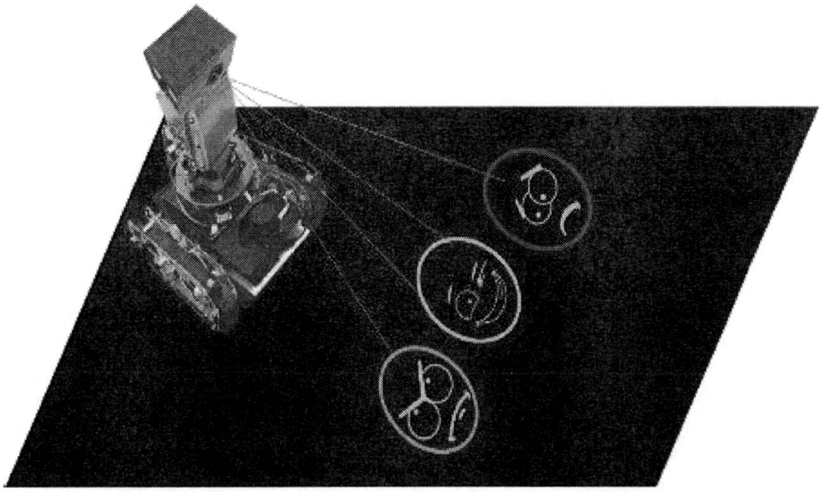

Fig. 4. Emotions laser projections via interaction between ManyCam and LaserCam applications

Developing mobile robots and their algorithms inevitably involves a lengthy debugging process, requiring the creation of integration software and a virtual development environment [7,10]. Simulators using virtual twins of robotic complexes for modelling sensory interaction with environmental objects have become widespread. They provide additional opportunities for debugging intelligent control systems and expanding functionality, including executing loosely defined tasks, orienting the robot using voice commands, moving objects based on indirect operator cues, and providing guidance or information about the equipment being interacted with. As part of the "Intelligent Robotronics" project, we developed robotic complexes for various purposes and implemented route-following algorithms using ArUco markers or external control scripts for logistics operations. Specifically, a universal sensor-control unit was created, that works with data from vision systems, lidars, ultrasonic rangefinders, odometers and features 6 PWM control channels at 30A each. The implementation is designed for use with an STM32-based microprocessor network and neural accelerators. Work is underway on force-torque sensing for dual-arm grippers. Tests are being conducted to assess the operator's ability to navigate using highly compressed images transmitted from the robotic complex, with simultaneous highlighting of target manipulation objects. Implementing such a system became possible on micro-PCs like Raspberry Pi 3/4 due to a new method of forming color clusters and histogram analysis for partial object recognition. Transmitting compressed images allows for a rate of 25 frames per second, even over RS-232 communication channels, due to significantly reduced data volume. A separate research direction involves using short-range microwave sensors to identify moving objects by comparing their spectra against background noise. For this purpose, specialized test benches were developed in our robotics lab to study multi-channel signals.

3. Conclusions

The adoption of collaborative robotics not only enhances productivity and quality, but also improves working conditions, reduces the risk of injury and makes work more productive and creative. Initial steps have been taken on this path towards deeper human-robot integration, creating methods for indicating mutual intentions using the laser, immersive, and other hardware-software technologies presented in this work. The discussed challenges of creating a software environment for developing sensor-control systems for mobile service robots are being actively addressed and applied in our developments [7,9,10,14]. The work was partially supported by the Ministry of Science and Higher Education of the Russian Federation (FFMN-2025-0022 project).

4. References

[1] Pryanichnikov V.Ye., Yastrebov V.V. Sposob svortki izobrazheniy dlya posleduyushchey identifikatsii ob"yektov servisnymi robotami / Zapis' i vosproizvedeniye ob"yomnykh izobrazheniy v kinematografe, nauke, obrazovanii, media i v drugikh oblastyakh: XVI yezhegodnaya nauchno-prakticheskoy konferentsiya. Materialy i doklady, chast' II // Pod obshchey redaktsiyey O.N. Rayeva. — Moskva : IPP «KUNA».(A method for convolving images for subsequent identification of objects by service robots), 2024 pp. 128-135, (UDC 681.5 BBK 32.965), ISBN 978-5-98547-148-9, https://www.rsuh.ru/education/minot/conf_3d-2024.php2024 S.128-135, (UDK 681.5 BBK 32.965), ISBN 978-5-98547-148-9, https://www.rsuh.ru/education/minot/conf_3d-2024.php

[2] Gost R 60.1.2.3-2021 /ISO/TS 15066:2016 Roboty i robototekhnicheskiye ustroystva. Trebovaniya bezopasnosti dlya robotov, rabotayushchikh sovmestno s chelovekom (Robots and robotic devices – Collaborative robots, IDT)

[3] Gost R 60.0.0.12– 2023 Ontologii robototekhniki.Ponyatiya i otnosheniya, opisyvayushchiye gruppovoye vzaimodeystviye (Concepts and relationships describing group interaction)

[4] Davydov O.I., Pryanichnikov V.Ye. Arkhitektura sistemy upravleniya mobil'nogo servisnogo robota. (Architecture of the control system of a mobile service robot) //Informatsionno-izmeritel'nyye i upravlyayushchiye sistemy, №7, t.13, 2015, str.41-50

[5] Davydov D.V., Eprikov S.R., Kirsanov K.B., Pryanichnikov V.E. (2017). Service Robots Integrating Software and Remote Reprogramming, Proceedings of the 28th DAAAM International Symposium, pp.1234-1240, B. Katalinic (Ed.), Published by DAAAM International, ISBN 978-3-902734-11-2, ISSN 1726-9679, Vienna, Austria DOI: 10.2507/28th.daaam.proceedings.172.

[6] Pryanichnikov V.Ye. Sovmeshcheniye fil'tratsii ul'trazvukovykh dannykh i programmnogo upravleniya lineynym privodom (Combination of ultrasonic data filtration and software control of a linear drive) / Pryanichnikov V.Ye.// Pribory+avtomatizatsiya. – 2008. – № 12. – S. 22–29

[7] V.Ye. Pryanichnikov, A.A. Aryskin, YU.S. Kolesov, A.YA. Ksenzenko, S.V. Kuvshinov, A.V. Plotnikov, M.D. Solov'yova, R.B. Tarasov, D.S. Telezhkin, D.K. Fedotov, K.V. Kharin, V.V. Yastrebov. Kompleks Bazovykh Sensorno-Upravlyayushchikh Sistem Mobil'nykh Servisnykh Robotov (A Complex of Basic Sensor-Control Systems of Mobile Service Robots.) - XVI Vserossiyskaya mul'tikonferentsiya po problemam upravleniya (MKPU-2023): materialy mul'tikonferentsii (g. Volgograd, 11–15 sentyabrya 2023 g.). V 4 t. T. 1. Robototekhnika i mekhatronika (RiM-2023) / redkol.: I. A. Kalyayev, V. G. Peshekhonov, S. YU. Zheltov [i dr.]. – Volgograd: VolgGTU, 2023. – 308 s., S.204-207, ISBN 978-5-9948-4703-9 ISBN 978-5-9948-4704-6 (T. 1) https://www.elibrary.ru/query_results.asp

[8] Pryanichnikov, V[alentin]; Mariia, S[oloveva]; Chernyshev, V[adim] & Kalinin, Y[aroslav] (2023). Analysis and Modernization of the Dual Manipulator of Service Robots, Proceedings of the 34th DAAAM International Symposium, pp.0324-0329, B. Katalinic (Ed.), Published by DAAAM International, ISBN 978-3-902734-41-9, ISSN 1726-9679, Vienna, Austria DOI: 10.2507/34th.daaam.proceedings.043.

[9] V.Ye.Pryanichnikov, A.A.Aryskin, A.A.Drachkova, K.E.Brundukova, O.I. Vlasova, A.A.Kartseva, YU.S.Kolesov, A.V.Snopkov, M.D.Solov'yova, A.I.Telegin, D.S.Telezhkin, A.N.Tikhomirov, A.V.Fedorov, D.K.Fedotov, V.V.Yastrebov. Eksperimenty po rasshireniyu funktsional'nykh vozmozhnostey mobil'nykh servisnykh robotov (Experiments to Expand the Functional Capabilities of Mobile Service Robots). «Ekstremal'naya Robototekhnika», Sbornik tezisov 34 -y Mezhdunarodnoy nauchno-tekhnicheskoy konferentsii. – Sankt-Peterburg: TSNII RTK - OOO «Tipografiya Fursova», 2023. – 306 s.: il., S.34-36, ISBN 978-5-6048032-9-5, https://er.rtc.ru/images/docs/2023/Sbornik-tezisov_ER-2023.pdf

[10] Pryanichnikov, V[alentin]; Kolesov, Y[uriy]; Tarasov, R[adomir] & Plotnikov, A[leksey] (2021). Integration Software for a Distributed Group of Mobile Service Robots, Proceedings of the 32nd DAAAM International Symposium, pp.0311-0315, B. Katalinic (Ed.), Published by DAAAM International, ISBN 978- 3-902734-33-4, ISSN 1726-9679, Vienna, Austria DOI: 10.2507/32nd.daaam.proceedings.045

[11] Vasilyev A.V, Lopota A.V. Utochneniye tiporazmernykh grupp nazemnykh distantsionno upravlyayemykh mashin dlya primeneniya v opasnykh dlya cheloveka usloviyakh (Clarification of standard-size groups of ground-based remotely controlled vehicles for use in conditions hazardous to humans) // Nauchno-tekhnicheskiye vedomosti SPbGPU. 2015. № 1 (214). S. 226–234

[12] Damic, V[jekoslav]; Cohodar, M[aida] & Muratovic, M[unir] (2017). Dynamic Modelling of Mobile Robots Based on its 3D CAD Model, Proceedings of the 28th DAAAM International Symposium, pp.0144-0149, B. Katalinic (Ed.), Published by DAAAM International, ISBN 978-3-902734-11-2, ISSN 1726-9679, Vienna, AustriaDOI: 10.2507/28th.daaam.proceedings.019

[13] A.A. Hatab and R. Dhaouadi, "Dynamic modelling of differential-drive mobile robots using Lagrange and Newton-Euler methodologies: A unified framework," Adv. Robot. Autom., vol. 02, no. 02, 2013.

[14] Plotnikov A.V. Kinetostaticheskaya model' zakhvata ob"yektov servisnymi robotami pri aktivnom peremeshchenii mobil'noy platformy (Kinetostatic model of object capture by service robots during active movement of a mobile platform) // Preprinty IPM im. M.V.Keldysha. 2023. № 14. 21 s., https://doi.org/10.20948/prepr-2023-14 , https://library.keldysh.ru/preprint.asp?id=2023-14

DOI: 10.2507/36th.daaam.proceedings.045

CELLULAR ENVIRONMENT AND 4-DIMENSIONAL CUBES FOR EFFECTIVE REPRESENTATION OF ROBOT IN LABYRINTH

Roman V. Khelemendik & Valentin E. Pryanichnikov

This Publication has to be referred as: Khelemendik, R[oman] & Pryanichnickov, V[alentin] (2025). Cellular Environment and 4-Dimensional Cubes for Effective Representation of Robot in Labyrinth, Proceedings of the 36th DAAAM International Symposium, pp.0321-0327, B. Katalinic (Ed.), Published by DAAAM International, ISBN 978-3-902734-47-1, ISSN 1726-9679, Vienna, Austria
DOI: 10.2507/36th.daaam.proceedings.045

Abstract

In robotics, one of the essential tasks is to simplify the representation of the operating environment in the form of a two-dimensional cellular field of dimension nxm. Let's consider the formalism of representing data about the operating environment in the robot's memory in the form of a labyrinth in which some neighbouring cells are separated by walls (partitions) and the robot cannot move from one such cell to the next. The main task is to determine the control of the robot (trajectory planning), which ensures the transition from the current position (i0, j0) - forming a sequence of transitions to neighbouring cells so that there is a path to the target cell (ig, jg). Let's divide the cellular field into enlarged 4x4 cells (or larger ones). For them, we will build correlated 4-dimensional cubes, that will set the numbering inside each 4x4 square in such a way, that the numbering of vertices in binary form allows you to move from one vertex (cell) to the neighbouring one by changing only one bit. The absence of a corresponding edge for the transition means, that these cells are separated by an obstacle. The path on such a field is constructed so that the entrances and exits of the paths on adjacent 4x4 cells coincide. The separation of neighbouring cells by an obstacle can be set both a priori (then this is the task of a known map), and due to the activation of the sensor system of near detection. The paper analyses the proposed method of constructing correlated four-dimensional cubes (graphs) on which the sequence of transitions and path search is carried out in a more efficient and economical way.

Keywords: Mobile service; collaborating robots; representation of environment information; n-dimensional cubes.

1. Introduction

The movement of objects across the neighbouring cells of a field on a plane is considered. We assume that the number of cells is 2^n, and the field has dimensions 2^m x 2^{n-m}. We study methods for numbering cells with Boolean sets of length n and numbers from 0 to 2^{n-1} such that adjacent (vertically or horizontally) fields differ in exactly one binary digit. This numbering helps save time and memory when formalizing movement on a discrete plane. This, in particular, allows us to change just one bit in software to move the robot one cell in a straight line; we also gain new capabilities for modelling robots in flat labyrinths. For problems where fast and efficient transitions between states of objects are important, Gray codes ([2], [3]) are used. Their distinctive feature is that only one position changes when transitioning from one code combination to another.

Gray codes have been studied in terms of Hamiltonian cycles in graphs and n-dimensional Boolean cubes ([2], [4], [5]), development of combinatorial algorithms ([6]), error correction in communication via a communication channel ([7]), and design of hard drives and databases ([7], [8]). In this article, we propose specialized truncated n-dimensional cubes, flat rectangular fragments and their combinations for new applications of n-dimensional cubes and Gray codes, in particular, for moving robots in cellular environments. In practice, this is important for representing the working environment in the robot's memory as a labyrinth and for planning the movements of a mobile service robot such as Amur. Based on these principles, it is possible to build a description of the efficient algorithm for by passing wards by a logistics robot in a hospital.

2. Definitions and designations

A simple path in a graph or *a simple chain* of length k>1 on a graph G will be called a sequence (set) $C=<v_1,u_1,\dots,u_{k-1},v_k>$, where v_i belongs to the set of vertices V, and u_i to the set of edges U, and none of the elements of the sequence is repeated (if there are repeating elements in this sequence, the word "simple" is omitted). Similarly, a simple cycle in a graph is a simple chain C for which $v_1 = v_k$.

The set of all sets $<\alpha_1...\alpha_n>$ of binary (Boolean) values is called an *n-dimensional (Boolean) cube* of size n and is denoted by B^n. For sets of values we use the following notations: $\boldsymbol{\alpha}^n=<\alpha_1...\alpha_n>$. *The distance* $r(\boldsymbol{\alpha}^n,\boldsymbol{\beta}^n)$ between vertices $\boldsymbol{\alpha}^n$ and $\boldsymbol{\beta}^n$ is the number of digits in which they differ: $r(\boldsymbol{\alpha}^n,\boldsymbol{\beta}^n)=\Sigma^n_{i=1}|\alpha_i-\beta_i|$. *The number* $\eta(\boldsymbol{\alpha}^n)$ *of a set* $\boldsymbol{\alpha}^n$ is the number whose binary representation is $\boldsymbol{\alpha}^n$: $\eta(\boldsymbol{\alpha}^n)=\Sigma^n_{i=1}2^{n-i}\alpha_i$. Geometrically, the Boolean cube B^n is a graph consisting of 2^n vertices, and any two of its vertices $\boldsymbol{\alpha}^n$ and $\boldsymbol{\beta}^n$ are connected by an edge if and only if $r(\boldsymbol{\alpha}^n,\boldsymbol{\beta}^n)=1$. *A Hamiltonian cycle* (HC) in a graph is a sequence of all its vertices, connected by edges and not repeating, in which the last vertex is connected by an edge to the first. If the vertices of B^n are denoted by binary words (sets) of length n, then the HC in B^n is also called the Gray code [2, 3]. Fig. 1 shows flat projections of n-dimensional cubes for n=1,2,3,4, as well as some HCs (arrows).

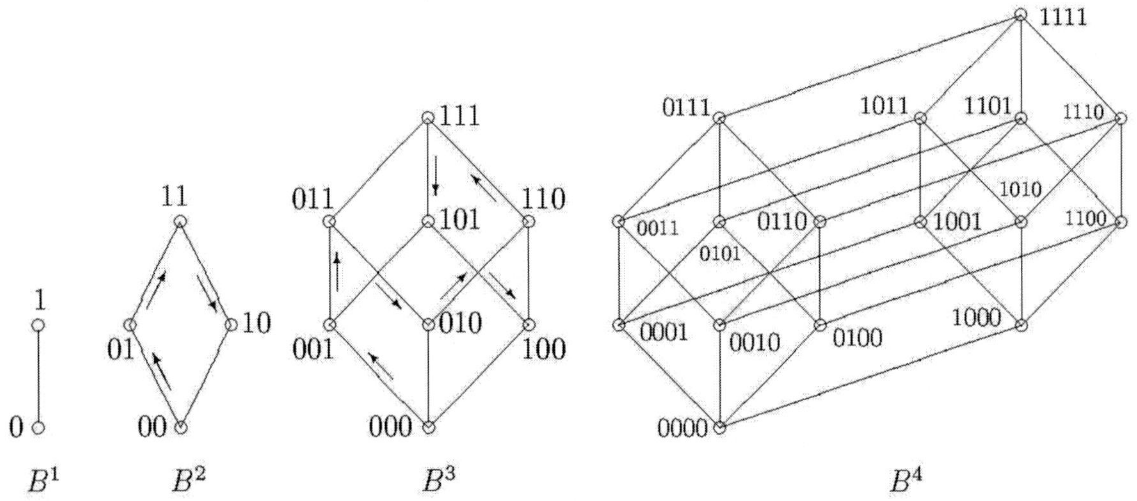

Fig. 1. n-dimensional cubes B^n, $n\leq 4$

A plane rectangular (m,n)*-fragment* ($PF_{(m,n-m)}$) of an n-dimensional cube is a subgraph B^n such, that its vertices are mapped to the cells of a rectangle of size 2^m x 2^{n-m}, the edges of the subgraph connect vertically and horizontally adjacent (neighbouring) cells of this rectangle, and the remaining edges from B^n are removed. In this case, the cells of a 2^m x 2^{n-m} rectangle can be numerated with numbers representing binary codes (of vertices). Thus, the HC in Fig. 1 for B^2 is the basis for $PF_{(0,2)}$ – a rectangle of size 1 x 4, the cells of which are sequentially numbered with the numbers 0, 1, 3, 2; and if we restore the edge <00>–<10> in this graph, we obtain $PF_{(1,1)}$ – a rectangle of size 2 x 2 (square), consisting of the rows [0,2], [1,3].

3. Construction of the cellular environment and PF (plane fragments) for the first values of n

Obviously, for n=1 there is only one PF, namely $PF_{(0,1)}$ with a corresponding rectangle of size 1 x 2, and "almost identically" (indistinguishably) numbered cells – in the order [0,1] or [1,0]. In the case of n=2, as shown above, there are only two PFs: $PF_{(0,2)}$ (strip) and $PF_{(1,1)}$ (square), for which the indistinguishable cell numberings are given above. We move on to the case of $n\geq 3$. The statement is obvious:

Statement 1. *Every* HC *in* B^n *uniquely corresponds to some* $PF_{(0,n)}$, *and vice versa.*
If in HC for B^n we can recover (n-1) edges, then we get $PF_{(1,n-1)}$.

Statement 2. *For* n≥3, *not every* HC *in* B^n *has a corresponding* $PF_{(1,n-1)}$.

Indeed, for n≥3 there is HC (2,0,4,5,1,3,7,6). However, to obtain $PF_{(1,2)}$, we can restore edges (2,6) and (5,1), but we cannot restore edges (0,7) and (4,3), since they are not in B^3. Similar, as well as more complex, situations arise when n>3. A sequence of numbers $\{a_k\}$ of length s is called ↑-unimodal (↓-unimodal) if there exists k such that $a_1 < a_k > a_{k+1} > a_s$ ($a_1 > a_k < a_{k+1} < \ldots < a_s$), 1<k<s. Thus, the sequences (0,1,3,2) and (6,7,5,4) are ↑-unimodal. Obviously, if the sequence $\{a_k\}$ ↑-unimodal (↓-unimodal), then its inverse (taken in reverse order) sequence is ↓-unimodal (↑-unimodal). By combining two oppositely directed unimodal sequences (0,1,3,2) and (4,5,7,6), we obtain $PF_{(1,2)}$, consisting of rows with these sequences (Fig. 2). Note that the ↑-unimodal sequence is simply unimodal ([1], p. 178), so in our definition, unimodality is somewhat specified (clarified) for this sequence, whether it is increasing or decreasing.

4	5	7	6
0	1	3	2

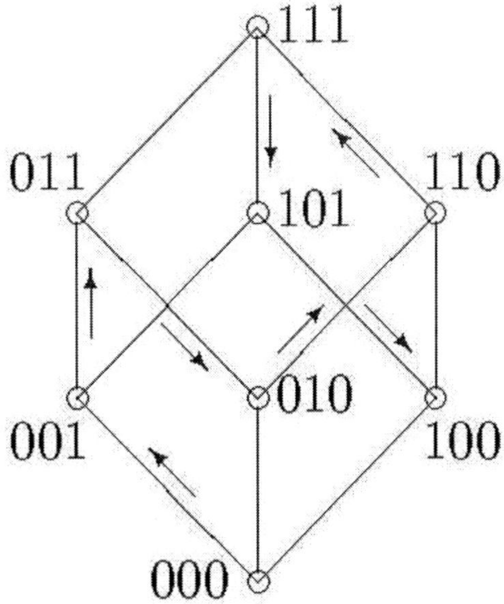

Fig. 2. $PF_{(1,2)}$, B^3 and HC in B^3

Further, unimodal sequences of numbers in rows (columns) will be called unimodal rows (columns). It is easy to see that both unimodal rows form a GC in the B^3 subgraphs, which are B^2 cubes (squares) and represent boolean functions depending only on the second and third variables.

Statement 3. *For* B^4 *there exists a* $PF_{(2,2)}$, *such that the sequences of all its rows and columns are unimodal.* One of these $PF_{(2,2)}$ and its corresponding B^4, from which the edges unnecessary for the 4x4 square (a truncated four-dimensional cube) are removed, are shown in Fig. 3.

8	10	14	12
9	11	15	13
1	3	7	5
0	2	6	4

1000	1010	1110	1100
1001	1011	1111	1101
0001	0011	0111	0101
0000	0010	0110	0100

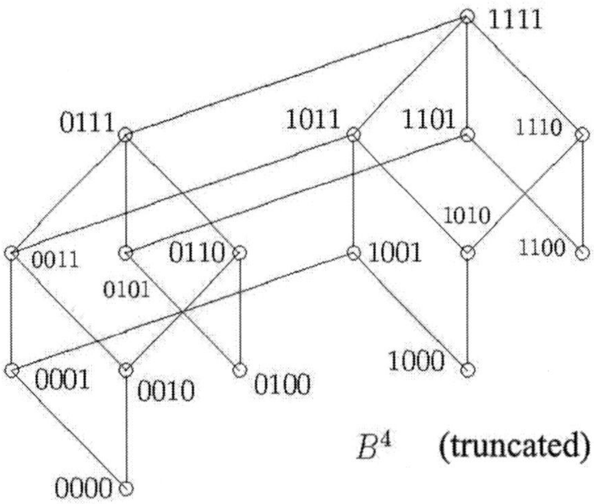

Fig. 3. $PF_{(2,2)}$ and truncated B^4

It is easy to see, that in this 4x4 square, all rows are unimodal in the second and third digits, and the columns are in the first and fourth. More specifically, from this we get, that the horizontal movement between the second and third verticals corresponds to a change in the second digit, and the movement between the first and second or third and fourth verticals corresponds to a change in the third digit. Similarly, a vertical movement between the second and third horizontals corresponds to a change in the first digit, and a movement between the first and second or third and fourth horizontals corresponds to a change in the fourth digit. *The weight* $\omega(\alpha^n)$ *of a boolean set* α^n is the number of its unit components: $\omega(\alpha^n) = \Sigma^n_{i=1} \alpha_i$. Then the weights in the unimodal rows (x_2 and x_3) and columns (x_1 and x_4) also form a unimodal sequence: 0, 1, 3, 2. Thus, instead of the $PF_{(2,2)}$ shown in Fig. 3, other 4x4 squares can be constructed using forward and reverse unimodal sequences of weights for various pairs of digits. Note that to obtain $PF_{(1,3)}$ from $PF_{(2,2)}$, it is sufficient to rotate 270 degrees counter clockwise relative to the midpoint of the left side of the 4x4 square of the two upper rows. Then in $PF_{(1,3)}$ we obtain rows consisting of the concatenation of the unimodal rows (13,15,11,9,1,3,7,5) and (12,14,10,8,0,2,6,4), and in the truncated cube B^4 we remove 4 edges corresponding to the "gaps" (due to rotation), and restore (from the full B^4) the edge (8,0). Continuing this process (cutting along the central horizontal line, rotating 270 degrees and restoring one edge), from $PF_{(1,3)}$ we obtain $PF_{(0,4)}$.

4. Construction of the cellular environment and PF for n≤6

To go from n=4 to n=5, simply duplicate the $PF_{(2,2)}$, mirroring it relative to the vertical centerline, increasing each number by 16, and positioning this PF to the right of the original. Then, for all adjacent cells in both $PF_{(2,2)}$, the condition $r(\alpha^n,\beta^n)=1$ remains satisfied, and for the fourth and fifth columns, horizontal adjacency means a difference of only one digit – the most significant one (Table 1).

Recording cells by natural numbers							
8	10	14	12	28	30	26	24
9	11	15	13	29	31	27	25
1	3	7	5	21	23	19	17
0	2	6	4	20	22	18	16
Recording cells as binary numbers							
01000	01010	01110	01100	11100	11110	11010	11000
01001	01011	01111	01101	11101	11111	11011	11001
00001	00011	00111	00101	10101	10111	10011	10001
00000	00010	00110	00100	10100	10110	10010	10000

Table 1. Cell numbering in $PF_{(2,3)}$ (rectangle of size 2^m x 2^{n-m}, m=2, n=5 , i.e. the size of rectangle is 4x8).

When moving from n=5 to n=6, we duplicate the constructed $PF_{(2,3)}$ and perform similar actions of mirror symmetry relative to the horizontal midline, increasing the numbers by 2^5=32, and positioning the new PF above the original. An example of such a $PF_{(3,3)}$ and the associated truncated cube B^6 are shown in Table 2 and Fig. 4. This $PF_{(3,3)}$ is not exactly an extension of the previously constructed $PF_{(2,3)}$, since in the starting $PF_{(2,2)}$ we chose oppositely directed unimodalities: rows are ↓-unimodal, and columns are ↑-unimodal.

Recording cells by natural numbers							
34	32	36	38	54	52	48	50
35	33	37	39	55	53	49	51
43	41	45	47	63	61	57	59
42	40	44	46	62	60	56	58
10	8	12	14	30	28	24	26
11	9	13	15	31	29	25	27
3	1	5	7	23	21	17	19
2	0	4	6	22	20	16	18
Recording cells as binary numbers							
100010	100000	100100	100110	110110	110100	110000	110010
100101	100001	100101	100111	111101	110101	110001	110011
101011	101001	101101	101111	111111	111101	111001	111011
101010	101000	101100	101110	111110	111100	111000	111010
001010	001000	001100	001110	011110	011100	011000	011010
001011	001001	001101	001111	011111	011101	011001	011011
000101	000001	000101	000111	011101	010101	010001	010011
000010	000000	000100	000110	010110	010100	010000	010010

Table 2. Cell numbering in $PF_{(3,3)}$

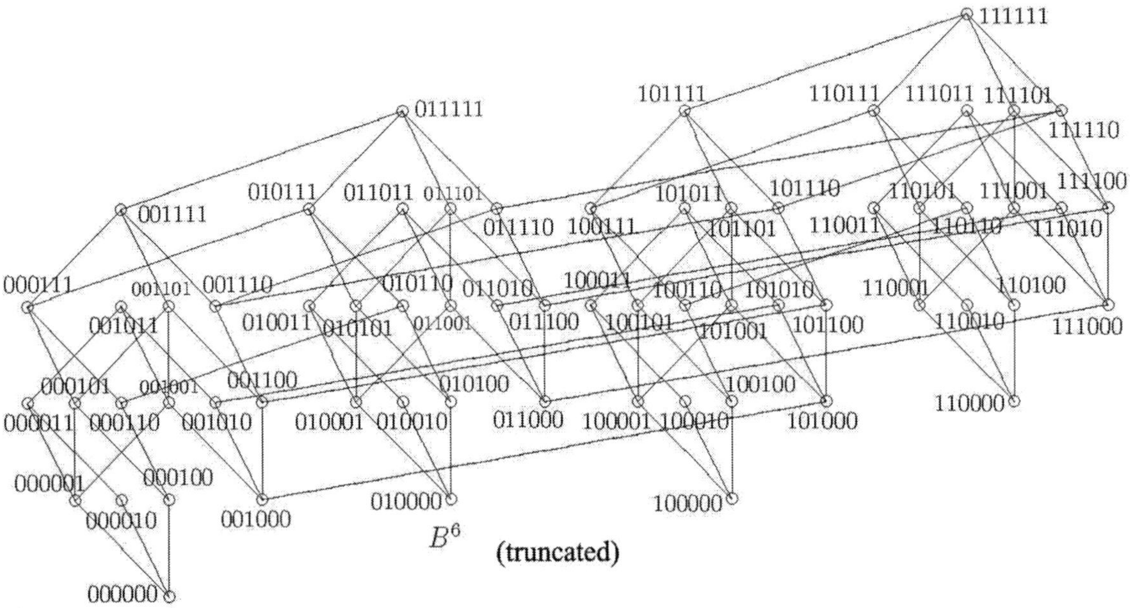

Fig. 4. Truncated six-dimensional Boolean cube

5. Construction of the cellular environment and PF for n>6

We will construct the $PF_{(4,4)}$ for n=8, and its lower half will also be the $PF_{(3,4)}$ for n=7. First, we will slightly modify the original $PF_{(3,4)}$ for n=4, according to which we will construct the $PF_{(3,3)}$ for n=6 (see Table 3) as was done above.

Recording cells by natural numbers							
32	33	37	36	52	53	49	48
34	35	39	38	54	55	51	50
42	43	47	46	62	63	59	58
40	41	45	44	60	61	57	56
8	9	13	12	28	29	25	24
10	11	15	14	30	31	27	26
2	3	7	6	22	23	19	18
0	1	5	4	20	21	17	16
Recording cells as binary numbers							
100000	100001	100101	100100	110100	110100	110001	110000
100010	100011	100111	100110	110110	110111	110011	110010
101010	101011	101111	101110	111110	111111	111011	111010
101000	101001	101101	101100	111100	111101	111001	111000
001000	001001	001101	001100	011100	011101	011001	011000
001010	001011	001111	001110	011110	011111	011011	011010
000010	000011	000111	000110	010110	010111	010011	010010
000000	000001	000101	000100	010100	010101	010001	010000

Table 3. Unimodal cell numbering in $PF_{(3,3)}$

The main difference between this PF and the previous $PF_{(3,3)}$ (see Table 2) lies precisely in the lower left 4x4 square: previously, the rows were unimodal in digits 2 and 3, and the columns in digits 1 and 4, but now the rows are unimodal in even digits (2, 4), and the columns in odd digits (1, 3). When moving to the new $PF_{(3,3)}$, unimodality in row concatenations is also preserved for even digits, and unimodality in column concatenations is preserved for odd digits. Continuing the transition from n=6 to n+1 in width, and then to n+2 in height, as was previously done for n=4, we arrive at $PF_{(4,4)}$ – a 16x16 square consisting of 16 4x4 squares (Table 4). The cells in this field contain two-digit numbers written in hexadecimal notation. The first number represents the position of a 4x4 square, and the second represents the cell's position within the 4x4 square (Table 5). The cells of this field contain two-digit numbers written in hexadecimal notation.

In this case, the first number characterizes the position of the 4x4 square, and the second characterizes the location of the cell already in the 4x4 square (Table 5, the symbols * denote arbitrary hexadecimal numbers, and $*^4$ – four arbitrary binary digits).

80	81	85	84	94	95	91	90	D0	D1	D5	D4	C4	C5	C1	C0
82	83	87	86	96	97	93	92	D2	D3	D7	D6	C6	C7	C3	C2
8A	8B	8F	8E	9E	9F	9B	9A	DA	DB	DF	DE	CE	CF	CB	CA
88	89	8D	8C	9C	9D	99	98	D8	D9	DD	DC	CC	CD	C9	C8
A8	A9	AD	AC	BC	BD	B9	B8	F8	F9	FD	FC	EC	ED	E9	E8
AA	AB	AF	AE	BE	BF	BB	BA	FA	FB	FF	FE	EE	EF	EB	EA
A2	A3	A7	A6	B6	B7	B3	B2	F2	F3	F7	F6	E6	E7	E3	E2
A0	A1	A5	A4	B4	B5	B1	B0	F0	F1	F5	F4	E4	E5	E1	E0
20	21	25	24	34	35	31	30	70	71	75	74	64	65	61	60
22	23	27	26	36	37	33	32	72	73	77	76	66	67	63	62
2A	2B	2F	2E	3E	3F	3B	3A	7A	7B	7F	7E	6E	6F	6B	6A
28	29	2D	2C	3C	3D	39	38	78	79	7D	7C	6C	6D	69	68
08	09	0D	0C	1C	1D	19	18	58	59	5D	5C	4C	4D	49	48
0A	0B	0F	0E	1E	1F	1B	1A	5A	5B	5F	5E	4E	4F	4B	4A
02	03	07	06	16	17	13	12	52	53	57	56	46	47	43	42
00	01	05	04	14	15	11	10	50	51	55	54	44	45	41	40

Table 4. Unimodal cell numbering in $PF_{(4,4)}$

8*	9*	D*	C*
A*	B*	F*	E*
2*	3*	7*	6*
0*	1*	5*	4*
Binary numbering			
1000$*^4$	1001$*^4$	1101$*^4$	1100$*^4$
1010$*^4$	1011$*^4$	1111$*^4$	1110$*^4$
0010$*^4$	0011$*^4$	0111$*^4$	0110$*^4$
0000$*^4$	0001$*^4$	0101$*^4$	0100$*^4$
Big square			

*8	*9	*D	*C
*A	*B	*F	*E
*2	*3	*7	*6
*0	*1	*5	*4
Binary numbering			
$*^4$1000	$*^4$1001	$*^4$1101	$*^4$1100
$*^4$1010	$*^4$1011	$*^4$1111	$*^4$1110
$*^4$0010	$*^4$0011	$*^4$0111	$*^4$0110
$*^4$0000	$*^4$0001	$*^4$0101	$*^4$0100
Little square			

Table 5. Numbering of cells of the large and small square in $PF_{(4,4)}$

Thus, any horizontal movements correspond only to a change in even digits, and any vertical movements correspond only to a change in odd digits. In this case, a change in the 1st (2nd) digit is a transition between horizontals (verticals) 8 and 9; a change in the 3rd (4th) digit is a transition between horizontals (verticals) 4-5, 12-13; a change in the 5th (6th) digit is a transition between horizontals (verticals) 2-3, 6-7, 10-11, 14-15; a change in the 7th (8th) digit is a transition between horizontals (verticals) 1-2, 3-4, 5-6, 7-8, 9-10, 11-12, 13-14, 15-16. The algorithm for constructing a PF from n=2k to n+1 in width, and then to n+2, also works in the general case n>8.

6. Comparison of traditional and multidimensional representations of the cellular environment

It is common to represent a robot's position in a cellular environment as a pair of coordinates <x,y>, where x is the vertical axis and y is the horizontal axis. The simplicity and clarity of this representation are beyond doubt. Moving to an adjacent cell then means adding or subtracting one from the x or y variable. It should be noted, that repeated execution of such simple arithmetic operations in computing devices with a multidimensional representation of the cellular environment requires significantly less time and memory. As an example, consider the case of n = 8 and a path from cell <1,1> to cell <16,16>. Storing the current position of an object (robot) requires one byte for both cellular representations. Any shortest path from <1,1> to <16,16> consists of 15 elementary vertical and 15 horizontal movements. In the multidimensional representation, an elementary movement results in a speed savings of some δ, since changing one bit is faster than adding or subtracting one. For 30 movements, this already yields a savings of 30δ. We achieve memory savings if we want to remember a specific route from <1,1> to <16,16>. With the traditional representation, we need to remember a sequence of coordinates (bytes), i.e., 30 bytes, whereas in the multidimensional representation, 30 bits are sufficient. At the same time, the hierarchy of bits by seniority allows us to obtain instant detailed information about the robot even when only one bit changes: for example, the inversion of one of the most senior bits immediately means that the robot has moved to the other half of the square (vertical or horizontal).

7. Formalism of representing a labyrinth in the robot's memory

A labyrinth in a cellular environment is a PF in which transitions between some adjacent cells may be forbidden. Let's first consider a field of size $2^{n/2}$ x $2^{n/2}$, which we represent as a $PF_{(n,n)}$, as shown above. A specific labyrinth is then defined by enumerating all pairs of adjacent cells between which transitions are forbidden. Since the total number of possible transitions already at n=8 amounts to hundreds, listing forbidden pairs of adjacent cells as pairs of their coordinates is only suitable for a small number of prohibitions; moreover, it is not visually clear. Using a multidimensional representation of the cellular environment, forbidden pairs are specified by specifying the bits responsible for the transitions. Thus, for n=8, a set of values $\beta^8=<\beta_1,\ldots, \beta_8>$ is first formed such that $\beta_j=1$ if and only if, for the case of at least one change in the j-th digit, there is a transition prohibition. Then, an additional set ξ is formed, allowing us to specify all such prohibitions. The representation of prohibitions becomes clear if the labyrinths are represented as truncated cubes in which prohibited edges are coloured red (or removed). Further refinement of prohibitions and formalization of routes can be carried out using Boolean functions of a small number of variables and propositional logic.

8. Conclusion

The key provision of the article is to propose a new logical representation for describing the operating environment of mobile service robots in the internal memory of a top-level control microprocessor. The practical significance of the results lies in the possibility of maximum data compression, close to the theoretical limit with a very promising method for determining the neighbourhood on a two-dimensional cellular field. The most important conclusion is that the approach found opens up new tools for creating effective on-board software for collaborative robots. It should also be noted, that the work was carried out mainly at the Keldysh Institute of Applied Mathematics of the Russian Academy of Sciences as part of research related to the "Intelligent robotronics" project (see also the works [9], [10], [11], [12]).

9. References

[1] Yablonskii S.V. (1986). Vvedenie v diskretnuiu matematiku / M.: Nauka, 1986. — 384s.

[2] Perezhogin A. L., Bykov I. S. Review of constructions and properties of Gray codes // Mathematical Issues of Cybernetics, no. 20, A collection of papers, ed. by O.B. Lupanov. Moscow: Fizmatlit, 2022, pp. 41-60., URL: http://library.keldysh.ru/mvk.asp?id=2022-41 DOI: 10.20948/mvk-2022-41

[3] Gray F. Pulse code communications. Patent no. 2632058 US.

[4] Gilbert E. N. Gray codes and paths on the n-cube //The bell system technical journal. 1958.—V. 37, No 3.—P. 815–826.

[5] Perezhogin A. L. On the spectrum of Hamiltonian cycles in the n-cube // Journal of Combinatorial Theory, Series B.—2021.—V. 151.—P. 435–464.

[6] Reingold E. M., Nievergelt J., Deo N. Combinatorial Algorithms: Theory and Practice. Prentice Hall College Div., 1977.

[7] Goddyn L., Lawrence G. M., Neme h E. Gray codes with optimized run lengths // Utilitas Mathematica. 1988.—V. 34.—P. 179–192.

[8] Faloustos C. Gray codes for partial match and range queries // IEEE Transactions on Software Engineering. 1988.—V. 14, No 10.—P. 1381–1393.

[9] Davydov D.V., Eprikov S.R., Kirsanov K.B., Pryanichnikov V.E. (2017). Service Robots Integrating Software and Remote Reprogramming, Proceedings of the 28th DAAAM International Symposium, pp.1234-1240, B. Katalinic (Ed.), Published by DAAAM International, ISBN 978-3-902734-11-2, ISSN 1726-9679, Vienna, Austria, DOI: 10.2507/28th.daaam.proceedings.172.

[10] Pryanichnikov, V[alentin]; Kolesov, Y[uriy]; Tarasov, R[adomir] & Plotnikov, A[leksey] (2021). Integration Software for a Distributed Group of Mobile Service Robots, Proceedings of the 32nd DAAAM International Symposium, pp.0311-0315, B. Katalinic (Ed.), Published by DAAAM International, ISBN 978- 3-902734- 33-4, ISSN 1726-9679, Vienna, Austria DOI: 10.2507/32nd.daaam.proceedings.045

[11] Aryskin A.A., Bogdanovich A.V., Davydov O., Khelemendik R., Petrakov M., Pryanichnikov V.E., (2019). Control and Interaction Algorithms for Industrial and Service Robots, Chapter 33 in DAAAM International Scientific Book 2019, pp.377-384, B. Katalinic (Ed.), Published by DAAAM International, ISBN 978-3-902734-24-2, ISSN 1726-9687, Vienna, Austria DOI: 10.2507/daaam.scibook.2019.33

[12] Khelemendik R., Pryanichnikov V. E. (2020). Contradictory Reasoning Connecting with the Problems of Intelligent Robotronics, Proceedings of the 31st DAAAM International Symposium, pp.0276-0281, B. Katalinic (Ed.), Published by DAAAM International, ISBN 978-3-902734-29-7, ISSN 1726-9679, Vienna, Austria DOI: 10.2507/31st.daaam.proceedings.038